Volume II
Infrastructure Health in Civil Engineering
Applications and Management

Volume II

Infrastructure Health in Civil Engineering

Applications and Management

Mohammed M. Ettouney
Sreenivas Alampalli

CRC Press
Taylor & Francis Group
Boca Raton London New York

CRC Press is an imprint of the
Taylor & Francis Group, an **informa** business

CRC Press
Taylor & Francis Group
6000 Broken Sound Parkway NW, Suite 300
Boca Raton, FL 33487-2742

© 2012 by Taylor & Francis Group, LLC
CRC Press is an imprint of Taylor & Francis Group, an Informa business

No claim to original U.S. Government works

Printed in the United States of America on acid-free paper
Version Date: 20110815

International Standard Book Number: 978-1-4398-6653-5 (Hardback)

Visit the Taylor & Francis Web site at
http://www.taylorandfrancis.com

and the CRC Press Web site at
http://www.crcpress.com

This book is dedicated to Mohammed A. Ettouney, Fatima A. Abaza, and William Zacharellis, may God be merciful on their souls.

Mohammed M. Ettouney

To my wife Sharada and son Sandeep for their love and patience.

Sreenivas Alampalli

Contents

Preface

There is a purpose for building infrastructure, and infrastructure owners are responsible for ensuring that the purpose is served while achieving maximum benefit at minimal costs. Taking appropriate and timely actions requires a good understanding of the infrastructure—its current and expected condition. This volume and its companion, *Infrastructure Health in Civil Engineering: Theory and Components* (CRC Press, 2012), are dedicated to discussions of these aspects (see Figure 0.1). The companion volume focuses on an overview of Infrastructure Health in Civil Engineering (IHCE) and associated theories, followed by a description of its four components: measurement, structural identification, damage identification, and decision making. Decision making is a unique feature and is introduced with the argument that any project that does not integrate decision making (or Cost–Benefit) ideas in all its tasks cannot be successful. This volume builds upon the ideas presented in the companion volume and deals with the application of IHCE and asset management.

PART III: APPLICATIONS OF IHCE

The aim of IHCE application is to prolong the beneficial life of a structure, recognizing the numerous hazards that can shorten its beneficial life. To effectively prolong beneficial life, these techniques should be well suited for addressing various hazards and their attributes. Thus, it is important to understand, within the IHCE context, the nature of hazards, the likelihood of their occurrence, how they affect the structure, and the impact of their occurrence. It is also equally important that IHCE techniques are also affected by the type of structure and its components. For example, temperature monitoring can be of paramount importance for bridge decks but not as important for scour effect on foundations.

This part of the volume is devoted to IHCE techniques as applied to some important hazards and applications as they affect bridge structures. Of all the hazards faced by structures, scour is recognized as the major cause of more bridge failures than any other. Earthquakes have caused much damage to infrastructures all over the world, with considerable causalities. Corrosion damage has been the leading cause of infrastructure deterioration over the years. Fatigue failures have received considerable attention recently owing to the sudden nature of failures with limited warning. Thus, we focus on scour, earthquake, corrosion, and fatigue in this part of the volume in Chapters 1 through 5. Corrosion effects on regular reinforced concrete systems (Chapter 3) and prestressed/posttensioned concrete bridges (Chapter 4) are treated separately.

Each hazard/deterioration has its own characteristics. For example, scour can affect a bridge from a slow/progressive manner to a fast/accelerated manner depending on hazard and structural/site characteristics, whereas earthquakes affect the bridge in a very short duration. Corrosion is a slow process; fatigue effects can be slow or sudden depending on the material, structural, or deterioration characteristics. Thus, the need for IHCE treatment is greatly dependent on the hazard and its characteristics. Our main interest in this volume is to assess the different uses of IHCE techniques in minimizing damage due to hazards.

In addition to the differences between hazard effects on structures, there are differences in applying IHCE techniques for different structural components. Note that we have not devoted separate chapters to other prominent components such as superstructures and substructures, since the IHCE treatment of those are covered throughout the volume. We felt that fiber reinforced polymer (FRP) bridge decks and FRP wrappings are different enough from other components to merit a special chapter for each. Thus, Chapter 6 acknowledges the emergence of FRP material as structural material, as evidenced by its use for bridge decks. Because of the special process required for manufacturing FRP bridge decks, and their geometric and material nature, they require special IHCE

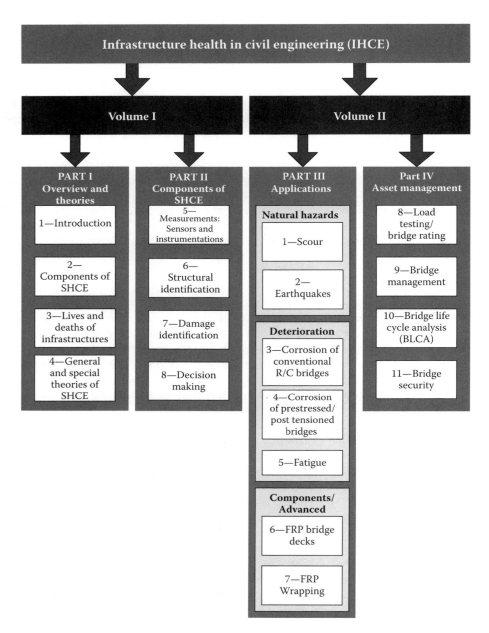

FIGURE 0.1 General layout of this volume and the companion volume, *Infrastructure Health in Civil Engineering: Theory and Components* (CRC Press, 2012).

treatment, which is the focus of Chapter 6. Similarly, Chapter 7 discusses the use of IHCE for FRP-wrapping applications

Although the subjects presented in Chapters 1 to 7 are varied, we have tried to handle each of the chapters in a consistent fashion, as shown in Figure 0.2. We start each chapter with a background discussion on the subject, emphasizing the unique aspects of the subject presented in that chapter. It is these unique aspects that influence IHCE treatment. We emphasize, where applicable, the economic importance of the subject and the various design and performance issues. In addition, we present the history and evolution of the subject matter.

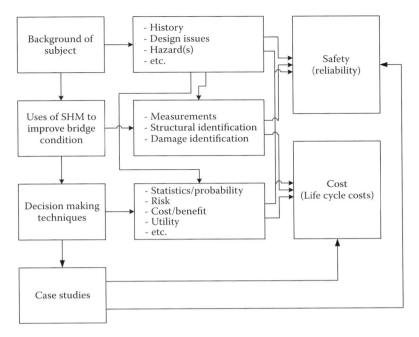

FIGURE 0.2   Composition of Chapters 1 through 7.

Next, we apply all four IHCE components to each of these chapters: measurement, structural identification, damage identification, and decision making. (These are discussed in detail in the companion volume.) For example, we apply some of the sensor technologies discussed in Chapter 1 of the companion volume. As needed, special purpose sensors used specifically for a given topic are discussed. e.g., Structural and damage identification techniques, as integral parts of the IHCE process, are also discussed as they apply to specific topics discussed in the chapters. For example, modeling or detecting delaminations in an FRP bridge deck is different from modeling and detecting corrosion in a prestressed/posttensioned concrete bridge. Decision making processes presented in these chapters are all based on the techniques of Chapter 8 of the companion volume: Infrastructure Health in Civil Engineering: Theory and Components (CRC Press, 2012). Relevant case studies are presented to illustrate the discussion.

Accurate and timely life cycle analysis (LCA) assists decision makers in making more informed decisions to ensure safety while reducing costs. Thus, in the end, a section on LCA is included to highlight the use of IHCE techniques in ensuring accurate assumptions and computations of LCA of the structure.

PART IV: ASSET MANAGEMENT

Management issues are global and wide ranging, and so these issues cover all technical aspects of IHCE as well as all potential hazards and components. In addition, the many goals, theories, and methodologies of IHCE are also a part of asset management practices. Thus, this part of the volume explores the IHCE techniques as applied to asset management that emphasis on bridge structures.

Chapter 8 discusses load testing and other rating techniques as applied to bridges. Load testing of bridges (or other types of infrastructures) is perhaps one of the most popular applications of IHCE. It has been used for bridges of all types, shapes, and sizes, that can be attributed to its simplicity and importance. Chapter 9 is devoted to the concepts of bridge management systems (BMS) and the different ways in which IHCE techniques can help bridge managers realize their objectives. Figure 0.3 shows the basic construct of Chapter 9. We discuss the tools and strategies of BMS. In addition, we

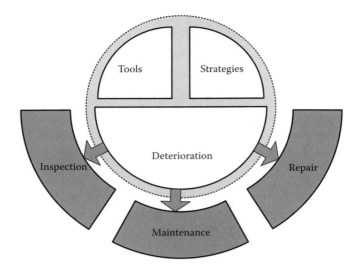

FIGURE 0.3 Elements of Chapter 9 on bridge management systems.

allocate a major part of Chapter 9 to deterioration and the three main management tools to fight deterioration: inspection, maintenance, and repair. The IHCE techniques that can help the manager in performing these tasks are covered in this chapter. Finally, the confluence of multihazards concepts, bridge management, and IHCE techniques is explored in this chapter.

Chapter 10 discusses bridge life cycle analysis (BLCA) methods in an integrated fashion. We emphasize the roles of different IHCE techniques in providing accurate parameters that the BLCA method needs. This use of SCHE techniques is often overlooked even though it could be one of the most valuable uses of IHCE methods.

Enhancing infrastructure/bridge security has emerged as a major asset management concern since the tragic events of 9/11. High costs prompted managers to seek IHCE techniques in improving bridge security at reasonable costs. Chapter 11 explores ways of using IHCE techniques for bridge security.

It should be noted that all the opinions and views expressed in this volume are those of the authors and not necessarily of the organizations they represent.

Acknowledgments

The life of every person is measured, among other things, by the special people who kindly affect such a life. This volume is a direct result of the many beautiful interactions I have had over the years with those special people. My late mentor, Dr. Mohammed S. Aggour, taught me the very foundations of structural engineering and design. My childhood friend, Dr. Elsayed A. Elsayed, has always been with me during good and bad times. My long friendships with Dr. Raymond (Ray) Daddazio, Michael (Mike) Udvardy, Amr Aly, Dr. Loraine Whitman, Ron (Ronnie) Check, Dr. Hamid Adib, and Christina (Tina) Plink have had a direct, as well as an indirect, influence on my writing this volume. I am grateful for the professional help, encouragement, and support given by Antranig (Andy) Ouzoonian, Norman (Norm) Glover, Gary Higbee, Dr. Amar Chaker, and Arturo (Artie) Mendez. The special friendship of, and the unparalleled technical and professional help and support given by Dr. Anil K. Agrawal will always be warmly remembered and deeply acknowledged. The advice, and patience of the editors of CRC Press helped immensely in making this project materialize. For that i am deeply grateful.

Over the millennia, when darkness descended, human beings used to look up to the heavens for help and guidance. The countless stars of the Milky Way never disappointed them. The combined belt of Orion and Sirius have always guided those who needed help through the darkest of the dark hours. I am no exception: without my own belt of Orion and Sirius this book could have never materialized. For that, I shall always be grateful to Milagros (Mila) Kennett and Dr. Sreenivas (Sreeni) Alampalli. Their friendship, kindness, help, and guidance will always be remembered as long as there are the belt of Orion and Sirius.

Mohammed M. Ettouney

It has been a great pleasure to have worked for the past 20 years with several of my hard-working and intelligent colleagues at the New York State Department of Transportation (NYSDOT), who strive to ensure that all users of transportation infrastructure are safe and secure every day. I learned a lot working with them, and several examples in this book are the direct result of working with many of them. One person requiring special acknowledgment is my mentor, the late Dr. Robert J. Perry, for his encouragement and support in advancing my professional interests.

During the past 20 years, I was very fortunate to have worked with a variety of people from various state and federal agencies, universities, private industry, and professional organizations. This volume has benefited from the practical and research experiences gained by working with them. I am specially grateful to Dr. Glenn Washer and Dr. Hamid Ghasemi for their professional help and support during these years. I thank my good friend Dr. Anil Agrawal for his support and for simply being there when needed. Several people have had an influence on me but none as much as Dr. Mohammed M. Ettouney. I met Dr. Ettouney about 15 years ago and found in him a great friend, an excellent colleague, and a mentor. I am always amazed at his professional competence and his ability to look beyond the normal. It has been a privilege writing this volume with him. I also thank CRC Press and Taylor & Francis Group for publishing this book and the support they provided during the publication process. Finally, and most important, many thanks to my wife, Sharada, and my son, Sandeep, for their patience, understanding, and good humor during the last few years it took to complete this book.

Sreenivas Alampalli

Authors

Dr. Mohammed M. Ettouney, MBA, PhD, PE, F AEI, Dist. M. ASCE, was conferred the Innovators Award in 2008 by the New Jersey Inventors Hall of Fame after he was nominated to receive such a great honor by the American Society of Civil Engineers (ASCE). Dr. Ettouney also received the Homer Gage Balcom lifetime achievement award by the Metropolitan Section of ASCE (2008). He won the Project of the Year Award, Platinum Award (2008) for the New Haven Coliseum Demolition Project (ACEC, NY). He is a fellow of the Architecture Engineering Institute (AEI) and a distinguished member in American Society of Civil Engineers. Among his other recent achievements are the pioneering work on "Theory of Multihazards of Infrastructures," "Theory of Progressive Collapse" pioneering work in multihazards/multidisciplinary evaluation of risk and resiliency of buildings, tunnels, bridges, and transit stations, and an innovative green design method for protecting utilities from demolition/blasting (City of New Haven, CT). He has professional interest in diverse areas of structural engineering as demonstrated by the list of his publications, invited presentations, seminars, and sessions organized during national/international conferences, besides his membership in different professional organizations.

Dr. Ettouney is a principal with Weidlinger Associates, based in New York City, NY. He received his Doctor of Science in structural mechanics from the Massachusetts Institute of Technology (MIT), Cambridge, MA, in 1976. Since then, his interest in structural engineering has been both as a practitioner and a researcher in multihazards safety of structures; probabilistic modeling of progressive collapse of buildings; uncertainties in structural stability; blast mitigation of numerous buildings around the world; innovative concepts such as "Probabilistic Boundary Element Method," "Scale Independent Elements," and "Framework for Evaluation of Lunar-Based Structural Concepts." He is a past president and past member of the board of governors of AEI, member of the Board of Directors of the BSC, member of several technical committees on building/infrastructure security, earthquake hazards, architectural engineering, and nondestructive testing and structural health monitoring. He was chair of the AEI National Conference, 2006 and 2008. He is a member of the NIBS Advanced Materials Council. He is also editor of the *Journal of Advanced Materials*.

Dr. Ettouney has authored or co-authored more than 325 publications and reports and contributed to several books. He has introduced several new practical and theoretical methods in the fields of earthquake engineering, acoustics, structural health monitoring, progressive collapse, blast engineering, and underwater vibrations. He co-invented the "Seismic Blast" slotted connection. More recently, he introduced the "Economic Theory of Inspection," "General and Special Theories of Instrumentation," and numerous principles and techniques in the field of infrastructure health—all pioneering efforts that can help in developing durable infrastructures at reasonable costs.

Dr. Sreenivas Alampalli, PE, MBA, is Director of the Structures Evaluation Services Bureau at the New York State Department of Transportation (NYSDOT). His responsibilities include managing structural inspection, inventory, and safety assurance programs at the NYSDOT. Before taking up the current responsibility in 2003, Dr. Alampalli was director of the Transportation Research and Development Bureau. In this position, he managed a targeted transportation infrastructure research and development program to enhance the quality and cost-effectiveness of transportation policies, practices, procedures, standards, and specifications. He taught at Union College and Rensselaer Polytechnic Institute as an adjunct faculty.

Dr. Alampalli obtained his PhD and MBA from Rensselaer Polytechnic Institute, his MS from the Indian Institute of Technology (IIT), Kharagpur, India, and his BS from S.V. University, Tirupati, India. His interests include infrastructure management, innovative materials for infrastructure applications, nondestructive testing, structural health monitoring, and long-term bridge performance. He co-developed the theory of multihazards and has been a great proponent of it to integrate all vulnerabilities, including security for effective infrastructure management. Dr. Alampalli is a Fellow of the American Society of Civil Engineers (ASCE), American Society for Nondestructive Testing (ASNT), and International Society for Health Monitoring of Intelligent Infrastructure (ISHMII). He has received several awards, including the prestigious Charles Pankow Award for Innovation, from the Civil Engineering Research Foundation in 2000; ASNT Mentoring Award in 2009; and Herbert Howard Government Civil Engineer of the Year Award from ASCE Metropolitan (NYC) section in 2009. He has authored or co-authored more than 250 technical publications.

Dr. Alampalli is an active member of several technical committees in TRB, ASCE, and ASNT, and currently chairs the ASCE Technical Committee on Bridge Management, Inspection, and Rehabilitation. He served as the Transportation Research Board representative for the NYSDOT and also as a member of the National Research Advisory Committee (RAC). He is an Associate Editor of the *ASCE Journal of Bridge Engineering* and serves on the editorial boards of the journal *Structure and Infrastructure Engineering: Maintenance, Management, Life-Cycle Design and Performance* and the journal *Bridge Structures: Assessment, Design and Construction.*

1 Scour

1.1 INTRODUCTION

1.1.1 FLOODS AND SCOUR

Floods can have devastating effects on bridges. They occur in most geographical regions, and their effects are felt by the structural system of the bridge, as well as the supporting soil and the approaching streams. Floods can affect the bridge system in five distinct ways as shown in Figure 1.1. Overtopping happens when the floodwater level becomes higher than the bridge superstructure and can lead to major bridge damage (Figure 1.2). Debris carried by floodwaters can impact the bridge structure and cause damage in the process (Figures 1.3 and 1.4). Harmful effects of the debris from hurricanes can be seen in Figure 1.4. Direct water pressure can affect the bridge and cause structural failure. Another form of flood effect is the potential failure of embankments, which in turn can damage the bridge. Finally, floodwater and stream flow can cause soil erosion, either under the foundation of the bridge or near the approaches to the bridge. This process is called "scour." Scour effects account for most failures due to floods (Table 1.1). The five modes of flood damage to bridges are summarized in Table 1.2.

1.1.2 SPECIAL NATURE OF SCOUR

Scour as one of the important bridge hazards has several special features that make it fairly unique and difficult to manage. It affects all foundation types. Deep foundations as well as shallow foundations are susceptible to scour effects. Scour is of several types: local, stream, etc. All are dangerous to bridge structures. Since it undermines the stability of the foundation, its effects can be global in nature: a failure of a single footing may lead to progressive failure of the overall bridge system, as what happened during the failure of the Schoharie Creek Bridge in 1987. Another special feature of scour is that it can happen in almost any type of soil or rock. The mode of scour failure can also be gradual erosion or sudden failure. One of the most difficult aspects of scour events is that they are mostly hidden under water. This particular fact makes monitoring of scour difficult. Conventional Structural Identification (STRID) methods need to be adjusted to accommodate fluid, soil, and structure interactions—a difficult and mostly unexplored issue. Because of the usually hidden nature of scour damage (under water), Damage Identification (DMID) is also a difficult task. Finally, because of the high degree of uncertainties of all the parameters associated with the scour problem, decision making regarding scour vulnerabilities and mitigations can be extremely difficult.

Because of the dangers that the scour problem poses, and because of the above-mentioned difficulties, structural health monitoring (SHM) application to the problem is particularly important. It offers accuracy and efficiency, which can be translated into cost-effectiveness and enhanced safety.

1.1.3 THIS CHAPTER

We discuss first different types of scour and how they affect bridges. To do so, we also need to discuss the causes of scour. Of special interest are different scour failure modes and how they can impact bridges. We then offer an overview of scour mitigation measures. In all, we observe how

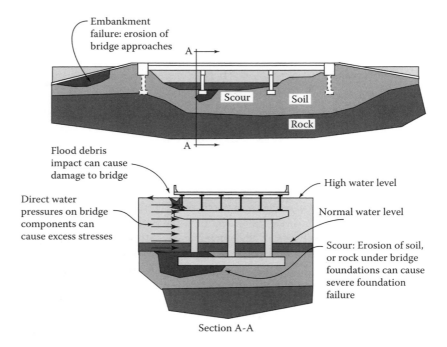

Section A-A

FIGURE 1.1   Flood-damaging modes.

those scour vulnerabilities and mitigation measures interact with SHM. We then offer a different vantage point: we present different SHM components (sensing, STRID, and DMID) and the role each plays in the problem. Three detailed case studies are offered: STRID applications to scour, fixed/continuous scour monitoring, and portable/intermittent scour monitoring. The last two sections of this chapter offer decision making examples of the scour problem, and we conclude the chapter with some scour management strategies. Figure 1.5 offers a schematic presentation of topics in this chapter. We note that much of the background material presented here is based on the landmark works by Richardson and Davis (2001), Lagasse and Richardson (2001), and Lagasse and Zevenbergen (2001). Those documents are also known as HEC-18, HEC-20, and HEC-23, respectively. We will be referring to them by their short names, for the sake of simplicity.

1.2   TYPES AND CAUSES OF SCOUR

1.2.1   SCALES OF SCOUR EFFECTS

We established the importance of scour to the health of bridges in the previous sections. In order to devise efficient strategies to ensure healthy protection of bridges against scour, we first need to identify the causes and types of scour. We observe that scour affects bridges on three spacial scales: (1) particles of earth material scale, (2) local scale, and (3) regional scale. Types of earth material include, but are not limited to, cohesive, noncohesive, and rock types. Local scour includes erosion under bridge foundations (abutments, piers, etc.) and scour from pressure flow during severe floods. Regional scour is the scour that occurs away from the bridge, but it can have major effects on the bridge system; it includes contraction scour, aggradations-degradation, and shifting of streams. Figure 1.6 shows the three scales of scour effects. We discuss each of the effects in more detail next. There are, of course, temporal scour scales, see Melville (1988 and 1999). These temporal scales are observed in several sections in this chapter.

FIGURE 1.2 Effects of overtopping from damaging hurricanes: (a) Hurricane Ike. (Courtesy of Eric Letvin.) and (b) Hurricane Katrina. (Courtesy of Milagros Kennett.)

FIGURE 1.3 Debris damage during flooding. (Courtesy of New York State Department of Transportation.)

(a)

(b)

FIGURE 1.4 Effects of debris from damaging hurricanes: (a) Hurricane Ike. (Courtesy of Eric Letvin.) (b) Hurricane Katrina. (Courtesy of Milagros Kennett.)

TABLE 1.1
Causes of Bridge Failure in the United States

Flood-Damaging Mode	Relative Contribution (%)
Overtopping	14
Debris	5
Direct structural failure	19
Embankment failure	22
Scour	40

Source: Annandale, G. W. Risk analysis of bridge failure, *Proceedings of Hydraulic Engineering* 93, ASCE, San Francisco, CA, 1993. With permission.

TABLE 1.2
Modes of Flood Damage

Flood-Damaging Mode	Effect on Bridges
Overtopping	Floodwater overtopping the bridge would cause abnormal hydrostatic and hydrodynamic pressures on the bridge. These pressures can dislodge the bridge from its bearings, leading to local or global bridge instabilities
Debris	The impact of floating debris on different submerged bridge components would result in an impacting force on these components These impacting forces can cause structural damage
Direct structural failure	High floodwater volume would result in increased hydrostatic and hydrodynamic pressures on submerged bridge components. At the limit, such increased pressures can cause structural damage to the bridge
Embankment failure	Embankment failure can have one or both of the following effects: (1) Failure of approaches, thus affecting traffic to and from the bridge, and/or (2) if embankment failure is close to the bridge support, the bridge superstructure can fail
Scour	Scour will cause soil erosion at the foundation interface. It also can cause soil erosion near the bridge, which can cause harmful water flow at the water-structure or soil-structure interfaces. Clearly, loss of soil mass at the foundation interface can cause instability, either local or global, to the bridge structure

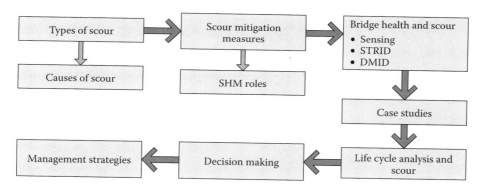

FIGURE 1.5   Scour chapter layout.

1.2.2 Material Response to Scour: Chemical Gel versus Physical Gel

Earth materials can be put into two categories: physical and chemical gels (Annandale 2006). The main difference between the two is in the manner the constituent particles connect together. In physical gels, the particles connect by simply touching each other. In chemical gels, the connections are formed by bonding; such bonds can be cohesion or cementation. When water pressure affects either of the two categories, the nature of the connection plays an important part in how scour might occur, the nature of scour, potential mitigation measures, and monitoring techniques.

The equilibrating forces in physical gel situation are (1) friction between particles, (2) own weight of the particles, and (3) disturbing hydraulic forces. When the disturbing hydraulic forces pass a particular threshold, the particle moves, thus causing scour. Annandale (2006) observed that such a threshold is lower during turbulent flow than during laminar flow. Physical gels include noncohesive silt, and gravel materials.

Chemical gels consist of two categories: (1) materials whose properties do not change when exposed to water, such as intact rocks, and (2) materials whose properties change when exposed to water, such as clay. Chemical gels experience flexure, brittle fracture, and fatigue. Depending on the

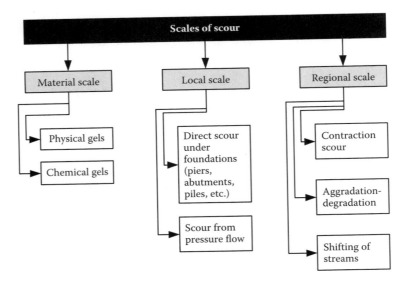

FIGURE 1.6   Scales of scour.

fracture toughness of the material, they can fail under hydraulic forces (scour) in one of two modes: brittle failure and fatigue.

Brittle failure: As flow fluctuations are applied to the material surface, fissures start forming on the surface; as the fissures increase, the stress intensity at the fissure edge increases. When the stress intensity reaches the material fracture toughness level, a brittle fracture occurs. This fracture, similar to all other brittle fractures, can be sudden with no previous warning. This form of scour occurs mostly in intact rocks.

Fatigue: When stresses due to hydraulic forces are much lower than fracture toughness, another form of failure can occur in chemical gels. As the stresses are repeated, slip surfaces develop and increase in length. With enough cyclic flow, the slip surfaces fail, causing scour. This form of scour occurs mostly in cohesive materials.

Figure 1.7 shows categories of material scour behavior. For a detailed discussion on the physics of scour, the reader is referred to Annandale (2006) and Richardson and Davis (2001).

1.2.3   LOCAL SCOUR AT PIERS OR ABUTMENTS

When the flow is interrupted by a pier, abutment, or other bridge substructure components, it becomes more complex in nature, since it tries to compensate for the rigid boundary of the substructure. This complex flow pattern will create vortices near the bed. The vortices formation will be accompanied by an increase of flow velocity that will cause a transport of the bed sediments, thus causing local scour. The degree of local scour is related to several factors, as shown in Figure 1.8. Note that local scour can also be either live bed (bed sediments transported from upstream) or clear water (no bed sediments are transported from upstream). We also observe that local scour is mostly a cyclic event, a factor that can affect its monitoring methodologies. Figure 1.9 shows local scour around piers in the floodplain of a bridge. Figure 1.10 shows a local scour around a bridge abutment. Note that the capacity of the footings under the abutment has been compromised by the scour events.

1.2.4   SCOUR FROM PRESSURE FLOW

If stream waters are blocked by the bridge superstructure, increased pressure under the bridge will result. The increased pressure may cause local pier/abutment scour or contraction scour (Figure 1.11).

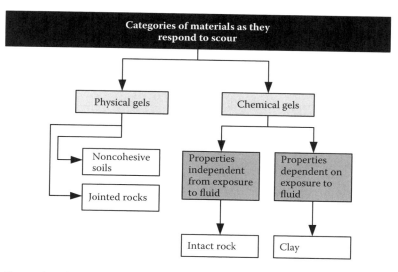

FIGURE 1.7 Categories of materials as they respond to scour.

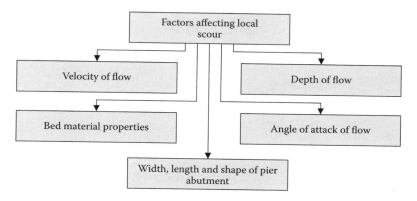

FIGURE 1.8 Factors affecting local scour.

FIGURE 1.9 Local scour at bridge piers. (From USGS site http://ky.water.usgs.gov/Bridge_Scour/ BSDMS.)

FIGURE 1.10   Local scour at abutment. (Courtesy of New York State Department of Transportation.)

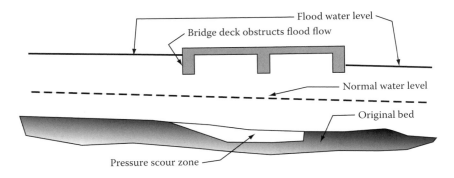

FIGURE 1.11   Scour due to pressure.

This situation can occur during severe floods. For a more in-depth discussion, see Richardson and Davis (1995) or Jones et al. (1993).

1.2.5 CONTRACTION SCOUR

Contraction scour occurs due to the reduction of the stream or river cross-sectional area. The reduction would increase the flow speed. The flow speed increase can lead to transport of the bed sediments in the contraction area, thus causing a scour. There are two types of contraction scour: live-bed contraction and clear-water contraction. Live-bed contraction scour happens when the flow speed is higher than a critical speed V_c. Above such a critical speed, transport of the sediments from upstream into the bridge area does occur. Clear-water scour occurs when the flow speed is lower than the critical speed. Below the critical speed, there is no net sediment transport from the upstream uncontracted area to downstream. Figure 1.12 shows some of the causes of contraction scour as described by Richardson and Davis (1995). Note that contraction scour is a cyclic event, not a long-term one. This can have implications for the monitoring methodologies of this type of scour.

1.2.6 AGGRADATIONS AND DEGRADATION

Man-made or natural events can cause long-term transport of sediment material from streambed/riverbed. Aggradations occur when the transported sediments move into the area, thus increasing the bed level, while degradation takes place when the transported sediments move away from the

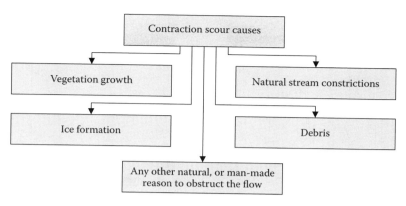

FIGURE 1.12   Contraction scour causes.

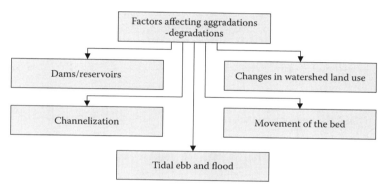

FIGURE 1.13   Some factors that affect aggradations and degradations.

area, thus lowering the bed level. Some factors affecting the long-term level of the bed are shown in Figure 1.13 (Richardson and Davis 2001).

1.2.7   SHIFTING OF STREAM

Richardson and Davis (1995) observed that streams are dynamic in nature—change in time and space. These changes can either be meandering or braided (Lagasse and Richardson 2001), as shown in Figure 1.14. These changes in the banks of streams can be both lateral movements and movements in the direction of downstream, as in Figure 1.14. When a bridge, which by definition is a static object, is located at the site of a continually moving stream, some severe scour can occur. Factors that affect the movement of the stream (thus causing scour) are shown in Figure 1.15. For a more in-depth discussion of stream shifting, the reader is referred to Lagasse and Richardson (2001).

1.3   SCOUR MITIGATION MEASURES

1.3.1   MITIGATION OF SCOUR: GENERAL

There are three basic measures to mitigate scour effects on bridges: soil, structural, and site mitigation measures (Figure 1.16). These methods illustrate the complexity of the scour hazard. The hydraulic, soil, and structural components of scour are evident in the three mitigation categories. Efficient methods need to account for more than one of those scour components. This section gives

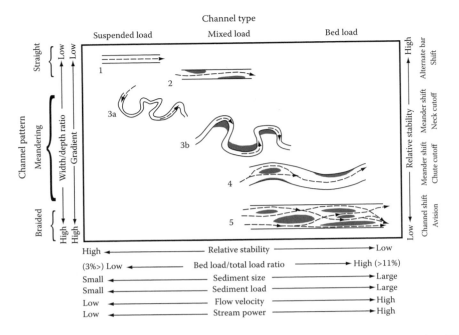

FIGURE 1.14 Channel classification and relative stability effects. (With permission from National Highway Institute.)

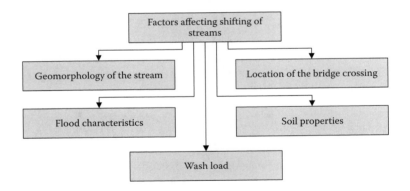

FIGURE 1.15 Some factors that affect stream shifting.

an overview of different mitigation methods, with emphasis on the SHM interrelationship with these methods. For comprehensive scour mitigation coverage, see HEC-23.

1.3.1.1 Reducing Risk versus Enhancing Safety

There are two types of scour countermeasures: those that reduce risk and others that enhance bridge safety. Risk-reducing countermeasures are used depending on (1) whether the bridge is scheduled for replacement within a few years, (2) bridge topology such as limited clearances and limited options for countermeasures, and (3) the type of bridge where consequences of failure are low, such as a bridge with low-volume traffic. Safety countermeasures will require advanced hydraulic studies. They include structural foundation strengthening of the bridge. See HEC-18 for a detailed discussion.

pier ($K = 1.7$ for rectangular piers, and $K = 1.5$ for round-nosed piers). Equation 1.1 shows what to monitor in case of riprap:

Monitor demands: Flow velocity V can be monitored as needed, both during normal flow and during flood conditions.

Monitor capacity: The geometry and condition of the riprap formation can be monitored visually to ensure adherence to design parameters.

Figures 1.17 and 1.18 show riprap use as a scour countermeasure.

Guide banks (spur dykes): (HEC-23, p. DG10.3)

Bridges crossing over openings from large floodplains are particularly vulnerable to scour damage due to turbulent flows approaching and crossing the opening (see Figure 1.23). By placing guide banks at the opening, a more streamlined flow is achieved, thus reducing scour risk (see HEC-23 for design details).

Check dams

When the drop of a streambed is steep near a bridge, it might result in high velocities, thus increasing scour potential at bridge site. Check dams are then constructed to control and stabilize water flow at bridge foundations (see Figures 1.24 and 1.25). Visual inspection and occasional flow velocity monitoring are sufficient in most situations.

Revetments (HEC-23, p. DG13.3)

These can be flexible revetments, such as rock riprap, rock, and wire, or precast concrete blocks. Old tires and natural vegetation can also be used. In addition to flow velocity and size of revetment particles, there are other factors that affect the design, such as the filtering system that would prevent soil erosion under the revetments. The edges of the revetment region should be designed and constructed carefully, since scour can occur around the edges. Visual inspection should include monitoring the conditions of the filter and the edges.

Rigid revetment includes soil cement and grouted/partially-grouted riprap (see HEC-23 for more details).

1.3.5 SHM AND SCOUR MITIGATION MEASURES

1.3.5.1 General Rules

Monitoring scour hazard depends on the phases of flooding/scour as follows:

Before floods/normal conditions: During normal conditions, a program for inspecting scour conditions and countermeasures needs to be implemented. Automatic scour-monitoring devices might

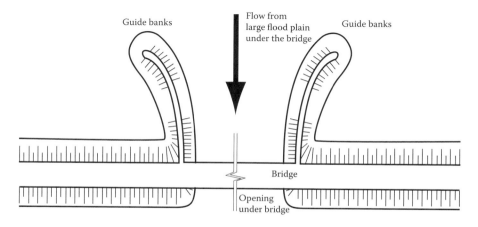

FIGURE 1.23 Geometry of guide banks.

FIGURE 1.24   Check dam (under bridge). (Courtesy of New York State Department of Transportation.)

FIGURE 1.25   Check dam on the bridge approach. (Courtesy of New York State Department of Transportation.)

also be installed for real-time observations. A stability analysis of critical bridge foundation should be performed a priori. The critical stability conditions should be monitored in the fields, with adequate safety factors. Also, decision making examples of this situation are given in the next section.

During flooding: Station inspectors at the flooding site should monitor conditions closely. If there are real-time monitoring devices, ensure that the monitored results are processed in real time and appropriate decisions made immediately.

After flooding: Bridge closing or immediate retrofit efforts should be planned in advance and kept ready for implementation. Readiness can improve scour resiliency. See Chapter 2 for more detailed discussions on resilience.

1.3.5.2   Capacity/Demand Monitoring

Monitoring techniques can be devised by inspecting the design procedures of each countermeasure. Parameters that control design would be good candidates for monitoring countermeasures and their adherence to design assumptions.

TABLE 1.3
Scour Countermeasures for New and Existing Bridges

Scour Mitigation Category	New	Existing
Site	Choose appropriate site	NA
	Improve stream performance to improve stream stability. Methods mentioned above are applicable to both new and existing bridges	
Bridge structure	Streamline bridge components (abutments, piers, etc.)	NA
	Advanced scour foundation design	Strengthen foundations
	Design foundations to resist scour without relying on riprap or similar material	Check dam construction
	Use deep foundations, if possible	Underpinning
	Check dam construction	Relief bridges
		Additional spans (lengthen bridges)
Improve soil conditions	Vegetation, soil cement, etc.	Vegetation, soil cement, etc.

Scour can be monitored by either observing the capacity of the systems or the demands of floods/scour. In most cases, it is advisable to monitor both capacity and demands. This is because both system capacities such as scour holes or deterioration of countermeasures change as time passes. Demands such as stream water volume or flow velocity may also change. Thus, safety margins can decrease suddenly, leading to unwanted failures. Capacity is monitored by observing the geometry and condition of the countermeasures. This observation can be performed either visually or by other devices such as optical lasers or displacement measurements. Demands are monitored by measuring flow at normal and flood conditions. Flow measurements include velocity, pressure (on foundations), and/or flow volume.

1.3.5.3 Existing versus New Bridges

Mitigating scour measures for new bridges are different from those for existing bridges. Table 1.3 shows the differences. Monitoring and inspection methods, however, are fairly similar and should be strictly enforced.

1.4 BRIDGE HEALTH AND SCOUR

1.4.1 Inspecting a Bridge for Scour

1.4.1.1 General

Because of the importance of scour damage to the well-being of bridges, there are several methods for evaluating scour effects on bridges. Most of them have a graduating degree of complexity. Figure 1.26 shows in a schematic manner the major steps that different scour evaluation schemes follow. The steps start with an in-office evaluation. Such an evaluation includes mainly the collection of information about the bridge under consideration. The information includes data from the National Bridge Inventory (NBI; see Section 1.9), if available, the history of any type of problems encountered since the bridge was commissioned, the importance of the bridge, and the consequences of any damage or failure of the bridge. A decision is then made as to whether there is a need for further evaluation. Such a decision is generally based on a ranking scheme.

The next step is manual inspection of the site with the aid of prepared forms. The forms are usually based on the observation of several factors known to promote scour. The observations are recorded with a rating system. At the end of the inspection, the ratings of all observations are added, and a general scour rating is obtained. If such a general rating crosses a specific acceptance limit, the evaluation process moves to the next step.

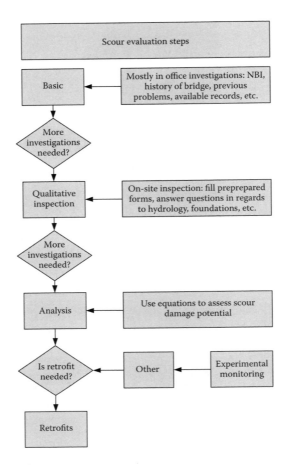

FIGURE 1.26   Scour evaluation steps.

The next step in evaluation is analysis. Several analytical techniques are available for estimating scour potential. The estimation of scour analytically is usually coupled with some *engineering judgment* to ascertain the severity of the scour potential. Such an *engineering judgment* is needed since many of the available equations are empirical equations derived by using many general assumptions; thus, personal expertise (engineering judgment) might be needed.

We now come to the main subject of our dissertation: the role of experimental monitoring in scour evaluation. In general, this is not used as of the writing of this document. With a few exceptions, there are no records of experimental monitoring being *prescribed* in most practice guides. In fact, there are several situations that would entail, even necessitate, the use of experimental monitoring for scour.* It would be natural to include such situations in scour evaluation guides.

1.4.1.2   HEC-8

One of the main reference standards in bridge scour field is the work by Richardson and Davis (1995). They supplied a concise checklist to qualitatively assess scour potential at bridge sites. Figure 1.27 shows the main subjects in their checklist.

In assessing the banks upstream or downstream, the inspector will look for signs of stability or instability of the bank. Presence of natural vegetation, riprap, pavements, or dykes would indicate

---

* We point at these situations throughout this chapter.

pile bents, spaced 9.1 m apart, and made up of a single row of 610-mm² precast concrete piles. Only the estimated pile lengths, with no pile driving records or actual installed lengths, were noted.

Borg (2000) concluded that the sonic echo (SE) method was not feasible for the experiment as none of the structures except one had clearance between the top of the pile cap and the superstructure. The steel pile jackets were expected to cause wave interference, and the likelihood of the piles being tapered would increase the error for length determination. If applicable, investigation of every pile bent using this method would have been feasible. In SE Method, the top of the pile (or top of the pile cap) is struck with a handheld hammer, generating a downward-traveling compression wave. An accelerometer measures the response of the wave reflections to factors that alter the acoustic impedance of the pile, such as stiffness, density, and cross-sectional area. Using the speed of the sound wave in the pile material and the observed pile toe reflector from the wave signature, the length can be calculated. In general, this method is limited to a pile embedment to diameter ratio of less than 30.

Bending wave (BW) tests were also tried at four pier bent locations. BW tests rely on reflections of BWs from changes in impedance such as that at the toe of the concrete piles. Reflections could not be identified in any of the BW tests and were therefore inconclusive.

Borg (2000) also used the borehole PS method. This requires that a cased or uncased borehole be installed adjacent to the existing pile (typically within 1–1.5 m) to some depth (at least 3 m) below the anticipated pile toe. A hydrophone is lowered to the bottom of the casing and raised incrementally after the pile is struck with a handheld hammer. The hammer impact induces a low-strain wave to travel down the pile and into the surrounding soil, thus triggering the hydrophone. Initial arrival time of the wave is plotted against the coinciding depth of the hydrophone in the casing. The slope of this curve indicates the speed of the sound wave in the material. Where the slope indicates a change from the wave speed in the pile material to the wave speed in the soil, the pile toe is indicated.

This determination is possible when the input wave is sufficiently strong to trigger the hydrophone, and the speed of sound in the pile material and the soil medium is sufficiently different. This method works well for purely columnar foundation elements and where the surrounding soil wave speed is constant, as is the case with fully saturated soils (soil below groundwater level). PS data presents itself to direct interpretation and is generally considered to be the most reliable low-strain NDT method for determining pile length.

PS tests were performed with a 1.4-kg hammer and an eight-channel hydrophone string. Twenty pile bents were tested, with one to three piles tested per bent. Due to the strong channel currents, steel casing was used. Because of low superstructure clearance, the casing could be installed only as close as 3.5 m. The pile toe was clearly indicated in most of the PS results. Of the 20 bents tested, PS results were inconclusive for only three bents due to the dominant waves traveling through the steel.

1.4.2.6 Time Domain Reflectometry

TDR application to scour detection was described by Dowding and O'Connor (2000). The system consists of a metallic coaxial cable placed in a drill hole and anchored to the walls by tremie placement of an expansive cement grout. When movements taking place in the rock or soil are sufficient to fracture the grout, cable deformation occurs. A voltage pulse is propagated along a cable grouted into place. When the pulse reaches a deformity in the cable, a portion reflects. Location of the deformity is calculated by the time of flight of the reflected pulse. It is possible to locate deformation zones but also to distinguish shearing from tensile deformation and to quantify the magnitude of deformation. The cable is crimped prior to placement in the hole to provide distance reference markers in the TDR records. Figure 1.33 shows the schematics of the TDR system.

Mine application

One application mentioned in this study is to monitor abandoned mines. TDR monitoring combined with phone lines can serve as a remote SHM. Data can be monitored continuously, providing

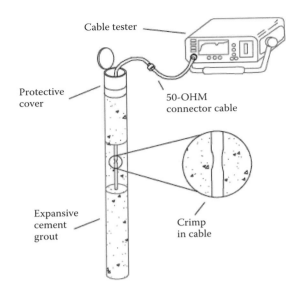

FIGURE 1.33   Schematics of cable installation and monitoring. (Courtesy of CRC Press.)

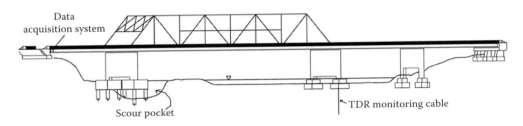

FIGURE 1.34   Bridge system. (Courtesy of CRC Press.)

continuous time history, and used on the basis of triggers to give alarms. This can be used to monitor slopes, scour applications, buildings on weak soils, etc.

Bridge Case Study

TDR was used for scour repair monitoring of a steel truss bridge supported on piers and spread footings over the Klamath River in Horse Creek, California, where a deep scour pocket beneath the pier on the west side of the river was discovered (Figure 1.34). The scour hole was repaired and the structure retrofitted appropriately. Coaxial cables were installed through the foundation of the pier on the east side end of the truss to measure footing movement that might result from scour of the graphitic schist supporting the spread footing. It was supplemented using water level "sensor" cables and tiltmeters. Monitoring data showed some initial movement but stabilized later with no further movement.

Considerations

To maximize sensitivity in soil, the shear capacity of the grouted cable should be less than the bearing capacity of the soil to facilitate grout fracture, so that the cable can be deformed as movement occurs within the surrounding soil.

TDR can also be used to measure water level. This offers other monitoring opportunities for trigger mechanisms.

1.4.2.7 Closing Remarks

We end this section with a comparison of different scour-sensing strategies as presented by HEC-23 in Table 1.9.

1.4.3 STRUCTURAL IDENTIFICATION

1.4.3.1 Conventional STRID Methods

Olson (2005) studied different STRID techniques on in-field–simulated scour damage. To perform the studies, the author first completed a set of dynamic tests on two bridge sites. The tests were performed for pristine and damaged conditions. The scour-damaged conditions were simulated by removing the soil surrounding foundation piles (Figure 1.35). Using the test results, modal identification of the structure was done. Differences in modal responses between pristine and damaged conditions were observed. Using dynamic test results for parameter identification did not produce a similar satisfactory answer. One potential reason might be the complexity of the structural system

TABLE 1.9
Advantages and Disadvantages of Scour-Monitoring Systems

Instrument Category	Advantages	Limitations
Fixed	Continuous monitoring, low operational cost, easy to use	Maximum scour not at instrument location, maintenance/loss of equipment
Portable	Point measurement or complete mapping, use at many bridges	Labor intensive, special platforms often required
Geophysical Positioning	Forensic investigations	Specialized training required, labor intensive

Source: HEC-23, with permission from National Highway Institute.

Excavating under foundation

Excavating around piles

FIGURE 1.35 Simulation of scour conditions. (Courtesy of Olson Engineering.)

and the interactions of different modes. The final phase of the research was the study of the use of Hilbert-Hwang transform (HHT) in trying to identify scour damage. The initial findings were encouraging, and further research was recommended.

Neural networks have also been used to identify complete structure-soil and water systems, Jeng et al. (2005). Since neural networks avert the complications of physical modeling of constituents, its use in scour situations is relatively simple. The disadvantage, of course, is that no physical insight can be had into the system behavior.

1.4.3.2 System Stability

Earlier, we presented earlier three modes of scour damage: gradual, brittle fracture, and fatigue. We hasten to add that even though the damage modes are different, the foundation system will usually behave in a similar manner: it will fail suddenly when the damage reaches a threshold of failure. Such a sudden failure, which is a form of structural instability, is usually catastrophic. We will discuss structural stability phenomenon as it relates to scour problem in this section.

Before we begin, we need to emphasize that the goal of STRID is to identify certain aspects of structural properties by using sensing and measurements. In a scour stability situation, the STRID goal is to identify, by some kind of scour field measurements, the propensity of the system to become unstable. We first investigate the suitability of conventional STRID to this task. We then offer a general STRID method that can be used.

1.4.3.2.1 Structural Stability and Conventional STRID Methods

Modal identification: Ettouney and Alampalli (2012, Chapter 8) showed a simple example of using measured local frequencies in identifying the potential loss of stability of truss bridges. The example shows the immense potential of using modal techniques in detecting the onset of structural instability. To utilize such a potential, the conventional modal techniques need to be readjusted to accommodate axial or second-order effects within the system, in a manner similar to the example given in Chapter 8 by Ettouney and Alampalli (2012).

Parameter identification: Olson (2005) studied the modeling of soil erosion around piles. Different STRID techniques were used, including parameter identification methods. Mixed results were reported. Perhaps parameter identification methods need to be developed specifically for considering stability. Currently, they are designed for parameter identification of the conventional force-displacement relationships. Formally, the current parameter identification model is

$$KU = P \qquad (1.10)$$

with K, U, and P representing the stiffness matrix, the displacement and the force vectors, respectively. And by measuring some U and P values, the parameters of K can be identified. We recall that structural stability model is

$$(K - lK_1)U = 0 \qquad (1.11)$$

The geometric matrix is K1 and the stability eigenvalue is l. Clearly, parameter identification of K and K_1 in the stability model needs techniques that are different from techniques used in the conventional case. More research in this area is needed.

Trial and error: Hunt (2005) used a trial-and-error technique to detect the onset of structural instability due to scour. The method is summarized as follows:

1. Model the structure, foundation, and soil system using a detailed finite-element method.
2. Apply design loads using a nonlinear incremental technique
3. Check the resulting nonlinear displacements of the structure

4. Remove some soil material from the model and repeat the analysis. Repeat steps 2 and 3.
5. Continue removing the soil, thus simulating the gradual scour effects, until the structure becomes unstable. The finite-element scour depth in this situation is the scour threshold that would cause structural instability

This method is simple and efficient. It succeeded in detecting the stability threshold by identifying the soil erosion geometry. By installing a monitoring mechanism, it was possible to warn against that stability threshold. We note that the method is applicable only for gradual scour. It is not applicable to brittle fracture-like or fatigue-like scour modes. However, the techniques can certainly be generalized as a basis for a stability-centric STRID method. We explore a generic scour-based technique later in this section.

1.4.4 DAMAGE DETECTION

1.4.4.1 General

The most popular metric of scour damage is the depth of scour, that is, the difference in elevation between the current soil level and the baseline soil level. In situations where the scour depth is measured directly, the DMID is direct. However, in many situations scour depth cannot be measured directly. It is estimated only by measuring or estimating other factors. For physical gel (noncohesive materials, for example), empirical damage expressions for many conditions were developed over the years. The expressions aim at relating scour depth with other variables that control scour. They were developed by field observations and/or laboratory testing. They have been used to evaluate the severity of scour damage. We immediately observe that DMID in this situation is another direct example of VSP. The scour damage is not measured directly; it is arrived at by measuring or estimating the environment that causes it. In the rest of this section, we summarize scour damage expressions for some conditions. HEC-23 and HEC-18 include more techniques of estimating scour damage.

The above definition of scour damage (as scour depth) does not include scour damage in situations where failure occurs in a brittle fashion. The scour damage expressions for this situation can be found in Annandale (2006).

1.4.4.2 Contraction Scour

Contraction scour is affected by two conditions: live bed and clear water. The empirical scour equations for each are presented next:

1.4.4.2.1 Live-Bed Scour

Live-bed scour occurs when there is transport of bed material from upstream to locations downstream. The damage is described by

$$\frac{y_2}{y_1} = \left(\frac{Q_2}{Q_1}\right)^{6/7} \left(\frac{W_1}{W_2}\right)^{k_1} \tag{1.12}$$

This equation is known as Laursen equation. The variables are defined as

y_1 = average depth in the upstream main channel
y_2 = average depth in the contracted section of the channel
Q_2 = flow in the upstream main channel (unit: volume/time)
Q_1 = flow in the contracted section of the channel (unit: volume/time)
W_2 = width of the bottom of the upstream channel
W_1 = width of the bottom of the contracted section of the channel
k_1 = exponent ranging from 0.59 to 0.69; this factor depends on shear stresses in the streambed and mass density of water.

The units of Equation 1.12 should be self-consistent. More information can be found in HEC-18.

1.4.4.2.2   Clear-Water Scour

Clear-water scour occurs when there is no movement of the bed material in the flow upstream. The damage is described by

$$y_2 = \left(\frac{K_u Q^2}{D_m^{2/3} W^2} \right)^{3/7} \tag{1.13}$$

The variables are defined as

y_2　= average depth in the contracted section of the channel
K_u　= 0.025 (SI units) and 0.0077 (English units)
Q　= flow in the contracted section of the channel (unit: volume/time)
D_m　= diameter of the smallest nontransportable particle in the bed material in the contracted section
W　= width of the bottom of the contracted section of the channel, less the total widths of piers.

The units of Equation 1.13 should be self-consistent. More information can be found in HEC-18.

1.4.4.3   Local Scour Piers

A general scour damage expression at the piers is expressed by

$$y_s = 2.0 \, y_1 \, K_1 \, K_2 \, K_3 \, K_4 \left(\frac{a}{y_1} \right)^{0.65} \mathrm{Fr_1}^{0.43} \tag{1.14}$$

The variables are defined as

y_s　= scour depth
y_1　= flow depth directly upstream of pier
K_1　= correction factor for pier nose shape (from 0.9 to 1.1)
K_2　= correction factor for angle of attack of flow (from 1.0 to 5.0)
K_3　= correction factor for bed condition (from 1.1 to 1.3)
K_4　= correction factor for armoring by bed material size
a　= pier width
$\mathrm{Fr_1}$　= Froude number directly upstream of the pier

Note that Equation 1.14 applies for both live-bed and clear-water scour. The units of Equation 1.14 should be self-consistent. More information can be found in HEC-18.

1.4.4.4   Local Scour Abutments

To detect scour effects at abutments for both live-bed and clear-water conditions, HEC-18 recommended the use of same equations. Two possible equations can be used. The Froehlich expression, based on an analysis of 170 live-bed scour conditions in the laboratory, is

$$y_s = 2.27 \, y_a \, K_1 \, K_2 \left(\frac{L'}{y_a} \right)^{0.43} \mathrm{Fr}^{0.61} + 1 \tag{1.15}$$

The HIRE equation for scour at abutments based on field data of scour in the Mississippi river (Richardson et al. 2001) is expressed as

$$y_s = 4.0 \, y_1 \, \mathrm{Fr}^{0.33} \frac{K_1}{0.55} K_2 \tag{1.16}$$

The variables in Equations 1.15 and 1.16 are defined as

y_s = scour depth
y_a = average depth of low in floodplain
K_1 = coefficient of abutment shape (from 0.55 to 1.00)
K_2 = coefficient depending on angle of attack of flow
Fr = Froude number of approach flow upstream of abutment
y_1 = depth of flow at the abutment

The units of Equations 1.15 and 1.16 should be self-consistent. More information can be found in HEC-18.

1.4.4.5 Other Empirical Formulations

There are several other empirical methods for evaluating scour local conditions near piers. For example, a simple expression by Laursen and Toch (1956) estimates scour as

$$Y_s = 1.35 D^{0.7} Y^{0.3} \tag{1.17}$$

The pier diameter is D, and the average depth of approaching flow is Y. Shen (1971) suggested the expression

$$Y_s = 0.00022 \left(\frac{UD}{v} \right)^{0.619} \tag{1.18}$$

The average velocity of approaching flow is U. For more complex pier geometries, Sheppard (2003) developed a method for predicting local scour damage.

1.4.5 BRITTLE BEHAVIOR TECHNIQUE

1.4.5.1 General

Scour geometry is basically an interaction of four components: bridge columns, bridge foundations, supporting soil (or rock), and flowing water. For an accurate evaluation of the problem, the intricacies of the interaction of the four components must be accommodated. In this section, we simplify the process as shown in Figure 1.36. The bridge columns are subjected to a set of vertical and lateral forces. The vertical forces are the gravity loads as well as any traffic or live loads. The lateral forces include any lateral design forces. The lateral forces also can include hydrostatic (or hydrodynamic) flood pressures, if applicable. The bridge foundation and the underlying soil (or rock) can be analytically simulated by a set of springs or more complex finite elements. The scour process degrades the properties of those analytical simulations.

Figure 1.36 poses several questions:

- What is the critical limit state of the soil foundation at which the whole system will become unstable?
- What are the effects of the uncertainties in the geometry of Figure 1.36 on those critical limit states?
- What is the role of SHM in aiding to improve the accuracies of the above two questions?

We present two methods of solving the problem. The first subdivides the problem into vertical and lateral loadings and then combines them using a simple interaction approach. The second is a unified approach.

1.4.5.2 Critical Stability Condition

We consider first the vertical loading condition of Figure 1.36. The problem is simplified even further in Figure 1.37. Note that the soil foundation system is modeled as a rotational spring with a stiffness of K_ϕ. The column is modeled as a homogeneous column. The system is subjected to a vertical loading P. The flexural stiffness of the column is EI, where E is the modulus of elasticity and I is the moment of inertia. The system will fail when the vertical load reaches or exceeds the critical stability load of the system P_{cr}. The critical load can be obtained analytically using Timoshenko and Gere (1961). The governing stability equation can be shown to be

$$k\cos(k\ell) + K_{\text{SCOUR}} \sin(k\ell) = 0 \tag{1.19}$$

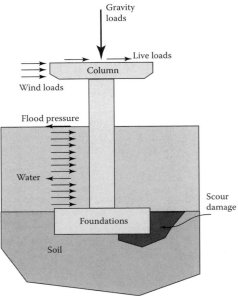

FIGURE 1.36 Simplified scour structural stability model.

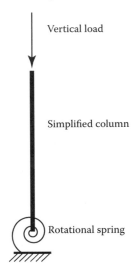

FIGURE 1.37 Simplified vertical loading model.

FIGURE 1.44 Vibroseis truck. (Courtesy of Olson Engineering.)

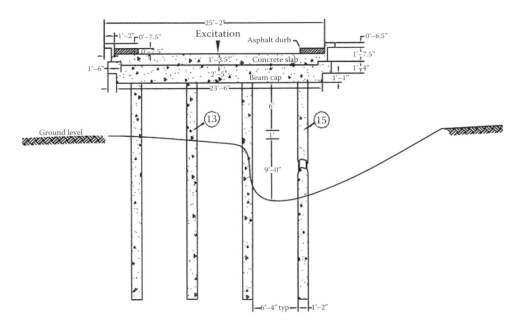

FIGURE 1.45 Typical tested bent. (Courtesy of Olson Engineering.)

authors applied parameter identification techniques to the tested bridges. Comparisons of the parameter identification results and field results did not prove to be satisfactory. A promising use of the dynamic test data was the application of HHT to the measured wave forms. The authors concluded that studying HHT shows that HHT-based signature recognition might be a means of identifying local dynamic properties. In addition, HHT analysis might be a means of determining differences in the signatures of local structural members. The HHT approach appears to identify local, nonlinear, and lower frequency responses of structural members to damage. Figure 1.48 shows a HHT of measured accelerations at a typical bent for baseline, excavated, and broken conditions. The effects of the changed states on the three spectra are evident.

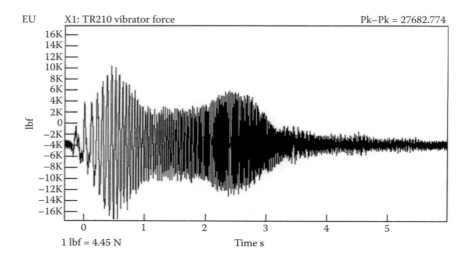

FIGURE 1.46 Typical generated vertical force. (Courtesy of Olson Engineering.)

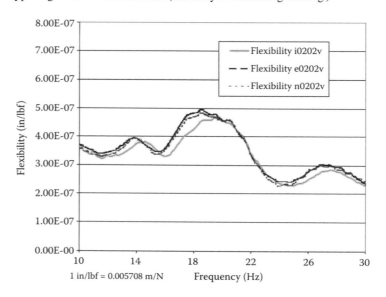

FIGURE 1.47 Comparison of flexibility of three bent states. (Courtesy of Olson Engineering.)

The study showed the promise and limitations of different STRID methods as applied to the scour problem. Certainly, more such studies and investigations are needed.

1.5.3 SCOUR MONITORING BY FIXED SONAR

Due to a pile failure resulting from scour at Wantagh Parkway over Goose Creek in Nassau County, New York, in 1998, NYSDOT initiated a program to investigate the cause in order to carry out appropriate repairs, identify scour-prone locations, and monitor key locations, using sonar devices to provide appropriate warning in case of an unexpected problem (see NYSDOT 1998). These locations have the ability to automatically collect periodic data and do subsequent analysis, routine monitoring, and continuous monitoring during scour critical events.

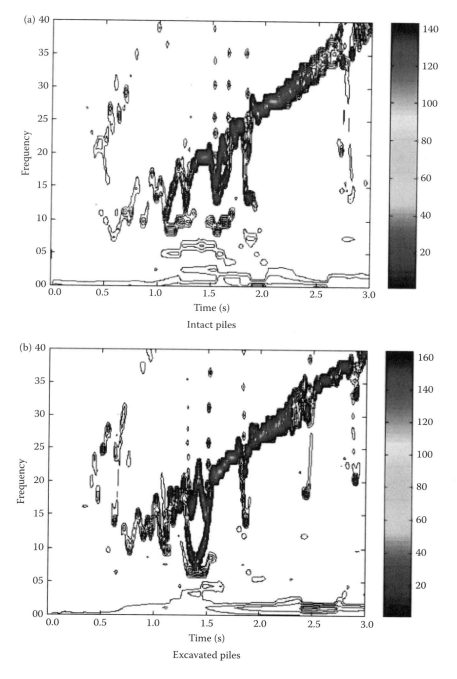

FIGURE 1.48 HHT for different scour-damaged States (a) intact piles, (b) excavated piles, and (c) broken piles. (Courtesy of Olson Engineering.)

The Wantagh Parkway Bridge was a 93-ft bascule bridge with concrete pile bent approach piers and 15 spans over waters subjected to tidal action. The streambed at one pier experienced about 29 ft localized scour since it was built in 1929 because of various storm events over the years and degradation due to daily tidal action. This caused pile settlement and fracture of pile cap. NYSDOT decided to replace the approach spans immediately, but bascule piers were to remain for several

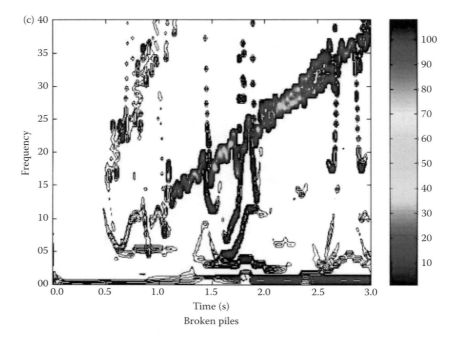

FIGURE 1.48   (Continued).

years. Thus, to ensure the safety of the bridge until it was completely replaced, a scour-monitoring system was installed after investigating several other options. Four scour monitors were installed at the bascule piers of Goose Creek, and they are still operational.

Scour monitors were also installed at Robert Moses Causeway over Fire Island Inlet in Suffolk County of New York as a long-term solution to the scour problems at the bridge. Due to the high flow rates expected at this bridge, the riprap needed monitoring according to HEC-23 when used as a countermeasure at piers. The scour monitors were placed on 13 piers considered to be most likely candidates to experience scour failure.

The system uses transducers to record streambed elevation measurements at designated piers. It includes one master control station per bridge, a remote station at each pier instrumented, a water temperature sensor, a water stage sensor, and an automatic alerting system (Figures 1.49 through 1.50). All the data gathered is transmitted to the master control station. A portable computer is set up at the designated office to retrieve the data from the systems. The scour monitor electronics measure the lapsed time an acoustic pulse takes to travel from the transducers to the streambed and back. They are converted into distances using the sound velocity, adjusted to the temperature. The transducers are programmed to take measurements at specified intervals. The intervals can be changed as needed. The program includes daily routine monitoring of these locations, including data acquisition and analysis, round-the-clock monitoring during scour critical events, preparation of weekly graphs of the streambed elevations and tidal gauge data, periodic data reduction analysis and graphs, and routine maintenance, inspection, and repairs (Figure 1.51). The condition of the scour monitors and the accuracy of the streambed elevation readings are checked during the regularly scheduled diving inspections at each bridge. During these inspections, debris and/or marine growth on the underwater components is also cleared. The program includes determining cautionary or critical scour elevations to trigger warning to the engineers as needed. Figure 1.52 shows the setup and steps used in monitoring the scour demands and scour foundation capacity. For a detailed discussion of this project, see Hunt (2005).

FIGURE 1.49   Sonar scour-monitoring system. (Courtesy of New York State Department of Transportation.)

FIGURE 1.50   Sonar scour monitoring installation. (Courtesy of New York State Department of Transportation.)

1.5.4   Scour Detection by Moving Sonar

Remote sensing using moving sonar sensing can also be performed. Stock (2008) reported a successful implementation of scour detection using boat-mounted sonar. The sonar used dual acoustic beams that operate at two different frequency ranges. First, a straight beam that aims in a vertical conical shape directly under the boat operates in the low-frequency range of 83–200 kHz. Such a beam can offer enough penetration of the bottom of the water body directly under the boat. The low-frequency range can distinguish soft and hard ground, as well as the depth of the water under the boat. A second higher frequency sonar beam that operates in the frequency range of 455–800 kHz

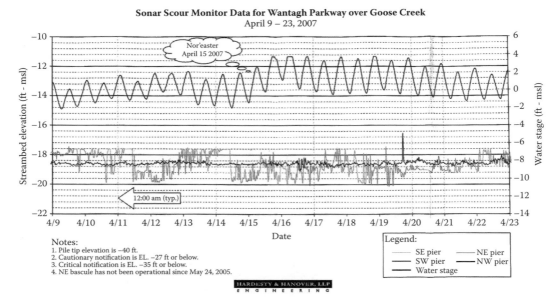

FIGURE 1.51 Typical sonar data. (Courtesy of New York State Department of Transportation.)

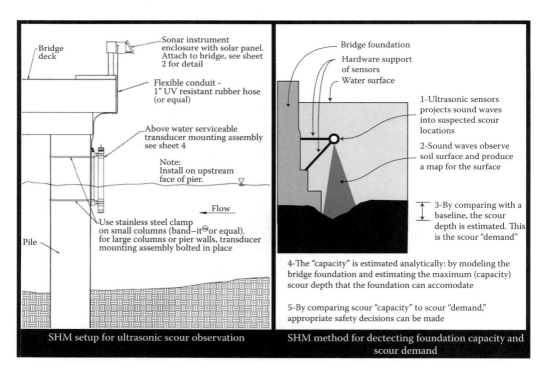

FIGURE 1.52 SHM instrumentation setup and steps for monitoring scour foundation capacity and scour demands.

1.7.1.2   Refining the Decision: Posterior Probability

The previous analysis was based on historically known probabilities of damaging scour situations at the bridge site. If the probabilities of scour situations $P(\theta_i)$ can be updated, it will lead to more accurate decisions. Such probabilities are known as posterior probabilities. The theoretical basis of posterior probabilities was covered by Ettouney and Alampalli (2012, Chapter 8). The question in the current example is how to obtain posterior probabilities.

There are, generally, two possible approaches to defining posterior probabilities. The first is to search for available data from similar sites and conditions. The second is to obtain site-specific data. Site-specific data can be numerically based, experimentally based, or a mix of the two. Figure 1.57 shows the different approaches to obtaining posterior probability information.

Let us assume that the bridge official did not find any data suitable for posterior probability information for the site. Remember, after all, that there is already a decision on hand, and the idea here is to improve on it. Therefore, the data needed for posterior probability analysis must exhibit at least some accuracy and relevance.

That leaves a site-specific data preparation. Pure analysis seems to be the obvious course of action. It is inexpensive, rapid, and does not require exposing the structure to unwanted scouring. Since, as was mentioned above, the goal of the official is to improve on the available decision, a purely analytical approach would not be appropriate in this situation. Similarly, pure monitoring would also present hurdles. For accurate experimental sampling, there should be some damaging scour for either of the two abutments and/or the foundation. Obviously, damaging the structure is not an acceptable option.

The way out of this dilemma is to adopt a hybrid analytical/experimental approach. It should be noted that this would entail perhaps one of the most important subjects of this volume: we just showed that health-monitoring experiments are needed, and sometimes should be used, in bridge scour investigations, especially those investigations that need to update the probabilities of occurrence. The complexity, duration, and extent of such an experimental monitoring would, obviously, depend on the particular situation.

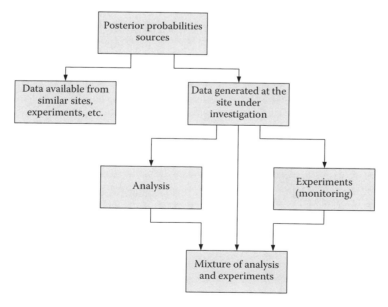

FIGURE 1.57   Posterior probability sources.

Let us suggest the following possible steps that the official might follow to obtain the posterior information data:

1. Evaluate the many scour evaluation equations that can help in investigating structural behavior and/or failures. Note that some of those equations have been discussed earlier in this chapter. Let us assume that the chosen equation is of the form $y_{sc} = Y_{sc}(x_1, x_2, \ldots x_n)$. Note that x_i represents either a constant or a variable in the chosen equation and Y_{sc} represents scour depth. The total number of constants and variables in the equation is n.

2. Recognizing that the chosen equation can lead to inaccurate results, the bridge owner will then identify the constants in the equation that need improvement in their values so that the equation would yield accurate results for the site. In other words, the equation would need validation and correction. Such validation and correction can be achieved by a monitoring experiment.

3. Design a monitoring experiment at the bridge site. The bridge official can use the information and the bibliography in this chapter. The monitoring can be so designed as to be of a reasonable length of time.

4. The data obtained from the monitoring experiment can now be used to validate and improve on the scour equation.

5. Using the variables in the equation as random variables, the Monte Carlo method can be used to evaluate the mean and variance of y_{sc} (see Ettouney and Alampalli 2012, Chapter 8, for a description of the use of the Monte Carlo method in a function of random variables situation).

6. Knowing the mean and variance of y_{sc}, the bridge official can now evaluate the mean and variance of the structural behavior. The required conditional probabilities of failure (or survival) of different solutions (S_1, S_2, and S_3) can then be established.

Let us assume that the conditional probabilities for the three alternatives S_1, S_2, and S_3 have been computed as a result of the above effort; they are shown in Table 1.20. The costs of the three options are then computed in Tables 1.21 through 1.23.

From Tables 1.21 through 1.23 we see that utilization of the posterior probabilities has paid off! The weighted costs are now much lower than before. The best decision remains to only retrofit the two abutments. However, the cost of doing nothing is now the second best decision. The major retrofit cost is now the least desirable decision. Figure 1.58 shows the logic of the different steps in this example.

TABLE 1.20
Conditional Probabilities $P(z|\theta_i)$

State of Nature θ_i	Prior $P(\theta_i)$	Conditional Probabilities $P(z|\theta_i)$		
		S_1	S_2	S_3
θ_1	0.82	0.70	0.75	0.76
θ_2	0.10	0.15	0.12	0.12
θ_3	0.06	0.10	0.11	0.11
θ_4	0.02	0.05	0.02	0.01
Total	1.00	1.00	1.00	1.00

TABLE 1.21
Weighted Costs for S_1: Do Nothing (Posterior Probabilities)

State of Nature θ_i	Prior $P(\theta_i)$	Posterior Probabilities $P(\theta_i|z)$	S_1 ($1,000)	Weighted Cost ($1,000)
θ_1	0.82	0.963	0	0
θ_2	0.10	0.025	500	12.58
θ_3	0.06	0.010	700	7.05
θ_4	0.02	0.002	1500	2.52
Total	1.00	1.00		22.15

TABLE 1.22
Weighted Costs for S_2: Retrofit both Abutments (Posterior Probabilities)

State of Nature θ_i	Prior $P(\theta_i)$	Posterior Probabilities $P(\theta_i|z)$	S_2 ($1,000)	Weighted Cost ($1,000)
θ_1	0.82	0.9701	15	14.25
θ_2	0.10	0.0189	0	0
θ_3	0.06	0.0104	0	0
θ_4	0.02	0.0006	1500	0.95
Total	1.00	1.0		15.50

TABLE 1.23
Weighted Costs for S_3: Retrofit Abutments and Foundations (Posterior Probabilities)

State of Nature θ_i	Prior $P(\theta_i)$	Posterior Probabilities $P(\theta_i|z)$	S_3 ($1,000)	Weighted Cost ($1,000)
θ_1	0.82	0.9707	100	97.07
θ_2	0.10	0.0187	0	0
θ_3	0.06	0.0103	0	0
θ_4	0.02	0.0003	0	0
Total	1.00	1.0		97.07

Note that the weighted costs in this example are not actual costs. They are the costs of particular decisions weighted by different probabilities of occurrences. The weighted costs are used only as a utility measure to help in ranking different decisions. See Ettouney and Alampalli (2012, Chapter 8) for further discussion on the use of cost as a utility measure in decision making. However, the use of experimental monitoring in combination with different decision making techniques has helped the bridge official to reach a clear decision in this otherwise complex situation.

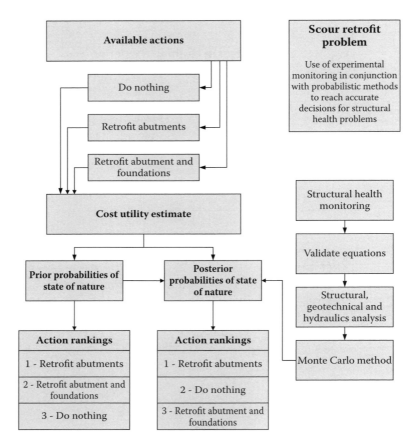

FIGURE 1.58   Decision making steps.

1.7.2   DECISION MAKING CASE STUDY: RISK-BASED METHOD

Scour hazard and bridge response to it involve numerous uncertainties. Because of this, risk-based decision making methods are naturally suited for use in these situations. A risk-based methodology to manage scour at bridges with unknown foundations was developed by Stien and Sedmera (2006). It is based on estimating failure probabilities and the consequences of such a failure.

A risk-based method for ranking scour countermeasures was proposed by Johnson and Niezgoda (2004), based on the failure mode effect analysis (FMEA) described by Ettouney and Alampalli (2012, Chapter 6). The method is based on ranking different countermeasures using three parameters: consequence of failure R_C, likelihood of failure occurrence R_L, and ease (difficulty) of failure detection R_D. The risk priority number (RPN) is then computed as

$$\text{RPN} = R_C\,R_L\,R_D \tag{1.49}$$

The reader is referred to Johnson and Niezgoda (2004) for several examples on the use of FMEA in scour-related decision making situations. For a detailed discussion of FMEA, the reader is referred to Ettouney and Alampalli (2012).

1.8 MANAGEMENT STRATEGIES FOR SCOUR HAZARD

1.8.1 GENERAL

We reasoned earlier that scour can affect the system in three distinct modes: gradual, brittle fracture, and fatigue modes. Each of these modes is a function of time: they behave differently as time progresses. Therefore, management strategies to assess and mitigate each mode should be different. We now offer different management strategies for each mode.

1.8.2 GRADUAL (DUCTILE) EFFECTS

Scour in noncohesive materials occurs in a relatively gradual manner. As such, it can be monitored, and advanced warning can result such that mitigating decisions can be made. The advent of soil erosion can be regarded as a gradual increase of damage as a function of time. The process is illustrated in Figure 1.59. Management strategies can follow one of the following methods:

Analytical solutions: Different analytical solutions that predict scour advance can be used to aid management decisions. Use of adequate safety margins is sufficient in most situations since a gradual increase in damage can be accommodated with an added safety margin: such a safety margin is not available in more sudden (brittle) scour situations.

Direct monitoring: Direct monitoring of scour erosion is possible. Correlating observed erosion to foundation behavior is then possible, and thresholds of structural failure can be established and observed in near real time.

Generally, two mitigation strategies can be followed:

- Increase threshold of accepted damage level. This can be done by decreasing demand, such as flood control measures.
- Increase capacity of scour resistance, such as soil improvements, or add riprap around the affected foundations.

1.8.3 SUDDEN (BRITTLE FRACTURE-LIKE) EFFECTS

Since soil/rock responses, such as erosion, do not reveal themselves in this situation, direct monitoring would not yield beneficial information. Two strategies can be used, as follows:

Virtual sensing: VSP is described at length by Ettouney and Alampalli (2012, Chapter 5). It is the mainstay of popular strategies for this situation. We recognize the empirical expressions used as

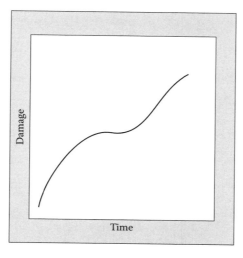

FIGURE 1.59 Damage accumulation for gradual scour.

a form of VSP. As usual, the accuracy of estimating the behavior depends mainly on the accuracy of the empirical expressions.

Probabilistic methods: In any structure-related brittle behavior, as shown in Figure 1.60, the response of the structure remains tolerable, until the loading reaches a certain level, at which the structure suddenly fails. It is reasonable to assume that the uncertainties (as represented by variances, for example) increase as the loading increases, as illustrated in Figure 1.60. This shows the limitations of the direct use of empirical formula in scour mitigation management. Since the formulas are generic, they do not accommodate all possible situations, including uncertainties of the site and loadings. Because of this, any built-in safety margin can be breached if the uncertainties are high in a particular site. Such a breach can lead to sudden brittle failure in some situations.

The above also points to a possible management strategy.

- Isolate locations where potential scour brittle fracture can occur. This can be done using any of the SHM techniques of this chapter or any of the cited references. In more complex situations, artificial neural network method can be used.
- Using pertinent scour prediction empirical formula, evaluate the probabilistic performance of the variables. This can be done by a semianalytical approach (Taylor series), or a simulation approach, or Monte Carlo analysis.
- The variances and means of the pertinent variables can now be used to estimate different probabilistic statements, such as nonexceedance probabilities for a given loading condition.
- Make mitigation decisions accordingly.

1.8.4 SUDDEN (FATIGUE-LIKE) EFFECTS

Scour and ultimately failure in this mode is similar to fatigue behavior. The damage accumulates as the demands are added at every flood occurrence until the damage reaches a failure threshold when the system fails. Figure 1.61 illustrates the damage accumulation behavior as a function of time. We

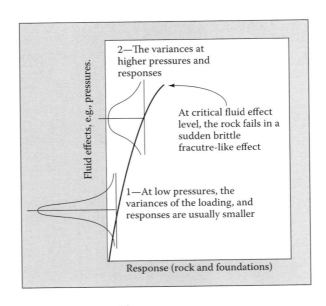

FIGURE 1.60 Brittle behavior in scour problems.

note that such a behavior is fairly similar to metal fatigue behavior. Of course, the monotonic damage accumulation in fatigue is expected. Scour damage accumulation is more likely to resemble a series of sudden increases, as in Figure 1.61. Those sudden increases in damage accumulation occur when severe floods affect the system. If the damage threshold is reached during one of those sudden increases, the system will fail suddenly.

Figure 1.61 shows that this particular scour mode of behavior has the features of both earlier scour types. It includes sudden brittle-like damage accumulation. It also includes monotonic

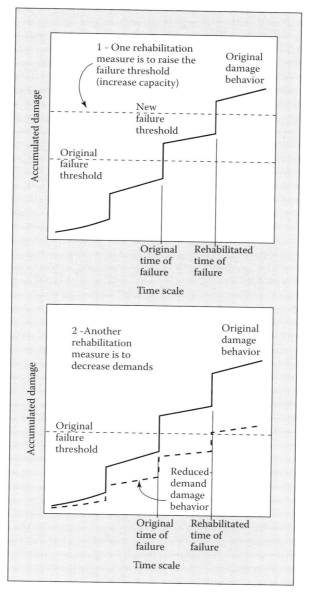

FIGURE 1.61 Damage accumulation for fatigue-like scour.

increase in damage accumulation. Because of this, management strategies include all previously discussed strategies:

- Monotonic/gradual damage accumulation: Can be managed by monitoring such damage accumulation. Different SHM-monitoring methods discussed earlier can be used.
- Brittle-like behavior can be managed only by including probabilistic techniques during the analysis, design, and decision making processes.

As before, the strategy can involve one of the following

- Increase damage threshold (increase capacity)
- Reduce demands
- Both of the above

The effects are shown in Figure 1.61. The efficiency of the mitigation solution depends on the same factors discussed earlier in this section.

From Figure 6.61, it is seen that the efficiency of mitigation solutions depends on several factors: (1) the shape of the accumulated damage versus time relationship, (2) available budget (which is translated to the degree of threshold or demand shifting), and (3) the level of accumulated damage at the time of implementing the project. Clearly, the earlier the implementation, the more efficient is the solution.

1.8.5 APPLICATION TO OTHER HAZARDS

The gradual, brittle fracture, and fatigue-like modes of failure as discussed earlier can occur in other hazards. The management strategies mentioned above for each mode are fairly applicable to other hazards, as shown in Table 1.24. Note that Table 1.24 addresses the way the hazard affects the system on a long time scale, which is needed for management strategies.

1.9 APPENDIX: NBI SYSTEM

The NBI system is an exhaustive database that contains useful information about the bridge system in the United States. The records (rows) in the database contain each bridge in the inventory. Each record contains a set of fields (items). The NBI items are shown in Table 1.25. The NBI database makes it easy to study and research individual bridges as well as to perform statistical investigations on any subset of the NBI. The details of each item in the NBI can be found in FHWA (1995).

TABLE 1.24
Modes of Applications for Other Hazards

Hazard	Gradual	Brittle	Fatigue Like
Earthquakes	Earthquakes occur suddenly, not gradually, in nature	This mode is similar to the scour mode since earthquakes occur suddenly	Since systems are usually retrofitted after each earthquake event, this mode is not applicable
Wind	Can occur	Can occur	Can occur
Corrosion of reinforcing rebars	Most applicable	Corrosion does not occur suddenly	Can occur
Damage of posttensioned tendons	Most applicable	Damage does not occur suddenly	Can occur

TABLE 1.25
NBI Record Format

Item No	Item Name	Item Position	Item Length/Type
1	State code	1–3	3/N
8	Structure number	4–18	15/AN
5	Inventory route	19–27	9/AN
5A	Record type	19	1/AN
5B	Route signing prefix	20	1/N
5C	Designated level of service	21	1/N
5D	Route number	22–26	5/AN
5E	Directional suffix	27	1/N
2	Highway agency district	28–29	2/AN
3	County (Parish) code	30–32	3/N
4	Place code	33–37	5/N
6	Features intersected	38–62	25/AN
6A	Features intersected	38–61	24/AN
6B	Critical facility indicator	62	1/AN
7	Facility carried by structure	63–80	18/AN
9	Location	81–105	25/AN
10	Inventory Rte, Min Vert clearance	106–109	4/N
11	Kilometerpoint	110–116	7/N
12	Base highway network	117	1/N
13	Inventory route, subroute number	118–129	12/AN
13A	LRS inventory route	118–127	10/AN
13B	Subroute number	128–129	2/N
16	Latitude	130–137	8/N
17	Longitude	138–146	9/N
19	Bypass/detour length	147–149	3/N
20	Toll	150	1/N
21	Maintenance responsibility	151–152	2/N
22	Owner	153–154	2/N
26	Functional class of inventory Rte	155–156	2/N
27	Year built	157–160	4/N
28	Lanes on/under structure	161–164	4/N
28A	Lanes on structure	161–162	2/N
28B	Lanes under structure	163–164	2/N
29	Average daily traffic	165–170	6/N
30	Year of average daily traffic	171–174	4/N
31	Design load	175	1/N
32	Approach roadway width	176–179	4/N
33	Bridge median	180	1/N
34	Skew	181–182	2/N
35	Structure flared	183	1/N
36	Traffic safety features	184–187	4/AN
36A	Bridge railings	184	1/AN
36B	Transitions	185	1/AN
36C	Approach guardrail	186	1/AN
36D	Approach guardrail ends	187	1/AN
37	Historical significance	188	1/N
38	Navigation control	189	1/AN
39	Navigation vertical clearance	190–193	4/N

continued

TABLE 1.25 (continued)
NBI Record Format

Item No	Item Name	Item Position	Item Length/Type
40	Navigation horizontal clearance	194–198	5/N
41	Structure open/posted/closed	199	1/AN
42	Type of service	200–201	2/N
42A	Type of service on bridge	200	1/N
42B	Type of service under bridge	201	1/N
43	Structure type, main	202–204	3/N
43A	Kind of material/design	202	1/N
43B	Type of design/construction	203–204	2/N
44	Structure type, approach spans	205–207	3/N
44A	Kind of material/design	205	1/N
44B	Type of design/construction	206–207	2/N
45	Number of spans in main unit	208–210	3/N
46	Number of approach spans	211–214	4/N
47	Inventory Rte total Horz clearance	215–217	3/N
48	Length of maximum span	218–222	5/N
49	Structure length	223–228	6/N
50	Curb/sidewalk widths	229–234	6/N
50A	Left curb/sidewalk width	229–231	3/N
50B	Right curb/sidewalk width	232–234	3/N
51	Bridge roadway width curb-to-curb	235–238	4/N
52	Deck width, out-to-out	239–242	4/N
53	Min Vert clear over bridge roadway	243–246	4/N
54	Minimum vertical underclearance	247–251	5/AN
54A	Reference feature	247	1/AN
54B	Minimum vertical underclearance	248–251	4/N
55	Min lateral underclear on right	252–255	4/AN
55A	Reference feature	252	1/AN
55B	Minimum lateral underclearance	253–255	3/N
56	Min lateral underclear on left	256–258	3/N
58	Deck	259	1/AN
59	Superstructure	260	1/AN
60	Substructure	261	1/AN
61	Channel/channel protection	262	1/AN
62	Culverts	263	1/AN
63	Method used to determine operating rating	264	1/N
64	Operating rating	265–267	3/N
65	Method used to determine inventory rating	268	1/N
66	Inventory rating	269–271	3/N
67	Structural evaluation	272	1/AN
68	Deck geometry	273	1/AN
69	Underclear, vertical, & horizontal	274	1/AN
70	Bridge posting	275	1/N
71	Waterway adequacy	276	1/AN
72	Approach roadway alignment	277	1/AN
75	Type of work	278–280	3/N
75A	Type of work proposed	278–279	2/N
75B	Work done by	280	1/AN
76	Length of structure improvement	281–286	6/N
90	Inspection date	287–290	4/N

TABLE 1.25 (continued)
NBI Record Format

Item No	Item Name	Item Position	Item Length/Type
91	Designated inspection frequency	291–292	2/N
92	Critical feature inspection	293–301	9/AN
92A	Fracture-critical details	293–295	3/AN
92B	Underwater inspection	296–298	3/AN
92C	Other special inspection	299–301	3/AN
93	Critical feature inspection dates	302–313	12/AN
93A	Fracture-critical details date	302–305	4/AN
93B	Underwater inspection date	306–309	4/AN
93C	Other special inspection date	310–313	4/AN
94	Bridge improvement cost	314–319	6/N
95	Roadway improvement cost	320–325	6/N
96	Total project cost	326–331	6/N
97	Year of improvement cost estimate	332–335	4/N
98	Border bridge	336–340	5/AN
98A	Neighboring state code	336–338	3/AN
98B	Percent responsibility	339–340	2/N
99	Border bridge structure number	341–355	15/AN
100	STRAHNET highway designation	356	1/N
101	Parallel structure designation	357	1/AN
102	Direction of traffic	358	1/N
103	Temporary structure designation	359	1/AN
104	Highway system of inventory route	360	1/N
105	Federal lands highways	361	1/N
106	Year reconstructed	362–365	4/N
107	Deck structure type	366	1/AN
108	Wearing surface/protective system	367–369	3/AN
108A	Type of wearing surface	367	1/AN
108B	Type of membrane	368	1/AN
108C	Deck protection	369	1/AN
109	Average daily truck traffic	370–371	2/N
110	Designated national network	372	1/N
111	Pier/abutment protection	373	1/N
112	NBIS bridge length	374	1/AN
113	Scour critical bridges	375	1/AN
114	Future average daily traffic	376–381	6/N
115	Year of future avg daily traffic	382–385	4/N
116	Minimum navigation vertical clearance vertical lift bridge	386–389	4/N
	Federal agency indicator	391	
	Washington headquarters use	392–426	
	Status	427	
n/a	Asterisk field in SR	428	1/AN
SR	Sufficiency rating (select from last 4 positions only)	429–432	4/N

Status field:
1 = Structurally deficient
2 = Functionally obsolete
0 = Not deficient
N = Not applicable
Source: http://www.fhwa.dot.gov/bridge/nbi/format.cfm, accessed September 5, 2007.

REFERENCES

Aguilar, R. and Kaslan, E. (2002). "Use of NDE in the California Bridge Inspection Program," *Proceedings, NDE Conference on Civil Engineering*, ASNT, Cincinnati, OH.

Annandale, G. W. (1993). "Risk Analysis of Bridge Failure," *Proceedings of Hydraulic Engineering '93*, ASCE, San Francisco, CA.

Annandale, G. W. (2006). *Scour Technology: Mechanics and Engineering Practice.* McGraw-Hill, New York City, NY.

Boehmler, E. and Olimpio, J. (2000). "Evaluation of Pier-Scour Measurement Methods and Pier-Scour Predictions with Observed Scour Measurements at Selected Bridge Sites in New Hampshire, 1995–98," Report No. FHWA-NH-RD-12323E. U.S. Geological Society, NH/VT District.

Borg, S. (2000). "Nondestructive Testing for Length Determination of Piles for Five Long Island Bridges," *Structural Materials Technology: An NDT Conference*, ASNT, Atlantic City, NJ.

Dowding, C. and O'Connor, K. (2000). "Real Time Monitoring of Transportation with TDR Technology," *Structural Materials Technology: An NDT Conference*, ASNT, Atlantic City, NJ.

Dowling, C. and Pierce, C. (1994). "Use of Time Domain Reflectometry to Detect Bridge Scour and Monitor Pier Movement." *Publications, Infrastructure Technology Institute*, Northwestern University, Evanston, IL.

Ettouney, M. and Alampalli, S. (2012). *Infrastructure Health in Civil Engineering: Theory and Components*, CRC Press, Boca Raton, FL.

FHWA. (1995). "Recording and Coding Guide for the Structure Inventory and Appraisal of the Nation's Bridges," Report No. FHWA-PD-96-001, Office of Engineering, Bridge Division, Washington, DC.

Holt, J. and Slaughter, S. (2000). "Mississippi's Approach to Unknown Bridge Foundations," *Structural Materials Technology: An NDT Conference*, ASNT, Atlantic City, NJ.

Hunt, B. E. (2005). "Bridge Health Scour Monitoring," *Third New York City Bridge Conference*, New York, NY, September 12–13.

Jeng, D., Bateni, S., and Lockett, E. (2005). "Neural Network assessment for scour depth around bridge piers," University of Sydney, Department of Civil Engineering Research Report No. R855, Sydney, Australia.

Johnson, P. and Niezgoda, S. (2004). "Risk-based method for selecting bridge scour countermeasures," *Journal of Hydraulic Engineering*, ASCE, 130(2), 121–128.

Jones, J. S., Bertoldi, D. A., and Umbreli, E. R. (1993). "Preliminary Studies of Pressure Flow Scour," *ASCE Hydraulic Engineering, Proceeding 1993 National Conference*, San Francisco, CA.

Lagasse, P. and Richardson, E. (2001). "Stream Stability at Highway Structures," 3rd Edition, FHWA Report No NH1 01-002, also HEC-20, Office of Bridge Technology, Washington, DC.

Lagasse, P., Richardson, E., Schall. J., and Price, G. (1997). "Instrumentation for Measuring Scour at Bridge Piers and Abutments," NCHRP Report 396, Transportation Research Board, National Research Council, National Academy Press, Washington, D.C.

Lagasse, P. and Zevenbergen, L. (2001). "Bridge scour and stream instability countermeasures," 2nd Edition, FHWA report No NH1 01-003, also HEC-23, Office of Bridge Technology, Washington, DC.

Laursen, E. M. and Toch, A. (1956). "Scour around bridge piers and abutments," Iowa Highway Research Board Bulletin No. 4, Iowa City, IA.

Marron, D. (2000). "Remote Monitoring of Structural Stability Using Electronic Clinometers," *Structural Materials Technology: An NDT Conference*, ASNT, Atlantic City, NJ.

Melville, B. W. and Chiew, Y. M. (1999). "Time scale for local scour depth at bridge piers." *Journal of Hydraulics Engineering*, ASCE, 125(1), 59–65.

Melville, B. W. and Sutherland, A. J. (1988). "Design method for local scour at bridge piers," *Journal of Hydraulics Engineering, ASCE,* 114(10), 1210–1226.

Mercado, E. and Woodroof, L. (2008). "The Pneumatic Scour Detection System; Development and Case History," *NDE/NDT for Highways and Bridges: Structural Materials Technology (SMT)*, ASNT, Oakland, CA.

Mueller, D. (1998). "Summary of Fixed Instrumentation for Field Measurement of Scour and Deposition." *Proceedings, Federal Interagency Workshop, Sediment Technology for the 21st Century*, St. Petersburg, FL.

Mueller, D. and Landers, M. (1999). "Portable Instrumentation for Real-Time Measurement of Scour at Bridges," *FHWA-RD-99-085*, Federal Highway Administration, McLean, VA.

NYSDOT. (1998). "Scour Program Monitoring Manual," New York State Department of Transportations Report, Albany, NY.

NYSDOT (2003) "Hydraulic Vulnerability Manual," New York State Department of Transportation, 50 Wolf Road, Albany, NY.

Oberg, K. and Mueller, D. (1994). "Recent Applications of Acoustic Doppler Current Profilers," In *Fundamentals and Advancements in Hydraulic Measurements and Experimentation*, Pugh, C. A., Ed, ASCE, Reston, VA.

Olson, L. (2005). "Dynamic Bridge Substructure Evaluation and Monitoring," FHWA Report No. FHWA A-RD-03-089, Federal Highway Administration, McLean, VA.

Richardson, E. and Davis, S. (1995). "Evaluating scour at bridges," 3rd Edition, Federal Highway Administration, Washington, DC.

Richardson, E. and Davis, S. (2001). "Evaluating scour at bridges," 4th Edition, FHWA report No NH1 01-001, also HEC-18, Office of Bridge Technology, Washington, DC.

Richardson, E.V., Simons, D.B., and Lagasse, P.F. 2001, "River Engineering for Highway Encroachments - Highways in the River Environment," FHWA NHI 01-004, Federal Highway Administration, Hydraulic Series No. 6, Washington, D.C.

Shen, H. (1971). "Scour near piers," In *River Mechanics*, Shen, H, Editor, Vol II, Chapter 23, Colorado State University, Fort Collins, CO.

Sheppard, D. (2003). "Scour at Complex Piers," Report No. FDOT BC354 RPWO 35, Florida DOT, Tallahassee, FL.

Stien, S. and Sedmera, K. (2006). "Risk-Based Management Guidelines for Scour at Bridges with Unknown Foundations," National Cooperative Highway Research Program, NCHRP Report No. 24-25, Transportation Research Board, Washington, DC.

Stock, T. (2008). "Use of a Hummingbird Model 997 Side Scan Sonar As a Bridge Inspection Tool," Report to Wisconsin Division of FHWA.

Timoshenko, S. and Gere, J. (1961). *Theory of Elastic Stability*, McGraw-Hill, New York, NY.

U.S. DOT. (1993). "Evaluating scour at bridges. Hydraul. Eng. Circular No.18," Federal Highway Administration, U.S. Dept. of Transp., McLean, VA.

2 Earthquakes

2.1 INTRODUCTION

2.1.1 OVERVIEW

Earthquake hazard is a major factor in considering the health of structural systems. The recent earthquakes in the United States and abroad showed that though seismic events are not frequent, they have enormous consequences. Therefore, decision makers have paid special attention to protecting infrastructures from damage due to earthquakes. We propose that approaching the operation and management of civil infrastructures from structural health monitoring (SHM) viewpoint would help in (1) reducing cost of ownership, and (2) ensuring safe system operations in the event of an earthquake.

Because of the intermittent occurrences of earthquakes, the initial costs of protection (which include increasing seismic worthiness and SHM costs) need to be justified. Such justification can be most convincing in a life cycle overview. Such an overview should consider all aspects of hazard-system response interrelationship. Such interrelationship is temporal, spatial, managerial, and economic. In the rest of this section, we discuss different interrelationships.

2.1.2 CATEGORIZATIONS OF EARTHQUAKE ISSUES

2.1.2.1 General

To use structural health in civil engineering (SHCE) techniques efficiently for handling earthquake hazards to bridges, we present several categorizations of the subject matter. The use of SHCE techniques changes drastically depending on these categories. The first of these is identified as the *phase* category. There are three phases for earthquakes and bridges: before-event phase, during-event phase, and after-event phase. The goals, techniques, and tools for SHCE for each of these phases differ immensely. Any SHCE project should account for these differences; otherwise, there is a risk of not meeting some of its stated goals. Another categorization is the *mode* category; capacity and demand modes. Another is the *goals* category; the single-hazard goal or the multihazard goal. The interrelationship between the phases, modes, and goals is shown in Figure 2.1.

2.1.2.2 Phases

The three phases of earthquake hazards are self-evident. The first phase, before event, is preparatory in nature. The bridge will be behaving in a linear fashion, with mostly small and elastic strains. The magnitudes of motion measurements are also expected to be small. The sensors must be online most of the time in order to be ready to measure an earthquake when/if it happens. This can result in a large amount of data, which must be handled properly. Structural identification and possible damage identification algorithms can be validated in this phase. This is the phase when the multihazard mode must be utilized for increased cost efficiency.

The second phase is the during-event phase. Earthquake motions usually last for few seconds, perhaps 5–20 seconds. This phase needs to sense pertinent motions of the bridge that can be utilized for both demand and capacity modes. The sensors, and their supporting instrumentation and

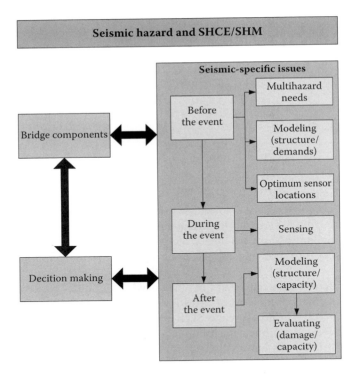

FIGURE 2.1   Categories of seismic hazard in SHCE.

networks should be resilient so as not to fail during unexpectedly large seismic motions. Data collection schemes must be consistent with the needed decision making algorithms in phase three. SHCE techniques in this phase serve research, as well as emergency management needs.

The third and final phase is the after-event phase. SHCE techniques can help in many different ways. By detecting any abnormal behavior or damage immediately after the event, emergency authorities can respond effectively, thus saving or reducing any casualty. In addition, traffic can be routed appropriately. Later on, inspection efforts can be aided by using the structural and damage identification algorithms that were validated during the pre-event phase. The same algorithms can be used for prioritizing retrofitting needs of the bridge.

2.1.2.3   Modes

There are two modes that SHCE can assist in for the seismic mitigation of bridges: demand and capacity modes. We identify the demand mode as that of the whole bridge seismic motion and behavior. In a way, we identify it as the structural identification step in SHCE. By identifying the structure, either analytically or experimentally, we can, for a given seismic motion, identify seismic *demands* from the structure. The *capacity* mode is closely linked to the damage identification step in SHCE. It involves studying the seismic damage in different parts of the bridge. The sensors needed for the capacity mode would be mostly localized (strains, for example) while the demands mode would rely more on displacement, velocities, or acceleration sensing. The capacity mode is localized, whereas the demand mode is global, by definition. Not only are the types of measurements different, but it is also clear that the two modes would require very different approaches in choosing the number and location of those sensors. The optimum sensor locations (OSL) problem described by Ettouney and Alampalli (2012, Chapter 5) can be used in the demand mode to evaluate the sensor layout. For optimal sensor layout for capacity mode, a mix of analysis and personal experiences is needed.

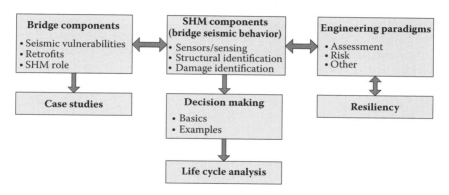

FIGURE 2.2   Organization of this chapter.

2.1.2.4   Goals

Two distinct goals can be mentioned here: a seismic-only SHCE goal or a multihazard SHCE goal. It is obvious that a multihazard approach to SHCE for bridges is preferred. A multihazard approach would potentially have a cost–benefit advantage. The multihazard SHCE goal can be achieved mostly in the pre-event phase. Potential hazards that can be monitored, while waiting for seismic events, include loss of capacity due to corrosion, misalignment due to impact and fatigue effects due to wind loading. A potentially great tool in sensor layout in a multihazard situation is the OSL method. However, considerable coordination must be done in choosing appropriate sensors for multihazard applications, as well as the underlying algorithms and decision making tools.

2.1.3   Organization of This Chapter

The next two sections explore, from two different vantage points, the structural health of bridges as related to earthquake hazards. First, we discuss vulnerabilities of bridge components as they respond to earthquakes. During this discussion, we examine the role of SHM in relation to different phases of the seismic event. The following section explores how different components of SHM can play a role in managing seismic hazards. We then offer some examples of current bridge seismic SHM applications. Decision making examples for seismic hazards and bridges are presented next. We are interested specifically in the cost–benefit implications of seismic monitoring and proving that, in a life-cycle overview, the benefits of SHM experiments can outweigh the costs. The health of bridges (or any other type of civil infrastructure) is a function of design engineering paradigms. We, therefore, devote the following two sections in discussing different engineering paradigms. Special emphasis is given to emerging engineering paradigms such as performance-based designs (PBD) and resiliency. We finally discuss some special factors that can affect seismic hazard effects on life cycle analysis (LCA) of bridges. Figure 2.2 shows the composition of this chapter.

Note that although this chapter will discuss mainly earthquake hazards to bridge structures, most of the concepts can be easily applied to other civil infrastructures, such as buildings, tunnels, manufacturing facilities, or dams.

2.2   BRIDGE COMPONENTS AND SEISMIC HAZARDS

2.2.1   General

Bridge structures deteriorate and degrade as time passes. The causes for deterioration are numerous: environmental (corrosion, freeze-thaw, temperature, etc.), loading (fatigue), natural hazards (wind, floods), and so on. Deterioration of mechanical properties of bridges and their components will result in lowering of as-designed seismic toughness. This concept was discussed in detail by Ettouney and Alampalli (2012, Chapter 2). The lesson is clear: we need to ensure that bridges, in their day-to-day state, have enough seismic toughness. In other words, for a desired level of seismic hazard, the bridge and its components should have enough capacity to resist the seismic demands. One way to ensure such capacity to demand balance, in a continuous or even in an intermittent manner, is by continued or intermittent monitoring. We immediately note that *monitoring* entails all four SHCE components: sensing, structural identification (STRID), damage identification (DMID), and decision making (DM). In what follows, we explore many of the bridge components. We discuss briefly their known seismic behavior and popular retrofit measures. We also suggest potential *before-event* monitoring techniques that can ensure adequate performance if seismic hazard occurs. We also suggest SHCE techniques during the seismic event. Finally, we suggest pertinent SHCE techniques that might help in saving *after-event* costs while ensuring safety.

2.2.2 Overall Geometry

Since seismic forces affect bridges in its entirety (as opposed, say, to fatigue or corrosion, where the initial effects are localized), the overall geometry significantly influences the seismic response. Hence, monitoring those responses should be ideally global in nature. Some of the global seismic effects and SHM roles are

Skew angle: Having a skew angle would force large three-dimensional responses that can lead to unseating of the bridge and the consequent failure of bridge spans (see Figure 2.3).

Curved spans: Similar to skewed spans, curved spans excite three-dimensional motions that can cause unseating and potential failures. One of the major bridge failures during the 2008 Wenchuan, China, earthquake was attributed to its curved spans (see Figure 2.4).

Large seat widths and seismic restrainers are two popular retrofitting schemes for skew or curved bridges. SHM roles for overall bridge behavior are

Before event: Monitor parameters of the bridge by conducting STRID to ensure that it performs as designed. Study causes of any abnormal modal behavior and correct them as needed. The effects of any nonseismic-related retrofits must be studied to ensure that no unwanted side effects (multihazard effects) occur. Moehle and Eberhard (1999) noted that a short channel reinforced concrete wall at Bull Creek Canyon shortened the pier columns, thus causing major damage during the Northridge earthquake.

During event: Well-placed accelerometers should monitor in real time the bridge behavior to detect any major damage.

After event: Strain gauges can reveal nonlinear behavior. STRID can help in detecting overall performance and aid in any needed retrofits.

FIGURE 2.3   Seismic damage of skew bridge. (Courtesy of MCEER, University of Buffalo.)

FIGURE 2.4   Seismic failure of curved bridge. (Courtesy of MCEER, University of Buffalo.)

2.2.3 BEARINGS

Bridge bearings play a crucial role during seismic events. They transmit seismic motion from the substructure to the superstructure. Because of the impedance (inertial and kinematic) mismatch between the superstructure and the substructure, the bearings must endure relative motions (linear and rotations) as well as reaction forces and moments. The consequences of a failure of a bearing during or after a seismic event can be catastrophic. The different methods of using SHM techniques to mitigate and address seismic behavior of bridge bearings are discussed in this section.

Table 2.1 shows the different types of bridge bearings currently in use. All those bearing types have one or more of the following components: sole plate, bearing/bearing surface, masonry plate, and anchorage. For a detailed description of different bearing components, the reader is referred to Hartle et al. (2002). To discuss the efficient use of SHM techniques on seismic bearing behavior, it is essential to reiterate the main functions of bridge bearings as follows:

- Transmit loads (forces and moments) between superstructures and substructures
- Allow for seismic demand motions of superstructures and substructures. Note that these motions can be in vertical, longitudinal, and/or transverse directions. Also, note that these motions include both linear and rotational types

We observed earlier in this chapter that SHM techniques can be used either before event (demand) or during and after events (capacity). We discuss below these two modes for seismic behavior of bearings.

Before event: Table 2.1 includes a description of vulnerabilities of each type of bearing that can affect its seismic performance. We note that one of the major vulnerabilities is the corrosion-induced freezing of bearing components, FHWA (1995). During seismic motion, this effect would prevent the superstructure from moving relative to the substructure, and the resulting internal seismic stresses can lead to an undesired damage or failure of the bearing. Regular visual corrosion inspection can help in initiating repairs if/when corrosion is observed. However, since bearings are usually difficult to inspect visually, and corrosive effects might not be obvious to visual inspection, an SHM solution can be of help in few cases. For example, a corrosion-rate sensor can be installed near potential frozen surfaces. We recognize the difficulty of such a solution, especially for large-size bearings, where corrosion can start on the other side of the bearing, *away* from the corrosion sensor. Perhaps a better solution is to monitor the frozen effects of the corrosion. This can be accomplished by sensing the relative motions of the potential frozen surfaces and comparing them with the

TABLE 2.1
Role of SHM and Different Types of Bearings (see Figure 2.5)

Type of Bearing	Description	Vulnerability	SHM Role		
			Before Event	During Event	After Events
Sliding plate bearings	Used with spans less than 40 ft. Two steel plates sliding against each other with adequate lubrication between them	Misalignment of the plates	Ensure that the two plates are not frozen or corroded	Record absolute and/or relative motion between the two plates (in all directions)	Analyze after-event motions. Compare with baseline (before event) and ensure that changes are within acceptable tolerances
Roller bearings	Single or multiple steel rollers that roll longitudinally to permit longitudinal bridge motion	Rollers should be centered and aligned properly	Monitor for frozen or corroded rollers		
Rocker bearings	Provide rolling surface at bottom. That permits rolling while providing for large curvature, thus allowing for longer spans	Loose or missing fasteners. Proper tilting (should not be excessive)	Ensure that pins and rolling surfaces are clean and corrosion free		
Pin and link	A steel link pinned at top and bottom. Can also be used as a restraining device that prevents uplifting of the support	Deck can move up and down during longitudinal motion. Risk of breakage if frozen due to corrosion. Proper tilting (should not be excessive)	Monitor 3-D strain states, especially near pin locations. Establish a baseline of vibration characteristics of the link	Record strain and relative displacement time histories	Analyze changes of vibration behavior of links
Elastomeric	Rectangular pad of neoprene that provides for lateral relative motion through shear deformation	Adhesion to bearing surfaces can be a problem. Excessive bulging, splitting, or tearing	Characterize force-displacement relationships of the elastomeric bearing. Ensure that such constitutive relationships are within design parameters	Record relative motions of the bearing surfaces	Ensure that the after-event absolute and relative motions of the bearing surfaces are within acceptable design parameters
Seismic isolation bearings	Special types of bearing that provide isolation from seismic motions; includes lead core, friction pendulum, and high-damping rubber bearings	Adhesion to bearing surfaces can be a problem. Excessive bulging, splitting, or tearing	Establish baseline for normal (nonseismic) behavior	Record absolute and/or relative motions	Ensure that no residual permanent displacements (rotations) have occurred. Also, ensure that changes to the before-event baseline are acceptable
Pot bearings	Allows for multidimensional rotations of the structure	Proper seating. Welds and cracks should be monitored			
Restraining bearings	Special bearings that can be used as a restraining device that prevents uplifting of the support	Uplift of bearing	Monitor for any potential uplift of bearings	Record relative motions. The SHM scheme should be designed so as not to fail if uplift does occur during a seismic event	Monitor relative motions. Ensure that changes in baseline behavior are acceptable

Note: The different bearings are after FHWA, BIRM (2002).

expected relative motions under normal conditions. If the sensed relative motions fall far below the expected relative motions, this would be a clear sign of a corrosion-freezing effect, and the needed repairs might be commenced. This technique is also valid when the freezing effects are due to factors other than corrosion, such as in Elastomeric-bearing types (see Figure 2.5).

Prevention (freezing) of relative motion is not the only major seismic bearing vulnerability. Misaligned bearings or loose bearings can also lead to unwanted damage during seismic events. Both types of vulnerabilities can be visually observed. However, SHM techniques, because of their more sensitive sensing nature, can help in detecting vulnerabilities that might otherwise go undetected visually. Loose bearings can be detected by installing acoustic sensors near the anchors of the bearing. The alignment, or lack of it, can be detected by sensing the relative motions between the superstructure and the substructure.

Note that all the above-mentioned SHM techniques can be designed as part of a more general SHM effort. Such a method, when designed carefully, can be used for seismic hazards as well as other types of hazards, such as bridge overload. It can also be very useful in providing information about normal day-to-day bridge behavior.

2.2.4 EXPANSION/MOVING JOINTS

These bridge components are needed to allow for movements that are generated by different sources, such as gravity loads, temperature, and creep effects, as well as wind and seismic loads. The expected movements can be as small as a few millimeters and as large as several inches (not counting seismic demands), depending on the size and type of bridge. The designs vary in detail, but they all have in common the need to allow for (1) stress-free (or stress-controlled) relative motions between the two sides of the bridge, (2) eliminate or reduce moisture leaks, and (3) reduce maintenance and long-term repair needs.

Two types of joints can be identified based on the stress paths between the two sides of the joint (item "1" above). They are

No stress continuation: In this configuration, the two sides of the joint do not have any stress continuation between them.

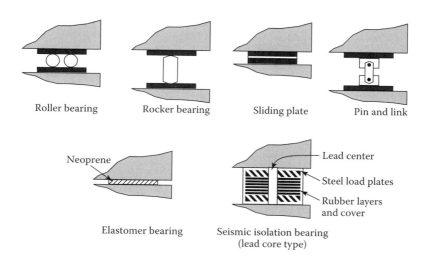

FIGURE 2.5   Typical bridge bearings.

Vertical shear transmission: The two sides of the joint can (1) move independently in the longitudinal direction, and (2) rotate independently about the horizontal transverse axis, but there is a vertical shear constraint between them. There are several configurations for this type, such as seated hinge, tension bar-type hinge, and shear keys.

The physical location of the joints makes them susceptible to traffic-induced stresses, especially with multiaxle vehicles. Thus the loosening of bolts, cover plates, welds, or other details of the joint are common. Also, because of potential weather related conditions (moisture and deicing salt) , corrosion damage is common. This creates the need for inspection, maintenance, and repair.

During seismic events, the motion of the two bridge sides of the joint can have two damaging effects: (1) if the relative seismic displacement demands are higher than the as-built gap in the joint, then the two bridge sides would impact each other, thus causing potential local damage to the bridge structure, and (2) the hinge at the joint might be unseated (see Figure 2.6). Note that the corrosion and/or normal wear-and-tear damage can result in a reduced seating capacity before the seismic event, so that the potential of unseating might be even larger than design conditions. The seating design requirements depend on the location and type of joints. These requirements are discussed in FHWA (1995) and NYSDOT (2004).

One of the more popular seismic retrofit strategies for joints is seismic restrainers. They are used to tie together the two sides of the joint. There are numerous detailing techniques for seismic restrainers that are beyond the scope of this chapter. However, the health and adequacy of restrainers must be always ensured in anticipation of seismic events.

SHM techniques can thus aid in several ways, before, during, and after seismic events, as follows.

Before event: SHM techniques can be used to ensure that the condition of the longitudinal joint is as designed. Any deterioration that might have undesirable seismic effects would be detected through SHM. For example, the moisture penetration that might cause corrosion within the joint can be monitored by using humidity sensors. A more direct approach would be to sense the rate of corrosion in appropriate locations in the joint. Unfortunately, these two approaches would require too many sensors in order to be effective.

Perhaps a better approach is to sense the motions at the joint, then estimate potential deterioration or departure from design conditions. When/if the seismic demands are judged as not having been met, maintenance or repair activity should be initiated. Figure 2.7 shows the schematics of the process.

After event: In addition to visual inspection, closer nondestructive testing (NDT) operations might be needed. Applicable NDT methods can be impact-echo, thermography, or penetrating radiation.

FIGURE 2.6   Unseating of joints during seismic events. (Courtesy of MCEER, University of Buffalo.)

2.2.5 PIERS: COLUMNS AND BENTS

Bridge piers can be solid wall, single columns, or multiple columns (In the latter situation they are called bents). Pier failure modes can be flexural and/or shear. Many of these failure modes were detected during seismic events: link beam failure, failure of connections with foundations (Figure 2.8), or failure of connections with superstructure (Figure 2.9). Sensing during seismic events should be near connections (top or bottom). For seated bearings, monitoring can be coupled with bearing investigation. Monitoring can also be coupled with corrosion monitoring in case of reinforced concrete construction. Monitoring according to seismic event phases can be as follows:

Before-event sensing can be used for validation (small strains) or other issues (traffic vibrations and loads)

During-event sensing can be used for validation (large strains)

After-event sensing can be used for emergency management services (EMS), damage assessment, traffic routing, monitoring retrofitted components, such as steel jackets or fiber reinforced polymers (FRP) wraps of columns, etc.

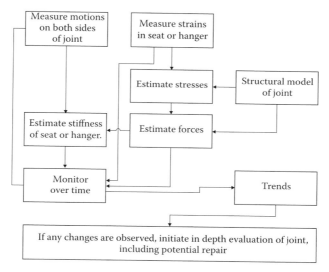

FIGURE 2.7 SHM process.

Baihua Bridge, China

Damaged column of pier 1 perpendicular to the direction of fault

FIGURE 2.8 Failure of bridge column at foundation connection. (Courtesy of MCEER, University of Buffalo.)

2.2.6 ABUTMENTS

The main modes of failure during seismic events (Figure 2.10) are loss of stability or excessive soil pressure that might lead to excessive abutment displacements, Wang and Gong (1999). Both modes of failure can lead to unseating of the supported superstructure. SHM roles can be summarized as

Before event: Measurements (strains, vibrations) can help in validating analytical models (albeit in small strain levels) and other normal conditions.

During event: Accelerations of abutment and back pressure measurements can give insights into the severity of the seismic motions. Also, both types of measurements can help in STRID and DMID efforts after event.

After event: To detect potential failure modes, sensing displacements or strains at appropriate locations are needed.

FIGURE 2.9   Failure of column-superstructure connection. (Courtesy of MCEER, University of Buffalo.)

FIGURE 2.10   Excessive seismic abutment motion. (Courtesy of MCEER, University of Buffalo.)

2.2.7 FOUNDATIONS AND SOIL CONDITIONS

Soil liquefaction and lateral spreading are the major observed causes of bridge foundation failures during earthquakes. Moehle and Eberhard (1999) observed that there were limited reports of foundation damage in situations other than liquefaction events. They opined that foundation damage might have been underreported because it is hidden below ground. This is an ideal situation for SHM applications as follows:

Before event: Monitoring for potential soil liquefaction or soil spreading near bridge foundations is essential. Also, accurate methods for monitoring status of as-built foundations (either surface or deep) must be devised.

During event: Accelerations at foundation and at free field need to be monitored.

After event: Validation of analytical models by measured accelerations might reveal any damage or abnormal behavior.

2.2.8 SHEAR KEYS

Shear keys provide lateral restraint of the bridge girders. Their seismic behavior and potential retrofit needs must be considered carefully. Keady et al. (1999) provided a detailed exploration of the ramifications of retrofitting shear keys. On the one hand, retrofitting shear keys to transmit seismic forces to substructures might create the need for an expensive substructure retrofit. On the other hand, if the shear keys failed during an earthquake, the performance of the superstructure might be severely impaired. The role of SHM of shear keys can be

Before event: Gaps between shear key and girders need to be inspected and adherence to design condition ensured.

During event: Lateral motions of girders and supporting piers/columns can reveal relative motion between both bridge components. In case of impact between the two, via closing the gap of the shear key, the severity of the impact might be detected using signal-processing techniques. Thus, the health of the shear key might be evaluated in real time.

After event: Visual inspection for visible damage. Accurate measurements, such as laser measurements, might be needed to estimate relative lateral motions, if any. In case of visible girder impacting shear key, advanced NDT methods might be needed. Different ultrasound techniques or infrared thermography can be used.

2.2.9 SUPERSTRUCTURE

Moehle and Eberhard (1999) observed that seismic damage or failure of superstructure is usually due to the damage or failure of bearings or substructures. However, there are situations where direct seismic damage to superstructures occurs. Some of these are

- Impacting of adjacent spans
- Lateral buckling or failure of light steel girders or trusses

Installing seismic restrainers can mitigate impacting of superstructures. Lateral buckling failures can be mitigated by appropriate seismic hardening. See Moehle and Eberhard (1999) for more details. SHM role for bridge superstructures include

Before event: Validation of bridge modeling using different STRID techniques. This involves modal identification, parameter identification, or neural networks, Jovanovic (1998). Measurement of displacements or accelerations is usually the main mode of sensing. Of particular interest is using modal identification to accurately estimate damping of lower natural modes that would verify seismic design assumptions. The models can be used as baseline models for eventual use to detect damage after seismic events.

During event: Measuring bridge response during the seismic event for a real- time or near real-time evaluation of severity of damage. This can aid emergency response and first responders (fire, police, etc.).

After event: Coupled with manual and visual inspection, different NDT methods can be used to assess damage. Appropriate methods include ultrasonic, thermography, acoustic emission (AE), laser-based methods, and penetrating radiation.

2.3   SHM COMPONENTS AND SEISMIC HAZARDS

2.3.1   GENERAL

2.3.1.1   Seismic Monitoring Philosophies

In this section, we concern ourselves with seismic health issues from a technical viewpoint, that is, from the SHM components viewpoint. Recall that we established three components of SHM, namely sensing, STRID, and DMID (the fourth component DM will be considered in Section 2.5). We look at the seismic applications of each of these components in detail. Considering that seismic effects on structures can be categorized according to their temporal phases and their spatial effects, each temporal phase and each spatial effect would have a different SHM role. We discuss the general temporal phases and the spatial effects first. In the rest of the section, we discuss, in more detail, the application of SHM components to seismic events.

2.3.1.2   Monitoring Temporal Phases

The importance of the roles of the four components of SHCE during each of the three phases of seismic events is shown in Table 2.2. When planned properly, SHCE can play a major role in seismic responses of bridges (and other structures). Specifically, Figure 2.11 shows the potential roles of monitoring during and after seismic events. We emphasize the importance of preplanning and integration of all the components of monitoring: sensing, STRID, DMID, and decision making. Without such planning, the efficiency and value of monitoring are reduced.

2.3.1.3   Monitoring Spatial Categories

One of the distinguishing features of seismic effects on structures is that its affects the system both globally and locally. The global effects are due to earthquake forces affecting the structure in a manner proportional to the structural mass. Because of this, most of the seismic forces are generated by the massive structural components, hence the global seismic effects. The local seismic effects are generated at locations where different structural components intersect (joints and connections). Because of the equilibrating and deformation compatibility needs between structural components, localized effects arise. Thus, it is essential that both global and local behaviors are monitored for a complete and accurate assessment.

In general, monitoring global structural seismic behavior includes global vibration and displacement monitoring. STRID tools for the whole system are used. Global damage considerations, including excessive global motions (displacements/accelerations), are an integral part of global

TABLE 2.2
Interrelationship between SHCE Components and Seismic Phases

SHCE Component	Before	During	After
Measurements	Medium (STRID)	High (emergency and rescue)	High (for damage detection)
Structural identification	Medium (design assessment)	High (assess failures in real time)	High (for damage detection)
Damage identification	Medium (other hazards)	High (assess failures in real time)	High (for estimating rating)
Decision making	Medium (all of the above)	Highest (rescue, closures, emergency)	High (cost/safety issues)

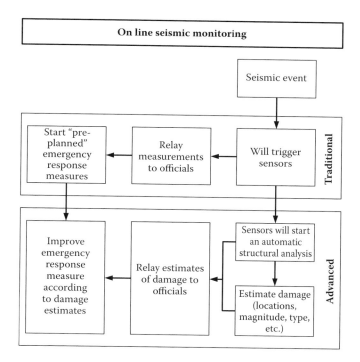

FIGURE 2.11 SHM role during and after seismic events.

monitoring. Monitoring local structural behavior includes using different NDT methods. Also, VSP can be used to detect local damage.

2.3.2 VISUAL INSPECTION AND EARTHQUAKE HAZARD

Visual inspection plays a major role before and after any seismic event. It should complement any automatic monitoring system. It has many unique attributes such as the ability to detect (1) qualitative parameters (overall conditions, surface cracks, nonstructural behavior, etc.), and (2) local components that are not monitored. Specifically:

Before event: The main goals of visual inspection before seismic events are (1) assessing the conditions of different components of the bridge, and (2) assessing potential onsite seismic vulnerabilities. Qualitative onsite reports and actual measurements can be of use in vulnerability assessment and post disaster evaluation.

After event: Overall response, local or global failures, extent of damage, support validate any monitoring results, qualitative reporting and assessments, and perform simple and advanced NDT.

2.3.3 SENSORS AND INSTRUMENTATION

2.3.3.1 General

We note that when a seismic event occurs, it affects the bridge in its entirety: its effects are global. However, initial damage, if it occurs, will occur locally first; in many situations, this damage can lead to global damage. For example, a shear failure in a column might be considered a local damage; however, such a failure can lead to the collapse of a whole span of the bridge. The concept is shown in Figure 2.12. Sensing techniques need to reflect this global/local/global phenomenon. Due to the high potential for nonlinearities, global seismic sensing is necessary but not sufficient. Thus there is a need for local sensing. This situation is unique for seismic hazard as compared to all other hazards. Table 2.3 shows potential local versus global techniques.

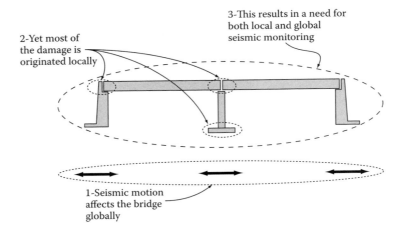

FIGURE 2.12   Local vs. global seismic behavior.

TABLE 2.3
Seismic Local versus Global Sensing

Global Sensing	Local Sensing
Thermography	AE
Load tests	Ultrasonic
Laser	Thermography
Remote sensing	Penetrating radiation
Other	Other

2.3.3.2   Optimum Sensor Locations—Multihazard Approach

One of the most difficult issues in sensing seismic behavior of bridges, especially large bridges, is locating a finite number of sensors on a bridge. This is especially difficult since earthquake motions affect the whole bridge. Thus, the bridge domain to be covered by the sensors is large, while the number of sensors is limited. If we recall that sensing overall seismic behavior would require sensing vibration modes of the bridge, in case of demand-type monitoring, it would be reasonable to expect that sensing vibration modes would also be useful for sensing bridge responses to some other conditions, such as wind loads. This is an ideal application of the concept of OSL discussed by Ettouney and Alampalli (2012, Chapter 5). The examples given in that reference show that OSL are sensitive to the assumed damage profile (location and magnitude) as well as the assumed loading combinations. The number of assumed damaged structural elements can have an effect on OSL, as demonstrated by the importance of damage indicator (stress ratio) and stress ratio threshold.

OSL concepts are still in need for further research. For example, more complex structural systems, different types of sensors (strains, velocity, etc.), different damage sources (corrosion, temperature, fatigue), and nonlinear effects are all among the important subjects that can have profound effects on any OSL conditions.

2.3.3.3   Principle of Sensor Interactions

The importance of monitoring seismic events at their three phases is well established, as evidenced throughout this chapter. There is one condition, though: such monitoring needs to be continuous. Since seismic events occur suddenly with no prior notice' the sensing and all necessary algorithms should to be functioning continuously. Continued seismic monitoring requires investment, and such an investment will be not functional unless an earthquake strikes the site. Questions about cost (a monitoring investment sitting idle while waiting for an earthquake to occur) and benefit (the value of information gathered during such an earthquake) will arise. Because of the infrequent nature of earthquakes, the value of earthquake monitoring does not appear to be as high as that of other monitoring goals.

We propose that this should not be so. The value of earthquake monitoring can be maximized by improving the efficiency of using earthquake sensing. Such an efficiency can be realized by recognizing that sensors interact in several ways. By optimizing such an interaction, we can enhance the value of seismic monitoring. These interaction modes are discussed briefly next:

Temporal interaction: Different hazards have different temporal distribution, as shown in Table 2.4. Intermittent hazards, by definition, will interact with continuous hazards. For example, seismic hazard will interact with traffic hazard. So, efficiency and optimal value can be realized if a monitoring scheme can be designed to continuously monitor traffic *and* seismic hazards.

Spatial interaction: Different hazards have different spatial distribution, as shown in Table 2.5. Globally applied hazards, by definition, will interact with local hazards. For example, the global seismic hazard will interact with the local fatigue hazard. So, efficiency and optimal value can be realized if a monitoring scheme can be designed to monitor fatigue *and* seismic hazards. An even greater value can be obtained by designing monitoring systems for two global hazards, such as wind and seismic hazards (see Celebi et al. 2004).

Type of sensing: Using an appropriate sensing method in a particular monitoring project for a particular hazard can have a great impact on monitoring efficiency. Table 2.6 shows relative effectiveness of sensing methods for different hazards. Using the same sensing method to monitor more than one hazard can improve monitoring efficiency.

We close this section by noting that there are two potential metrics for evaluating monitoring efficiency: cost and quality of information. Some rational methods of evaluating these metrics are discussed by Ettouney and Alampalli (2012, Chapter 4).

TABLE 2.4
Temporal Interaction of Sensors

	Hazard					
	Seismic	Wind	Scour/Flood	Traffic	Fatigue	Corrosion
Frequency	I	I	I	C	C	C

TABLE 2.5
Spatial Interaction of Sensors

	Hazard					
	Seismic	Wind	Scour/Flood	Traffic	Fatigue	Corrosion
Spatial effects	G	G	G	G	L	L

G = Global, L = Local

2.3.4 STRUCTURAL IDENTIFICATION/MODELING

2.3.4.1 General
The seismic behavior of a system spans wide ranges and categories. For example, structural responses can be linear or nonlinear. The behavior also can be dynamic or quasi-static (very low-frequency response). Structural modeling techniques depend on specific behavior. The same can be said for STRID techniques. Table 2.7 shows some structural modeling considerations. We also observe that the purpose of analysis can affect modeling techniques. For example, Figure 2.13 shows different modeling methods for estimating seismic capacity or demand. We note that if the ultimate goal of the models is to estimate damage, the modeling should be geared to estimate capacity. This means that the model should be of high resolution (component or frame level). It also, it should have some degree of nonlinearity. Figure 2.14 shows a typical use for seismic STRID modeling effort.

2.3.4.2 Seismic Testing
Seismic testing can be done on component or multicomponent levels. A popular form of testing is by placing the test article on a shake table (Figure 2.15). The shake table is then subjected to time-dependent motions that simulate earthquake shaking. Measuring the responses of the test article can then be investigated as needed. The test articles can be full scale or reduced scales. When this type of testing is done, the potential STRID method would be the single input (earthquake motion), multiple output technique.

TABLE 2.6
Type of Sensing Interaction

Type of Sensing	Hazard					
	Seismic	Wind	Scour Flood	Traffic	Fatigue	Corrosion
Acceleration	H	H	M	M	L	L
Displacement	H	H	H	H	L	L
Strains	H	H	H	H	H	M
Load Cells	H	H	H	H	L	L
Vibration	H	M	M	H	L	L

H, high; M, medium; L, low.

TABLE 2.7
Matching Analysis Techniques with Analysis Goals

Type	Linearity	Detail	Method
Static	Linear, within elastic limits	Only global displacements needed	Semianalytical or *smeared* FE model
	Linear, within elastic limits	Detailed strains in core and joints	FE: explicit modeling of core and surfaces
	Nonlinear/failure	Failure modes	FE: explicit modeling of adhesives, core, surfaces and other complex details
Dynamic	Vibrations/noise	Only global displacements needed	Semianalytical or *smeared* FE model
	Earthquakes, blast/impact	Behavior beyond elastic limits	FE: explicit modeling of adhesives, core, surfaces and other complex details

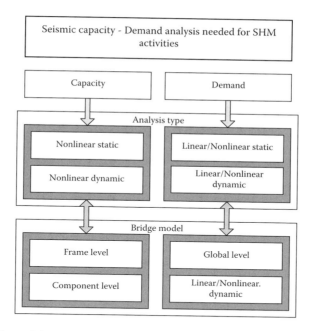

FIGURE 2.13   Capacity and demand needs for seismic structural modeling.

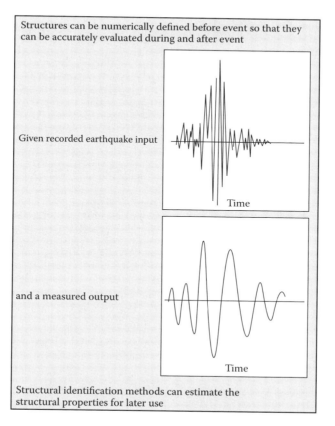

FIGURE 2.14   STRID role: before event.

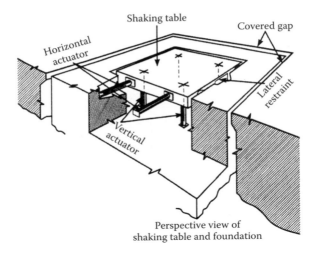

FIGURE 2.15   Details of typical shake table. (Courtesy of MCEER, University of Buffalo.)

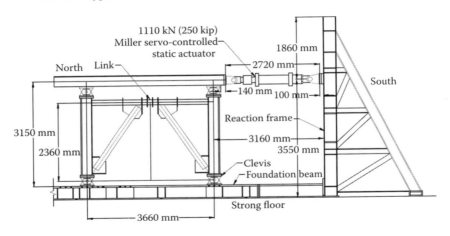

FIGURE 2.16   Test setup for bridge steel-braced frame. (Courtesy of MCEER, University of Buffalo.)

Another method of seismic testing is to subject the test article to lateral forces that simulate the seismic inertia forces while either keeping the base stationary (to simulate a rigid structural base) or placing the whole system on simulated springs (to simulate potential soil motion). Figure 2.16 shows a test setup for a bridge steel-braced frame. Figure 2.17 shows a seismic testing of a shear wall subjected to quasi-static forces via a set of mechanical actuators (Figure 2.18). A typical force-deformation result is shown in Figure 2.19.

Another conceptual vibration-based test system using automatic laser for armored vehicle launched bridge (AVLB) was developed by Chen et al. (2000). The AVLB is a mobile folding scissors-type assault bridge, which is hydraulically operated and is designed to be light and flexible (see Figure 2.20 for AVLB on top of an armored vehicle launcher). Due to very high-stress/low-cycle type of loading accompanied by large deflections, AVLBs have been found to fail under high-stress fatigue.

The conceptual system is presented in Figure 2.21, where AVLB is shaken with shakers in a pure bending mode and an automated laser system will capture modal data. A software system was used to obtain modal data and associated derived parameters to compare with expected data, and to detect safety by using a web-based expert system that gives red, yellow, green conditions based on integrity checks. This concept shows the potential for using laser in vibration monitoring. It can be

FIGURE 2.17   Seismic testing of shear wall. (Courtesy of MCEER, University of Buffalo.)

FIGURE 2.18   Simulating actuators. (Courtesy of MCEER, University of Buffalo.)

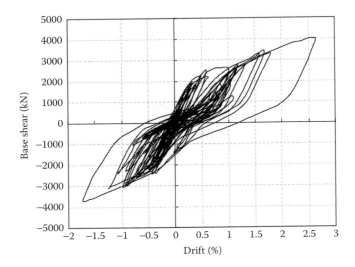

FIGURE 2.19   Typical seismic force-deformation relation. (Courtesy of MCEER, University of Buffalo.)

FIGURE 2.20   Armored vehicle launched bridge. (Courtesy of CRC Press.)

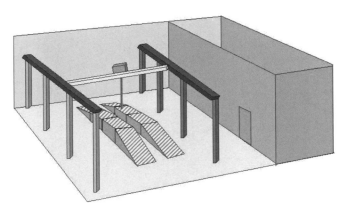

FIGURE 2.21   Shaker system. (Courtesy of CRC Press.)

generalized for an SHM application to detect, for example, the alignment of seats in an NDE setting or an SHM setting.

2.3.4.3 Parameter Identification

Parameter identification approach can be extremely valuable as an analytical tool during the three phases of the earthquake. It is based on building a suitable finite-element model for the bridge or any other structure. The finite element (FE) model should represent all important features of the bridge. The model can be fairly simple or very complex. For example, the simple 3-spans bridge of Figure 2.22 can be modeled for seismic investigations as shown in Figure 2.23. Note the special attention in the model for the gaps and supports. Similarly, the smaller bridge of Figure 2.24 and

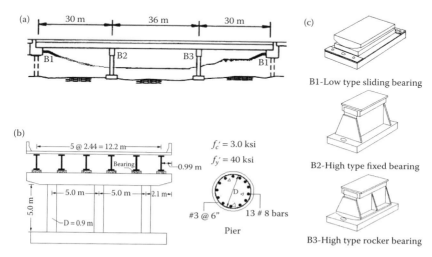

FIGURE 2.22 Typical steel bridge with three spans: (a) elevation, (b) side view, (c) bearing details. (Courtesy of Dr. Anil Agrawal.)

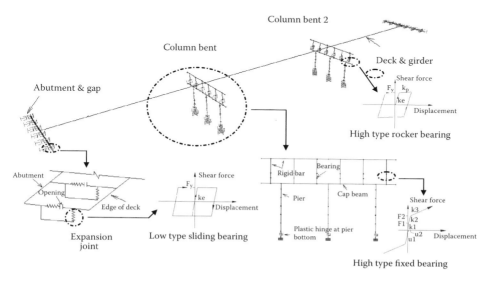

FIGURE 2.23 Finite elements modeling details of the typical steel bridge with three spans. (Courtesy of Dr. Anil Agrawal.).

its model in Figure 2.25 show different styles of modeling of column bents. Complex bridges, as in Figure 2.26, require complex finite element (FE) models (Figure 2.27).

After the FE model of the bridge of interest is built, the important parameters need updating. This can be done via bridge testing, for example, diagnostic testing (Chapter 8). The parameter updates

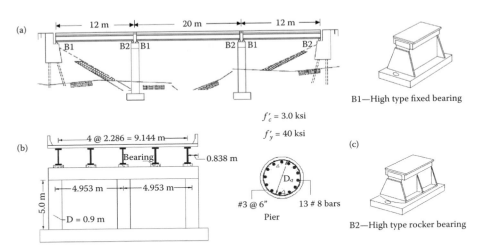

FIGURE 2.24 Typical smaller steel bridge with three spans: (a) elevation, (b) side view, (c) bearing details. (Courtesy of Dr. Anil Agrawal.)

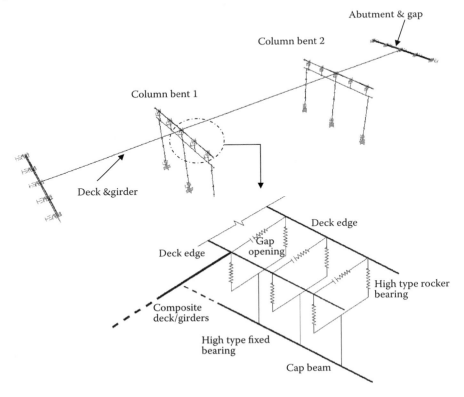

FIGURE 2.25 Finite elements modeling details of the smaller typical steel bridge with three spans. (Courtesy of Dr. Anil Agrawal.)

FIGURE 2.26 Complex bridge system.

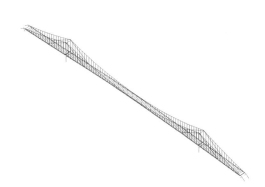

FIGURE 2.27 Finite elements modeling details of a complex bridge.

can be achieved using an appropriate parameter update method (see Ettouney and Alampalli 2012, Chapter 6). Now the model can be used as follows:

Before event: Accurate design assessments can be made using the model.

During event: The model can be used to evaluate, in real time, seismic damage. Such information can be invaluable for emergency and rescue operations.

After event: The model can be used to estimate the extent of damage and the needed repair levels.

2.3.4.4 Neural Network and Structural Identification

Neural networks, see Amini et al. (1997), can be applied to the earthquake problem in several modes. These modes differ in the types of input and output, as shown in Figure 2.28. One approach is shown in Figure 2.29. In this approach, the measured earthquake records are used as an input layer. The measured outputs (can be displacements or any other type of output measure) are used in the output layer. The use of numerous input-output sets can completely define the neural network. The network can then be used to predict outputs for any given earthquake input history.

Another form of neural network use in seismic applications was introduced by Gonzalez and Zapico (2007). The method uses modal data as an input to the network. The modal data includes pertinent frequencies and mode shapes. The output consists of different structural stiffness and mass. The network is trained twice: initially, before any damage, and after the system is damaged by an earthquake. By comparing the stiffness before and after the earthquake, the damage can be estimated.

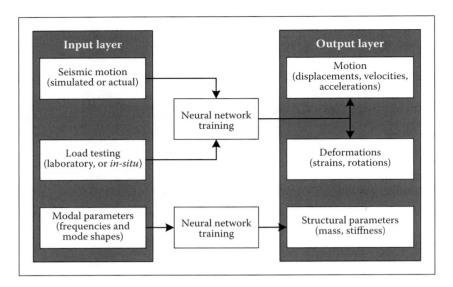

FIGURE 2.28   Modes of neural networks in seismic STRID.

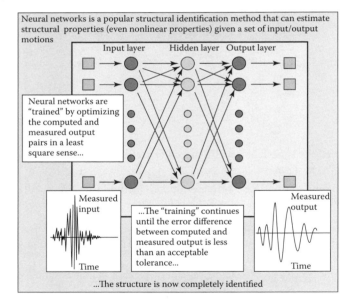

FIGURE 2.29   Neural networks in earthquake STRID.

Neural networks are well suited for earthquake STRID since they are applicable to nonlinear behavior. However, since there is no clear physical interpretation of the values of the neurons in the resulting network, it is difficult to develop an exact correlation between the network and the physical structure.

2.3.4.5   Nonvertically Propagating Waves

2.3.4.5.1   Introduction

Nonuniform seismic motion affects the seismic behavior of long-span bridges. Three main reasons for this nonuniformity have been identified. They are local soil conditions, wave passage, and incoherency effects. Other effects, such as extended source and attenuation, are relatively small. The importance of nonuniform seismic motions, especially for sensitive and important structures, has led to the development of several methods of analysis. These methods can be subdivided into two general categories: deterministic and stochastic. Due to the inherent uncertainty of the nonuniform seismic motions, deterministic methods, mainly time integration methods, can be computationally inefficient. Stochastic methods have been based mainly on modal analysis methods. The input/output are described in terms of either power spectral density (PSD) or response spectra. This chapter presents a direct frequency domain method based on formulating the whole soil-structure problem. The use of this direct frequency domain method for solving nonuniform seismic support motions is shown. The application of the proposed method to a simple two-dimensional bridge and a long-span suspension bridge is presented. Several observations are made. It is observed that nonuniform support motions may result in a large shifting of resonant frequencies of the structure. A large redistribution of bridge responses and internal forces were also observed while using nonuniform seismic motions, when compared with the uniform seismic motions case.

Several spatial seismic recordings have been performed and analyzed (Harichandran and Vanmarcke 1986; Hao 1989; Abrahamson et al. 1991). The results of these recordings have helped in gaining an understanding of the nature of this problem. Generally, there are two basic methods to account for the nonuniform seismic motions: deterministic and stochastic. Deterministically, the basic approach is to estimate a set of compatible time histories that might affect the support of the system. The system is then analyzed using a time domain approach (Clough and Penzien 1975). The main disadvantage of the deterministic method is that several sets of compatible support motions might be needed to ensure realistic and conservative results (Ettouney, Brennan, and Brunetti 1979). This makes the deterministic approach inefficient and less desirable for practical applications.

The stochastic method for nonuniform seismic motion is more appealing. It accounts for all the important parameters of motion nonuniformity in an efficient and accurate manner. It relies on the use of the PSD of the input motions; also, the output is produced in terms of PSD. The differences between the different support motions are accounted for by the use of an incoherence function based on empirical or field measurements. Some stochastic methods utilize the PSD as the only measure of the input and output parameters, (Zerva 1990), while some other researchers utilize the more practical response spectra as the input/output measures (Der Kiureghian and Neuenhofer 1992). In most cases, the analysis is based on using the mode shapes of the structure.

This chapter looks at the analysis of nonuniform support motions, using stochastic methods. Instead of using a modal approach to analyze the structure, we will use a direct frequency domain approach. To investigate the trends and sensitivities of structures to different aspects of nonuniform seismic motions, two case studies are presented. A simple two-dimensional bridge and a large-size complex suspension bridge are studied. Different aspects of the input motions, the methods of considering the nonuniform motions, the material damping, and the soil impedances (springs) considerations are studied. Different output measures are also studied, namely, PSD frequency spectra, average responses, and statistics of responses. In all, we studied internal forces and displacements of both bridges.

2.3.4.5.2 *Theoretical Background*

Equations of motion in the frequency domain: The general equation of motion of a structure subjected to seismic motion is

$$M\ddot{u} + C\dot{u} + Ku = 0 \tag{2.1}$$

where M, C, and K are the total mass, damping, and stiffness matrices of the structure, respectively. These matrices include the effects of both structural and support degrees of freedom, while $\ddot{u}$, $\dot{u}$, and u vectors represent the total acceleration, velocity, and displacements of both the structural and support degrees of freedom. The order of Equation 2.1 is $N = n + m$, where n is the number of unsupported degrees of freedom, while $m \geq 1$ is the number of supports in the structure subjected to seismic motion. Note that this definition is general and can accurately account for the existence of soil spring as well as the soil-structure interaction effects. If we assume a steady-state motion in the form $u = Ue^{-i\Omega t}$, with $i = \sqrt{-1}$, then the structural motions can be expressed as

$$D(\Omega)U(\Omega) = 0 \qquad (2.2)$$

The vector $U(\Omega)$ represents the steady-state complex amplitude of vibration, while the matrix $D(\Omega)$ is the frequency-dependent impedance matrix of the structure; both are functions of the driving frequency Ω. The matrix $D(\Omega)$ can be expressed as

$$D(\Omega) = K^* - i\Omega C - \Omega^2 M \qquad (2.3)$$

Note that the matrix $K^* = K + iC_1$, where C_1 is the hysteretic damping matrix and can be represented as

$$C_1 = \sum_j 2\beta_j k_j \qquad (2.3a)$$

where k_j is the stiffness matrix of the jth finite element and β_j is the hysteretic (material) damping ratio of the jth element. Note that the summation in Equation 2.3a is performed over all the elements in the structure. As such, the appropriate degrees of freedom for matrices k_j and C_1 should be observed during the summation procedure. Equation 2.2 is now capable of utilizing both viscous damping, C and element-specific hysteretic (material) damping, C_1.

The structure is subjected to multiple-support motions, $U_s(\Omega)$ where the order of the vector $U_s(\Omega)$ is m, such that

$$U_s(\Omega) \subset U(\Omega) \qquad (2.4)$$

The multiple-support seismic motions, $U_s(\Omega)$, can then be easily prescribed in Equation 2.2, and the total structural motions can be obtained by solving Equation 2.2, using any available complex equation solver. Note that by utilizing a steady-state (frequency domain) formulation, Equation 2.2 offers a simple and accurate form when compared with the modal approach (Zerva 1990; Der Kiureghian and Neuenhofer 1992).

Multiple-support motion: When a structure is subjected to seismic nonuniform multiple-support motions, it was shown that three main factors must be accounted for (Der Kiureghian and Neuenhofer 1992). These are (1) local soil conditions, (2) wave passage effects, and (3) incoherency of the seismic waves. Wave passage effects are those that result from the propagation of seismic waves in the soil media. Wave (shear, normal, etc.) velocities and their corresponding wavelengths can result in nonuniformity of the structural support motions. Wave passage effects can be represented either deterministically or stochastically. The incoherency of the seismic waves results from the fact that soil properties are not uniform, including the elevations of soil layers, as well as the nonhomogenous nature of the soil material. Due to this, the propagation of seismic waves in the soil media, is not coherent. The local soil condition effects result from the differences in the local soil conditions at each structural support. For example, for a long-span bridge, one end of the bridge can be supported on a stiff soil, while the other end can be supported on a landfill. This dissimilarity will result in large differences of the support motions. The local soil effects are traditionally computed in a deterministic fashion. However, these effects have also been used in studies of suspension bridges that are subject to nonuniform seismic motions (see Abdel-Ghafar and Rubin 1982, and Shrikande 1999). The reader is referred to Gazetas and Mylonakis (1998) for an overview of local soil condition effects during seismic events.

Let us assume that only the jth support degree of freedom of the structure is excited, where $1 \leq j \leq m$. The rest of the supports are assumed to be stationary. The scalar displacement of this jth support is $a_j(\Omega)$. Equation 2.2 can be solved for this single support excitation case, and the total structural displacement vector $U_{sj}(\Omega)$ can be obtained. The relationship between the jth support-prescribed displacement, $a_j(\Omega)$, and its displacement PSD, $g_{jj}(\Omega)$, is well known (Bendat and Piersol 1971). When all the m structural supports are excited, the total resulting structural displacement vector is

$$U(\Omega) = \sum_{j=1}^{j=m} U_{sj}(\Omega) \tag{2.5}$$

Note that the summation in Equation 2.5 is a complex summation; thus, it preserves any phase differences between the support motions. Expressions similar to Equation 2.5 can be derived for other desired response measures, such as internal forces, reactions, or internal stresses. For example, if we define the internal strain vector that will result from the structural displacement vector $U_{sj}(\Omega)$, to be $\varepsilon_{sj}(\Omega)$, the structural strains can be obtained by replacing $U_{sj}(\Omega)$ by $\varepsilon_{sj}(\Omega)$ in the summations in Equation 2.5 through 2.7. The values of $\varepsilon_{sj}(\Omega)$ can easily be obtained from $U_{sj}(\Omega)$ and the knowledge of different element impedance matrices. The same back-substitution steps can be used for other structural design parameters, such as internal stresses or reactions. For more details on this well-known back-substitution steps, the reader is referred to finite-element books (for example Zienkiewicz 1971). We will use the total structural displacement vector $U(\Omega)$ as an example of structural output throughout this work.

The PSD vector of the structural displacements can be expressed as

$$G(\Omega) = \sum_{k=1}^{k=m} \sum_{\lambda=1}^{\lambda=m} U_{sk}(\Omega) \, U_{s\lambda}(\Omega) \, G_{U_{sk}U_{s\lambda}}(i\Omega) \tag{2.6}$$

The cross-PSD, $G_{U_{sk}U_{s\lambda}}(i\Omega)$, is a complex scalar that can be expressed as

$$G_{U_{sk}U_{s\lambda}}(i\Omega) = \gamma_{k\lambda}(i\Omega)[G_{U_{sk}U_{sk}}(\Omega) \, G_{U_{s\lambda}U_{s\lambda}}(\Omega)]^{1/2} \tag{2.7}$$

The quantity $G_{U_{sj}U_{sj}}(\Omega)$ is the PSD of the input seismic motion at support j.

Gazetas Mylonakis (1998) showed that local soil conditions can be subdivided into kinematic and inertial effects. The kinematic effects are included in $G_{U_{sj}U_{sj}}(\Omega)$. The inertial effects of the local soil conditions are included in the structural impedance matrix, $D(\Omega)$. In addition, the parameter $G_{U_{sj}U_{sj}}(\Omega)$ includes the frequency (and temporal) characteristics of the incoming seismic motions. The wave passage effects and the incoherency effects are expressed in the parameter $\gamma_{k\ell}(i\Omega)$ that can be defined as (Der Kiureghian and Neuenhofer 1992)

$$\gamma_{k\ell}(i\Omega) = \exp\left[-\left(\frac{\alpha \Omega d_{k\ell}}{v_s} \right)^2 \right] \exp\left(i \frac{\Omega d_{k\ell}^L}{v_{\text{app}}} \right) \tag{2.8}$$

where v_s and v_{app} are the effective shear wave velocity and the apparent surface wave velocity of the site. For a plane seismic wave, the relationship between $d_{\lambda k}$ and $d_{\lambda k}^L$ can be expressed as

$$d_{\lambda k}^L = d_{\lambda k} \sin(\theta) \tag{2.9}$$

Where $d_{\lambda k}$ is the longitudinal distance, between support λ and support k. Factor $d_{\lambda k}^L$ is the projected distance in the direction of the wave propagation between support λ and support k. Factor α is an incoherence factor, which can be found from field measurements. Angle θ is the angle between a line extending from support λ and support k and the direction of the propagating wave.

Finally, upon solving for $G(\Omega)$, the vector of the mean squares of the displacements σ^2 can be found by using the well-known integral:

$$\sigma^2 = \int_{-\infty}^{+\infty} G(\Omega)\, d\Omega \tag{2.10}$$

Practical considerations: The method outlined so far is based on the use of PSD seismic input motions, $G_{U_{s\lambda}U_{s\lambda}}(\Omega)$ (Equation 2.7). The method produces root mean squares of any design parameter σ^2 (Equation 2.10). In practice, the seismic input is usually expressed in response spectra measures, and the maxima of design parameters are needed.

For input measures, the methodology that evaluates seismic PSD input for a given response spectra is well known. Several authors have studied the subject and presented concise expressions to evaluate PSDs from known response spectra. For example, see Der Kiureghian (1980), Der Kiureghian and Neuenhofer (1992), and Christian (1989). For completion, we report here the relation reported by Der Kiureghian and Neuenhofer (1992):

$$G_a(\Omega) = \frac{\Omega^{p+2}}{\Omega^p + \Omega_f^p}\left(\frac{2\zeta\Omega}{\pi} + \frac{4}{\pi\tau}\right)\left[\frac{D(\Omega,\zeta)}{p_s(\Omega)_0}\right]^2, \quad \Omega \geq 0 \tag{2.11}$$

$$G_d(\Omega) = \frac{G_a(\Omega)}{\Omega^4} \tag{2.12}$$

The variable $D(\Omega, \zeta)$ represents the input seismic acceleration spectrum at oscillator damping of ζ. $G_a(\Omega)$ and $G_d(\Omega)$ are the required seismic acceleration and seismic displacements PSDs, respectively. The physical and mathematical interpretations of the parameters $p, \tau, p_s(\Omega)_0$ are explained in Der Kiureghian and Neuenhofer (1992).

The mean and the standard deviation of the response maxima, μ_{max} and σ_{max} of a stationary process, needed for design, were studied and utilized by several authors (e.g., Der Kiureghian 1980; Die Kiureghian and Neuenhofer 1992; Dumanoglu and Stevern 1990). It was shown that the means and standard deviation of the response maxima are related to the response root mean squares σ by peak factors, as follows (using the expressions found in Dumanoglu and Stevern 1990):

$$\mu_{max} = \rho\sigma \tag{2.13}$$

$$\sigma_{max} = q\sigma \tag{2.14}$$

where ρ and q are peak factors. The evaluation of these peak factors, as well as their behavior, is well explained in Der Kiureghian (1980). For any design parameter (displacement, forces, etc.), the use of Equations 2.13 and 2.14, coupled with any design-specified nonexceedance probability (see the methods of Benjamin and Cornell 1970), would produce the desired final design value.

We note that Equations 2.11 through 2.14 were displayed in this section for completion only. They are not used in the case studies that will be presented later.

Comparison with normal modes-based solutions: It is of interest to compare the methodology as presented here with normal modes-based methodologies for considerations of multiple-support motions. This approach was introduced by Mindlin and Goodman (1950). Clough and Penzien (1975) generalized the formulation for application to matrix analysis. The basic approach is to use the conventional fixed-base modal analysis. A pseudostatic term is added to account for the differential base motions. The method was used in nonuniform seismic motions by several authors (e.g., Der Kiureghian and Neuenhofer 1992; Zerva 1990). The method represents the solution of the normal modes-based multiple-support problem utilizing two terms, in the form

$$f = \sum_{j=1}^{j=m} f_{j1} + \sum_{j=1}^{j=m} \sum_{i=1}^{i=n} f_{ij2}, \tag{2.15}$$

where f represents the desired response and n represents the total number of normal modes in the analysis. The expression f_{j1} contains the dynamic responses due to jth support, while the expression f_{ij2} contains the pseudostatic effects of the interaction through the structure of the ith normal mode due to the motion of the jth support. The PSD version of Equation 2.11 is expressed as

$$p = \sum_{k=1}^{k=m} \sum_{\lambda=1}^{\lambda=m} p_{k\lambda1} + \sum_{k=1}^{k=m} \sum_{\lambda=1}^{\lambda=m} \sum_{i=1}^{i=n} p_{k\lambda i2} + \sum_{k=1}^{k=m} \sum_{\lambda=1}^{\lambda=m} \sum_{i=1}^{i=n} \sum_{j=1}^{j=n} p_{k\ell ij3} \tag{2.16}$$

The expressions of p_{kl1}, p_{kil2}, and p_{klij3} are dependent on f_{j1}, f_{ij2}, the input PSDs, and the coherence functions. For our immediate purpose, however, we note that Equation 2.5 is similar in function to Equation 2.15, while Equation 2.6 is similar in function to Equation 2.16. In the current method, there is no need for a second (pseudostatic) term since the impedance matrix, $D(\Omega)$, contains all the structural interaction effects that necessitated the use of the second, doubly summed term of Equation 2.15. This, in turn, eliminated the need for the second, triply summed, and third, quadruply summed term in Equation 2.16. Basically, the direct solution of Equation 2.2 and the subsequent summations of Equations 2.5 and 2.6, in the current study, have replaced the conventional modal analysis and the subsequent summations of Equations 2.15 and 2.16.

Advantages and limitations of the proposed method: Frequency domain methods have several advantages and some limitations for practical use. Some of the advantages are the following: (1) damping of different material can be expressed on an element level (in the finite-element method) in a straightforward manner. Other methods of analysis can certainly handle different material damping by specifying a nonproportional damping matrix C. However, in a modal analysis technique, nonproportional damping might require the solution of a quadratic eigenvalue problem, thus leading to complex frequencies and complex mode shapes. Similar to the frequency domain method, step-by-step time integration can be used to account for different material damping; and (2) Simple handling of soil-structure interaction. Popular soil-structure interaction methods, such as those of SASSI computer code (Lysmer et al. 1981), produce SSI effects as complex-valued, frequency-dependent parameters.

There are some well-known limitations of the frequency domain. First, the solution of Equation 2.2, which is the basic equation in any direct frequency domain analysis, is based on solution at discrete frequencies, Ω. Careful choice of those discrete frequencies is needed to ensure accurate representation of all resonances in the structure. For example, in the suspension bridge case study to be presented later in this work, a frequency step of 0.0115 Hz was needed to capture all closely spaced natural modes of the bridge. Second, frequency domain methods are applicable only for linear systems. For systems with large nonlinearities, a nonlinear step-by-step time integration approach is needed. Currently, most of the design engineering community is not familiar with this method.

2.3.4.5.3 Case Studies

General: The methods of this study are applied to two bridges. In analyzing these two bridges, the effects of multiple-support seismic motions are considered. For both bridges, the sensitivities of the response to different parameters that affect the nonuniform support excitation are studied, using the same five cases reported by Der Kiureghian and Neuenhofer (1992). The reason for this is to provide uniformity for future researchers and practitioners who will be studying this class of problems. For convenience, we cite the five cases as follows

Case 1: Fully coherent (uniform) motions at all supports of the structure
Case 2: Only wave passage effect included (i.e., $\alpha = 0$) and $v_{app} = 400$ m/s
Case 3: Only incoherence effect included (i.e., $v_{app} = \infty$) and $v_s/\alpha = 600$ m/s
Case 4: Both wave passage and incoherence effects are included, with $v_{app} = 400$ m/s and $v_s/\alpha = 600$ m/s
Case 5: Mutually statistically independent support motions, that is, $\gamma_{k\lambda}(\Omega) = 0$ for $k \neq \lambda$

Input motions: The input seismic motion for both structures is a plane seismic wave with constant displacement unit amplitude at all frequencies of interest. For simplicity, the kinematics parts of the local soil conditions at each of the four support groups are assumed to be identical. The sensitivities of the responses to inertial effects of local soil conditions will be investigated in the second case study.

Let us assume that the input plane seismic displacement motion is $u_g = 1.0$. The general coordinate system of the two cases is x_1 (longitudinal), x_2 (vertical), and x_3 (lateral), as shown in Figure 2.30. The angle between x_1 and the propagation direction of the seismic wave is θ_1°. The angles θ_2° and $\theta_3^{\circ} = 180° - \theta_1^{\circ} - \theta_2^{\circ}$ are the angles between the propagation direction and the x_2 and x_3 axes, respectively. The seismic input displacement amplitudes in the three principal directions, u_{g1}, u_{g2}, and u_{g3} can be expressed as

$$u_{g1} = u_g \cos(\theta_1) \tag{2.17}$$

$$u_{g2} = u_g \cos(\theta_2) \tag{2.18}$$

$$u_{g3} = u_g \cos(\theta_3) \tag{2.19}$$

Using these relations, the input seismic displacement PSDs in each of the three principal directions can be evaluated (Bendat and Peirsol 1971). All the needed input parameters for Equations 2.2 through 2.10 are now defined.

Organization of results: One of the aims of the case studies is to explore the sensitivities and trends of the two structures to different parameters, when nonuniform support motions are accounted for. A frequency spectrum of different PSDs of different responses can be evaluated and displayed. These spectra can be beneficial in displaying the dynamic characteristics of different responses. For practical design purposes, however, it is more important to study the root

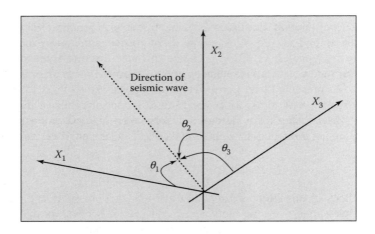

FIGURE 2.30 Coordinate system and direction of seismic wave.

mean square of the responses, as computed by Equation 2.10, or the maxima of the responses, as computed by Equations 2.13 and 2.14. For the purpose of the current work, we will consider only the square root of the mean squares of the responses, σ, as the basic tool to study the trends and sensitivities. Following Equations 2.14 and 2.15, it is straightforward to conclude that the observed trends and sensitivities using square root of mean squares should be similar to those trends and sensitivities of the response maxima.

The basic problem here is to try to simplify the observation methodology for the two case studies. We note that these two case studies, as well as most structures (bridges) susceptible to nonuniform seismic motions, are extremely complex structures. As such, there are many output measures (displacements, forces, etc.) that result from each situation. Coupled with the five uniform/nonuniform case possibilities, as well as other variables, the need for a systematic and simplified method(s) of studying the output is clear. Following is an outline of such a methodology.

For each problem, we will compute all square roots of the mean squares of the response measure of interest, using Equation 2.10. For example, σ_{ui}, the square root of the mean squares of the ith displacement measure, $i = 1, 2 \ldots n_1$. The total number of displacement measures of interest is n_1. For force measures, we define σ_{fi} as the square root of the mean squares of the ith force measure, where $i = 1, 2 \ldots n_2$. The total number of force measures of interest is n_2. In simple and small structures, the trends and sensitivities of individual σ_{ui} and σ_{fi} can be studied. Unfortunately, for larger structures, the values of n_1 or n_2 can be in thousands, making it impractical to study individual responses and reach accurate conclusions.

A simple way out of this dilemma is to consider σ_{ui} as a statistical sample. We can then easily evaluate the statistical average and standard deviation of that sample. We will also evaluate the maximum and minimum values $\max(\sigma_{ui})$ and $\min(\sigma_{ui})$, respectively. This procedure can be followed for the force measure sample σ_{fi}. The four parameters (average, standard deviation, maximum, and minimum) can provide a clear view of the overall behavior of the particular structure under consideration and make it easy to compare with other cases, thus determining the required trends and sensitivities.

Finally, in our presentation of different results, we will display normalized responses whenever deemed appropriate. In normalizing responses, we will use conventional analysis methods as a baseline. This way it becomes easy to note at a glance, the trends and sensitivities of the structural responses to different parameters.

2.3.4.5.4 *Steel Girder Bridge on Reinforced Concrete Towers*

Let us consider the simple two-dimensional bridge of Figure 2.31. The structure is a three-span steel girder supported by two intermediate piers. The piers and the two outer girder supports are assumed

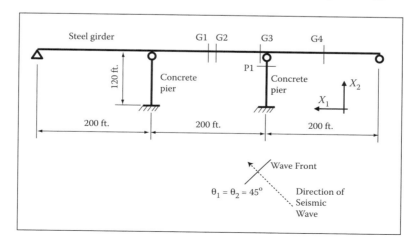

FIGURE 2.31 Two-dimensional bridge problem.

to be directly supported by the underlying soil. No-soil springs are included in this case. This system is modeled using a set of beam finite element. The total number of girder nodes is 30, and the total number of the nodes in each of the piers is 10. The scale independent element (SIE) approach was used for formulating the beam finite elements, so that the number of nodes in this model is sufficient to simulate the behavior at all the frequency ranges of interest. For more descriptions of the SIE methodology, see Ettouney et al. (1997). The input motion is assumed to be a seismic plane wave that is propagating in the $x_1 - x_2$ plane. The angles of propagation are $\theta_{1}^{\circ} = 45°$, $\theta_{2}^{\circ} = 45°$ and $\theta_{3}^{\circ} = 90°$. The damping ratio of the steel girders is assumed to be 2%, while the damping ratio of the concrete piers is assumed to be 5%. The damping is assumed to be hysteretic and frequency-independent. The output locations were chosen as follows: point G1 is at the center of the central girder, point G2 is 6.096 m (20 ft) off center, point G3 is directly to the left of the pier, and point G4 is at the center of the right outer girder. Point P1 is located on the right pier, 6.096 m (20 ft) below the girder support.

Frequency spectra of PSDs of internal forces: Let us consider the PSD frequency spectra of the internal forces at some of the output points. Figures 2.32 and 2.33 show the effects of nonuniform multiple-support motions on the PSD frequency spectra. Figures 2.32 and 2.33 compare case #1 (uniform seismic motion at all supports) with cases #2 and #3 (wave passage effects only and incoherence effects only). Note that the frequency axes are normalized to the first natural flexural frequency of a simply supported two-dimensional beam that has the same properties as the steel girder and a span of 60.96 m (200 ft). Figures 2.32a and 2.32b show the PSD spectra for axial

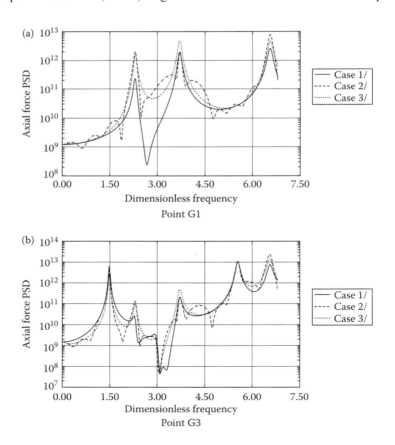

FIGURE 2.32   Comparison of axial force PSDs (a) Point G1 and (b) Point G3.

forces at points G1 and G3, respectively. Figures 2.33a and 2.33b show the PSD spectra for bending moments at points G4 and G2, respectively.

It is important to note that the internal forces in Figures 2.32 and 2.33 portray considerable sensitivity to the uniform and nonuniform support assumptions. First, we note that several of the magnitudes of the peaks and valleys of the PSD are different. Second, we observe considerable shifting in response resonance frequencies, especially in the case of bending moments (Figure 2.33a). A bending resonance at a dimensionless frequency of 3.25 for the uniform support motion case is shifted to a dimensionless frequency of 3.6 for both nonuniform motion cases. Thirdly, and most interestingly (Figure 2.33a), we observe that the nonuniform support motions actually excite a bending resonance at a dimensionless frequency of about 2.0. This resonance does not exist in the uniform support motion case. One of the reasons for both resonance shifting and new resonances may be that nonuniform multiple-support motions produce different modal participation factors from those participation factors produced by the uniform motions situation. Another possible reason is that the pseudostatic terms (second term in Equation 2.15) might have a larger effect on the responses than the convention modal terms (first term in Equation 2.15) at some driving frequencies.

The implications of the above observations are important; more studies of this phenomenon are advised. Such studies should account for the frequency-dependency of the input earthquake motions.

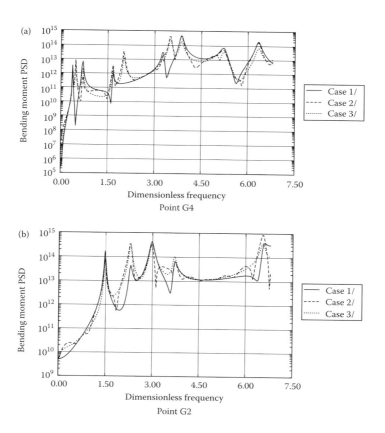

FIGURE 2.33 Comparison of bending moment PSDs. (a) Point G4 and (b) Point G2.

2.3.4.5.5   Closing Remarks

This section presented a method to account for nonuniform seismic support motions for large structures. The method relies on random vibration techniques, and solves for structural motions using a direct frequency domain method. The frequency domain approach can easily account for several aspects of the physical behavior of the bridge, for example, frequency-dependent soil springs and variable material damping. The method is simple to implement and computationally efficient.

A simple two-dimensional plane bridge was investigated. The internal forces of the two-dimensional bridge case displayed large sensitivities of the computed average forces to the nonuniformity of the support motions. More importantly, the frequency spectra of the PSDs of the bending moments at some cross sections displayed large resonant frequency shifts, and in some cases it displayed new resonances that were not observed in the uniform support motions case. This was explained by a redistribution of the modal participation factors and the presence of the pseudostatic term in the nonuniform case, which is not present in the uniform motions case. The implications to design are important, and more studies of this phenomenon are recommended.

A detailed and more complex suspension bridge case was discussed by Ettouney et al. (2001). The results of the study are the following. (1) Large response redistribution occurs when using soil springs compared with no-soil springs condition; (2) Using frequency-dependent springs results in a higher maximum response and a lower minimum response than using the fixed-support case or the constant soil spring case; (3) Accurate modeling of soil springs, through the use of frequency-dependent soil springs, is recommended; (4) Sensitivity and redistribution of responses to the method of specifying material damping were observed. Accurate specification of material damping is recommended; (5) In all studied cases, large sensitivities of the responses to seismic wave direction, θ_1, and multiple-support motion considerations were observed; and (6) Response redistribution was observed in most cases. The authors recommended careful consideration of the variability of multiple-support motions for large and complex suspension bridges.

2.3.4.5.6   SHM Role for Arbitrary Seismic Waves

The sensitivity of bridge responses to arbitrary seismic waves, as shown above, points to the importance of including such a factor in seismic analysis of bridges. We observe that such an analysis is based on three factors: (1) the incoherence factor α; (2) the effective shear wave velocity v_s; and (3) the apparent surface wave velocity of the site v_{app}. The accuracy of these factors controls the accuracy of the results. A systematic seismic monitoring effort to compute those three parameters during seismic events can help in establishing an accurate database that can be used to account for arbitrary seismic waves during any seismic analysis of bridges.

2.3.5   DAMAGE DETECTION

2.3.5.1   DMID Methods

The objective of any seismic consideration of bridges is to minimize damage during seismic events. Thus, stakeholders need always to identify potential seismic damage. This needs to be done in any/all of the three phases of the seismic events as follows:

Before event: Seismic damage can be identified by structural simulations where conventional analysis or more formal StrId can be used. When using StrId methods, some form of testing (*in situ* or laboratory, full size or component) needs to be done.

During event: Signal processing is an essential activity. In addition, real-time StrId or DMID efforts can be performed.

After event: DMID can be local, using any of NDT methods such as ultrasonic, optical, thermography, and so on. It can also be global, using vibration or load-testing techniques.

2.3.5.2 Signal Processing

The dynamic behavior of bridges during earthquakes is the basis of most seismic-based structural identification or damage identification efforts. Signal-processing techniques are of course used heavily during such efforts. The relative importance of signal processing during different phases of seismic events is shown in Table 2.8. Traditionally, Fourier transform and response spectra have been the main signal-processing methods in earthquake considerations (Boggess and Narcowich 2001). The two methods relate the frequency content of the signal to relative amplitudes. More recently, methods such as wavelet transforms (WT) (Gurley and Kareem 1999) proved to be beneficial in providing more information regarding time-frequency interrelationships (Figure 2.34). Note that the frequency level in the time signal increases as time increases. The WT of the signal reflects this time-frequency relationship. Figure 2.35 shows typical measured earthquake data and its WT. Heidari and Salajegheh (2008), investigated the use of FT and WT for the Taft Earthquake record and showed that important information can be deduced from WT analysis.. Hilbert-Huang transform (HHT) also provides time-frequency information that can be useful in analyzing earthquake records. Both WT and HHT have additional advantages: both are applicable to nonstationary events, such as earthquakes, and to nonlinear events, which are usually how systems respond to earthquakes. Figure 2.36 through 2.38 show processing of several earthquake records using both continuos (Morlet) and discrete (Haar) functions. The three earthquake records represent longitudinal, transverse and vertical motions at the same location in a moderate seismic zone. Note that the high frequency power of the longitudinal earthquake (Figure 2.38) arrives earlier than the lower frequency power of horizontal earthquakes (Figures 2.36 and 2.37).

2.3.5.3 Damage Types: Vulnerabilities, Mitigation, and SHM

Throughout this chapter, we have pointed out that damage to structures occurs in different ways. This is true of seismic damage to bridges. O'Connor (2004) summarized some of potential seismic damage to bridges and illustrated the potential retrofit or mitigation measures. Table 2.9 shows that summary. It also illustrates different health-monitoring techniques appropriate for detecting such damage. Note that the SHM techniques vary from global (displacement, load testing, etc.) to local (impact-echo, optical, etc.) Also, note that many of the SHM detection methods are analytical (STRID techniques) in addition to sensing and testing methods.

TABLE 2.8
Importance of Signal-Processing Techniques During Phases of Seismic Events

SHM Component	Before	During	After
Measurements	Needed	Needed	Needed
Str. Id.	Yes	NA	Yes
Dm. Id.	Yes (hazard)	Yes	Yes
Decision making	Yes	Yes	Yes

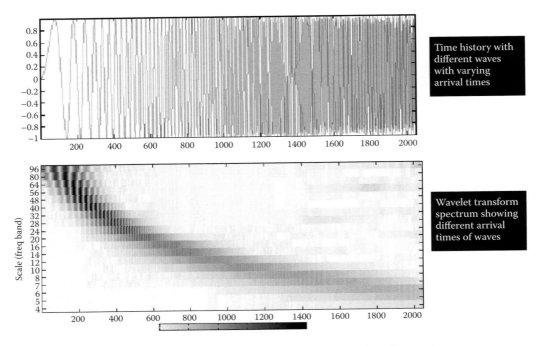

FIGURE 2.34   Wavelet transforms and arrival times. (Courtesy of Dr. Ahsan Kareem.)

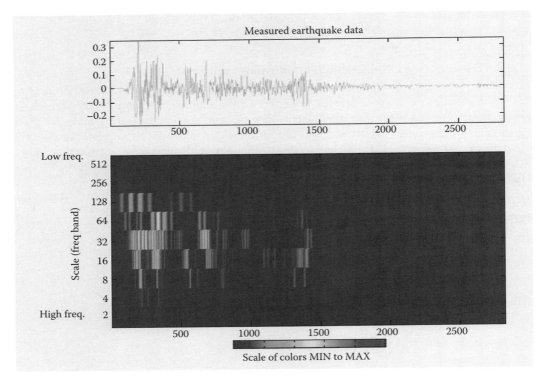

FIGURE 2.35   Wavelet transform of a typical earthquake record. (Courtesy of Dr. Ahsan Kareem.)

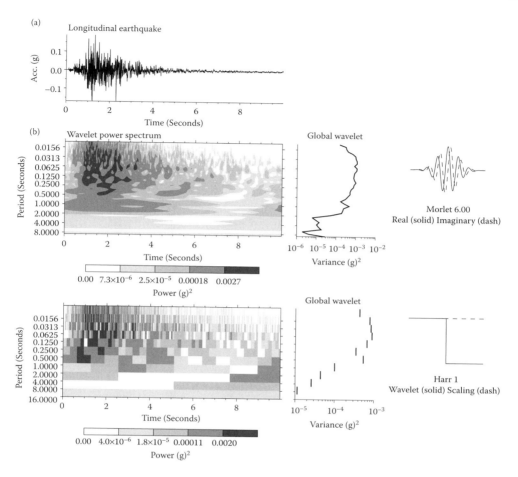

FIGURE 2.36 Wavelet transforms for longitudinal earthquake record using continuos function (Morlet) and discrete function (Haar).

2.4 CASE STUDIES

2.4.1 GENERAL

NDT and SHM techniques have been used extensively to monitor seismic damage and potential seismic damage. For example, Fujino and Abe (2002) presented an overview of the infrastructure situations in Japan. To show the importance of keeping the infrastructures healthy, they used the monitoring of Hakucho as an example of SHM application. They used laser Doppler vibrometer (LDV) for monitoring relative motions, which can be useful during all phases of seismic events. They also presented the theoretical basis of the LDV in their work. Sun and Change (2002) presented theory and experiment for damage identification, using wavelets analysis and statistical-based approach. This is applicable when damage scenarios can be approximated and the motion is a seismic-type motion. There is no need for structural modeling. The authors observed that the method is sensitive to damage but insensitive to measurement noise. Glisic and Inaudi (2002) discussed the use of fiber optics sensors in measuring cracks in concrete. This approach can be used during and after seismic events.

We present more in-depth SHM seismic-specific case studies in the rest of this section.

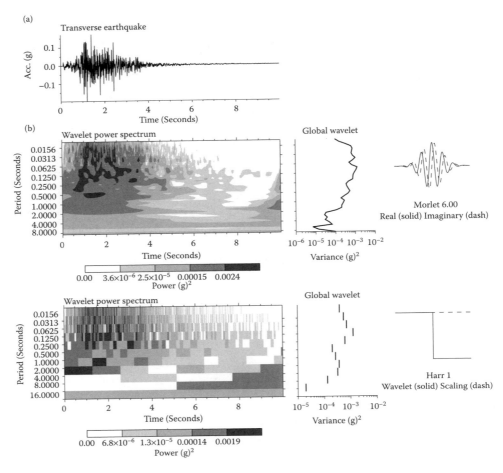

FIGURE 2.37 Wavelet transforms for transverse earthquake record using continuos function (Morlet) and discrete function (Haar).

2.4.2 LASER VIBROMETER DURING AND AFTER SEISMIC EVENTS

Scanning LDV was used to detect damage of different specimens by Sundaresan et al. (2000). Of particular interest was the study of an aircraft wing with stiffeners. The study included a pristine wing and a damaged wing (the damage was simulated by a crack in the stiffener). The experiment included the use of two surface-mounted piezoelectric patches to generate an in-phase excitation. Using an SLDV, line and surface scans of the wing were performed. The frequencies of the excitations were in the range of 45 KHz to 50 KHz. The resulting SLDV surface measurements clearly localized the damage. Since the stiffener is a linear element, both the linear and surface scans were adequate to identify the damage. For damage in two-dimensional plates, a surface scan might be needed. The effectiveness of this method in detecting damage that are deep in a three-dimensional system, such as reinforced concrete connections, is not clear.

Surface scans using laser-based technology can be applied to detect damage in steel connections during or after earthquakes. The use of conventional strain gauges will of course show the state of strains in the connection. However, one of the limitations of conventional strain gauges is that they offer only local sensing. Thus, if they strain localization during or after an earthquake occurs away from a strain gauge, the strain measurements by that gauge will not be too beneficial. Employing

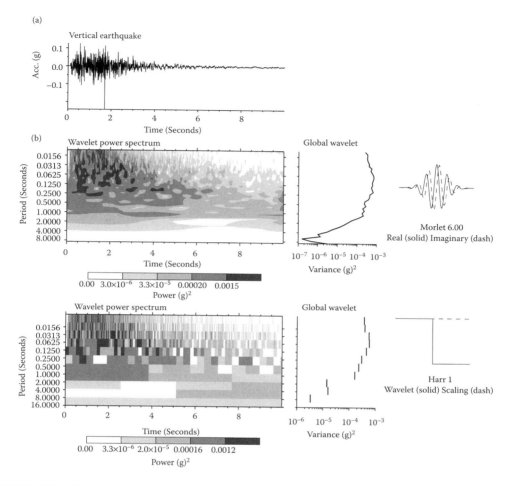

FIGURE 2.38 Wavelet transforms for vertical earthquake record using continuos function (Morlet) and discrete function (Haar).

a laser-based technology might overcome such a limitation. As shown by Sundaresan et al (2000), the range of detection can be large enough to cover a typical steel connection. If the scanning is performed on appropriate surfaces and locations of the connections, the surface vibrations of the connection during and after earthquakes can be used to detect the behavior of such connections, including any damage. Such a system might be instrumented so that it remains dormant until triggered by an earthquake.

2.4.3 AFTER SEISMIC EVENT DETECTION BY ACOUSTIC EMISSION

AE technique can be used for detecting post event damage in different bridge components. During an earthquake, several components will deform beyond elastic limit (see Paulay and Preistly 1992). The resulting stress distributions are complex and extremely difficult to measure accurately. Moreover, if there was no strain gauges mounted at the exact location of the nonlinearly deformed spot, mounting strain gauges *post event* will reveal neither the strain history *during* the event nor the state of strain. The reason for this is simple: the post event strain state has already occurred; the strain is now an initial strain as far as the bridge component is concerned, which cannot be measured by conventional strain sensors. Fortunately, the extent of damage can be evaluated by measuring the current stress

TABLE 2.9
Seismic Vulnerability, Mitigation, and Role of SHM Methods

Vulnerability (From O'Connor 2004)	Typical Retrofit Measure (From O'Connor 2004)	SHM Application
Lack of confinement in reinforced concrete column	Steel jackets, tensioned steel strands, or composite fiber wraps	Strains, impact-echo, thermography
Loss of superstructure support at pier	Longitudinal restraint cables	Strains, optical
Excessive lateral movement of superstructure	Install lateral restraint device, cables or shear blocks	Strains, optical
Uplift	Secure bearings	Displacements, optical
Toppling of bearings and dropping of superstructure	Replace bearings with elastomeric or other low bearing	Visual, optical
Lack of superstructure continuity on a multispan bridge	Join the spans to provide live load continuity	Conventional load-testing techniques
Excessive load transfer to substructure	Install isolation bearings	STRID techniques
Loss of support at abutment	Widen bridge seat	STRID techniques
Insufficient lap splices between column and footing or piles and pile cap; lack of top mat tension reinforcement in footing	Footing-column modification; additional piles and extended pile caps; micropiles	Ultrasound, STRID techniques
Unequal bent stiffness	Detailed analysis and retrofit	STRID techniques
Masonry substructure construction	Exterior reinforcement and confinement	Ultrasound
Liquefaction	Silty soil remediation by compaction grouting, stone columns, or micropiles	STRID techniques
Excessive longitudinal forces	Install load transfer devices	Optical, STRID techniques
Single column bridge piers	Provide redundancy by adding columns or outriggers	STRID techniques
Poor reinforcement detailing	Reconstruct joint	Ultrasound, STRID techniques

state, that is, the post event stresses. This can be accomplished if we view the current state of stresses as initial stresses, or in other words, residual stresses. If these residual stresses have reached yield stage, then it would be producing AE. By measuring these emissions, a reasonable estimation of state of stresses can be made. At the least, it would be possible to ascertain if the state of stresses as the point of interest has reached an inelastic stage. Riahi and Abdi (2005) presented a method for using AE in detecting residual stresses. Such an application can be extended easily to the post event evaluation of seismically damaged bridges or any other abnormal hazard.

2.4.4 Bill Emerson Memorial Bridge

Celebi et al. (2004) described an instrumentation system developed for Bill Emerson Bridge monitoring as this major cable-stayed bridge is located in a seismic region where the New Madrid earthquake of 1811–1812 occurred.

The instrumentation planning was initiated before the construction started as the bridge was designed for a strong earthquake of magnitude 7.5 or higher. The system is triggered by a prescribed threshold motion with the aim of recording free-field motions at the surface and at downhole locations reaching competent rock, the overall motion of the cable-stayed bridge, and the motions of the extreme ends of the bridge and the immediate pier locations. A total of 84 channels of accelerometers were under consideration with estimated hardware costs of about $335,000 (see Figures 2.39 and 2.40). The system functionality can be verified remotely. The system can also record low-amplitude motion to facilitate assessment of the dynamic characteristics of the structure and provide a

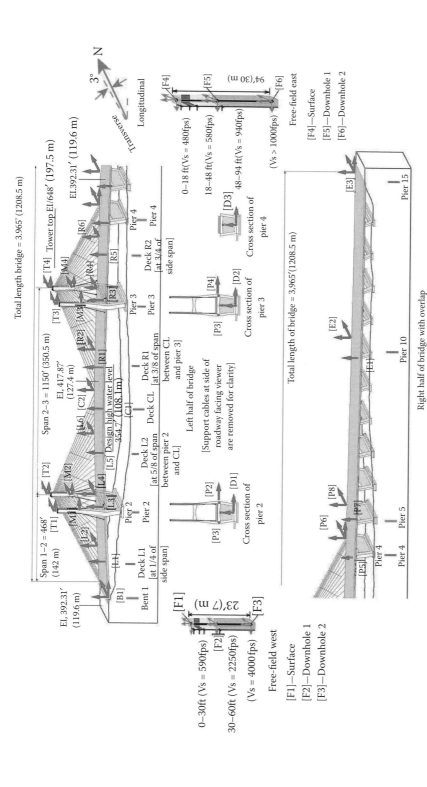

FIGURE 2.39   Sensor locations. (Courtesy of Dr. Mehmet Celebi.)

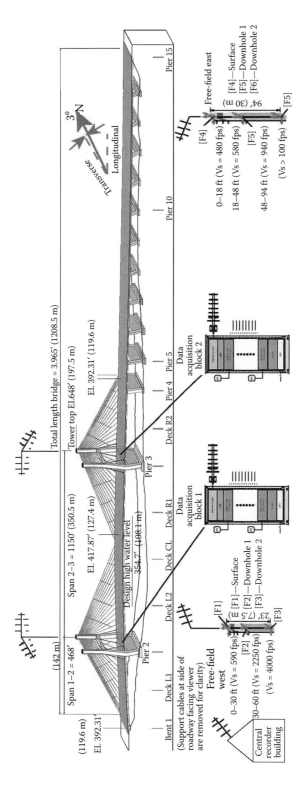

FIGURE 2.40 Data and wireless systems. (Courtesy of Dr. Mehmet Celebi.)

basis for estimating levels of shaking during stronger events, the return periods of which are longer than smaller events.

Some lessons noted are the following: (1) Planning should account for physical and scheduling constraints, which will limit where sensors and instrumentation can be placed, when it can be placed, and when response can be collected. Costs also depend on when these can be done; and (2) Accounting for multihazard considerations can save the owner resources by avoiding redundant efforts. Weather conditions can impose special requirements for instrumentation protection (in this case due to the high probability of high winds and thunderstorms). Wind was also an important factor, but in view of the financial constraints, it was realized that both cannot be done initially. Hence, the design considered this carefully so that it could be added in the future, but some sensors were accommodated in the current scheme through optimization.

2.4.5 Vincent Thomas Bridge

A STRID project utilizing SHM for Vincent Thomas Bridge, in San Pedro, CA, was performed by Smyth et al. (2000). The Vincent Thomas Bridge, which is situated in a seismic region, was instrumented with a limited number of accelerometers due to financial restrictions. Through parametric linear system identification approaches, damping estimates and other critical dynamic influence coefficients were estimated for each earthquake. Changes in the identified models were studied to determine the amount of damage incurred during the events. The results of the new global system identification techniques based on vibration data collected from the sensor array on the Vincent Thomas Bridge were compared. Such estimates of critical structural dynamic properties can be extremely useful for retrofit plans and risk assessment.

The bridge was completed in 1964, and in 1980 it was instrumented with 26 accelerometers as part of a seismic upgrading project. Ten accelerometers measured motion at the superstructure footings, and fifteen accelerometers were distributed at various locations and in lateral, longitudinal, and vertical directions about the superstructure itself (see Figure 2.41).

Since its installation, the instrumentation network has been triggered twice during large seismic events in southern California. The first was for the 1987 Whittier-Narrows earthquake (M = 6.1)

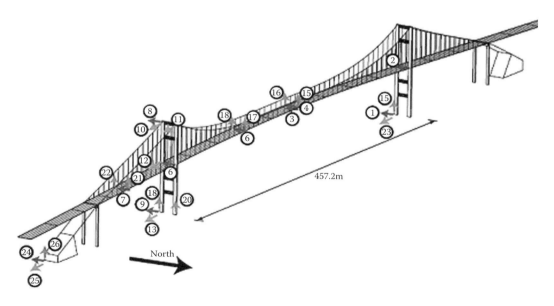

FIGURE 2.41 Accelerometer locations and directions for the instrumentation network on The Vincent Thomas Bridge. (Courtesy of Dr. Raimondo Betti.)

and the second for the 1994 Northridge earthquake (M = 6.7). Time domain SI procedures were used. A brief sample of the identification results are given in the paper, and they confirm that time domain–based identification procedures can capture, in a reduced order model, the essence of the response dynamics of this highly complex structural system. The modal frequencies obtained for the two earthquakes are similar but indicate some system changes. These types of results can be compiled to track structural changes during the excitation process.

2.5   DECISION MAKING AND EARTHQUAKE HAZARD

2.5.1   General

Decision making tools can be used in numerous situations that involve seismic events. We explore first a simple and popular situation: decision making based on seismic vulnerability rating. We make a qualitative case for improving safety and reducing costs by using SHM techniques in rating evaluations. We next provide some decision making examples for pre, during, and postevent situations. The purpose of all of those examples is to show how decision making tools can save costs, while maintaining or improving safety levels.

2.5.2   Example: Seismic Vulnerability Rating

Seismic vulnerability rating aims at qualitatively assigned values that describe seismic vulnerability. For example, Table 2.10 shows an NYSDOT seismic vulnerability rating system. It is a qualitative scale in the range of 1 to 6. The results of each of the ratings are also shown in Table 2.10. As can be seen from Table 2.10, lower ratings can result in costly repairs. So, it is desirable to produce as accurate a rating as possible. By improving the accuracy of rating, improved safety and reduced costs are achieved. The sources of reduced costs are

1. Direct: reduced labor
2. Immediate: by reducing cost of retrofits/repairs
3. Long term, by reducing life cycle costs

SHM methodologies can have a role in improving accurate seismic vulnerability ratings. In general, they provide a time advantage over lesser accurate ratings. Consider Figure 2.42 for example. It

---

TABLE 2.10
Seismic Vulnerability Rating, NYSDOT (2004)

VR	Vulnerability Rating Definitions
1	Safety priority action—This rating designates a vulnerability to failure resulting from loads or events likely to occur. Remedial work to reduce the vulnerability must be given immediate priority
2	Safety program action—This rating designates a vulnerability to failure resulting from loads or events that may occur. Remedial work to reduce the vulnerability does not need immediate priority, but waiting for the Capital Program action would be too long
3	Capital program action—This rating designates a vulnerability to failure resulting from loads or events that are possible but are not likely. The risk can be tolerated until a normal capital construction project can be implemented
4	Inspection program action—This rating designates a vulnerability to failure presenting minimal risk, provided that anticipated conditions or loads on the structure do not change. Unexpected failure can be averted during the remaining life of the structure by doing the normal scheduled bridge inspections with attention to factors influencing the vulnerability of the structure
5	No action—This rating designates a vulnerability to failure which is less than or equal to the vulnerability of a structure built to the current design standards. Likelihood of failure is remote
6	Not applicable—This rating designates that there is no exposure to a specific type of vulnerability

shows a deteriorating system. Since automatic evaluation offers more accurate damage detection capability than manual evaluation, it can result in a time advantage: the automatic evaluation can detect damage long before manual evaluation. This can improve safety and/or reduce costs. For the special case of rating before event, Figure 2.43 shows the qualitative value of automatic (SHM-aided rating) versus manual rating. The same conclusion can be drawn for after-event condition assessment. Figure 2.44 shows qualitatively how accurate assessment after an event can save costs. Alternatively, Figure 2.45 shows qualitatively how accurate assessment after an event can improve safety. The previous sections discussed several SHM-aided methods that can be used to improve seismic rating both before and after the event.

2.5.3 BEFORE-EVENT DECISION MAKING CASE STUDY: DAMAGE TRENDS

2.5.3.1 Overview

Placing sensors at locations where seismic damage is expected to occur during a seismic event is obviously a prudent decision. For, such a location will help in locating the damage and assessing if the damage warrants emergency action after the event. It will also help in pursuing any retrofit

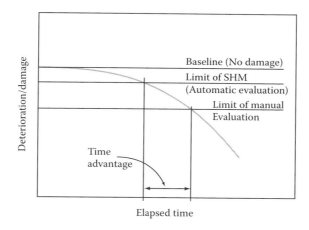

FIGURE 2.42 Value of automatic vs. manual sensing (time value).

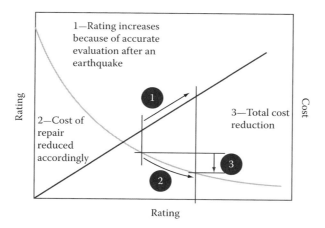

FIGURE 2.43 Value of automatic vs. manual sensing (cost value).

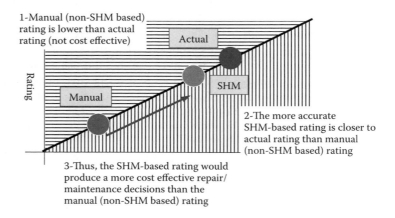

FIGURE 2.44 Value of accurate rating (after event)—cost savings.

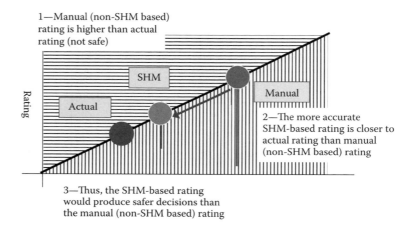

FIGURE 2.45 Value of accurate rating (after event)—safety.

actions. However, we must ask if placing sensors at expected damage locations can also be of benefit before the event. If another utilization of the sensing capability at such locations while waiting for the seismic event can be found, then the value of the experiment is improved immensely. We recall that the duality principle of the multihazard theory has implied the benefits of such dual use. Table 2.11 shows some potential dual use of seismic sensors.

Let us consider the case of the seat alignment of Table 2.11. The seat is manufactured of steel material. It is judged by the bridge official to be the most seismically vulnerable point in the bridge. However, since the bridge is located in a moderate seismic zone, the official is hesitant to commit funding for long-term seismic monitoring of the seat. The SHM consultant brought it to the attention of the official that the adoption of the duality principle can result in several by-products of long-term monitoring. For example, by designing a laser alignment experiment between the two sections of the bridge, and measuring the long-term alignment between the two sections, the benefits in Table 2.12 can be achieved.

TABLE 2.11
Dual Use of Seismic Sensing

Location	Type of Sensor	Seismic Use	Dual Use
Bearings/seats	Alignment	Unseating of bearings/seats during event	Long-term corrosion under bearing/seat; misalignment due to vibration, impact, wear, and tear
Hinging in bents	Strains	Damage at column-beam connections in bents	Validation of design or analysis assumptions; if placed appropriately, can be of benefit in monitoring any long-term corrosion effects
Midspans	Accelerometers	Measure motions during events	Validation of design or analysis assumptions; can be used during load-rating tests; wind response of bridge; long-term wear and tear, including deterioration measures
Soil-foundation interfaces	Pressure/strains/ accelerometers	Measure motions, liquefaction, and deformation during events	Validation of design or analysis assumptions; can be used during load-rating tests; scour potential, if applicable; deterioration (wear and tear) of foundations, piles; any unexpected foundation problems (see serendipity concept in Ettouney and Alampalli 2012, Chapter 4, Section 7.)

TABLE 2.12
Benefits of Long-Term Monitoring of Seat Alignment

	Benefit
Before seismic event	The temporal alignment trends can reveal any misalignments due to impact, accidents, or any other reason
	Alignments can validate any load-rating testing
	Alignments can validate any analysis assumptions
During seismic event	Detects behavior in real time and alerts if there is damage or unseating
After seismic event	By detecting damage level, can help in guiding emergency efforts
	Can guide any retrofit efforts

2.5.3.2 When Does a Change Occur?

Let us assume that the horizontal alignment measurements from the above experiment is $a(t)$, where t is the time variable and $a(t)$ is the measured relative motion between the two bridge sections. Normally, the average of $a(t)$ can be expressed as

$$E[a(t)] = \frac{1}{T}\int_T a(t)\,\mathrm{d}t \approx 0 \tag{2.20}$$

Let us assume that at time $T = T_1$, a sudden misalignment, δa, occurred between the two sections of the bridge (due to an accident, for example); then the measured average becomes

$$E[a(t)] = \frac{1}{(T-T_1)}\int_{T_1}^{T} a(t)\,\mathrm{d}t \approx \delta a \tag{2.21}$$

The new root mean square μ can be computed, using the methods of Ettouney and Alampalli (2012, Chapter 8).

The decision maker is faced with a question: How serious is the misalignment? There are two ways of answering this. A qualitative way to answer the question is to compare the new misalignment average, δa, with the old root mean square μ. If the ratio $\delta a / \mu$ is within a reasonable tolerance, say $\leq 5\%$, then the decision maker might ignore the recorded misalignment; otherwise, further investigations and remedies might be warranted. A more accurate and quantitative way is to investigate the effects of such misalignment on the state of stresses, strains, and forces in the seat as well as in the pertinent bridge sections. Such an effort might be costly, but it will result in more safety assurances.

Let us assume that the misalignment occurred in a very slow and gradual manner, for example, in the case of corrosion under the seat that could not be detected by visual inspection. Such corrosion will gradually affect the stiffness of the seat supports and induce slow misalignment. The use of Equation 2.20 to compute the average misalignment will be slow in revealing the gradual and slow changes. A better approach is to use a moving window-averaging technique.

2.5.4 During- and Immediately After-Event Decision Making Case Study: Risk Management

2.5.4.1 Overview

Seismic monitoring techniques have one major goal in common: the during- and after-event estimation of bridge behavior and helping the decision maker choose the course of action needed as a response to the seismic event. Let us assume that a particular bridge has been instrumented to monitor seismic events. Let us further assume that an earthquake did actually occur and affect the bridge. All the instrumentations did function properly, and the data were collected in near real time. The decision maker has now some important decisions to make. These involve invoking emergency response teams, managing traffic, alerting the public, and so on. All these decisions depend, in general, on two factors. The first is the level and type of the bridge damage. The second is the relative importance of the affected bridges within the bridge network; if the earthquake affected a large number of bridges within the network, then a prioritization of responses within the network might be needed. It is obvious that the decision making tools and algorithms for these two factors must be in place before the event. In this section, we present decision making methods for these two important factors.

2.5.4.2 Performance Levels

Let us assume that the bridge under consideration is monitored by N_1, N_2, and N_3 displacements, acceleration, and strain sensors, respectively. The monitored data during the earthquake is $u_i(t)$, $\ddot{u}_j(t)$, and $\varepsilon_k(t)$, representing displacements, acceleration, and strain measurements, respectively. Note that the subscripts denote the particular sensor of interests. Thus, $i = 1, 2, \ldots, N_1, j = 1, 2, \ldots, N_2$, and $k = 1, 2, \ldots, N_3$. In this example, for the sake of simplicity, we will not concern ourselves with soil failure modes such as landslides or liquefaction problems. We note that the decision making techniques presented below apply in a similar fashion to soil failure modes.

Damage levels: Deciding on the damage levels is perhaps the most important first step during and after an earthquake event. Preparation for such a decision should start during the installation of the SHM sensors and equipment, well before the earthquake. One way is to adopt the performance level approach that was first presented for existing buildings. Celebi et al. (2004) discussed utilization of the concept for health monitoring of buildings. We extend the method for bridges. As such, we place the seismic performance of the bridge at several discrete levels: continued operations, damage control, life safety, collapse prevention, and complete collapse. We define these levels as follows:

- Continued operations
- Damage control

- Life safety
- Collapse prevention
- Complete collapse

These discrete performance levels are qualitative; they can certainly be improved by using continuous, rather than discrete, performance levels. However, this is beyond the current scope of this example. For now, we define the five performance levels as PL_i with $i = 1, 2, 3, 4, 5$.

The crucial step now is to relate the performance levels of the bridge to the measured data. A simple approach would be to assign predetermined threshold values for performance levels and test the condition:

$$\text{If } f_u(u_t(t)) > U_j, \text{ then performance level } PL_j \text{ is realized} \tag{2.22}$$

Equation 2.22 is tested for each performance level, $j = 1, 2, 3, 4, 5$, thus establishing the performance level attained by the bridge as a result of the earthquake. Similar conditions can be established for accelerations and strain measurements as

$$\text{If } f_{\ddot{u}}(\ddot{u}_i(t)) > \ddot{U}_j, \text{ then performance level } PL_j \text{ is realized} \tag{2.23}$$
$$\text{If } f_\varepsilon(\varepsilon_i(t)) > E_j, \text{ then performance level } PL_j \text{ is realized} \tag{2.24}$$

The functions $f_u(u_t(t))$, $f_{\ddot{u}}(\ddot{u}_i(t))$, and $f_\varepsilon(\varepsilon_i(t))$ depend on the location of the sensors and the logic used in locating the sensors during the design phase of the SHM project. For example, if $u_i(t)$ and $u_{i+1}(t)$ are lateral displacements at top and bottom of a particular column in a bridge, then

$$f_u(u_k(t)) = \text{Max}\left|u_i(t) - u_{i+1}(t)\right| \tag{2.25}$$

Strain and acceleration functions are usually simple functions in the form

$$f_{\ddot{u}}(\ddot{u}_i(t)) = \text{Max}\left|\ddot{u}_i(t)\right| \tag{2.26}$$

$$f_\varepsilon(\varepsilon_i(t)) = \text{Max}(\varepsilon_i(t)) \tag{2.27}$$

Note that the strain functions allow for separate checking of compressive or tensile strains.

The thresholds U_j, $\ddot{U}_j$, and E_j can be established either qualitatively or quantitatively. Qualitative performance thresholds are established on the basis of the personal experiences of the design team. They also can be established by comparisons with known performances of similar bridges. Quantitative performance thresholds are established using analytical techniques of structural identification and damage detection, some of which are described in detail earlier in this chapter and in Chapters 6 and 7 by Ettouney and Alampalli (2012).

There is an inherent problem in the above approach: the potential for conflicting results. This conflict is possible because of (1) the thresholds being estimated qualitatively, (2) the likelihood of the quantitative methods for estimating the thresholds not being accurate, or (3) the qualitative nature of the performance levels. To illustrate such a situation, consider an SHM experiment where the bridge is instrumented with a total of 90 sensors: 30 displacement sensors, 30 acceleration sensors, and 20 strain sensors. First, let us assume that a moderate earthquake resulted in performance level estimations as shown in Table 2.13. A qualitative study of Table 2.13 would indicate that the performance level of the bridge is damage control. Some life-safety issues must be tended at three locations. If the earthquake is more severe, the performance levels might look similar to those in Table 2.14.

TABLE 2.13
Performance Levels, Moderate Earthquake

Performance Level	Displacement	Acceleration	Strain
Continued operations	12	5	10
Damage control	16	20	7
Life safety	2	5	3
Collapse prevention	0	0	0
Complete collapse	0	0	0
Total sensors	30	30	20

TABLE 2.14
Performance Levels, Severe Earthquake

Performance Level	Displacement	Acceleration	Strain
Continued operations	2	0	3
Damage control	8	7	5
Life safety	15	18	7
Collapse prevention	5	5	3
Complete collapse	0	0	2
Total sensors	30	30	20

Again, a qualitative evaluation of the results of Table 2.13 shows that the performance level of the bridge is life safety. This is fairly clear from the results of the displacement and acceleration sensors. The strain sensors show more disperse results where a distinct performance level is not obvious. This is to be expected from strain sensors during higher intensity earthquakes. Some locations might experience severe deformations while some deformations at other locations might not even exceed the elastic limit.

This example leads to an interesting conclusion: for evaluating the overall performance level of the bridge, only displacement or acceleration sensors can lead to a definite supposition. Strains sensors might lead to a diluted result.

On the other hand, we must note that a displacement-only or acceleration-only experiment might be successful in defining a global definition of the performance level of the bridge during and immediately after the earthquake; however, it is exactly because of the general nature of the performance definition that it becomes very difficult for the decision maker to arrive at a definite decision about the condition and safety of the bridge. Consider, for example, the case of the moderate earthquake of Table 2.13. It shows three strain sensors that indicate a life-safety performance. This is definitely in contrast to the conclusions that might be drawn from the displacement and acceleration sensors. Important decisions such as traffic routing, immediate dispatch of repair crews, or even bridge closings are difficult to ascertain. Similar inconsistent conclusions can be reached from Table 2.14 in the case of a high-intensity earthquake. In other words, the ability of the displacement/acceleration sensors to detect demands is necessary, but not sufficient, to make informed decisions during or immediately after the event. The capacity of the bridge must be monitored: this can be measured only by strain sensors.

From the above, it is clear that any seismic SHM must include an appropriate mix of displacement, acceleration, and strain sensors. This leads us to one shortcoming in monitoring the capacity of the bridge via strain sensors: such sensors are capable of monitoring only local damage. To overcome this limitation, there can be two approaches. The first approach, obviously, is to install strain monitors at all sensitive locations within the bridge $N_{3\text{-MAX}}$. This is costly and impractical since $N_{3\text{-MAX}} \gg N_3$. The other approach is to rely on structural and damage analytical/numerical modeling. We discuss this approach next.

Extent of damage: We just established the need for strain sensors in any seismic SHM experiment. We also observed that strain sensors have a localized range beyond which they can not detect damage. Let us assume that for a completely informed decision on the state of a bridge during and immediately after an earthquake, the damage, or remaining capacity, at $N_{3\text{-MAX}}$ locations must be known. Thus the ratio

$$\eta_S = \frac{N_3}{N_{3-\text{MAX}}} \tag{2.28}$$

Represent the efficiency of the SHM experiment in estimating the extent of the earthquake damage. If $N_3 = N_{3\text{-MAX}}$ ($\eta_S = 1.0$), then the SHM results will completely identify the extent of the earthquake damage. If $N_3 = 0$ ($\eta_S = 0$), then the SHM results will not be able to identify the earthquake damage in a direct fashion.

How about the case when

$$0 < \eta_S < 1.0 \tag{2.29}$$

It is clear that Equation 2.29 represents the vast majority of SHM situations, so we must find a satisfactory resolution to this situation.

2.5.5 After-Event Decision Making Case Study: Prioritization of Retrofits within Bridge Network

One of the characteristics of earthquakes is that it affects a wide area. Because of this, a single earthquake can damage a large number of bridges—a bridge network. In general, the amount of resources to repair/retrofit particular earthquake damage within a bridge network is limited. Thus, the important question of how to prioritize the repair/retrofit efforts within the bridge network needs to be answered. Traditionally, such prioritization is generally done in a qualitative or semiqualitative fashion. Qualitatively, the prioritization of repair/retrofit of bridge network is made by evaluating seismic vulnerability ratings. FHWA (1995) and NYSDOT (2004) seismic vulnerability rating guides discuss such qualitative techniques.

Our immediate concern is to discuss the potential of using SHM methods to prioritize repair/retrofit efforts within a bridge network in a more quantitative fashion. The above-mentioned vulnerability rating methods relates seismic vulnerability to L, V, and I. With L, V, and I, representing likelihood, vulnerability, and impact, are mostly qualitative; hence the semiqualitative nature of such vulnerability methods. Note that the likelihood, L, is an estimation of the probability of occurrence of the earthquake. The vulnerability, V, is also an estimation of the performance of the bridge as a result of the earthquake. When an earthquake does occur, both L and V are no longer needed, for the event has already occurred, and the bridge responses have already been estimated, in a global sense, by estimating the performance level PL_i (see above). Armed with this information, we need another analytical tool. Consider the cost function:

$$C_{EQ}^i = I(PL_i) \tag{2.30}$$

where
C_{EQ}^i = Cost of repair/retrofit of the ith performance level
$I(PL_i)$ = Cost function

The cost function includes all costs of repair/retrofits, bridge closure, traffic routing, and any other social or economic impact for the ith performance level of the bridge. The discrete function $I(PL_i)$ should be evaluated as an integral part of any seismic SHM project since it should be available in real time the instant an earthquake strikes. We note that such a function should be reviewed and revised, if needed, periodically, say once a year. An outdated cost function can lead to inaccurate decisions. A sample of a cost function is shown in Table 2.15.

Cost function is an important tool in decision making. Thus, understanding its behavior can help in constructing it, as will be seen later. First we note that in addition to the performance level, the cost function is dependent on two additional parameters: topology of bridge TP and social/traffic conditions ST. Thus, Equation 2.30 can be generalized as

$$C_{EQ}^i = C_{EQ}(PL_i, TP, ST) \tag{2.31}$$

We see that the cost function would be similar for similar bridges (similar topologies, TP), if the social and traffic conditions ST are similar. This means that for the same earthquake the cost of repair/retrofit will be the same. We can use the previous statement and the basis Equation 2.31 to make some logical observations if we allow for some perturbations in any of the three main factors PL_i, TP, or ST. For example:

- If we allow only for a change in soil conditions between two bridges (representing a slight change in TP), the limiting costs C_{EQ}^1 will remain the same, since the cost of continued operation conditions is independent of soil condition. However, the other limit, the cost of complete collapse C_{EQ}^5, will be higher for the bridge with weaker soil. Of course, if the same earthquake affected both bridges, the performance levels PL_i would be different, leading to higher repair/retrofit costs for the bridge on weaker soil, as seen in Figure 2.46.
- If the two bridges are identical, with one of them being located in an urban area and the other in a rural area (this represents a difference in the social/traffic conditions ST), then both the limiting costs will be different. The urban bridge location will cost more than the rural location at all performance levels. The performance levels for a given earthquake will be identical for the two bridges; however, the resulting costs will be different (Figure 2.47).
- If the two bridges are identical in every measure, except that one of them has been seismically retrofitted (this represents differences in TP), the two limiting costs will be similar; however, the intermediate costs will be higher for the seismically retrofitted bridge. This will be due to the additional costs of repairing the seismic details. The performance level for the seismically fit bridge will be lower than for the other bridge, if both are subjected to the same earthquake (Figure 2.48).

The function C_{EQ}^i is bridge and location specific. It is highly nonlinear and fairly difficult to compute. By some deductions and with limited information, we can arrive at an accurate cost function.

TABLE 2.15
Qualitative—Quantitative Performance Levels

Counter, i	Performance Level PL_i	$I(PL_i)$—$1,000
1	Continued operations	2.00
2	Damage control	10.00
3	Life safety	30.00
4	Collapse prevention	300.00
5	Complete collapse	1500.00

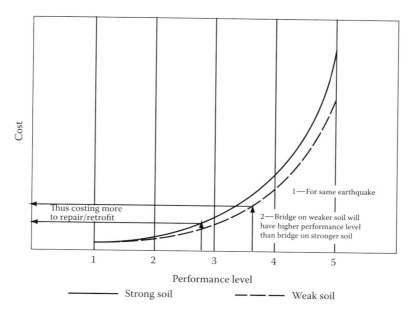

FIGURE 2.46 Cost functions: effects of soil conditions.

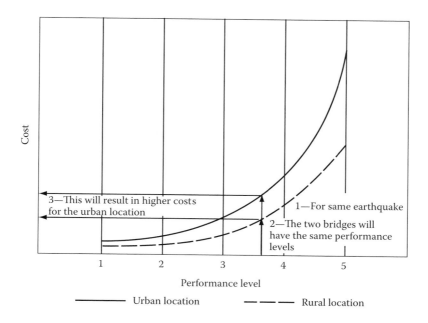

FIGURE 2.47 Cost functions: effects of bridge location.

We now present a simple method for evaluating the function, in lieu of a detailed study. We make the following observations:

- Continued operations, C_{EQ}^1 cost is fairly low; perhaps it is only the cost of routine mainte-nance. As such it is easy to estimate.

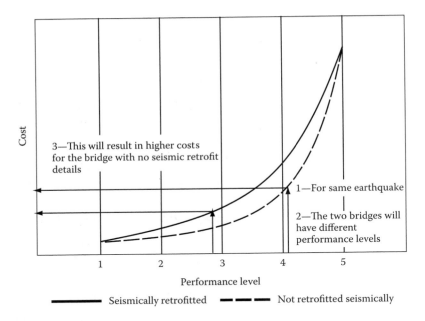

FIGURE 2.48 Cost functions: effects of seismic retrofitting.

- Cost of complete collapse C_{EQ}^5 is easier to compute than the remaining costs. The cost does not include structural evaluation of the bridge. It includes economic and social costs. It may also include the cost of replacement.
- The function C_{EQ}^i is highly nonlinear.

On the basis of the above, we propose

$$C_{EQ}^i = a_1 + a_2 (i-1)^M \tag{2.32}$$

Simple deduction would lead to

$$a_1 = C_{EQ}^1 \tag{2.33}$$

$$a_2 = \frac{C_{EQ}^5 - C_{EQ}^1}{(4)^M} \tag{2.34}$$

We only require that $M \geq 2$. The situations where $M \leq 1$ are not realistic. The example of Table 2.15 is plotted against Equation 2.32 with different values of M as shown in Figure 2.49. It is clear that the cost function is sensitive to M. As more accurate cost function studies are developed for different bridge types, different seismic zones, and different communities (rural, urban, etc.), more accurate values of M can be established.

Returning now to the prioritization issue, assume that for a bridge network composed of N_{BR} bridges the retrofit/repair cost of the jth bridge in the network is $C_{EQ}^i\big|_j$. Prioritization is possible now by simply sorting the N_{BR} components $C_{EQ}^i\big|_j$ in an ascending order.

As an example of repair/retrofit prioritization, using SHM techniques, consider the four-bridge network in Table 2.16. For this example, we will assume that the complete cost function is available only for bridge A, as shown in Table 2.17. Using the regression Equation 2.32, we find that a best fit is accomplished by

$$a_1 = 5.0 \qquad (2.35)$$

$$a_2 = 2.97 \qquad (2.36)$$

$$M = 5.1 \qquad (2.37)$$

A comparison between the regression and actual relationship is shown in Figure 2.50.

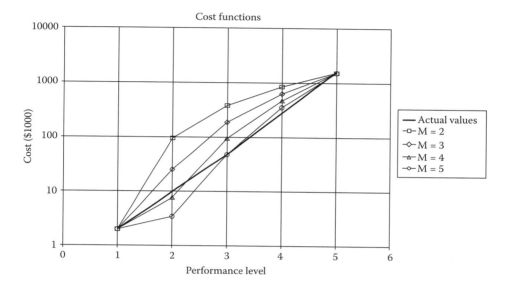

FIGURE 2.49 Accuracy of cost function equation.

TABLE 2.16
Description of the Four-Bridge Network

Bridge	Size	Location	Soil	Seismically Retrofitted?
A	Medium	Urban	Medium	No
B	Medium	Rural	Medium	No
C	Large	Urban	Medium	No
D	Large	Urban	Soft	No

TABLE 2.17
Cost Function for Bridge A

Counter (i)	Performance Level (PL_i)	$C_{EQ}^i - \$1,000$
1	Continued operations	5.00
2	Damage control	25.00
3	Life safety	150.00
4	Collapse prevention	750.00
5	Complete collapse	3500.00

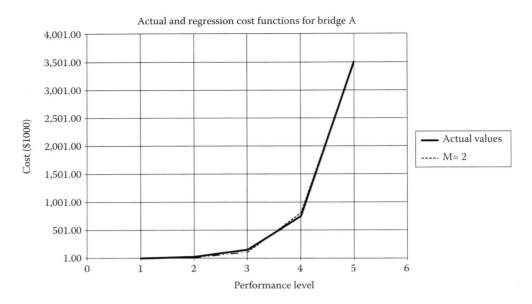

FIGURE 2.50   Comparison between regression and actual cost functions of bridge A.

We can now estimate the cost function regression equations of the rest of the bridges, using the regression equation of bridge A and the deductions based on the relative information in Table 2.16. Note that bridge type B is identical to bridge type A, except for its being in a rural area. As mentioned earlier, the two limiting costs will be different, since urban costs would be much higher than rural costs. Estimating the two limiting costs, relative to the costs of bridge A, should be easy for the decision maker. Let us assume that the lower and upper limiting costs would be about one-half and one-third of their urban counterparts; thus, $C_{EQ}^1 = 2.5$ and $C_{EQ}^5 = 1166.7$. We will assume the exponent of the regression equation, M to be unchanged at 5.1. Since research results are lacking, a constant exponent in this situation seems to be a reasonable assumption. Turning to bridge C, we note that it is larger than bridge A, which indicates that the limiting costs are larger than for bridge A. Estimating these limiting costs, either relative to A or by conducting a detailed analysis should be fairly easy tasks. We will assume that for bridge C, $C_{EQ}^1 = 10$ and $C_{EQ}^5 = 17500$. Note that the low-end cost increase ratio is 4, while the high-end cost increase ratio is 5. This is a logical nonlinear effect. At the low end, the costs involve only simple inspections, while at the high end it involves replacement of bridge. The exponent in this case can also be kept at 5.1, since there is no obvious reason to change it. Finally, for bridge D, we note that it is founded on soft soil; otherwise, it is identical to bridge C. Following the above discussion, the lower cost limit will remain same at $C_{EQ}^1 = 10$. The upper cost limit will increase for weaker soil. The higher cost is estimated at $C_{EQ}^5 = 19000$. Again, we will keep the exponent at 5.1. The resulting four-cost regression equations for different bridge costs are shown in Figure 2.51.

With the cost functions prepared, assume that a moderate earthquake affected the network. The four bridges within the network are equipped with SHM monitoring techniques capable of measuring the overall performance levels of each bridge, as explained in the previous section. The measured performance levels are shown in Table 2.18. Note that the performance levels of bridges A and B are the same. The performance level of bridge C is better than that of bridge D, showing the effects of weak soils.

The next step in this exercise is to use the measured performance levels as an input to the cost functions of Figure 2.51, obtaining the estimated retrofit/repair costs for each bridge, as shown in Table 2.18. These costs include both direct construction costs as well as any social/traffic costs, thus

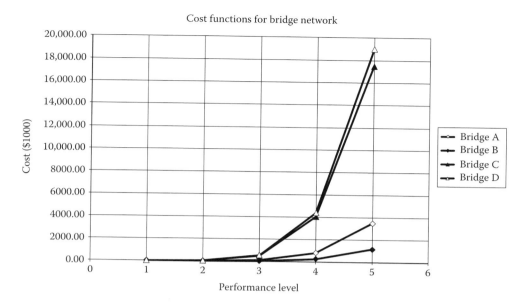

FIGURE 2.51   Comparison of cost functions for the bridge network.

TABLE 2.18
Prioritization Results

Bridge	Performance level	Repair/Retrofit Cost $1,000	Prioritization order
A	3	106.9	2
B	3	36.4	1
C	3	519.96	3
D	4	4388.6	4

making prioritization easy: from lowest to highest cost. The resulting prioritization order is also shown in Table 2.18. For a different earthquake, another prioritization order might result, which would show the potential of this prioritization technique. In addition, this prioritization will become available to the decision maker immediately after the ground shaking ends—a near real-time result.

2.6 GENERAL ENGINEERING PARADIGMS, EARTHQUAKES, AND STRUCTURAL HEALTH

2.6.1 GENERAL

Healthy performance of bridges during seismic events is of importance, from the point of view of both public safety and economic impact. Because of this, several engineering methodologies that address both safety and costs during seismic events have emerged. Of these, prioritization of strategies using risk-based techniques has become a very attractive tool. This addresses all elements of behavior: hazard level, bridge vulnerability, and consequences. Another equally popular design tool is PBD. It is popular precisely for the same reasons: it addresses hazard level, bridge vulnerabilities, and consequences. In what follows, we explore briefly both methods as applied to seismic behavior of bridge systems.

2.6.2 Methods of Risk-Based Earthquake Assessments of Bridges

Retrofitting existing bridges to meet current seismic demands is an expensive task. There are more than 600,000 bridges in the United States. Thus, it is prudent for owners to rate or prioritize the need for seismic rehabilitation of bridges. Most of the prioritization methods rely on risk-based concepts in prioritization. As such, the hazard, vulnerability, and consequences are usually the main parameters in different methods. The details of different methods vary according to the local seismic hazard as well as engineering judgment. We summarize two such prioritization approaches next. One of the approaches represents a high seismic area (California) and the other a moderate seismic area (New York). Because of the largely qualitative nature of the weight assignment in risk prioritization schemes, there is a potential for producing unrealistic results. To ensure realistic results for such methods, we offer a simple analytical procedure to evaluate these weights in a more quantitative fashion.

2.6.2.1 Caltrans

One approach for defining seismic risk to bridges is based on the formula (see Roberts and Maroney 1999)

$$R = H(0.6C + 0.4V) \tag{2.38}$$

$$H = \sum C_{Hi} W_{Hi} \tag{2.39}$$

$$C = \sum C_{Ci} W_{Ci} \tag{2.40}$$

$$V = \sum C_{Vi} W_{Vi} \tag{2.41}$$

The hazard, consequence, and vulnerability are H, C, and V, respectively. The weights are preset numbers based on experience. They are shown in Table 2.19. The factors C_{Hi}, C_{Ci}, and C_{Vi} depend on the bridge under consideration. Thus, by examining each bridge, the risk factor R can be computed. By comparing the risk factors for several bridges, efficient prioritization strategies for seismic rehabilitation and maintenance can be planned.

2.6.2.2 NYSDOT

NYSDOT has a simple seismic evaluation system, (NYSDOT 2004) that covers all important aspects of bridge behavior. It is subdivided into three cascading tiers: screening, classifying, and rating.

Screening: This process screens the bridges into one of four groups according to their seismic vulnerabilities: high, moderate-high, moderate-low, and low. The screening process depends on the geometry and location of the bridge. Among information used to screen the bridge seismically are

- Date of construction
- Importance of bridge
- Single or multiple spans
- Simple or continuous girders
- Bearing type
- Number of girders per span (girder redundancy)
- Skew
- Pier/footing type

In addition to the above factors, special conditions can also affect classifications, such as

- Type of structural system (arch, movable, suspension, stayed)
- Railroad, pipeline
- Culvert, tunnel, span length

TABLE 2.19
Risk Components and Their Weights

Risk Component	Constituent	Weight (%)
Hazard	Soil condition	33
	Peak rock acceleration	28
	Seismic duration	29
Impact (consequences)	Average daily traffic on structure	28
	Average daily traffic under/over structure	12
	Detour length	14
	Leased airspace (residential office)	15
	Leased airspace (parking, storage)	7
	RTE type on bridge	7
	Critical utility	10
	Facility crosses	7
Vulnerability	Year designed (constructed)	25
	Hinges (drop-type failure)	16.5
	Outrigger, shared columns	22
	Bent redundancy	16.5
	Skew	12
	Abutment type	8

Source: Roberts, J. and Maroney, B., Seismic retrofit practice, In *Bridge Engineering Handbook*, Chen, W.F. and Duan, L., Editors, CRC Press, New York, NY, 1999. Courtesy of CRC Press.

Classification: The seismic classification is fairly similar to the FHWA (1995) approach. It is based on evaluating the score

$$CS = V \cdot E \tag{2.42}$$

CS = Classification score ($0 \leq CS \leq 100$)
V = Structural vulnerability rating ($0 \leq V \leq 10$)
E = Seismic hazards rating ($0 \leq E \leq 10$)

The structural vulnerability rating, V, is a function of two independent vulnerabilities as follows

- Vulnerability for connections, bearings, and seat widths V_1
- Vulnerability for piers, abutment, and liquefaction V_2.

Finally,

$$V = f(V_1, V_2) \tag{2.43}$$

Function $f()$ depends on the geographic location and the criticality of the bridge.
The seismic hazard score, E, is expressed by

$$E = a S \tag{2.44}$$

Factor a depends on the geographic location, while factor S depends on the local soil condition.
The seismic classification score, CS, is used for identifying vulnerability class as in Table 2.20.

TABLE 2.20
Seismic Vulnerability Classes

Classification Score CS	Vulnerability Class
>70	High
25–75	Medium
<30	Low

Source: NYSDOT, Seismic Vulnerability Manual, New York Department of Transportation, Albany, NY, 2004. With permission.

The overlapping between the scores in Table 2.20 is meant to allow for some engineering judgment in the final vulnerability classifications.

Vulnerability rating (VR): This rating is meant to provide a uniform measure of the bridge's vulnerability to failure, on the basis of the three classical risk components: hazard, vulnerability, and consequences. The VR is evaluated as

$$\text{VR} = L + C + E_\text{T} + E_\text{F} \tag{2.45}$$

L = Likelihood
C = Consequence of structural failure
E_T = Exposure (traffic)
E_F = Exposure (functional)

The components and the relative weights of L, C, E_T, and E_F are shown in Table 2.21. The implications of the similarity of structure of Table 2.19 and Table 2.20 will be discussed later.

The final vulnerability rating can now be used by decision makers to assign a rating scale (1–6) as shown in Table 2.22. The rating is used for prioritization and rehabilitation efforts, as needed. See NYSDOT (2004) for more details.

2.6.2.3 Optimization Technique

Generally, many of the evaluation methods in this section and in other hazard evaluation methods have the form

$$\{A\}^T \{W\} = r \tag{2.46}$$

The vector $\{W\}$ contains preassigned weights that reflect the strengths of different controlling parameters. Vector $\{A\}$ contains scores that reflect the conditions of the different controlling parameters as assessed by the users. The order of the vectors in Equation 2.46 is the number of the controlling parameters, $N_\text{PARAMETER}$. The final evaluation score is r. Obviously, the final score depends on two factors

- Controlling parameters
- Preassigned weights $\{W\}$

In most situations, these two factors are chosen in a qualitative fashion. In addition, the scores, $\{A\}$, are usually assigned in a qualitative fashion. Because of this, the final score r is qualitative. When Equation 2.46 is used over several systems in our current situation, seismic vulnerability of bridges, the resulting set of r would form a vector $\{R\}$. The size of $\{R\}$ is the number of samples.

TABLE 2.21
Vulnerability Rating Weights

Category	Description	Weight
Likelihood	Vulnerability class (Table 2.20)	
	High	10
	Medium	6
	Low	2
	Not vulnerable	0
Consequence (structural failure)	Failure type	
	Catastrophic	5
	Partial collapse	3
	Structural damage	1
Exposure (traffic)	Traffic volume	
	>25000 AADT	2
	4000–25,000 AADT	1
	<4000 AADT	0
Exposure (functional)	Functional classification	
	Interstate and freeway	3
	Arterial	2
	Collector	1
	Local road and below	0

TABLE 2.22
Seismic Rating

VR	Rating
>15	1
13–16	2
9–14	3
<15	4
<9	5
NA	6

Even though the constituents of $\{R\}$ are qualitative, they *should* exhibit two important properties:

1. Their relative magnitudes should be reasonable. For example, the need that the relative vulnerabilities, or seismic risks, for each bridge should be reasonably represented within $\{R\}$.
2. The absolute values of the constituents should also be well represented. For example, a bridge exposed to very high risk should have a high-risk score, while minimal risk should also be represented. This requirement also implies that the risk scores within $\{R\}$ should cover a wide range, or a narrow range, as needed. Koller (2000) observed that many conventional risk methodologies fail to provide such a needed range spread.

The highly qualitative nature of Equation 2.46 makes it particularly difficult to produce methodologies that satisfy the two above requirements. One obvious solution is to validate the problem by a trial-and-error procedure that covers a reasonable set of samples and to fine-tune the weight set and the parameter space until reasonable solutions are found. Another more formal approach is to use an SVD (see Ettouney and Alampalli 2012, Chapter 8) as follows:

- For a given number of sample bridges, N_{SAMPLE} assign a reasonable score vector $\{\bar{R}\}$; the size of $\{\bar{R}\}$ is N_{SAMPLE}
- For each of the bridges in the sample, estimate the scoring vector $\{A\}_i$, with $i = 1, 2, \ldots$ N_{SAMPLE}
- Form the matrix $[\bar{A}]$. The ith row of the matrix is the transpose of $\{A\}_i$

Equation (2.47) is the risk equation for all the samples.

$$[\bar{A}]\{W\} = \{\bar{R}\} \tag{2.47}$$

Note that the weights $\{W\}$ are unknown. If we can find these unknown weights, they can be used to evaluate any other similar evaluation problem while satisfying the two above conditions.

Solving Equation 2.47 in the form

$$\{W\} = [\bar{A}]^{-1}\{\bar{R}\} \tag{2.48}$$

will produce the unknown weights. We note that since, in general,

$$N_{\text{PARAMETER}} \neq N_{\text{SAMPLE}} \tag{2.49}$$

A SVD solution process is needed. This means that the resulting weights are going to satisfy Equation 2.47 in the least square sense.

2.6.3 Performance-Based Design

We presented the emerging PBD paradigm as described by Ettouney and Alampalli (2012, Chapter 2). We now discuss specific application to bridges as offered by Roberts and Maroney (1999). Two performance levels can be considered:

- Immediate service performance level: normal traffic resumes almost immediately after the earthquake.
- Limited service performance level: Limited access, for example, reduced lanes, within days and full service within months.

Coupled with these performance levels, three types of damage states are defined:

- Minimal damage: Elastic or near-elastic performance
- Repairable damage: Damage repaired with minimal loss of functionality
- Significant damage: Minimal risk of collapse, might have to close the bridge for repair

Deciding which performance level should be assigned to which bridge is based on the importance of the bridge. Two types of bridges are defined: (1) important bridge performance level, and (2) minimum performance level (see Roberts and Maroney 1999 for detailed description of the two levels).

Table 2.23 shows the different performance levels, damage levels, and the functional and safety requirements consistent with them.

TABLE 2.23
Performance-based Design Criteria for Bridges

Ground Motion at Site	Design Levels	Minimum Performance Level	Important Bridge Performance Level
Functional evaluation	Service level	Immediate	Immediate
	Damage level	Repairable	Minimal
Safety evaluation	Service level	Limited	Immediate
	Damage level	Severe	Repairable

Source: Roberts, J. and Maroney, B., Seismic retrofit practice, In *Bridge Engineering Handbook*, Chen, W.F. and Duan, L., Editors, CRC Press, New York, NY, 1999. Courtesy of CRC Press.

The different performance levels can be achieved through strict analysis and design criteria. We also observe that SHM can play a major role in PBD, as discussed next.

2.6.3.1 Role of SHM

All components of SHM can play a major role in PBD before and after the seismic event as follows:

- **Sensors/sensing:** Sensing structural and load (traffic) behavior can help in STRID and DMID before and after the events. Several types of sensors can be used: accelerometers, strain gauges, displacements, or load cells. In addition, several NDT technologies can be used before and after the events.
- **STRID:** Accurate STRID can be of help during design stages (for new bridges). For existing construction, STRID (modal identification, for example) can help in defining the as-built structural properties, such as stiffness, damping, or material properties. After the event, STRID methods can help in defining damaged structural properties, if any.
- DMID: After the event, DMID methods can be used to identify levels of damage and verify PBD assumptions.

2.7 RESILIENCE OF INFRASTRUCTURES

2.7.1 Overview

Infrastructure resiliency is an important concept for both existing and new infrastructure. A decision maker needs to have an accurate estimate of the resiliencies of different infrastructures in the network. This is needed for appropriate emergency planning as well as for prioritization of projects at the local, state, and national levels. Resiliency also offers a simple way of relating varying types of infrastructures as well as comparing the effects of different hazards of varying magnitudes.

2.7.2 Resiliency of Bridges

Resiliency is defined in the dictionary as "the power or ability to return to the original form, position, etc." or "the ability to recover readily from adversity or the like." Bridge resiliency is not any different. Bridge resiliency can be defined as its ability to resume its level of service after it is subjected to an (un)expected or (un)common hazard during its service life. Thus it can be defined in terms of the time required to resume the load and/or number of vehicles it is designed to carry. The resiliency can be individual bridge resiliency (i.e., local in nature) or network resiliency. Since a network is comprised of a number of bridges, the resiliency of a network can be totally dependent, or totally independent, or in between, depending on the network characteristics. The big difference is that for local resiliency, the structure characteristics become very important whereas for network resiliency they may not be important.

The resiliency of a given bridge can be characterized by using three components: required level of service, redundancy (structural vs. nonstructural), and the time required to bring it back to the level of service. Hence, as shown in Figure 2.52, for two bridges with the same level of service expected, resiliency *for a given hazard* can be known by comparing the shaded area. The smaller the area, the more resilient the bridge is.

One should note the following:

1. For the same bridge/structure, resiliency varies depending on the type and magnitude of a hazard. Hence, the resiliency is associated with the level of service and hazard. Thus, for risk management, one has to look into the probability of a hazard of a given magnitude and the subsequent consequence (i.e., risk) before estimating the resiliency of a bridge for design or evaluation.
2. One way of maintaining resiliency is by using both structural and nonstructural redundancy. For example, if you need a two-lane bridge to accommodate traffic and expect a security risk associated with a given blast event, you can provide structural redundancy by hardening the structure to ensure that the event does not impact capacity. Alternatively, you can have a four-lane bridge designed in such a way that only a portion of the structure carrying two lanes will be impacted by a given event (or have two parallel bridges each with two lanes).
3. The other way to improve redundancy is to design in such a way that the bridge can be restored in a short time by choosing appropriate materials and other details for rapid construction.
4. A combination of both redundancy, and selective materials and details for accelerated construction can be used to achieve the required level of resiliency.

Network resiliency depends on bridge resiliency and redundancy in the network. For example, consider a bridge with a half-a-mile detour in a rural area. If traffic volumes are such that they do not impact travel times and loads, then bridge resiliency becomes irrelevant to network resiliency irrespective of the hazard type or magnitude. But for the same bridge with the same detour length in an urban environment, the network can be affected significantly due to high traffic volumes as the network level of service can decrease. So the resiliency of the bridge becomes very important in this case.

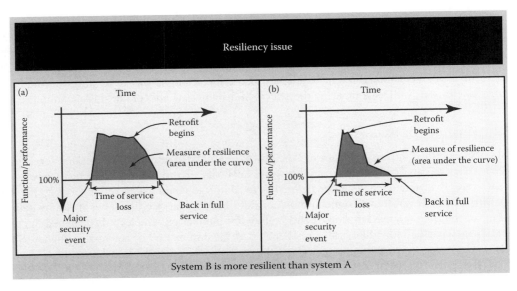

FIGURE 2.52   Concept of infrastructure resiliency. (a) high resiliency, (b) low resiliency

Bridge owners consider these issues on the basis of experience and make decisions day-to-day, but having an effective bridge management tool to perform comparisons will be valuable. At the same time, since resiliency depends on the magnitude and type of hazard, system resiliency should consider all hazards that the bridge is subjected to during its service life and use a risk-based approach from design stages. Designers should try to improve the resiliency of the bridge to more than one hazard in order to add value by selecting appropriate materials and details, when more than one option is available, in the case of a no-increase or a slight increase in cost and design times. For example, if bridges are being retrofitted for seismic reasons, there may be details that can offer similar seismic resistance but also can improve flood and wind resistance at similar costs; they should be investigated to reduce the overall risk, that is, increase overall resiliency. Fragility data should be developed for bridge details such that owners can make informed decisions, using performance-based criteria and a multihazard approach. Figure 2.53 shows several of the issues regarding resiliency and infrastructures.

2.7.3 COMPONENTS OF RESILIENCY

Infrastructure resiliency contains three essential components: hardening, redundancy, and time of recovery. They are explained below:

Hardening: Hardening is the capacity of the infrastructure to withstand the direct effect of the calamity or hazard. Traditionally, hardening is a property of the infrastructure itself.

Redundancy: Redundancy is the capacity of the infrastructure itself, or the local, regional, or national network, to act as alternative paths or routes to fill in the gap of damage due to calamity or hazard.

Time of recovery: Time of recovery is defined as the time it takes to retrofit the damage due to the calamity or hazard until the functionality of the infrastructure is fully restored.

In many ways, there is an analogy between the resiliency of infrastructures and human health. Qualitatively, resiliency can be expressed as

$$\text{resiliency} = (1 - (\text{hardening} + \text{redundancy})) \times (\text{time of recovery})$$

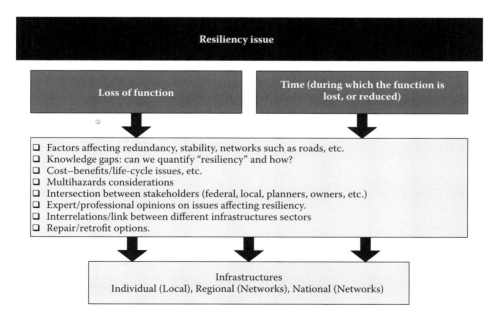

FIGURE 2.53 Resiliency issues in infrastructures.

Figure 2.52 shows a schematic presentation of resiliency. Clearly, there are many issues that can affect infrastructure resiliency; some of these are shown in Table 2.24

2.7.4 SEISMIC RESILIENCY OF A SINGLE BRIDGE

2.7.4.1 Method

Bruneau and Reinhorn (2007) introduced a wide-ranging study of the quantification of resilience of acute healthcare facilities. They formalized resilience as

$$R = \int_{t_0}^{t_1} (100 - Q(t)) \, dt \tag{2.50}$$

The dimensionless quality of the infrastructure (QIF), is $0 \le Q(t) \le 100$. Note that the QIF in the Bruneau and Reinhorn (BR) model includes the combined effects of redundancy and hardening. The initial time when the hazard (earthquakes or any other suddenly applied hazard) strikes is t_0; the time when the infrastructure quality returns to its initial state is t_1. We observe that the resilience in Equation 2.50 has a dimension of time. The BR conceptual resilience model is shown in Figure 2.54. For completeness, we require

$$\Delta T > 0 \tag{2.51}$$

and

$$\int_{t_0}^{t_1} (Q(t)) \, dt > 0 \tag{2.52}$$

TABLE 2.24
Parameters Affecting Infrastructure Resiliency

Resiliency Component	Parameters Affecting Resiliency
Hardening	Advanced materials, construction techniques, detailing, design, and analysis issues, deterioration of systems/materials, nature of hazard, considerations, environmental, energy considerations, and climate change
Redundancy	Design/type of infrastructure system, infrastructure network (local, regional, and national), interaction between stakeholders, communication technologies
Time of recovery	First responders, retrofit techniques and materials, interaction between stakeholders, other social and economical issues, nature, and type of hazard

---

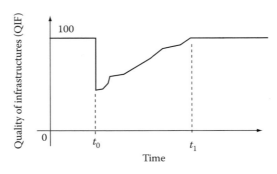

FIGURE 2.54 Bruneau-Rienhorn resiliency concept. (Courtesy Dr. Michel Bruneau.)

with the down time defined by

$$\Delta T = t_1 - t_0 \qquad (2.53)$$

The conditions 2.51 and 2.52 are designed to satisfy the trivial state of no down time, $\Delta T = 0$, and no damage, $Q(t) = 0$. We now discuss the quantification of resiliency. First, we discuss the initial quality of the bridge immediately after seismic events, $t = t_0$. We then discuss the temporal characteristics of $Q(t)$. Finally, we explore the role of SHM in estimating R.

2.7.4.2 Initial Damage State

After a seismic event, the usability of the bridge becomes the main concern. One way of quantifying this usability is through a qualitative rating system. FHWA and all state governments have a well-established rating system that can estimate bridge usability. The QIF, immediately after an earthquake, $t = t_0$, can then be estimated as

$$Q(t = t_0) = \frac{R_{\text{AFTER}}}{R_{\text{BEFORE}}} \cdot 100 \qquad (2.54)$$

where
  R_{AFTER} = Bridge rating immediately after a seismic event
  R_{BEFORE} = Last bridge rating before the seismic event
  Since it is reasonable to expect that $R_{\text{AFTER}} \leq R_{\text{BEFORE}}$, condition 2.52 is satisfied.
  Bridge rating in the immediate aftermath of an earthquake, or for simulation purposes, can be established analytically or experimentally (see Chapter 8).

2.7.4.3 Temporal Characteristics of QIF

When QIF is reduced by an earthquake, it is expected that efforts will be taken to bring it back to it the preearthquake level. Rehabilitation measures are performed as quickly as possible. As such efforts progress, QIF is improved. For planning purposes, the decision maker would need a simple temporal expression for QIF. We offer the following analytical expression:

$$Q(t) = A + Be^{\alpha t} \qquad (2.55)$$

The values of A and B are constants, which are functions of the following conditions:

$$Q(t = t_1) = Q_0 \qquad (2.56)$$

and

$$Q(t = t_2) = Q_1 \qquad (2.57)$$

With
  Q_0 = QIF immediately after the seismic event (or any hazard of interest)
  Q_1 = QIF at the end of rehabilitation effort
  The constant α controls the shape of QIF. It can indicate fast, slow, or gradual rehabilitation. It can be considered a measure of the rehabilitation rate. For example, if after a seismic event, Q_0 and Q_1 are 50 and 100, respectively, the different possible time-dependent QIF values are shown in Table 2.25 and Figure 2.55. The start and end times are assumed to be $t_0 = t$ and $t_1 = 1.0$, respectively. Note that for high negative values of α, the rehabilitation is fast, as $\alpha \to 0$ the rehabilitation efforts are gradual. As the values of α become positive, the rehabilitation rate becomes slower. As

TABLE 2.25
Effects of Rate of Rehabilitation on QIF

Time	Value of α						
	30	5	1	0.1	−1	−5	−30
0.00	50.00	50.00	50.00	50.00	50.00	50.00	50.00
0.05	50.00	50.10	51.49	52.38	53.86	61.13	88.84
0.10	50.00	50.22	53.06	54.78	57.53	69.81	97.51
0.15	50.00	50.38	54.71	57.19	61.02	76.56	99.44
0.20	50.00	50.58	56.44	59.60	64.34	81.82	99.88
0.25	50.00	50.84	58.26	62.04	67.50	85.92	99.97
0.30	50.00	51.18	60.18	64.48	70.50	89.11	99.99
0.35	50.00	51.61	62.19	66.93	73.36	91.59	100.00
0.40	50.00	52.17	64.31	69.40	76.08	93.53	100.00
0.45	50.00	52.88	66.54	71.88	78.66	95.03	100.00
0.50	50.00	53.79	68.88	74.38	81.12	96.21	100.00
0.55	50.00	54.97	71.34	76.88	83.46	97.12	100.00
0.60	50.00	56.47	73.92	79.40	85.69	97.83	100.00
0.65	50.00	58.41	76.64	81.93	87.81	98.39	100.00
0.70	50.01	60.89	79.50	84.47	89.82	98.82	100.00
0.75	50.03	64.08	82.50	87.03	91.74	99.16	100.00
0.80	50.12	68.18	85.66	89.60	93.56	99.42	100.00
0.85	50.56	73.44	88.98	92.18	95.29	99.62	100.00
0.90	52.49	80.19	92.47	94.77	96.94	99.78	100.00
0.95	61.16	88.87	96.14	97.38	98.51	99.90	100.00
1.00	100.00	100.00	100.00	100.00	100.00	100.00	100.00

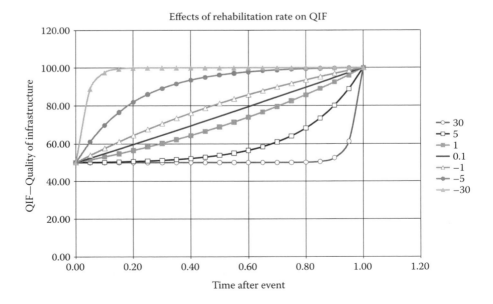

FIGURE 2.55   Rehabilitation effects on QIF.

the positive values of α become large, the rehabilitation slows down. At the theoretical limit $\alpha \to \infty$, the rehabilitation effort comes to a complete stop. Conversely, at the theoretical limit $\alpha \to -\infty$, the rehabilitation is instantaneous.

The effects of temporal behavior of rehabilitation on the resiliency of the previous example, using Equation 2.50, are shown in Table 2.26. We note that Equation 2.50 produces lower values for more resilient systems. A perhaps more consistent resilience expression is

$$R = \int_{t_0}^{t_1} (Q(t)) \, dt \tag{2.58}$$

The results are also shown in Table 2.26. The resiliency using the latter expression is shown in Figure 2.56. Clearly, it is of benefit to rehabilitate QIF as soon as possible. However, it seems that a rehabilitation rate of $\alpha \approx -5$ offers an optimum resilience value beyond which a diminishing rate of return occurs.

2.7.4.4 Role of SHM

Notice that four parameters are needed to completely evaluate resilience quantitatively. They are

- Initial QIF Q_0
- Time length of recovery ΔT

TABLE 2.26
Effects of Rehabilitation Performance on Resiliency

Governing Equation	Value of α						
	30	5	1	0.1	−1	−5	−30
Equation 2.50	48.03	40.29	29.09	25.42	20.91	9.71	1.97
Equation 2.58	51.97	59.71	70.91	74.58	79.09	90.29	98.03

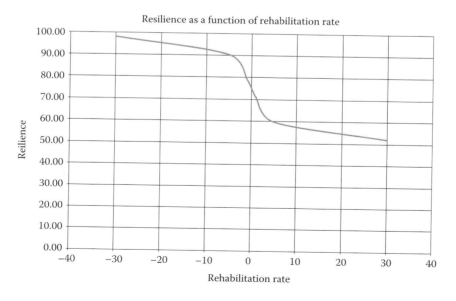

FIGURE 2.56 Rehabilitation rate and resiliency.

TABLE 2.27
SHM Role in Estimating Resiliency Parameters

Parameter	Role of SHM
Q_0	Using structural analysis and STRID tools, an accurate prediction of Q_0 can be computed numerically. Using load-rating techniques can be an alternative route for computing Q_0
ΔT	Decision making techniques that rely on evaluating past experiences in length of recovery after a seismic event with a particular level can help in estimating ΔT
Q_1	Decision making tools that evaluate costs-benefits can be of use in determining Q_1
α	Cost–benefit analysis that studies cost/benefit implications for choosing different rehabilitation rates

- Desired final QIF Q_1. Note that in some situations the decision maker might want the final QIF to be either higher or lower than 100
- Rate of rehabilitation α

Table 2.27 shows how these values are estimated using SHM.

2.7.4.5 Closing Remarks

We showed the interrelationship between resiliency and rehabilitation rates. We also offered a simple expression for quantifying seismic resilience of bridges. Finally, we highlighted the important role SHM can play in estimating resiliency. These tools can be expanded for estimating resiliency in a multihazard environment. The resiliency of bridge networks is also an important subject; the just-introduced tools can be used to evaluate such resiliency. Such a topic is beyond the scope of this chapter.

2.7.5 RESILIENCY AND LCA

Resiliency has direct use in estimating LCA. If we assume that the cost per unit time of complete loss of function of a particular bridge due to a seismic event or any other type of mishap is c_{LOSS}, then the total cost of loss of function is

$$C_{LOSS} = c_{LOSS}(100 - R) \tag{2.59}$$

The effects of loss of function can be used within expressions of LCA as appropriate. Additional costs, as explained elsewhere in this chapter, can be the costs of rehabilitation, replacement, management, and so on.

When we reflect on Equation 2.59, it becomes apparent that rehabilitation techniques can have as much of an effect as hardening and redundancy on the overall LCA. Rehabilitation techniques depend on two main factors: rate of rehabilitation α, and down time ΔT. By controlling these two factors effectively, the cost of hazard can be reduced.

2.7.6 RESILIENCY AND RISK

We showed that one of the forms of risk is expressed by relating hazard level, the vulnerability of the system to that hazard, and the consequences of the system response to the hazard (Ettouney and Alampalli 2012). A close inspection of Equation 2.50 or 2.58 shows that resiliency is also a function of the three basic risk components, as follows:

- **Hazard level:** The level of hazard affects QIF immediately after the event, $Q(t = t_0)$. For low-level hazard, the value of $Q(t = t_0)$ will be close to 100, for large hazards, the value of $Q(t = t_0)$ will be close to 0.

- **Vulnerability:** The hardening and redundancy of the system will also affect its QIF immediately after the event, $Q(t = t_0)$. For hardened and/or highly redundant systems the value of $Q(t = t_0)$ will be close to 100; for highly vulnerable systems the value of $Q(t = t_0)$ will be close to 0.
- **Consequences:** The down time ΔT is a direct consequence of the hazard event.

Thus, it is logical to conclude that resiliency is a measure of risk for a particular hazard level and a particular system (see Figure 2.57). We note that the consequences can be much more than just a down time. For example, cost of replacement/rehabilitation and other economic and/or social consequences can occur. Those consequences are not directly included in the down time. Because of this, resilience should be considered only a part of the overall risk of a hazard.

2.7.7 UNCERTAINTY OF RESILIENCY

The interrelation between risk and resilience brings to our attention another important aspect of resiliency: the uncertain foundations of resiliency. Bruneau and Reinhorn (2007) first observed that the components of resiliency are mostly uncertain. The response of the system to a particular hazard level is uncertain. The down time and rehabilitation rates are also highly uncertain. The use of fragility curves was suggested by Bruneau and Reinhorn (2007) to account for the uncertainties of the initial QIF, $Q(t = t_0)$. The uncertainty of rehabilitation can be estimated using the Monte Carlo simulation. However, given the closed-form approximation of Equation 2.55 to the rate of rehabilitation, a Taylor series approach (see Chapter 8 by Ettouney and Alampalli 2012) can be used in conjunction with system fragilities to estimate uncertainties in resilience.

If we assume that α is a random variable, the derivatives of QIF become

$$\frac{\partial Q}{\partial \alpha} = A + B t \, e^{\alpha t} \tag{2.60}$$

$$\frac{\partial^2 Q}{\partial \alpha^2} = A + B t^2 e^{\alpha t} \tag{2.61}$$

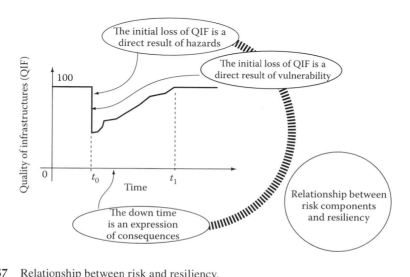

FIGURE 2.57 Relationship between risk and resiliency.

Applying the Taylor series method, we evaluate the function Q at $\bar{\alpha}$, such that

$$\bar{Q} = Q\big|_{\alpha=\bar{\alpha}} \tag{2.62}$$

The mean of α is $\bar{\alpha}$. We then expand Equation 2.55 around $\bar{\alpha}$ with

$$\varepsilon_1 = \alpha - \bar{\alpha} \tag{2.63}$$

the expectations of QIF can be shown to be

$$E(Q) = \bar{Q} + \frac{1}{2}\left(E(\varepsilon_1^2)\frac{\partial^2 \bar{Q}}{\partial \alpha^2} \right) \tag{2.64}$$

$$E(\bar{Q}^2) = \bar{Q}^2 + \left(E(\varepsilon_1^2)\left(\bar{Q}\frac{\partial^2 \bar{Q}}{\partial \alpha^2} + \left(\frac{\partial \bar{Q}}{\partial \alpha}\right)^2 \right) \right) \tag{2.65}$$

Note that, by definition

$$E(\varepsilon_1) = 0 \tag{2.66}$$

And

$$E(\varepsilon_1^2) = V_\alpha \tag{2.67}$$

The variance of Q can be expressed as

$$V(Q) = E(Q^2) - E(Q)^2 \tag{2.68}$$

The standard deviation of Q is

$$\sigma(Q) = \sqrt{V(Q)} \tag{2.69}$$

To illustrate the use of the above equations, consider a situation where the professional estimated that in the previous example the random variable rate of rehabilitation has a mean of $\bar{\alpha} = 1$. The coefficient of variation is 0.3%. Applying the above equations, the expected value and the uncertainty bounds at $\pm\sigma$ and $\pm2\sigma$ can be evaluated as shown in Table 2.28. The resiliency and the uncertainty bounds are also shown in Figure 2.58.

In the above example, we assumed that the variables A and B to be deterministic. These variables are functions of the hazard and vulnerability levels. Their uncertainties can be included using component or system fragilities, as suggested by Bruneau and Reinhorn (2007). The above Taylor series approach can easily be extended to account for this situation; the Taylor series will include three random variables. Such a development is beyond the scope of this chapter.

2.8   LCA AND EARTHQUAKE HAZARDS

We discuss seismic-specific bridge life cycle Analysis (BLCA) issues in this section. As noted elsewhere in this chapter, BLCA includes cost, benefit, and lifespan analysis. Each of these subjects is discussed next, with emphasis on the role of SHM tools.

TABLE 2.28
Uncertainty Bounds of Resiliency

	-2σ	$-\sigma$	0	$+\sigma$	$+2\sigma$
0.00	43.75	46.87	50.00	53.13	56.25
0.05	44.58	48.04	51.49	54.95	58.40
0.10	45.40	49.23	53.06	56.89	60.72
0.15	46.20	50.46	54.71	58.96	63.21
0.20	46.98	51.71	56.44	61.17	65.91
0.25	47.72	52.99	58.26	63.54	68.81
0.30	48.41	54.29	60.18	66.07	71.96
0.35	49.03	55.61	62.19	68.78	75.36
0.40	49.58	56.95	64.31	71.68	79.04
0.45	50.04	58.29	66.54	74.79	83.04
0.50	50.38	59.63	68.88	78.13	87.37
0.55	50.58	60.96	71.34	81.71	92.09
0.60	50.62	62.27	73.92	85.57	97.22
0.65	50.47	63.56	76.64	89.72	102.81
0.70	50.10	64.80	79.50	94.20	108.90
0.75	49.46	65.98	82.50	99.02	115.55
0.80	48.52	67.09	85.66	104.23	122.80
0.85	47.23	68.10	88.98	109.86	130.74
0.90	45.53	69.00	92.47	115.95	139.42
0.95	43.37	69.75	96.14	122.53	148.92
1.00	40.68	70.34	100.00	129.66	159.32

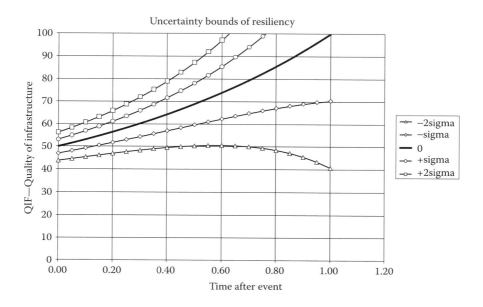

FIGURE 2.58   Uncertainty bounds of resiliency.

2.8.1 Bridge Life Cycle Cost Analysis (BLCCA): Cost

2.8.1.1 Methods for Seismic BLCCA

Computing the costs of seismic events during the lifespan of the bridge involves two parameters: probabilities of occurrence of different hazard levels and cost of damage of those occurrences. The bridge life cycle cost of seismic hazard, $BLCC_{Seismic}$ is expressed as

$$BLCC_{Seismic} = \int_S \int_h p(h) C_h \, dh \, dS \tag{2.70}$$

The probability of occurrence of seismic hazard of level h is $p(h)$. The cost of such an occurrence is C_h. The first integral in Equation 2.70 is over the lifespan, S. The second integral in the equation is over all reasonable seismic hazard levels, h. The expression in Equation 2.70 can be simplified to the discrete form

$$BLCC_{Seismic} = \sum_{i=1}^{N_S} \sum_{j=1}^{N_h} p_j C_{ij} \tag{2.71}$$

The lifespan is now subdivided into N_S steps, usually on an annual basis. Each ith step represents time span

$$T = \frac{S}{N_S} \tag{2.72}$$

The hazard level space is now subdivided into N_h steps. The probability that a seismic event with an jth strength during time span of T is p_j. Finally, the cost of such an occurrence is C_{ij}.

The probabilities $p(h)$ or p_j depend on the location and the ground conditions of the structure of interest. The costs of seismic occurrence C_h, or C_{ij}, are a bit more difficult to compute. We first have to decide on the components of such costs. Components of cost include, but are not limited to, (1) direct cost of repairing/retrofitting the damage, (2) direct costs of special inspection, and (3) direct costs of management during the postevent. There are numerous indirect costs such as (1) costs of detours, (2) costs of bridge closings or partial closings, and (3) other indirect social and economic costs. Note that at the high end of the seismic hazard level, the bridge might fail, so the above-mentioned direct costs need to be modified to direct costs of bridge failure; instead of repairing/retrofitting, the bridge may have to be demolished and rebuilt.

The use of Equation 2.70 or 2.71 to estimate seismic hazard costs is popular due to their simplicity. Unfortunately, these equations do not account for any interaction with other future events, such as the potential of other retrofits or other accidents that can change the costs. One possible way of including such interactions is by using a true-time Monte Carlo simulation. Examples of this method are given in Chapter 10 of this volume and Chapters 8 and 10 by Ettouney and Alampalli (2012).

In the rest of this section, we concern ourselves only with the process of estimating costs of repair/retrofit. We note that estimating the costs of repair/retrofit depends mainly on the degree of damage due to an earthquake. Thus, for a given earthquake level, the process can be summarized as in Table 2.29. Note that Table 2.29 also shows the roles of SHM techniques in enhancing the estimation process. We also observe that these steps are fairly similar to the steps of PBD of Chapter 2 by Ettouney and Alampalli (2012).

2.8.1.2 Fragility Computations and SHM

As noted in Table 2.29, system or component fragilities can be used during the process of estimating seismic costs. Table 2.30 summarizes the steps in computing seismic fragilities and the different roles that analysis and experiments can play in such computations.

TABLE 2.29
Seismic Repair/Retrofit Costs

No.	Comment	Analytical	Experimental—SHM?
1	Seismic input	In recognition of the potential nonlinear behavior of responses, as well as the nonstationary nature of seismic input, time frequency-processing techniques (such as Wavelet or Hilbert-Hwang) should be used	Place adequate seismic sensors (e.g., accelerometers) at the site
2	Estimate statistical properties of damage	Use detailed modeling/analysis combined with simulation (Monte Carlo); use fragility (Table 2.30)	Using STRID techniques to estimate changes in structural properties due to seismic event. Modal identification methods can identify global changes (damage), whereas parameter identification methods can help in identifying local changes (damage)
3	Estimate repair/retrofit strategy	Use available historical cost measures for different strategies and their potential statistical properties	Relates specific SHM-sensing measurements, STRID models, and DMID techniques to establish optimal decisions. See Chapter 8 by Ettouney and Alampalli (2012)
4	Estimate cost of strategy of # 3		
5	Estimate statistical properties of repair/retrofit costs		

TABLE 2.30
Fragility Computations

No.	Description	Analytical	Experimental—SHM?
1	Seismic input	Same as in Table 2.29	
2	Identify system or components statistical properties	Use conventional analyses methods such as finite-elements method	Use different STRID methods such as modal identification, parameter identification, or neural network. See Chapter 6 by Ettouney and Alampalli (2012)
3	Estimate damage	Use nonlinear analysis. The analysis can be static or dynamic. In some cases, linear analysis can be used to infer damage	Use *in situ* damage detection methods; see Chapter 7 by Ettouney and Alampalli (2012). In many situations, combinations of STRID and DMID techniques can yield even more accurate and cost-efficient methods for damage detection
4	Estimate statistical properties of damage. This will form a PDF (or histogram) of damage level for the specified input at step 1	Use methods of #2 and #3 above in combination with simulation techniques to determine needed statistical properties	Use adequate samples during the SHM experiment in combination with pertinent STRID methods to determine more realistic statistical properties

2.8.2 BRIDGE LIFE CYCLE BENEFIT ANALYSIS (BLCBA): BENEFITS

Seismic benefit analysis is computed indirectly. Benefits can occur when a seismic retrofit is performed. Such a retrofit can be used in a BLCA as a benefit or a partial benefit if it can be shown that such a retrofit can have a beneficial effect on mitigating another hazard. For example, if the bridge is located in a flood-prone area, the addition of seismic restraints might be beneficial in reducing cost of flood hazard. We note that for accurate computations of this type of benefit, the analysis must use a method that can accommodate the interrelationships between hazards, such as Monte Carlo simulation, as mentioned above.

2.8.3 BRIDGE LIFE SPAN ANALYSIS (BLSA): LIFESPAN

We note the dependence of $BLCC_{Seismic}$ on the estimated bridge lifespan S. An accurate estimation of S is paramount for accurate estimation of $BLCC_{Seismic}$. In other words, it is not too efficient to spend too many resources on estimating p_j or C_{ij}, while having an inaccurate estimate of S. In this section we consider two possibilities:

S **is independent of seismic events:** In this situation, the estimate of S should be based on other factors that affect bridge health, (see Chapter 10).

S **is dependent on seismic events:** This can occur in one of two scenarios. The most obvious is if the seismic event caused the bridge failure. This can result analytically only during a Monte Carlo simulation method (Equation 2.70 or 2.71 has a predetermined lifespan). The other potential occurrences can take place only if interrelations between hazards are used during cost computation. In such a case, if another hazard causes a bridge failure during the simulation process, the lifespan S is determined.

REFERENCES

Abdel-Ghafar, A. M. and Rubin, L. I. (1982). "Suspension bridge response to multiple-support excitations," *Journal of Engineering Mechanics Division*, ASCE, 108(2), 419–435.

Abrahamson, N., et al. (1991). "Empirical spatial coherency functions for applications to soil-structure interaction analyses," *Earthquake Spectra*, 7(1), 1–28.

Amini, F., Chen, H. M., Qi, G. Z., and Yang, J. C. S. (1997). "Generalized Neural Network Based Model for Structural Dynamic Identification, Analytical and Experimental Studies," iis, pp.138, 1997 IASTED International Conference on Intelligent Information Systems (IIS '97), 1997

Bendat, J. and Piersol, A. (1971). *Random Data: Analysis and Measurement Procedures*, Wiley-Interscience, New York, NY.

Benjamin, J. and Cornell, A. (1970). *Probability, Statistics and Decision for Civil Engineers*, McGraw-Hill, New York, NY.

Boggess, A. and Narcowich, F. J. (2001). *A First Course in Wavelets with Fourier Analysis,* Prentice Hall. Upper Saddle River, NJ.

Bruneau, M. and Reinhorn, A. (2007). "Exploring the concept of seismic resilience for acute care facilities," Earthquake Engineering Research Institute, *Earthquake Spectra*, 23(1) .

Celebi, M., Purvis, R., Hartnagel, B. A., Gupta, S., Clogston, P., Yen, P., O'Connor, J., and Frank, M. (2004). "Seismic Instrumentation of the Bill Emerson Memorial Mississippi River Bridge at Cape Girardeau (MO): A Cooperative Effort," *Proceedings, NDE Conference on Civil Engineering*, ASNT, Buffalo, NY.

Chen, S. E., Petro, S., Venkatappa, S., Ramamoody, V., Moody, J., Gangarao, H., and Culkin, A. (2000). "Automated Full-Scale Laser Vibration Sensing System," *Structural Materials Technology: An NDT Conference*, ASNT, Atlantic City, NJ.

Christian, J. (1989). "Generating seismic design power spectral density functions," *Earthquake Spectra*, 5(2), 351–368.

Clough, R. and Penzien, J. (1975). *Dynamics of Structures*, McGraw-Hill, New York, NY.

Der Kiureghian, A. (1980). "Structural response to stationary excitation," *Journal of Engineering Mechanics,* ASCE, 106(6), 1195–1213.

Der Kiureghian, A. and Neuenhofer, A. (1992). "Response spectrum method for multi-support seismic excitations," *Earthquake Engineering and Structural Dynamics*, 21(8), 713–740.

Dumanoglu, A. and Stevern, R. (1990). "Stochastic response of suspension bridges to earthquake forces," *Earthquake Engineering and Structural Dynamics*, 19(1), 133–152.

Ettouney, M. and Alampalli, S. (2012). *Infrastructure Health in Civil Engineering: Theory and Components*, CRC Press, Boca Raton, FL.

Ettouney, M., Brennan, J., and Brunetti, J. (1979). "Seismic Design Method for Arbitrary Propagating Waves," *Transactions of the 5th International Conference on Structural Mechanics in Reactor Technology*, Commission of European Communities, Berlin, Germany, K (5/7), 1–8.

Ettouney, M., Daddazio, R., and Abboud, N. (1997). "Some practical applications of the scale independent elements for dynamic analysis of vibrating systems," *Computers and Structures*, 65(3), 423–432.

Ettouney, M., Daddazio, R., and Hapij, A. (1999). "Optimal Sensor Locations for Structures with Multiple Loading Conditions." *Proceedings, International Society of Optical Engineering Conference on Smart Structures and Materials*, San Diego, CA.

FHWA. (1995). "Seismic Retrofitting Manual for Highway Bridges," Federal Highway Administration Report FHWA-RD-94–052, McLean, VA.

FHWA (2002). *Bridge Inspector's Reference Manual (BIRM)*, Vols. I and II, FHWA-NHI-03-001, Federal Highway Administration, U.S. Department of Transportation, Washington, DC.

Fujino, Y. and Abe, M. (2002). "Vibration-Based Structural Health Monitoring of Civil Infrastructures," *Proceedings of 1st International Workshop on Structural Health Monitoring of Innovative Civil Engineering Structures*, ISIS Canada Corporation, Manitoba, Canada.

Gazetas, G. and Mylonakis, G. (1998). "Seismic Soil-Structure Interaction: New Evidence and Emergin Issues," In *Geotechnical Earthquake Engineering and Soil Dynamics III*, Volume Two, Dakoulas et. al., Ed, ASCE Geotechnical Special Publication No. 75.

Glisic, B. and Inaudi, D. (2002). " Crack Monitoring in Concrete Element using Long-Gauge Fiber Optic Sensors," *Proceedings of 1st International Workshop on Structural Health Monitoring of Innovative Civil Engineering Structures*, ISIS Canada Corporation, Manitoba, Canada.

Gonzalez, M. and Zapico, J. (2007). "Seismic damage identification in buildings using neural networks and modal data," *Computers and Structures*, 86(3-5), 416-426.

Gurley, K. and Kareem, A. (1999). "Application of wavelet transforms in earthquake, wind and ocean engineering," *Engineering Structures,* 21(2), 149–167.

Hao, H., Oliveira, C., and Penzien, J. (1989). "Multiple-station ground motion processing and simulation based on SMART-1 array data," *Nuclear Engineering and Design*, 111(3), 293–310.

Harichandran, R., Hawari, A., and Sweidan, B. (1996). "Response of long-span bridges to spatially varying ground motion," *Journal of Structural Engineering*, ASCE, 122(5), 476–484.

Harichandran, R. and Vanmarcke, E. (1986). "Stochastic variations of earthquake ground motion in space and time," *Journal of Engineering Mechanics*, ASCE, 112(2), 154–174.

Heidari, A. and Salajegheh, E. (2008). "Wavelet analysis for processing of earthquake records," *Asian Journal of Civil Engineering (Building and Housing)*, 9(5).

Jovanovic, O. (1998). "Identification of dynamic system using neural network," *University of NIS, The Scientific Journal, Series: Architecture and Civil Engineering*, 1(4), Nis, Montenegro.

Keady, K., Alameddine, F., and Sardo, T. (1999). "Seismic Retrofit Technology," In *Bridge Engineering Handbook*, Chen, W-F. and Duan, L., Ed, CRC Press, New York, NY.

Lysmer, J., (1981), "*SASSI – A System for Analysis of Soil-Structure Interaction*," Report No. UCB/GT/81–02, University of California, Berkeley, CA.

Mindlin, R. and Goodman, L. (1950). "Beam vibrations with time-dependent boundary conditions," *Journal of Applied Mechanic,* Transactions of ASME, 17(4), 377–380.

Moehle, J. and Eberhard, M. (1999). "Earthquake Damage to Bridges," In *Bridge Engineering Handbook*, Chen, W-F. and Duan, L., Ed, CRC Press, New York, NY.

NYS-DOT. (2004). *Seismic Vulnerability Manual*, New York Department of Transportation, Albany, NY.

O'Connor, J. (2004). "Seismic Risk Assessment Procedures and Mitigation Measures for Highway Bridges in Moderate Earthquake Zones of the Eastern U.S.," *Proceedings of the 3rd US-PRC Workshop on Seismic Behavior and Design of Special Highway Bridges*, Shanghai, China. Also available in *Technical Report MCEER-05–0003*, MCEER, University at Buffalo, Buffalo, NY.

Paulay, T. and Preistly, M. (1992). *Seismic Design of Reinforced Concrete and Masonry Buildings*, John Wiley & Sons, New York, NY.

Priestley, M. J. N., Seible, F., and Calvi, G. M. (1996). *Seismic Design and Retrofit of Bridges*. John Wiley & Sons, Inc., New York, NY.

Riahi, M. and Abdi, M. (2005). "Residual Stress Analysis of Steel Parts via Utilization of Acoustic Emission (AE) Testing Method," *Proceedings of the 2005 ASNT Fall Conference*, Columbus, OH.

Roberts, J. and Maroney, B. (1999). "Seismic Retrofit Practice," In *Bridge Engineering Handbook*, Chen, W-F. and Duan, L., Eds, CRC Press, New York, NY.

Shrikande, M. and Gupta, V. K. (1999). "Dynamic soil-structure interaction effects on the seismic response of suspension bridges," *Earthquake Engineering and Structural Dynamics*, 28(11), 1383–1403.

Smyth, A., Betti, R., Lus, H., and Masri, S. (2000). "Global Health Monitoring and Damage Detection of the Vincent Thomas Bridge," *Structural Materials Technology: An NDT Conference*, ASNT, Atlantic City, NJ.

Sun, Z. and Chang, C. (2002) "Statistical-Based Structural Health Monitoring Using Wavelet Packet Transform," *Proceedings of 1st International Workshop on Structural Health Monitoring of Innovative Civil Engineering Structures*, ISIS Canada Corporation, Manitoba, Canada.

Sundaresan, M., Ghoshal, A., Schulz, M., Ferguson, F., Pai, P., and Chung, J. (2000). "Crack Detection Using a Scanning Laser Vibrometer," *Proceedings of 2nd International Workshop on Structural Health Monitoring*, Stanford University, Stanford, CA.

Wang, L. and Gong, C. (1999). "Abutments and Retaining Structures," In *Bridge Engineering Handbook*, Chen, W-F. and Duan, L., Eds, CRC Press, New York, NY.

Wolf, J. and Song, C. (1996). *Finite-Element Modeling of Unbounded Media*, John Wiley, New York, NY.

Koller, G. (2000), *Risk Modeling for Determining Value and Decision Making*, Chapman & Hall / CRC Press, Boca Raton, FL.

Zerva, A. (1990). "Response of multi-Span beams to spatially incoherent seismic ground motions," *Earthquake Engineering and Structural Dynamics*, 19(6), 819–832.

Zienkiewicz, O. C. (1971). *The Finite Element Method,* McGraw-Hill, New York, NY.

Hartle, R., Ryan, T., Mann, J., Danovich, L., Sosko, W., Bouscher, J. (2002), "Bridge Inspector's Reference Manual," *Federal Highway Administration, FHWA, Report No, FHWA NHI 03-001, Vol 1*, McLean, VA.

3 Corrosion of Reinforced Concrete Structures

3.1 INTRODUCTION

Corrosion poses considerable cost and safety threats to infrastructure. Yunovich et al. (2005) observed that the direct annual corrosion costs were about $276 billion as of the time of writing. They also observed that out of that amount, 16.4% is the estimated direct costs of corrosion of infrastructure. That amounts to almost $45.3 billion annually, again in 2005 dollars. They also reported that the indirect corrosion costs are almost 10 times the direct costs. For infrastructure alone, in 2005 dollars, that can amount to nearly half a trillion dollars, annually! Clearly, corrosion is a serious problem than requires serious consideration by all stakeholders.

Because of the seriousness of the problem, corrosion was/is the subject of intense interest by researchers, practitioners, and owners of infrastructure. Almost all types of civil infrastructure are affected by corrosion (bridges, roads, tunnels, dams, etc.). There are more than 600,000 bridges in the United States (see Ettouney and Alampalli 2012, Chapter 3). Of these, almost 41% are reinforced concrete bridges, 31% steel bridges, and 22% prestressed concrete bridges. Corrosion affects all of those bridges one way or the other. We concentrate in this chapter on corrosion of reinforced bridges. The deterioration of prestressed concrete bridges is discussed in the next chapter.

3.1.1 THIS CHAPTER

This chapter examines the problem of corrosion hazard (corrosion of steel rebars) in reinforced concrete structures, with emphasis on bridges, and the different roles that structural health monitoring (SHM)/structural health in civil engineering (SHCE) techniques can play in monitoring, mitigating, and managing the effects of corrosion. There are seven parts in this chapter, as shown in Figure 3.1. First, we present different aspects of corrosion as it affects reinforced concrete bridges. Understanding the causes of corrosion is essential for an efficient SHCE strategy. In addition, we discuss how corrosion affects different reinforced concrete components in a bridge since this can influence the SHCE strategy. Sections 3.2 through 3.5 are devoted to the corrosion problem from the viewpoint of the four components of SHCE field (see Ettouney and Alampalli 2012). Corrosion monitoring/detection and corrosion mitigation schemes are both functions of sensing and damage identification components. Structural identification as related to corrosion damage is shown to scale local to global structural geometry. Decision making applications are presented, with emphasis on the costs of potential corrosion-related decisions. We then offer some SHCE case studies for corrosion monitoring/study applications. We finally explore some corrosion-specific bridge life cycle analysis (LCA) issues.

3.2 CORROSION: THE PROBLEM

3.2.1 CAUSES OF CORROSION

3.2.1.1 Chloride and Carbonation

Chloride (Cl) content in concrete is one of the major reasons for corrosion damage. Yunovich et al. (2005) proved that chloride is one of the two main sources of corrosion damage due to sea water and deicing salt. When a reinforced concrete material is subjected to moisture that has some chloride in

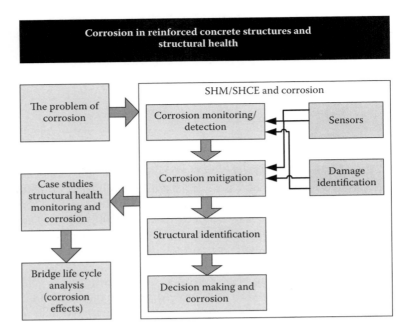

FIGURE 3.1   Overview of chapter on reinforced concrete corrosion.

it, the chloride starts penetrating the concrete cover in one of two ways: diffusion, and/or concrete cracks, if present. The chloride level in the concrete increases until it reaches a threshold level. At such a level, rusting starts in steel reinforcement. Rusting continues at a rate related to the chloride level in the concrete material. Trejo (2002) called the chloride threshold level that would initiate rebar corrosion as the critical chloride threshold value (CTV). Bentur et al. (1997) estimated that the value of CTV is in the range of 0.9–1.2 kg/m³. Trejo (2002) evaluated CTV for different steel rebars, as shown in Table 3.1.

The process of corrosive damage due to chloride is shown in Figure 3.2.

3.2.1.2   Moisture, Humidity, and Water

When concrete is produced, water is used to form the concrete composite. As the water evaporates, the moisture level in the concrete becomes uneven. This uneven moisture level continues throughout the lifespan of the structure, especially when it is subjected to external sources of water, such as rain, sea waves, tidal action, or snow. The uneven moisture levels can generate different internal stresses that may cause cracks in the concrete material. Such moisture can result in visible cracks that penetrate the concrete surface as much as 300 mm. For the sake of simplicity, we let us call this type of moisture "moisture cracking." As mentioned earlier, moisture that includes chloride content and cracking can both cause corrosion damage in reinforcing bars. Figure 3.3 shows the moisture interrelationship with corrosion.

3.2.1.3   Cracks

Concrete has a large compressive strength. But, it has two vulnerable material properties: weak tensile strength and brittle tensile behavior. Whenever the state of stress in concrete reaches the ultimate tensile strength, it will crack due to the lack of ductility. The use of steel reinforcement with concrete is meant to augment this vulnerability and make use of the other beneficial qualities of concrete material.

TABLE 3.1
CTV Levels for Different Materials

Material Type	CTV (pounds/y)
ASTM A615	1.1
ASTM A304	8.4
Microcomposite material	8.3

Source: Trejo, D. *Evaluation of the Critical Chloride Threshold and Corrosion Rate for Different Steel Reinforcement Types.* Texas A&M University, College Station, TX, 2002. With permission.

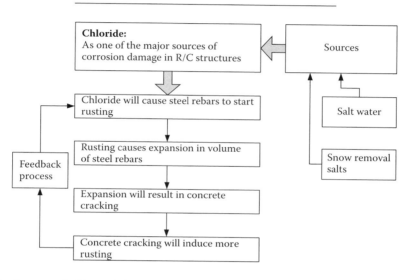

FIGURE 3.2 Source and effects of chloride.

Cracking in concrete can occur for many reasons. Moisture as a cause of moisture cracking was discussed earlier. However, perhaps the most important cause of concrete cracking is the formation of tensile stresses due to the different loadings that affect the structure: *mechanical cracking.* For example, positive or negative bending moments may induce concrete cracking (see Figure 3.5). Concrete joints and connections can be subjected to stress concentrations at the corners. If any of these concentrated stresses reach the ultimate tensile concrete stress, a crack will form.

When the reinforcing bars start corroding, it forms rust. The volume of rust can be several times the volume of steel. Such an increase of volume will create an outward normal compressive pressure on the surrounding concrete. Such compressive radial pressures will require equilibrating circumferential tensile stresses. These tensile stresses will cause the surrounding concrete to crack: *rust cracking.*

Cracking of concrete plays a major role in corrosion of reinforcing bars. Whenever a crack is formed in the concrete, it becomes a venue for increase of the chloride and carbon levels around the steel rebars, and this will increase the rate of the corrosion process.

As was mentioned earlier, the volume increase would promote more concrete cracking. This feedback would continue, and the corrosion damage would continue harming the structure at an increased rate. Figure 3.4 shows cracking and corrosion interrelationships of reinforced concrete systems.

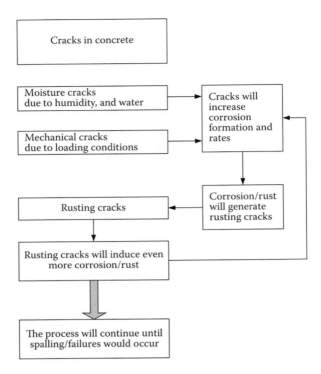

FIGURE 3.3   Corrosion and crack interrelationships.

3.2.1.4 Other Sources

There can be other sources for corrosion in reinforced concrete systems. Some of these are local defects and random-type sources. Local defects can occur either in the concrete or in the steel rebars. They can originate during the manufacturing, or construction processes. Random effects occur due to the inherent uncertainties in the aggregate material constituents of reinforced concrete. There are also uncertainties in all construction steps. Both local and random effects can be reduced by enforcing higher quality control (QC) and quality assurance (QA) standards. Manual inspection as well as SHM processes can also help in monitoring and reporting corrosion at early stages.

3.2.2 CORROSION, STRUCTURAL COMPONENTS, AND HEALTH EXPERIMENTS

3.2.2.1 Reinforced Concrete Columns

Columns in a structural system have several unique properties that, with close observation of the environment of the column, can help in making efficient corrosion monitoring decisions. The ACI (2008) defines structural column as any structural element that has

$$\frac{P}{P_U} > 0.10 \tag{3.1}$$

where P is the axial load in the column and P_U is the axial strength of the column. Figure 3.5 shows a typical axial load distribution in a column: (1) the effects of the column's own weight are neglected, and (2) the column axial force is assumed to be a compressive force. There can be bending moments

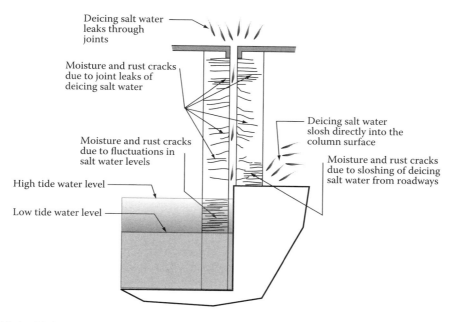

FIGURE 3.4   Moisture and rust cracks: patterns and sources.

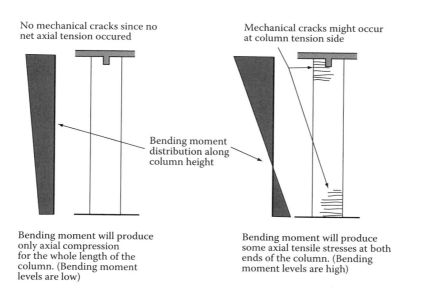

FIGURE 3.5   Column mechanical crack patterns and sources.

along the length of the column, depending on the type of loading applied to the structure as a whole. For the sake of simplicity, we will consider only the bending moment distributions shown in Figure 3.5. Of course, other bending moment distributions are possible, but their occurrences will not change the conclusions of this section.

The expected station locations of corrosion damage along the column length are affected by several factors: location of moisture source, cracks in concrete cover, and random blemishes in the concrete and/or steel rebars. Let us take a closer look at each of these factors.

3.2.2.1.1 Moisture Source

The source of moisture plays a major role in locating the corrosion damage distribution along the column height. Because of this, it is important that any corrosion monitoring experiment is preceded by an accurate estimation of the source(s) of moisture. For example, columns that support a bridge might be subjected to water leaks from the bridge joints at the top of the column. A partially submerged column in seawater subjected to tidal fluctuations would be susceptible to corrosion around the stations of the column that are subjected to the rising and falling tide water levels. Columns that have their bases at a street level might be subjected to water sloshing, especially during snow-removing seasons. Water sloshing near the column base can cause corrosion near the column base.

3.2.2.1.2 Cracks in Concrete

Reinforced concrete cracks encourage moisture to penetrate the steel rebars. It was also observed that there is a direct relation between concrete cracks and corrosion rates. The location and size of concrete cracks can be easily attributed to two factors: the level of concrete tensile stresses and the quality of concrete material.

Tensile stresses

Concrete material has a limited tensile strength. As soon as the tensile strength is reached, the concrete material will crack. The nominal concrete strength in tension is proportional to $\sqrt{f_c'}$ with compressive strength of concrete being f_c' as measured in psi (see ACI 2002).

Tensile concrete strength in columns can easily be reached near columns joints, depending on the level of end-bending moments of the column (Figure 3.5). Since columns are subjected, in general, to axial compressive forces, there exists a middle part of the column where no axial tension is expected, that is, no concrete cracking is expected. Figure 3.5 shows this cracking-free zone in a typical concrete column. Estimation of this zone can be of help in predicting if corrosion damage in columns can be traced to tensile cracking. If corrosion damage is observed within the zone, it can be safely assumed that the corrosion is caused by factors other than concrete tension cracks.

Quality of concrete material

Concrete strength f_c' is related to the quality of concrete material. This means that, for poor material quality, tensile cracks will be expected to be more prevalent than in better quality concrete material. Corrosion damage can then be expected in poorly manufactured concrete material. This indicates that QA and conformity to specification of concrete material can lessen corrosion damage.

Concluding remarks

In addition to the above, we recall that another corrosion source can be random blemishes/effects. The uncertain nature of this effect makes it difficult to prevent it completely. Only strict inspection/monitoring practices can reduce its damaging results.

3.2.2.2 Other Structural Components

Corrosion can also occur in other structural components such as beams (shown in Figure 3.6), joints/connections, foundations, slabs/decks, and walls/abutments. As in the case of columns, component-specific corrosion analysis and behavior can be studied. As in columns, some important factors that can affect corrosion occurrences in these components are

- Degree of exposure to moisture or salt (deicing or seawater)
- Patterns of freeze/thaw
- Drainage pathways
- QA/QC in manufacturing, transportation, and erection steps

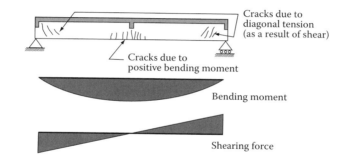

Simple beam mechanical crack patterns and sources

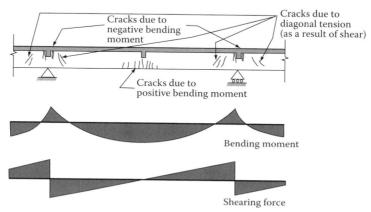

Continuous beam mechanical crack patterns and sources

FIGURE 3.6 Beam mechanical crack patterns and sources.

- Mechanical cracks, including deformation patterns
- Accidental impacts

The above factors can guide manual inspection processes as well as different SHM corrosion-related projects.

3.3 CORROSION MONITORING

3.3.1 LABORATORY TESTING

Several procedures are available to monitor corrosion in laboratories. Among these are

Half-cell potential: The method measures possible active corrosion areas in steel reinforcements (see Ettouney and Alampalli 2012, Chapter 7). It does not compute the corrosion rate. Also, it cannot be used in concrete fully submerged in water.

Detection of crack pattern: This method, used first by Goto (1971), is based on using ink solution to show crack patterns, which can also indicate corrosion damage.

Electropotential system: This method is used to produce a desired accelerated rate of corrosion. It is performed by immersing the specimen in a solution. With the aid of an electrical circuit, the metal ions can be transferred between the solution and the steel, thus accelerating the corrosion process.

Measuring chloride content: The test is performed by adding silver nitrate solution and thiocyanate solution to the specimen until a prescribed change in its properties occurs. The chloride content as a percent of the cement used is computed by (see Fazio 1996)

$$J = \left(V_5 - \frac{V_6 \, m}{0.1} \right) \left(\frac{0.3545}{M_c} \right) \left(\frac{100}{C_1} \right) \tag{3.2}$$

where

J = The chloride ion content by percentage of the weight of cement

M_c = The mass of sample used (in g)

V_5 = The volume of 01 M silver nitrate solution added (in mL)

V_6 = The volume of thiocyanate solution used (in mL)

m = The molarity of the thiocynate solution (in mol L)

C_1 = The cement content of the sample used (in %)

3.3.2 POPULAR METHODS OF FIELD MONITORING

3.3.2.1 Tests and Sensors

There are several types of tests for monitoring corrosion damage. Each test might need a special type of sensor to reach its goal. Tables 3.2 and 3.3 show different tests for corrosion damage detection, description of the experiment, and the sensors that can be used for each test.

3.3.2.2 NCHRP 558 Procedures

Several other types of field evaluation procedures that relate to corrosion damage are detailed in Sohanghpurwala, A. A. (2006). Table 3.4 illustrates some important details of those procedures.

For completion, we note that in addition to the methods of Table 3.4, Broomfield (1998) described additional methods as in Table 3.5.

We note that the methods of Tables 3.4 and 3.5 are mostly manual. They are performed at a given instant of time. They are also performed at limited locations. This brings up the usual question of how many locations (samples) should be used. Of course, there is a need for compromise between

---

TABLE 3.2
Testing for Corrosion Damage

Damage Classification	Damage Type
Time-dependent concrete damage (2a)	Time-dependent concrete deterioration. Crack growth due to corrosion.
Concrete damage (both internal and external) (1,2,3,4,5)	– Surface cracks
	– Internal cracks
	– Degraded concrete engineering properties such as modulus of elasticity or compressive strength.
	– Homogeneity or voids in concrete.
	– Engineering properties of concrete (modulus of elasticity, specific weight, etc.)
Identification of geometrical properties (6, 7, 8, 9)	– Rebar location
	– Prestressed or posttensioned tendon or duct locations
	– Rebar size
	– Concrete cover thickness
Prestressed or posttensioned tendon damage (10, 11)	– Presence of voids in ducts
	– Damage (corrosion, loss of area, etc.) in tendons
Corrosion process of reinforcements (12, 13)	– Corrosion progression
	– Corrosion rate
Concrete chemical and electrical properties (14,15)	– Concrete resistivity
	– Dielectric and conductive properties of concrete

TABLE 3.3
Field Testing Methods

Damage Classification	SHM Method	Comments
Time-dependent concrete damage (2a)	Ultrasound methods are popular for these damage mechanisms. See Chapter 5, Ettouney and Alampalli (2012)	Care is counseled, since environmental conditions might affect reading and interpretation of results.
Concrete damage (both internal and external) (1,2,3,4,5)		
Identification of geometrical properties (6, 7, 8, 9)s	See Chapter 5, Ettouney and Alampalli (2012)	
Prestressed or posttensioned tendon damage (10, 11)	See Chapter 4 of this volume	
Corrosion process of reinforcements (12, 13)	Half-cell potential	Measure electrochemical potential of steel in corroding concrete. Not applicable for water-saturated concrete. See description in Section 3.4.
	GPM	Measure polarization properties of rebars. Can be used in water-saturated concrete.
Concrete chemical and electrical properties (14,15)	See Chapter 5, Ettouney and Alampalli (2012)	

TABLE 3.4
Summary of Corrosion Field Procedures

Method	Method	Comment
Visual survey	Qualitative	Relies on visual assessment of damage (ACI 201.1 R-92)
Delaminating survey	Semiquantitative	Estimates surface delimitation. Computes percentage of delimitation of the surface (ASTM D-4580-86). Note that delimitation survey can be performed by one of several methods:
		Sounding: can be performed by a hammer or dragging chain along the surface
		Impact-echo (see Ettouney and Alampalli 2012, Chapter 7)
		Ultrasonic pulse velocity
		Infrared thermography (see Ettouney and Alampalli 2012, Chapter 7)
		Ground-penetrating radar
Cover depth measurements	Computes cover of rebars	Drill into the concrete until encountering of rebars
Direct current method (resistivity)	Measure electric resistance and voltage to estimate corrosion extent	The method would rely on exposing rebars at an adequate number of locations
Core sampling	Petrographic core extraction Epoxy-coated rebars Chloride ion distribution	Samples are collected according to ASTM C42/C42M-99 Three different procedures can be used as appropriate
Corrosion potential survey	Half-cell potential method	This method is performed according to ASTM C-876. The corrosion potential is computed on a grid basis of 2 ft intervals. In the longitudinal and transverse directions
Corrosion rate survey	Computes corrosion rate at a given location	Several devices are available for corrosion rate measurements. The process is mostly manual. Broomfield (1997) observed that this method is fairly suited for automatically determining deterioration rate in bridges and other infrastructure
Carbonation testing	Measures depth of carbonation	The procedure would require the use of 0.15% solution of phenolphthalein in ethanol to be sprayed into the cut concrete sample

Source: Sohanghpurwala, A.A., *Manual on Service Life of Corrosion-Damaged Reinforced Concrete Bridge Superstructure Elements*, National Cooperative Highway Research Board (NCHRP), Report No 558, Washington, DC, 2006.

TABLE 3.5
Summary of Additional Corrosion Field Procedures

Method	Method	Comment
Chloride testing	Measure chloride profile at different depths	The method is either field or laboratory based. It can show chloride concentration, thus revealing risk of corrosion (>0.4% chloride). The test can also reveal whether the chloride was included in the initial concrete casting or was diffused after the initial construction
Permeability testing	Measures diffusion of concrete	This is usually used to investigate the improvements in concrete status after a rehabilitation work. Can be field or laboratory based

Source: Broomfield, J., *Corrosion of Steel in Concrete*, E & FN Spon, New York, NY, 1998. With permission.

TABLE 3.6
Minimum Sampling Size for In-Depth Evaluation (NCHRP 558)

Test Method	Minimum Sampling Size
Clear Concrete Cover	
(Using nondestructive test methods). Several actual CCC measurements should be collected to calibrate nondestructive test methods equipment	30 measurements per span. If cover measurements from a previous PCCE are available, they can be used instead of collecting the data again in the in-depth evaluation
Visual survey	Entire surface of the concrete element
Delamination survey	Entire surface of the concrete element
Chloride profile analysis	1 location per 1000 square feet
Electrical continuity testing	5 reinforcing steel bars in each span. Must include both transverse and longitudinal bars
Epoxy-coated rebar cores	Minimum of 5
Pertrographic analysis	1 location per 3000 square feet or a minimum of 5, whichever is higher

Source: Sohanghpurwala, A.A., Manual on Service Life of Corrosion-Damaged Reinforced Concrete Bridge Superstructure Elements, National Cooperative Highway Research Board (NCHRP), Report No 558, Washington, DC, 2006. With permission from NCHRP.

the number of samples and the cost involved for collecting and testing the samples. NCHRP 558 recommended the number of samples as shown in Table 3.6.

3.3.3   SHM Role and Field Monitoring

3.3.3.1   SHM and Corrosion

The continued monitoring (SHM) of state of corrosion in reinforced concrete civil infrastructure was discussed by Broomfield et al. (1987). Since then, it has been addressed by many authors and researchers. Some of the objectives of corrosion SHM projects can be summarized as follows:

*   Detect/monitor corrosion-related structural damage such as surface defects, delaminations, spalling, or internal cracking due to rust
*   Detect corrosion rate. This can be used in an automated procedure to quantify deterioration rate of systems (see Agrawal et al. 2009a)
*   Detect risk of corrosion

TABLE 3.7
SHM Role in Corrosion Detection

Corrosion Issue	SHM Role and Methods
Structural damage	Can detect damage at earlier stages. Different nondestructive test methods are fairly suited for such detection. GPR, impact-echo, acoustic emission, or thermography can all detect corrosion damage with varying degrees of success
Corrosion rate	Many of the methods discussed elsewhere in this chapter can estimate corrosion rate
Corrosion risk	Environmental condition, e.g., moisture and site condition, e.g., drainage efficiency, are good indicators of potential corrosion formation. Thus, monitoring these conditions might give an early indication of corrosion

TABLE 3.8
SHM Concepts as Applied to Corrosion

SHM Concept	Comment
Virtual sensing	Virtual sensing paradigm (see Ettouney and Alampalli 2012) would measure indirectly corrosion in a system. By measuring different parameters that would cause corrosion, its presence and magnitude can be inferred. Examples of these parameters are moisture, temperature, concrete chemical composition, drainage conditions, and presence of salt in moisture
STRID	Simulating corrosion-caused degradations in structural properties (such as loss of area or delamination) in a structural model can help in estimating the safety effects of corrosion. We note that complete reliance on analytical models in this situation might lead to erroneous results. This is due to the high degree of uncertainty in modeling the corrosion process and its effects. It is recommended that such a STRID approach needs to be based on the results of in-field testing
DMID	Many of the nondestructive test methods would measure either the corrosion process itself (acoustic emission) or the damage resulting from corrosion (impact-echo or thermography)
Wireless sensing	Wireless sensing is emerging as an invaluable tool in SHM sensing. Lack of wires would enable wireless-sensing systems to use more sensors in an SHM system, thus covering a larger sensing range
Remote sensing	Remote-sensing techniques can help in detecting corrosion damage and its extent in situations where close sensing is not possible. Examples of such situations are hard-to-reach bridge superstructures and post-tensioned concrete systems where tendons are embedded deep inside the concrete systems
Cost benefit	See next section
LCA	Corrosion can have an immense effect on LCA. Thus, an accurate estimate of corrosion and its extent, as well as corrosion mitigation measures and their costs, are important components in an LCA effort

Table 3.7 explores different SHM methods in achieving these three objectives. Some basic concepts of SHM and corrosion are discussed in Table 3.8.

3.3.3.2 SHCE and Corrosion

One of the basic premises of this chapter is that the main difference between SHCE and SHM is that the former includes both SHM and decision making (DM) processes. We would like to highlight a subtle but major DM issue in the corrosion detection of civil infrastructure. We need first to carefully reexamine Table 3.6. After a bit of study, we can propose that Table 3.6 relates a particular corrosion detection method, i, with a minimum sampling size, N_i. We recognize that the units of N_i are method dependent; however, such dependence will not limit the accuracy of the following developments. Further, we make the following propositions:

- The sampling and the whole experiment will show the state of corrosion at a single instant of time. Since corrosion is a continuous event in time, such an experiment needs to be repeated at a time interval of Δt.
- The cost of each sample in the experiment is C_{ei}

The total cost of the intermittent corrosion monitoring of the system during a time span of T becomes

$$C_E = N_i C_{ei} \frac{T}{\Delta t} \qquad (3.3)$$

For continuous corrosion monitoring, the cost can reasonably be expressed as

$$C_C = C_{\text{initial}} + cT \qquad (3.4)$$

The initial cost of placing the SHM setup is C_{initial}, while the continuous operational costs per unit time is c. It is reasonable also to assume that

$$c \ll \frac{N_i C_{ei}}{\Delta t} \qquad (3.5)$$

Condition 3.5 is accurate since the continuous monitoring does not include any overhead costs such as labor, cost of installation, and data collection every time the experiment is performed. Figure 3.7 shows a schematic comparison of the intermittent and continuous corrosion monitoring costs. The cost of continuous monitoring will be higher than that of intermittent monitoring at the earlier stages. However, the trend will be reversed after a time period of

$$T_0 = \frac{C_{\text{initial}}}{(N_i C_{ei} / \Delta t) - c} \qquad (3.6)$$

In addition to direct cost savings, continuous monitoring has also the advantage of detecting corrosion damage in real time, rather than at discrete time intervals.

3.3.4 ADVANCED MONITORING OF CORROSION

Advanced corrosion monitoring systems should have some of the following attributes: (1) they should permit remote sensing and recording, (2) they should operate without major human supervision, and

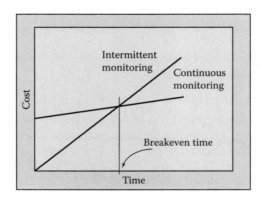

FIGURE 3.7 Relative costs of intermittent and continuous corrosion monitoring.

(3) they should be environmentally friendly. Agrawal et al. (2009b) provided a summary of many of the advanced corrosion monitoring techniques.

3.3.4.1 Fiber Optic Sensors (FOS)

There are many recent advances in FOS-specific corrosion monitoring in reinforced concrete. A study on the applicability of fiber optics in sensing pH levels was offered by Ghandehari and Vimer (2002). They showed that measuring other environmental effects are possible (moisture, corrosion, etc.). Also, Micheal et al. (2004) gave details of the use of Carbon Fiber Reinforced Polymers for flexural strengthening and Glass Fiber Reinforced Polymers spray-up for shear strengthening of a T-beam bridge in Florida. The beams were instrumented with fiber optic strain gauges and with some corrosion sensors (to sense rebar corrosion). In another work, Wiese et al. (2000) observed that embedding a special dye (pyridinium-N-phenolate betaine) in a polymethacrylonitrile polymer matrix would result in an absorption spectral (wavelength) shift that is dependent on the degree of water concentration. Thus, a calibrated fiber optic sensor can measure the water moisture content in concrete by measuring the shift of wavelength. The dry condition would produce a wavelength of 602 nm, while a water concentration of 28% (by weight) would produce a wavelength shift of about 40–562 nm. Figure 3.8 shows the water-content-wavelength spectrum. Embedding such a fiber optics sensor in reinforced concrete systems would make it possible to continuously monitor moisture inside the concrete.

3.3.4.2 Acoustic Emission and Corrosion Monitoring

The process of corrosion of metals can be considered a slow damaging process of the material. As such, it can be considered a source of acoustic emission (AE). Many researchers have studied the possibility of using AE to estimate corrosion damage. For example, Riahi and Khosrowzadeh (2005) have researched the subject. The main thrust of their effort was to investigate the correlation, if any, between AE count rate and corrosion damage (as measured by loss of weight of the base metal). They found a clear correlation between the two. On the basis of their finding, it is clear that AE can be used to estimate corrosion damage.

As was mentioned earlier, one of the main disadvantages of the AE technique is that the AE signals can be so low that the ambient noise might increase the error of measurement above any acceptable tolerance. Riahi and Khosrowzadeh (2005) found that the measured amplitude of the AE

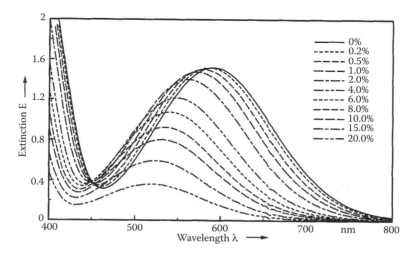

FIGURE 3.8 Absorption spectra of an indicator dye at different water concentration levels. (From Wiese, S. et al., Fiberoptical Sensors for on-line Monitoring of Moisture in Concrete Structures, *Proceedings of 2nd International Workshop on Structural Health Monitoring,* Stanford University, Stanford, CA, 2000. With permission from Dr. Fu-Kuo Chang.)

signals in their experiment was about 40 dB, while the ambient noise amplitude was about 26 dB. The 14 dB difference is large enough to distinguish the AE corrosion signals from the ambient noise. They showed that, with appropriate tools, the AE corrosion signals can be measured accurately. This opens the door for using AE technique in detecting corrosion damage.

3.3.4.3 Linear Polarization Resistance

So, Millard, and Law (2006) investigated the effect of environmental conditions (temperature, humidity, and rainfall) by the linear polarization resistance (LPR) method to measure the corrosion rate of steel-reinforcing bar embedded in concrete (see Figure 3.9). They noted that LPR is a well-established method for instantaneous corrosion rate of steel rebars in the concrete; the approach has been developed from the Stern-Geary Theory where the corrosion current is calculated as a ratio of Stern-Geary constant and polarization resistance. The polarization resistance was measured using a

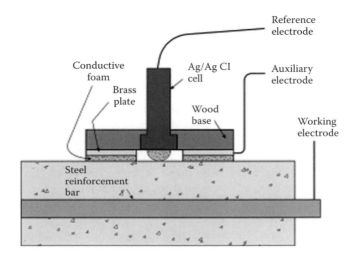

FIGURE 3.9   Three-electrode LPR measurement. (Reprinted from ASNT publication.)

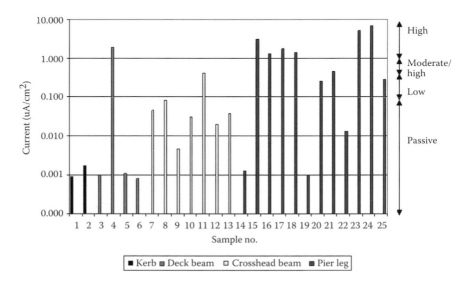

FIGURE 3.10   Evolution of steel corrosion in different environmental conditions. (Reprinted from ASNT publication.)

three-electrode system, as shown in Figure 3.9, by coupling reference and auxiliary electrodes to a concrete surface that acts as a working electrode. On the basis of this study, the authors concluded that the corrosion rate increased with temperature and rainfall, that these are more significant for active corrosion than for passive corrosion, and that a single data point does not indicate corrosion activity but long-term monitoring is needed (see Figure 3.10).

3.3.4.4 Smart Pebbles

With the recent advances in miniaturized electronic devices, several miniaturized corrosion sensors have been developed. For example, a new type of sensor called "smart pebbles" has been developed recently (Watters et al. 2001). Smart Pebbles is long-life wireless sensor that continually monitors the health of the bridge from deep inside the bridge deck. In less than a few seconds, the sensor checks chloride ingress and relays the information instantly and wirelessly, providing an early warning system for assessing damage before safety issues arise. It is powered remotely, so no lifetime-limiting batteries are required. It is inserted in the bridge deck either during initial construction, or during rehabilitation, or in existing structures through a back-filled drilled core hole. These sensors are called pebbles since they are roughly the size and weight of a typical piece of rock aggregate used in concrete. They are smart in that each contains a chloride sensor and radio frequency identification, or RFID, a chip that can be queried remotely to identify it and that can indicate chloride concentration levels in that part of the bridge. The Smart Pebble reader can be either hand-held or mounted on the underside of a vehicle. While driving over the bridge, the reader picks up information from the smart pebbles embedded in the bridge deck and sends them to a collection point. These sensors are in their early stages of development and embedded in bridge decks by the California Department of Transportation for monitoring chloride levels in bridge decks. Figures 3.11 and 3.12 show smart pebbles and chloride sensor concepts, respectively (Watters et al. 2001).

3.3.4.5 Embedded Combined Type of Measurement (ToM)

Ong and Grimes (2002) provided another advanced corrosion sensor research. The sensor is embedded in concrete and can sense stresses, temperatures, and chlorine (corrosion) levels. It is a sensor array that works with harmonic motions, in a wireless fashion. The sensor can be embedded in concrete while the cement is curing, and the response can be found by using a portable detection coil that does not require contact with the sensor through wires. The sensor cost is estimated at 0.5 cents.

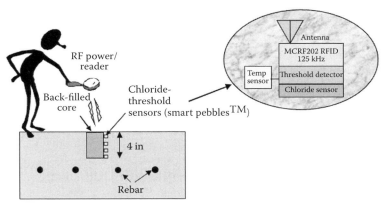

Simplified bridge-deck cross section

FIGURE 3.11   Smart pebbles concept. (From Watters, D.G. et al., Design and performance of wireless sensors for structural health monitoring, Review of Progress in Quantitative Nondestructive Evaluation, Brunswick, ME, July 29–August 3, 2001, 969–976, D. O. Thompson and D. E. Chimenti, Eds. (*Am. Inst. Phys.*, Melville, NY, 2001). Courtesy of Caltrans.)

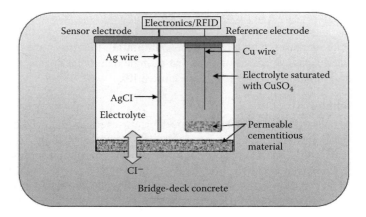

FIGURE 3.12 Chloride sensor used inside smart pebbles (From Watters, D.G. et al., Design and performance of wireless sensors for structural health monitoring, Review of Progress in Quantitative Nondestructive Evaluation, Brunswick, ME, July 29–August 3, 2001, 969–976, D. O. Thompson and D. E. Chimenti, Eds. (*Am. Inst. Phys.*, Melville, NY, 2001). Courtesy of Caltrans.)

Thus, it is an inexpensive, passive sensor with a lot of potential for SHM and nondestructive evaluations (NDE). This sensor actually is a combination of three sensors:

Temperature sensor: A commercially available alloy (Honeywell-brand Metglas 2826MB) or Metglas 2605SC is used, where its harmonic response (amplitudes) changes with the change in temperature. The accuracy is increased by monitoring more harmonics and taking an average. In the lab test described, the temperature measured by considering three harmonics was about ±3% compared to thermocouples. The sensor showed a reversible linear response.

Stress sensor: The above also showed that it can be used to monitor stress level in concrete through lab experiments. It also showed that the response is linear and reversible.

Chlorine sensor: These were fabricated by coating Metglas 2605SC with 200-μm-thick polyurethane followed by a 50-μm-thick alumina. The swelling of the coating in response to chloride ions creates a stress on the sensor, which in turn changes the harmonic signature.

3.3.4.6 Laser Surface Measurements

Fuchs et al. (2000) experimented with a laser system for measuring the surface profile of objects during large structural testing. The laser system is capable of making distance measurements over a maximum range of 100 ft with an accuracy of less than 0.03 in over the entire range. The system contains a scanner used for directing the laser measurements to various locations without the use of any special targets, and valid measurements can be made on both steel and concrete specimens.

The above test bridge using geosynthetic reinforced soil abutments was constructed (see Figure 3.13). The bridge superstructure consists of 70-ft-long concrete box girders provided by the New York State Department of Transportation. The structure serves as a test platform for NDE instrumentation research as well as for geotechnical research. The data collected using the scanner will serve as a baseline measurement of the original position and shape of the test bridge abutments. Periodic remeasurement of the structure could then provide information on the long-term changes.

The NDEVC laser system was also used as a part of the curved girder bridge testing (Figure 3.14). Global deflection and rotation of the entire 90-ft-long, three-girder bridge was measured during the test, using 124.5-in-diameter targets fixed to the bridge girders to allow an exact point on the structure to be tracked as the structure moves. A detailed scan of a section—the web and top flange of the component specimen at various load steps during the test—was also done to determine the out-of-plane distortions of the web and top flange (Figure 3.15). Tools were developed to postprocess the laser data immediately after data collection and produce plots representing the component out-of-plane distortions (Figure 3.16).

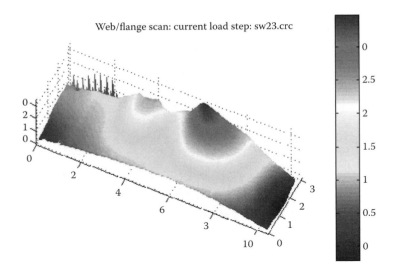

FIGURE 3.16   Laser system data from component specimen web showing buckling. (Courtesy of CRC Press.)

3.4   CORROSION MITIGATION METHODS

3.4.1   GENERAL

Decision making process and corrosion problems have been documented for many years. Many owners have their own road maps for responding to different corrosion problems. This poses two important questions:

- Can the tools and methods of structural health experiments aid in the decision making process?
- In view of the many choices available to the decision maker, what is the most beneficial decision to take?
- After a particular mitigation process decision is made, is there a role for structural health techniques during the mitigation process and after it is completed?

To answer these questions, it is important to remember that, for an appropriate decision making process, a full knowledge of available mitigation/protection schemes is necessary. In addition to the technology itself, there are several aspects of each method that must be known, as shown in Table 3.9.

Numerous methods have been applied over the years for corrosion protection of steel reinforcement in concrete. They have produced mixed results. Perhaps the most difficult part in discussing and presenting these methods is the way they are categorized. For example, Bentur et al. (1997) placed corrosion mitigation methods in three categories: Carbon control methods, chloride control methods, and sever chloride penetration methods. Virmani and Clemena (1998) and Yunovich et al. (2004) put the mitigation methods into two general categories: new construction and existing construction. They proceeded to subdivide each of the categories into several subcategories. Tables 3.10 and 3.11 summarize their corrosion mitigation/control methods. Note that prestressed concrete bridge cases are discussed in Chapter 4.

In what follows we discuss the above mitigation methods. Different factors that affect decision making, as highlighted earlier, will also be discussed. The section will conclude by introducing a general risk-based decision making process that can be used in most situations.

TABLE 3.9

Important Aspects of Corrosion Mitigation Methods

Basic information on technology

How it works

Advantages and disadvantages of the method

Is it applicable to new, existing or both types of bridges?

Lifespan of method

Present cost of method

Risk-based life cost of method

Potential for monitoring the performance of the method: visual, short term, long term?

Any other pertinent qualitative information

TABLE 3.10

Methods for Mitigating/Controlling Corrosion for New Construction

Method

Adequate concrete cover	
Quality concrete	
Alternative reinforcement	Steel bars with organic coatings
	Steel bars with metallic coatings
Corrosion-inhibiting admixtures	

Source: Courtesy of Yash Virmani.

TABLE 3.11

Methods for Mitigating/Controlling Corrosion for Existing Construction

Method

Surface barriers	
Cathodic protection (CP)	Impressed current CP
	Galvanic CP
Chloride removal	

Source: Courtesy of Yash Virmani.

3.4.2 NEW CONSTRUCTION

Three obvious methods to mitigate or reduce the effects of corrosion in newly constructed reinforced concrete structures were described in detail by Virmani and Clemena (1998):

Adequate concrete cover: An optimum concrete cover that is large enough so that the time needed for the chloride corrosion threshold to reach the steel is beyond the required service life. Such an optimal cover length is computed using the diffusion equation (see Crank 1983). Unfortunately, even with an optimal concrete cover, chloride will still penetrate the concrete due to the inevitable cracking in the concrete components such as beams, columns, or slabs.

Quality concrete: Concrete type and mix can affect the propensity of concrete to corrosion—an obvious conclusion. Several researches have been conducted over the years to identify different sensitivities of concrete mix to corrosion attack. See for example, Thompson and Lankard (1995) and Thompson et al. (1996).

Alternative reinforcement: The third obvious method that would mitigate corrosion addresses the steel rebars. Two known approaches can be used: (1) coat the rebars with adequate barrier that would resist chloride, moisture, or oxygen effects; and (2) use a material different from the conventional steel rebars, a material with better corrosion resistance properties. A good discussion of these alternatives can be found in the report by Virmani and Clemena (1998).

3.4.3 Concrete Replacement/Patching

When corrosion damage is extensive and/or deep, the damaged concrete needs to be removed and replaced. The removal can be achieved by pneumatic hammers, hydraulic jets, or milling machines. Vorster et al. (1992) compared the costs of the three methods and concluded that the use of mechanical hammers is much more expensive per unit area than the other methods. Milling machines cost the least per unit area. After the damaged concrete is removed, it is replaced by patches of special concrete mixes. Care must be taken in preparing the concrete mix as well as in the method of applying the patches. Broomfield (1998) observed that, without careful application of the patches, corrosion damage will continue to occur in the rehabilitated area. Obviously, this method is used for existing constructions.

3.4.4 Concrete Spraying/Encasement

In addition to patching, the damaged concrete can be replaced by overlays, spraying, or encasement. In all cases, the efficiency of the retrofit depends on the skill of the workers, the mixture of the replacement concrete, and the method of application.

3.4.5 Corrosion Inhibitors

Corrosion inhibitors are a class of chemicals applied to reinforcing steels that should inhibit cathode and/or anodic reactions, thus slowing or preventing the corrosion process, Broomfield (1998). There are many inhibitor products in the market, which have been used in several projects with varying degrees of success. This method can be used in both new and existing constructions.

Another class of corrosion inhibitors is applied directly to the concrete mix for new construction or the patch mix for rehabilitation projects (Virmani and Clemena 1998). They subdivided the inhibitors into three major classes: Anodic, cathodic, and organic inhibitors. There are in the market many commercial inhibitors that have been used with varying degrees of efficiency.

We need to mention that concrete inhibitors have also been used in the grout mixture for posttensioned bridges.

3.4.6 Surface Barriers

Water proofing and other types of barriers that protect concrete from chloride attack are used extensively as a corrosion mitigation scheme. They need to be used carefully to ensure long-life service. Their average lifespan is 10–15 years (Broomfield 1998). Cost–benefit and reliability studies need to be performed to ensure maximum efficiency. Table 3.12 shows a cost comparison for different surface barrier systems. Such information can be used in cost–benefit analysis or, along with lifespan information, in life cycle cost analysis (LCCA).

A passive barrier system is the *deflection* approach. In this approach a good drainage is provided. The drainage would deflect the water away from the susceptible concrete, thus reducing the chances of corrosion.

3.4.7 Cathodic Protection

Since corrosion is an electrochemical process, where an electric current is formed from the anode (corroding steel) to the cathode (oxygen and water are interacting chemically), it is logical that

TABLE 3.12
Cost and Prolonged Life for Different Surface Barrier Mitigation Methods

Method	Average Cost ($/m²)	Average Prolonged Life (years)	Range of Cost ($/m²)	Range of Prolonged Life (years)
Portland cement concrete overlay	170	18.5	151–187	14–23
Bituminous concrete with membrane	58	10	30–86	4.5–15
Polymer overlay/sealer	98	10	14–182	6–25
Bituminous concrete patch	90	1	39–141	1–3
Portland cement concrete patch	395	7	322–469	4–10

Source: Courtesy Yash Paul Virmani.

TABLE 3.13
Methods of Cathodic Protection

	Method	
Characteristic	Immersed Current CP	Galvanic CP
Power requirements	External power required	Requires no external power
Driving voltage	Driving voltage can be varied	Fixed driving voltage
Current	Current can be varied	Limited current
Current level	Can be designed for almost any current requirement	Usually used where current requirements are small
Resistivity	Can be used in any level of resistivity	Usually used in low-resistivity electrolytes

Source: Courtesy Yash Paul Virmani.

TABLE 3.14
Cost and Prolonged Life for Different Cathodic Protection Mitigation Methods

Method	Average Cost ($/m²)	Average Prolonged Life (years)	Range of Cost ($/m²)	Range of Prolonged Life (years)
Impressed current CP (Deck)	114	35	92–137	15–35
Electrochemical removal (Deck)	91	15	53–129	10–20
Impressed current CP (Substructure)	143	20	76–211	5–35
Sacrificial anode CP (Substructure)	118	15	108–129	10–20
Electrochemical removal (substructure)	161	15	107–215	10–20

Source: Courtesy Yash Paul Virmani.

electrochemical processes are used to counteract corrosion. One of the most popular methods is cathodic protection (CP), used mainly for existing constructions. There are many techniques for providing CP. They were divided by Virmani and Clemena (1998) into two main categories: immersed current and galvanic. The authors provided a comparison of two CP methods, as shown in Table 3.13. A cost comparison of different CP techniques is shown in Table 3.14.

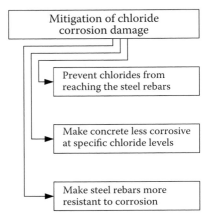

FIGURE 3.17   Chloride mitigation techniques.

3.4.8   CHLORIDE CONTROL

Corrosion mitigation through control of chloride attack can be achieved as shown in Figure 3.17. In addition, chloride can be removed by using an electrochemical process described by Broomfield (1998).

3.5   STRID AND CORROSION HAZARD

3.5.1   GENERAL

Corrosion damage is a slow, time-dependent process that starts locally. After enough damage has occurred, the global effects of corrosion start manifesting. Any Structural Identification (STRID) studies must recognize this local-to-global scaling. Next we discuss the two procedures and their interrelationships with SHM.

3.5.2   LOCAL CORROSION STRID

Corrosion process is initiated locally; as such it is of interest to model the local corrosion effects analytically. Local corrosion analysis can be subdivided into four steps as follows.

Time-to-corrosion initiation: Analytical expressions of time-to-corrosion initiation, T_{CI}, are based on diffusion equations (see Section 3.9). The accuracy of the solution depends on deterministic and probabilistic parameters, as follows:

- **Deterministic:** The diffusion coefficient of concrete, D, needs to be estimated accurately. In addition, the solution of the diffusion equation needs to account for more realistic boundary conditions, Carslaw and Jaeger (1947).
- **Probabilistic:** The uncertainties in the parameters of the diffusion equation result in a random T_{CI}. Many authors recognized the need to solve the diffusion equation probabilistically. For example, Thoft-Christensen (2001) used Monte Carlo techniques to produce a Weibull distribution function of T_{CI}.

Crack initiation: As soon as T_{CI} is reached, the steel rebar material will start an electrochemical process in which rust products will start forming. Since the volume of those rust products is larger than the volume of the steel material, the ratio of this increase ranges from about 2.0 to about 6.5. This increase of volume will exert internal stresses that will eventually cause concrete cracking.

Liu and Weyers (1998) investigated the crack initiation time, T_{CRI}. Thoft-Christensen (2001) modeled this process probabilistically using the Monte Carlo technique.

Crack evolution: As corrosion progresses and continues to increase the loss of the rebar diameter, Δd, the volume of the rebar increases. The volume increase further results in an increase in the crack diameter, Δw. An empirical relationship was suggested by Thoft-Christensen (2001) as

$$\Delta w = \gamma \Delta d \qquad (3.7)$$

The empirical parameter γ is estimated to be in the range of 1.4 to 4.2.

Spalling: Time for spalling in different reinforced concrete components such as slabs, beams, or columns can now be modeled using an appropriate analysis method (finite-element or boundary element methods, for example). The modeling will use the relationship (3.7). The analytical model should be with enough resolution to model individual rebars, the geometry of the concrete cover, and the distances between rebars. Because of the random nature of the material properties at such a small scale, some kind of stochastic analysis might be needed. See Ettouney et al. (1989) for a description of probabilistic boundary element method, which can be formulated and used in such a task.

The above procedure, summarized in Figure 3.18, can be used for understanding the process of corrosion. Evaluation of potential corrosion effects in complex geometries, and for more sensitive and important structures requires a detailed probabilistic/deterministic analysis. This is becoming increasingly possible with the rapid advances in computational capabilities. We note, however, that to execute such an advanced analysis for corrosion with accuracy, there is a need for accurate in-field information. Such information can be provided by monitoring at different stages of the processes of Figure 3.18.

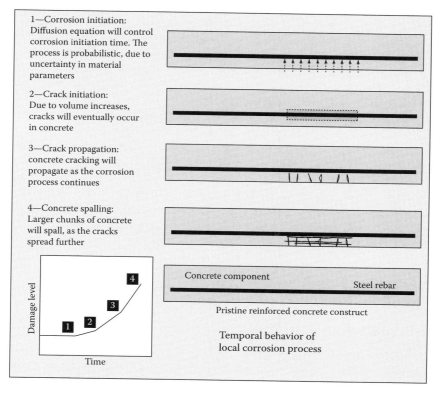

FIGURE 3.18 Local behavior of corrosion process.

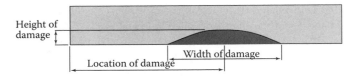

Reinforced concrete beam model for
corrosion parameter identification

FIGURE 3.19 Corrosion parameter identification of reinforced concrete beams.

3.5.3 GLOBAL CORROSION STRID

When corrosion degrades the properties of the structure in a global sense, the degraded structural properties can be identified following a parameter identification technique. The process follows these steps:

- Make an assumption on the general location, size, and shape of the degraded area
- Build an analytical finite-element model using one of the parameter identification methods (see Ettouney and Alampalli 2012, Chapter 6 for more)
- Perform dynamic or static tests of the subject structure and measure pertinent responses
- Use the test results to minimize errors in the finite-element model by optimizing (updating) the location, size, and shape of the assumed corrosion-degraded area

Jahn and Mehlhorn (1998) used a minimization procedure to optimize a finite-element model that identifies the degradations in reinforced concrete beams due to errors in construction planning, material defects, or corrosion. The finite-element model consisted of plane strain elements. They modeled the beam using isotropic material properties in the undamaged areas and orthotropic material properties in the assumed damaged area. The degradation of the concrete was assumed to be present in a parabolic area as shown in Figure 3.19. In addition to the material properties of the beam, the length, height, and longitudinal location of the damaged area were assumed to be unknown and were to be optimized using test results.

Dynamic tests were performed, and the first seven modes (frequencies and mode shapes) were identified. The mode shapes were used to optimize the model and update the assumed damage area. The authors reported that after 29 iterations, the differences between the measured and computed results dropped from a range of 14.22%–7.83% to a range of 0.14%–1.77%. There were considerable differences between initial and final properties. For example, the cracked modulus of elasticity was reduced from 30.00 to 17.10 kN/mm^2. The location and the size of the damaged area were also changed considerably.

The resultant model of the above example, or any other similar method, with its updated properties can be used for (1) accurate structural assessment, (2) decision making about retrofits or maintenance, and (3) evaluation of retrofit projects by using the model as a baseline to compare with future changes of structural properties.

3.6 DECISION MAKING AND CORROSION HAZARD

3.6.1 DECISION MAKING: SURFACE BARRIERS

3.6.1.1 General

In this section, we investigate some decision making techniques as they apply to corrosion mitigation methods that use surface barrier technologies. The decision maker is faced with choosing a method of mitigation out of the several methods of Tables 3.10 and 3.11. To facilitate the decision, the information in Table 3.15 will be relied upon. It will be shown that the final decision will depend mainly on the level of analysis that might be used in arriving at the decision. The level of analysis, in turn, will depend on the accuracy of information available. This information can be improved by additional experimentation. This example will use only the probabilistic/quantitative methods. In the next example the utility methods will be explored.

3.6.1.2 Description of the Problem

Table 3.12 shows the needed data for the retrofit methods. The decision maker needs to choose a single method for retrofit. One simple way is to follow some of the simple decision making methods of Ettouney and Alampalli (2012, Chapter 8). Let us first note that each of the methods has prior data information. This information includes the average and expected limits of cost of each method and the average and expected limits of the effective lifespan of each method.

3.6.1.3 Perfect Information Approach

Let us first consider only the perfect information situation, that is, let us assume that the averages reported in Table 3.12 are the accurate values of both cost and prolonged life. Implicitly, this indicates that there are no uncertainties in the information; thus, the ranges reported in Table 3.12 are ignored. Analytically:

$$\mu_1 \approx x_{1\max} \approx x_{1\min} \tag{3.8}$$

$$\mu_2 \approx x_{2\max} \approx x_{2\min} \tag{3.9}$$

Here we assume that μ_1 and μ_2 are the averages of expected cost and prolonged life. Whereas $x_{1\min}$, $x_{1\max}$, $x_{2\min}$ and $x_{2\max}$ are the range limits of cost and prolonged life, respectively.

One of the decision making tools for perfect (no uncertainty) information, given in Chapter 8 by Ettouney and Alampalli (2012), is the cost per unit time measure. Table 3.15 shows the cost per unit time for each of the methods of Table 3.12.

It is simple now to rank each of the methods using the cost rate as a criterion. It is clear that if the information about each method is perfect, following Equations 3.8 and 3.9, the bituminous concrete with membrane method is the most cost effective. The Portland cement concrete overlay and the polymer overlay/sealer rank very close to each other in the second and third places. The Portland cement concrete patch and the bituminous concrete patch methods product are ranked next; the rates for each of those two methods are much higher than for the first three methods.

If the perfect information assumption is correct, then the decision would be to employ one of the first three methods, with the first choice given to the bituminous concrete with membrane. However, neglecting the uncertainties in cost and prolonged life estimates, as well as the utility of each method, renders the rankings of Table 3.15 at best questionable. We include the uncertainties in the decision making process next.

TABLE 3.15
Cost Per Year for Each Mitigation Method

Method	Perfect Cost ($/m²)	Perfect Prolonged Life (years)	Cost Rate ($/m²/year)	Ranking
Portland cement concrete overlay	170	18.5	9.19	2
Bituminous concrete with membrane	58	10	5.80	1
Polymer overlay/sealer	98	10	9.80	3
Bituminous concrete patch	90	1	90.00	5
Portland cement concrete patch	395	7	56.43	4

3.6.1.4 Probabilistic Approach

The limits of Table 3.15 of both cost and prolonged life indicate uncertainty. Such uncertainties must be taken into consideration in making decisions about the choice of method of retrofit/mitigation. We will assume, for the time being, that the two parameters (cost and prolonged life) in Table 3.15 are independent. We will examine this assumption later. Let us also assume that the probability distribution for each of the two variables is not known. The decision maker will have to assume a probability distribution. For the sake of simplicity, a uniform probability distribution can be assumed. Perhaps a more accurate assumption is a truncated normal distribution.

We will use the Taylor series approach, as defined in Chapter 8 by Ettouney and Alampalli (2012), to evaluate the expectations of the cost per year y. Let us assume that y is defined as

$$y = \frac{x_1}{x_2} \tag{3.10}$$

In anticipation of the Taylor series parameters, we define

$$\frac{\partial y}{\partial x_1} = \frac{1}{x_2} \tag{3.11}$$

$$\frac{\partial y}{\partial x_2} = \frac{-x_1}{x_2^2} \tag{3.12}$$

$$\frac{\partial y^2}{\partial x_1 \, x_2} = \frac{-1}{x_2^2} \tag{3.13}$$

$$\frac{\partial y^2}{\partial x_1^2} = 0 \tag{3.14}$$

$$\frac{\partial y^2}{\partial x_2^2} = \frac{2 \, x_1}{x_2^3} \tag{3.15}$$

Further, we define the mean and the standard deviations of both the cost and the prolonged life as μ_1, σ_1, μ_2, and σ_2, respectively, by utilizing Equations 3.10 through 3.15, respectively in the Taylor series method proposed by Ettouney et al. (1989). The computed values are shown in Table 3.16 for all corrosion mitigation methods.

TABLE 3.16
Statistical Values of Uniform Assumption

Method	Uniform Distribution			
	μ_1	σ_1	μ_2	σ_2
Portland cement concrete overlay	169.00	10.39	18.50	2.60
Bituminous concrete with membrane	58.00	16.17	9.75	3.03
Polymer overlay/sealer	98.00	48.50	15.50	5.48
Bituminous concrete patch	90.00	29.44	2.00	0.58
Portland cement concrete patch	395.50	42.44	7.00	1.73

Applying the Taylor series, we evaluate the mean μ_y and variance Var_y of the cost per year as

$$\mu_y = \frac{\mu_1}{\mu_2} + \left[\left(\frac{-1}{\mu_2{}^2} \right) (\rho\, \sigma_1\, \sigma_2) + \left(\frac{\mu_1}{\mu_2{}^3} \right) \sigma_2{}^2 \right] \qquad (3.16)$$

$$\text{Var}_y = \left(\frac{1}{\mu_2} \right)^2 \sigma_1{}^2 + \left(\frac{-2\,\mu_1}{\mu_2{}^3} \right) (\rho\, \sigma_1\, \sigma_2) + \left(\frac{\mu_1}{\mu_2{}^2} \right)^2 \sigma_2{}^2 \qquad (3.17)$$

$$\sigma_y = \sqrt{\text{Var}_y} \qquad (3.18)$$

$$\text{CoefV}_y = \frac{\sigma_y}{\mu_y} \qquad (3.19)$$

Where σ_y and CoefV_y are the standard deviation and the coefficient of variation of the cost per year, respectively.

We note that the expectations of the above equations are dependent on the coefficient of correlation, ρ. Remember that by definition, $-1 \le \rho \le 1$. This brings up an interesting dilemma for the decision maker. It is clear that ρ needs to be accounted for in the current evaluations; unfortunately, the correlation between X_1 and X_2 is not available. On the basis of this, the decision maker might opt for one of two approaches. The first is to assume that $\rho = 0$, an assumption that is not fairly accurate since it is obviously against common sense. The second approach is to use a reasonable value for ρ. To do so, let us remember that it is common sense to assume that $0 < \rho$ since it is expected that the higher cost would result in a better product and hence a longer life for the retrofit. A negative ρ should not be the case. A limit to the dilemma is to assume that $\rho = 1$. Such a perfect correlation is an upper limit, of course. A more realistic assumption is that ρ is in the range of 0.6 to 0.9.

Using Equations 3.16 through 3.19, and different values for ρ, the expectations for y can be evaluated, as shown in Table 3.17 through Table 3.19. Note that the ranking of different methods, based on the mean μ_y is now different from the rankings based on the perfect information assumption of Table 13.5. The close tie between polymer overlay/sealer method and the Portland cement concrete overlay method no longer exists. The polymer overlay/sealer method has now a clear second ranking. Also, the two least- ranking methods in Table 3.15 have switched positions, with the Portland cement concrete patch holding the last rank. The first method, bituminous concrete with membrane of Table 3.15, still holds the top rank. The implication of this change of rankings is that the perfect information assumption is not adequate for this situation: a probabilistic approach is needed. It is also interesting to see that the results of Table 3.17 through Table 3.19 indicate little effects of the assumed correlation coefficient ρ.

From the information of Table 3.17 through Table 3.19, we can now establish the confidence intervals of the cost rate. In the current problem, we first establish that the decision maker wants to evaluate the different corrosion mitigation techniques at a nonexceedance probability level of 84%. This value is chosen since it is one standard deviation away from the mean, and such values can easily be computed, as shown in Table 3.20. The table also shows the rankings of different methods for different correlation coefficient values. The bituminous concrete with membrane method still has the highest ranking in all cases. However, accounting for the uncertainties switched back the ranking of the next two methods. The obvious reason for this rank switching is that the standard deviation of the polymer overlay/sealer method is much higher than the standard deviation of the Portland cement concrete overlay method. This affected the ranking at 84% nonexceedance level. Such an

TABLE 3.17
Statistical Values of Uniform Distributions Case ($\rho = 0$)

Method	μ_y	Var_y	σ_y	$CoefV_y$	Ranking Based on μ_y
Portland cement concrete overlay	9.32	562.89	23.73	2.55	3
Bituminous concrete with membrane	6.52	324.80	18.02	2.76	1
Polymer overlay/sealer	7.11	1205.37	34.72	4.88	2
Bituminous concrete patch	48.75	700.50	26.47	0.54	4
Portland cement concrete patch	59.96	9528.75	97.62	1.63	5

TABLE 3.18
Statistical Values of Uniform Distributions Case ($\rho = 1$)

Method	μ_y	Var_y	σ_y	$CoefV_y$	Ranking Based on μ_y
Portland cement concrete overlay	9.24	562.17	23.71	2.57	3
Bituminous concrete with membrane	6.01	321.74	17.94	2.99	1
Polymer overlay/sealer	6.01	1198.37	34.62	5.76	2
Bituminous concrete patch	44.50	509.25	22.57	0.51	4
Portland cement concrete patch	58.46	9444.00	97.18	1.66	5

TABLE 3.19
Statistical Values of Uniform Distributions Case ($\rho = 0.5$)

Method	μ_y	Var_y	σ_y	$CoefV_y$	Ranking Based on μ_y
Portland cement concrete overlay	9.28	562.89	23.73	2.56	3
Bituminous concrete with membrane	6.27	324.80	18.02	2.88	1
Polymer overlay/sealer	6.56	1205.37	34.72	5.29	2
Bituminous concrete patch	46.63	700.50	26.47	0.57	4
Portland cement concrete patch	59.21	9528.75	97.62	1.65	5

TABLE 3.20
Nonexceedance Level of 84% and Ranking of Mitigation Methods

Method	$\rho = 0$		$\rho = 0.5$		$\rho = 1$	
	84%	Ranking	84%	Ranking	84%	Ranking
Portland cement concrete overlay	33.04	2	33.00	2	32.95	2
Bituminous concrete with membrane	24.55	1	24.29	1	23.95	1
Polymer overlay/sealer	41.83	3	41.28	3	40.62	3
Bituminous concrete patch	75.22	4	73.09	4	67.07	4
Portland cement concrete patch	157.57	5	156.82	5	155.64	5

observation calls for adequate QC of different mitigation methods so as to reduce the uncertainties of performance. It is also interesting to note that the rankings are not sensitive to the assumed value of ρ. This can have an effect on budgetary decision as well as on the bridge life cycle cost analysis.

The ranking methods discussed so far treated all the corrosion mitigation methods objectively. No place has been given for the qualitative preferences and utility of the professional. The next section illustrates the use of utility in a decision making process.

3.6.2 Decision Making: Cathodic Protection

3.6.2.1 General

This section will apply another approach to the decision making process. The concept of a payoff table for making the decision with the least risk will be applied to a case where the decision maker is required to choose between several corrosion mitigation methods. The three CP methods for substructure from Table 3.13 are analyzed for the most appropriate method to use in a particular situation. We will use the cost and prolonged life limits and averages of Table 3.14 to help in reaching the decision. First, we will present the payoff table approach for the best decision. We then use the utility theory to improve the analysis. In using the utility theory, we will recognize that cost is not the only utility of importance in the decision making process. We will also use the experimental utility function in the analysis.

To construct the payoff table, we have to define the natural events θ as well as the costs that correspond to those events for each of the three substructure CP methods (Table 3.21). We will start by assuming a joint probability function for the cost and the prolonged life as the random variables. Since we do not have any information about such joint distribution, except the limits of each random variable, we can make one of several assumptions:

1. A uniform joint probability distribution as in Figure 3.20
2. A unit joint distribution as in Figure 3.21
3. A more general joint distribution as in Figure 3.22

The first distribution will have uniform marginal distributions for the cost and the prolonged life. All of the distributions will have different correlation coefficients. Let us consider the distribution of Figure 3.20. If we subdivide the prolonged life range into several ranges, we can assign each range

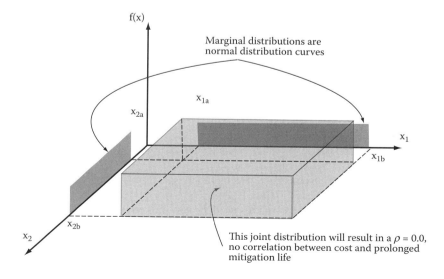

FIGURE 3.20 Uniform joint probability distribution.

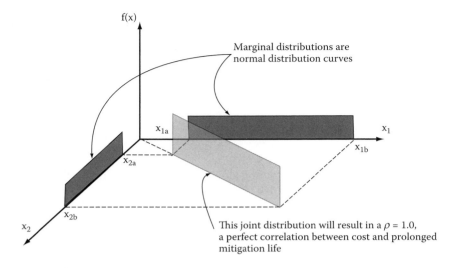

FIGURE 3.21   Unit joint probability distribution.

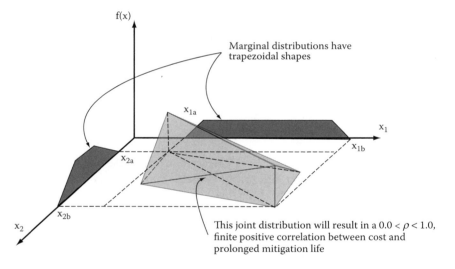

FIGURE 3.22   General joint probability distribution.

as the natural event θ_i. Then by a simple integration we can find the probability of the occurrence of θ_i and the cost that corresponds to such an event. The results of this process are shown in Table 3.21 for all the three CP mitigation methods.

It should be noted that the use of the other two joint probability distributions would have resulted in a similar payoff table.

The expected costs can now be constructed as in Table 3.22 through Table 3.24 for all the methods under consideration.

The sacrificial anode CP method seems to produce the most desirable (least) cost of the three methods. We turn our attention now to using different utilities in choosing the mitigation method.

3.6.2.2   Utility Functions

The decision making process so far utilized the weighted cost and prolonged lives probabilities to decide on the least costly mitigation, given all possible prolonged lives for all methods. The method

TABLE 3.21
Simple Payoff Table for Substructure Mitigation Techniques

	Decision		
	V_1 Impressed Current Cost ($/m²)	V_2 Sacrificial Anode Cost ($/m²)	V_3 Electrochemical Removal cost ($/m²)
Event (years)			
$\theta_1 = 5{-}10$	87.25	None	None
$\theta_2 = 10{-}15$	109.75	113.25	134
$\theta_3 = 15{-}20$	132.25	123.75	188
$\theta_4 = 20{-}25$	154.75	None	None
$\theta_5 = 25{-}30$	177.25	None	None
$\theta_6 = 30{-}35$	199.75	None	None

TABLE 3.22
Expected Costs for Impressed Current cp Method

Event	V_1: Impressed Current Cost ($/m²)		
	Probability	Cost ($/m²)	Weighted Cost ($/m²)
$\theta_1 = 5{-}10$	0.167	87.25	14.54
$\theta_2 = 10{-}15$	0.167	109.75	18.29
$\theta_3 = 15{-}20$	0.167	132.25	22.04
$\theta_4 = 20{-}25$	0.167	154.75	25.79
$\theta_5 = 25{-}30$	0.167	177.25	29.54
$\theta_6 = 30{-}35$	0.167	199.75	33.29
Weighted cost			143.50

TABLE 3.23
Expected Costs for Sacrificial Anode cp Method

Event	V_2: Sacrificial Anode Cost ($/m²)		
	Probability	Cost ($/m²)	Weighted Cost ($/m²)
$\theta_1 = 5{-}10$	0	None	0.00
$\theta_2 = 10{-}15$	0.5	113.25	56.63
$\theta_3 = 15{-}20$	0.5	123.75	61.88
$\theta_4 = 20{-}25$	0	None	0.00
$\theta_5 = 25{-}30$	0	None	0.00
$\theta_6 = 30{-}35$	0	None	0.00
Weighted cost			118.50

relied completely on the joint distribution probability function for cost and prolonged lives. There are several other factors that go beyond such a purely quantitative approach. Two of the most important factors are the decision maker's attitude toward monetary risk and the qualitative attributes of each mitigation solution. We recognize each of these factors as a *utility* that affect the decision. Remember that the mathematical basis of utility was discussed in Chapter 8 by Ettouney and Alampalli (2012). We will use this example to explore further those utilities and their uses.

Let us define the monetary utility function $u(x_1)$, where x_1 is the cost. It was shown that there are three monetary utility types. The concave utility function represents the conservative user, the

TABLE 3.24
Expected Costs for Electrochemical Removal CP Method

Event	V_3: Electrochemical Removal Cost ($/m²)		
	Probability	Cost ($/m²)	Weighted Cost ($/m²)
$\theta_1 = 5\text{–}10$	0	None	0.00
$\theta_2 = 10\text{–}15$	0.5	134	67.00
$\theta_3 = 15\text{–}20$	0.5	188	94.00
$\theta_4 = 20\text{–}25$	0	None	0.00
$\theta_5 = 25\text{–}30$	0	None	0.00
$\theta_6 = 30\text{–}35$	0	None	0.00
Weighted cost			161.00

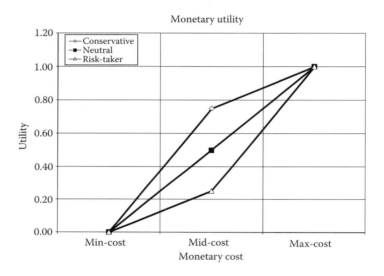

FIGURE 3.23   Monetary utility.

straight line function represents the neutral user, and the convex function represents the risk-taking user (see Figure 3.23). The first step in establishing the monetary utility function is to assign the limits of the experimental costs. These limits will denote the limits of the horizontal axis. The two limits will then be assigned a utility of 0 and 1.0. There are several methods to establish midpoints on the curve. It is important to note that those methods are either approximate or qualitative. For our purpose, for the sake of simplicity, we will merely assign a single midpoint based on the user's attitude toward monetary risks.

1. For a conservative user, assign $u(x_1) = 0.75$ for midpoint on horizontal (monetary) axis
2. For a neutral user, assign $u(x_1) = 0.50$ for midpoint on horizontal (monetary) axis. Note that this situation will have identical results to a cost–benefit analysis of the earlier section
3. For a risk-taking user, assign $u(x_1) = 0.25$ for midpoint on horizontal (monetary) axis

We will then use linear interpolation between the three points to completely define the function. Figure 3.23 shows the resulting monetary utility function.

The next step will be to relate the monetary utility to the actual costs of the three methods. This is shown in Table 3.25

In addition to $u(x_1)$, we also define the utility of experiment as $u(x_{2j})$. Note that x_{2j} is a set of discrete numbers, with $j = 1, 2, \ldots, K$. The number of different methods is K.

TABLE 3.25
Monetary Utility

Monetary Utility at Different Cost Levels					
Impressed Current		**Sacrificial Anode**		**Electrochemical Removal**	
Cost ($/m²)		Cost ($/m²)		Cost ($/m²)	
87.25	0.125	None	None	None	None
109.75	0.375	113.25	0.375	134	0.375
132.25	0.625	123.75	0.875	188	0.875
154.75	0.792	None	None	None	None
177.25	0.875	None	None	None	None
199.75	0.958	None	None	None	None

TABLE 3.26
Parameters that Affect Experimental Utility

Counter	Parameter	Impressed Current	Sacrificial Anode	Electrochemical Removal
1	Contractor's experience	30	60	50
2	Total construction time	70	90	50
3	Ease of inspection	90	70	90
4	Ease of posterior automatic monitoring	50	30	70
5	Ease of quality assurance	30	25	50
6	Propensity to accidental damage	90	50	90
7	Negative effects on other components?	100	70	100

We can express the utility as

$$u\left(x_{2j}\right) = \sum_{i}^{N_E} U_{Eij}$$

$$(3.20)$$

Vector U_E includes weighted values assigned to qualitative parameters that affect the experiment. The qualitative parameters are defined by the decision maker for each experiment. They reflect the nature of the experiment, the experience of the decision maker, and other specifics of the projects (the contractor, proposed schedule, and any limitations). The number of these parameters is N_E. Table 3.26 shows an example of the qualitative parameters of the CP corrosion mitigation methods. Note that $N_E = 7$. Table 3.26 also shows sample weighted values that the decision maker might assign to each of the three mitigation techniques. These values are for illustration only, and more realistic values should be considered in practical situations. For simplicity, each of the numbers in Table 3.26 ranges from 0 to 100. Vector U_E can then be computed by normalizing each column in the table to $100N_E$. The weighted values are shown in Table 3.27. Equation 3.20 can now be executed, and the resultant utility $u(x_{2j})$ is shown in Table 3.28.

Finally, we choose the total utility function to be

$$u(x_1, x_{2j}) = \alpha_1 u(x_1) u(x_{2j})$$

$$(3.21)$$

The factor α_1 is a weighting parameter that should be adjusted such that the bounding values of $u(x_1, x_{2j})$ would be consistent with the physical problem.

TABLE 3.27
Normalized Experimental Utility Vector

Counter	Parameter	Impressed Current	Sacrificial Anode	Electrochemical Removal
1	Contractor's experience	0.04	0.09	0.07
2	Total construction time	0.10	0.13	0.07
3	Ease of inspection	0.13	0.10	0.13
4	Ease of posterior automatic monitoring	0.07	0.04	0.10
5	Ease of quality assurance	0.04	0.04	0.07
6	Propensity to accidental damage	0.13	0.07	0.13
7	Negative effects on other components?	0.14	0.10	0.14

TABLE 3.28
Estimated Experimental Utilities

Method	Impressed Current $u(x_{21})$	Sacrificial Anode $u(x_{22})$	Electrochemical Removal $u(x_{23})$
Experimental utility value	0.66	0.56	0.71

TABLE 3.29
Total Utility Function

Monetary Utility $u(x_1)$	Mitigation Method		
	Impressed Current $u(x_1, x_{21})$	Sacrificial Anode $u(x_1, x_{22})$	Electrochemical Removal $u(x_1, x_{23})$
0 (minimum cost)	0.00	0.00	0.00
0.5 (midlevel cost)	0.69	0.59	0.75
1 (maximum cost)	0.92	0.79	1.00

For purposes of illustration, let us assume that the decision maker is conservative in monetary decisions. In addition, it seems appropriate to normalize the total utility function to have a value of 1.0 as its maximum possible value. This means that $\alpha_1 = 1/0.71$.

Table 3.29 shows the values and shape of $u(x_1, x_{2j})$. Note that the function is continuous for the monetary cost variable x_1 but discrete for the method of retrofit variable x_2.

The total utility function can now be established at different cost levels by interpolating Table 3.28 with the values from Table 3.29. The results are shown in Table 3.30.

We can now utilize the total utility values in the payoff computations. This is done by replacing the cost in Table 3.30 by the total utilities $u(x_1, x_{2j})$ from Table 3.29. The payoff tables are shown in Table 3.31 through Table 3.33.

Consideration of the total utilities of the three methods has changed the rankings of the three methods. Now the highest ranking method (highest utility) is the electrochemical removal CP. The method with the least ranking is the sacrificial anode CP method. Consideration of the monetary attitude of the decision maker and the qualitative experiment have a major effect on the decision making process.

TABLE 3.30
Total Utility Function at Different Cost Levels

		Mitigation Method			
Impressed Current		Sacrificial Anode		Electrochemical Removal	
$u(x_1)$	$u(x_1, x_{21})$	$u(x_1)$	$u(x_1, x_{22})$	$u(x_1)$	$u(x_{1,} x_{23})$
0.125	0.17	None	None	None	None
0.375	0.52	0.375	0.44	0.375	0.56
0.625	0.75	0.875	0.74	0.875	0.93
0.792	0.82	None	None	None	None
0.875	0.86	None	None	None	None
0.958	0.90	None	None	None	None

TABLE 3.31
Expected Total Utility for Impressed Current CP Method

	V_1: Impressed Current Total Utility		
Event	Probability	Total Utility	Weighted Total Utility
$\theta_1 = 5$–10	0.167	0.17	0.02839
$\theta_2 = 10$–15	0.167	0.52	0.08684
$\theta_3 = 15$–20	0.167	0.75	0.12525
$\theta_4 = 20$–25	0.167	0.82	0.13694
$\theta_5 = 25$–30	0.167	0.86	0.14362
$\theta_6 = 30$–35	0.167	0.90	0.1503
Weighted total utility			0.67

TABLE 3.32
Expected Total Utility for Sacrificial Anode CP Method

	V_2: Sacrificial Anode Total Utility		
Event	Probability	Total Utility	Weighted Total Utility
$\theta_1 = 5$–10	0	None	0.00
$\theta_2 = 10$–15	0.5	0.44	0.22
$\theta_3 = 15$–20	0.5	0.74	0.37
$\theta_4 = 20$–25	0	None	0.00
$\theta_5 = 25$–30	0	None	0.00
$\theta_6 = 30$–35	0	None	0.00
Weighted total utility			0.59

3.7 CASE STUDIES

3.7.1 SHM OF ADVANCED BRIDGE AND MATERIALS CONCEPTS

Lewis and Weinmann (2004) offered an in-depth philosophical discussion of the use of innovative technologies for future bridges and materials with some applications to reinforced concrete systems. In addition, they presented examples of the use of SHM for enhancing the performance and health

TABLE 3.33
Expected Total Utility for Electrochemical Removal CP Method

	V_3: Electrochemical Removal Cost ($/m^2$)		
Event	Probability	Total Utility	Weighted Total Utility
$\theta_1 = 5–10$	0	None	0.00
$\theta_2 = 10–15$	0.5	0.56	0.28
$\theta_3 = 15–20$	0.5	0.93	0.465
$\theta_4 = 20–25$	0	None	0.00
$\theta_5 = 25–30$	0	None	0.00
$\theta_6 = 30–35$	0	None	0.00
Weighted total utility			0.75

of infrastructure. They showed how the confluence of the four SHCE components can result in successful projects. We summarize some of their case studies next.

H3 viaduct: This is meant as an R&D project, where one of the largest sections of the H-3 Freeway connects the Leeward and Windward sides of Oahu. This is a segmental cast-in-place post-tensioned concrete box girder bridge. It was constructed using the cantilever construction method. Instrumentation was used to measure its long-term bridge performance in order to monitor creep and shrinkage strains so as to predict the future performance more reliably, to get a better assessment of the remaining structural capacity, and then to improve the design method of future bridges by fine-tuning the computer model for this type of structure.

Performance of new materials: Instrumentation was used to evaluate performance of double tee (DT) beams reinforced and prestressed with CFRP rods and strands. A full-scale DT girder was first tested in the lab to ultimate load, and then 12 DT beams were constructed. The bridge was load tested for 5 years at various times. A 400-sensor array with remote data collection capability was used to monitor the long- term bridge performance. Concrete strain and camber, and pretensioning forces at both dead and live ends of the beam, were measured during the fabrication and pretensioning process, respectively. Permanent load cells were also used to verify forces, force distribution, and long-term losses due to creep and shrinkage. Additional sensors were installed, after erection, to measure deck strains, temperature, and beam deflections.

SAFECO field: This application verifies the operation of seismic dampers on the structure located in an area subjected to seismic activity and high winds. The monitoring system also captured the actual earthquake response to the Nisqually earthquake in 2001. The field has a retractable roof, and the north side panels have eight 800-kip viscous dampers mounted between the horizontal trusses and the down-turned north legs to maintain the roof stability. Normal long-term verification and ensuring proper operation of the dampers included replacing one every year. Hence the continuous monitoring option was used instead. All dampers were instrumented with a wireless LAN remote data collection system to continuously monitor forces, displacements, acceleration along with wind speed to correlate with roof operation.

3.7.2 CONGRESS STREET BRIDGE

The use of FRP composite materials for bridge repair and rehabilitation, especially for bridge decks, concrete strengthening, and column wrapping, has emerged in recent years as a cost-effective solution for much of the damage that affects bridges and their components. Realizing the unavailability of the durability data of the structural components retrofitted with these materials and the lack of standards, numerous monitoring experiments were initiated to monitor and record the effectiveness of several of these applications. One of those experiments (Alampalli 2005) was done by NYS DOT at the Congress Street Bridge in Albany, NY, to evaluate alternative corrosion-damaged concrete removal strategies. One layer of FRP wrapping was then used to confine the future delaminations by sealing the concrete surface, thus extending the service life of the concrete columns in bridges. This has been discussed in detail in the FRP-wrapping applications in Chapter 7, with a brief explanation provided here to illustrate the various phases of SHCE.

Measurements phase

To assess the durability of repairs, data required included corrosion rate, humidity, temperature, and condition of bond between the concrete and FRP surfaces. The corrosion rates of the longitudinal rebar in the column were measured using the corrosion probes. Humidity/temperature probes were used to measure humidity and temperature inside the columns. Visual inspection was done to assess the condition of the bond between the concrete and the FRP surfaces.

Structural identification

This experiment did not include any formal structural identification phase. The reason is that there was no need/attempt to quantify the corrosion damage or the effectiveness of the retrofit solutions. In this experiment, measured quantities were utilized directly to identify damage, reach appropriate conclusions, and make decisions.

Damage identification

The damage identification in this experiment was based on measured/observed corrosion rates at different probes. The effectiveness of the mitigating measures was qualitatively judged by comparing the temporal behavior of corrosion rates and comparing the corrosion rates at each of the three probes.

Results

Corrosion rates, humidity, and temperature were collected from the sensors periodically and analyzed. Visual inspections were also done every time the data was collected.

Conclusions and decision making

The results indicated that FRP wrapping was effective in controlling the delaminated concrete columns, and the concrete removal strategies did not influence durability during the limited 4-year monitoring period. The inspections indicated that the bond between the FRP wrapping and the concrete surface had not deteriorated except for the failure of the paint used on the FRP wrapping. Such conclusions encourage further use of FRP wrapping to control corrosion damage in reinforced concrete columns.

3.8 BRIDGE LIFE CYCLE ANALYSIS AND CORROSION MONITORING

3.8.1 General

This section offers a simple method for estimating life cycle cost (LCC) of corrosion hazard. Several components are needed for this method:

- Corrosion damage as a function of time between retrofits
- Cost of corrosion retrofits

Other parameters such as the total lifespan of a bridge or discount rate are discussed elsewhere in this chapter.

3.8.2 Extension of Service Life after Corrosion Mitigation

Sohanghpurwala, A. A. (2006), Table 3.34, showed a useful relationship between extensions of the service lives of components and the corrosion retrofit system. They also developed a method for estimating service life based on measurements (laboratory or fields). Such a relationship can be used in estimating LCC of corrosion hazard for bridges.

TABLE 3.34
Extensions of Service Life for Different Corrosion Mitigation Systems

Corrosion Control System	Service Life	Comments
Patching	4–10, 4–7	Patching with Portland concrete cement and mortar
Reinforcing bar coating		No information available in literature
Repair of epoxy-coated rebar	>3	Study did not monitor the repair procedure beyond 3 years; therefore, it is difficult to predict its service life
Corrosion inhibitor surface application	4–6	Service life is based on application of the inhibitor in the test patches in highly contaminated concrete
Corrosion inhibitor plus patching	4–6	Service life is based on application of the inhibitor in the test patches in highly contaminated concrete
LMC overlay	20	Based on study of several bridges in the state of Virginia
LSDC overlay	20	Numerous studies corroborate the findings of this study
HMAM overlay	<10, 25	Less than 10 years is based on the failure of the HMA overlay, which would also mean the end of service life of the waterproofing membrane
Penetrating sealers	5–7	The service of 7 years for penetrating sealers is generally accepted
Surface coatings		There are numerous kinds of coatings, and sufficient information is not available to define this category
Corrosion inhibitor overlays		No information available in literature
CP	5 to > 25	There are several types of CP systems, and service life varies from one to another
ECE	10–20	Service life of ECE-treated concrete element is governed by ingress of chloride ions after the treatment. The service life quoted herein is based on no chlorides migrating into the concrete element

CP, cathodic protection; ECE, electrochemical extraction; HMAM, hot mix asphalt with a preformed membrane; LMC, latex-modified concrete; LSDC, low-slump dense concrete

Source: Sohanghpurwala, A.A., Manual on Service Life of Corrosion-Damaged Reinforced Concrete Bridge Superstructure Elements, National Cooperative Highway Research Board (NCHRP), Report No 558, Washington, DC, 2006. With permission from NCHRP.

3.8.3 Method and Examples

We provide a simple method for computing LCC of corrosion of reinforced concrete bridges. Our method is based on a corrosion damage model as in Figure 3.24. We assume that corrosion damage starts increasing at corrosion initiation time T_{CI}. For the sake of simplicity, we assume that the corrosion damage will increase until it passes through two thresholds, D_{TARGET} and D_{MAX}. The maximum allowable corrosion damage is D_{MAX} and this is when corrosion retrofit is initiated. The damage level D_{TARGET} is the desired damage level after retrofitting. Our model assumes that the damaged area is a function of time elapsed when the damage level is D_{TARGET} and D_{MAX}, that is, time between retrofits. A nonlinear quadratic relationship is assumed for this example as shown in Figure 3.25. If there is an accurate estimate of rehabilitation costs per unit area, as in Table 3.12 or Table 3.14, the LCC of corrosion can be estimated. In the following examples, we show how to use the method. In the examples, we make the following assumptions for the sake of simplicity

- No discount rate is included
- Lifespan of the bridge is known in advance to be 26 years from the time of corrosion initiation T_{CI}
- Cost of retrofit is $1.00 per unit area

These assumptions can be released, if desired, without any impact on the accuracy of the method.

3.8.3.1 Nonlinear Deterioration Behavior

We consider the case of Figure 3.24. The time between retrofits is constant at 4 years. Using the relationship of Figure 3.25, the damaged area can be estimated at the beginning of each retrofit. The cost of retrofit and the cumulative cost can then be easily computed. Table 3.35 shows the analysis details. The LCC of corrosion is $12.80.

To evaluate the effects of variable time between retrofits, let us consider a case when the first retrofit is performed after 14 years after T_{CI}, instead of after 10 years, as shown in Figure 3.26. Table 3.36 shows the details of the computations. The LCC of corrosion is $19.20. The variable retrofit schedule LCC is 50% higher than the uniform retrofit schedule LCC.

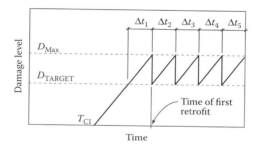

FIGURE 3.24 Corrosion damage for constant rehabilitation schedule.

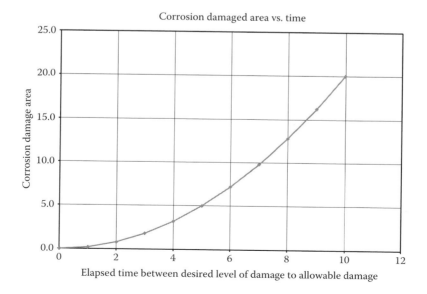

FIGURE 3.25 Corrosion damage as a function of elapsed time between rehabilitations.

TABLE 3.35
Corrosion LCCA—Constant 4-year Rehabilitation Schedule with Visual Inspection

Period	Elapsed Time	Delta-T (years)	Area (m²)	Retrofit Cost	Cumulative Cost
0	0.00	0.00	0.00	0.00	0.00
1	10.00	4.00	3.20	3.20	3.20
2	4.00	4.00	3.20	3.20	6.40
3	4.00	4.00	3.20	3.20	9.60
4	4.00	4.00	3.20	3.20	12.80
5	4.00				
Total	26.00			12.80	

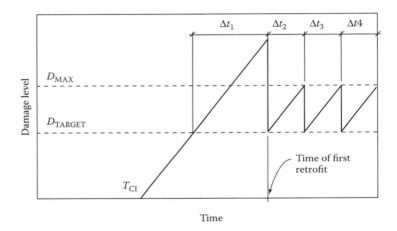

FIGURE 3.26   Corrosion damage for variable rehabilitation schedule.

TABLE 3.36
Corrosion LCCA—Variable Rehabilitation Schedule with Visual Inspection

Period	Elapsed Time	Delta-T (years)	Area (m²)	Retrofit Cost	Cumulative Cost
0	0.00	0.00	0.00	0.00	0.00
1	14.00	8.00	12.80	12.80	12.80
2	4.00	4.00	3.20	3.20	16.00
3	4.00	4.00	3.20	3.20	19.20
4	4.00	4.00			
Total	26.00			19.20	

3.8.3.2   Linear Deterioration Behavior

Next we investigate the sensitivity of LCC to the damage area-time relationship. Let us assume that such a relationship is linear, instead of quadratic, as was assumed in Figure 3.25. The results for both uniform and variable retrofit schedule are shown in Table 3.37 and Table 3.38. Evidently, Life Cycle Cost (LCC) of corrosion is independent of the retrofit schedule for linear damage progression.

This shows the need for accurate estimation of damage progression as a function of time to ensure an accurate Life Cycle Cost Analysis (LCCA) of corrosion.

TABLE 3.37

Corrosion LCCA—Constant 4-year Rehabilitation Schedule with Visual Inspection

Period	Elapsed Time	Delta-T (years)	Area (m²)	Retrofit Cost	Cumulative Cost
0	0.00	0.00	0.00	0.00	0.00
1	10.00	4.00	0.80	0.80	0.80
2	4.00	4.00	0.80	0.80	1.60
3	4.00	4.00	0.80	0.80	2.40
4	4.00	4.00	0.80	0.80	3.20
5	4.00				
Total	26.00			3.20	

TABLE 3.38

Corrosion LCCA—Variable Rehabilitation Schedule with Visual Inspection

Period	Elapsed Time	Delta-T (years)	Area (m²)	Retrofit Cost	Cumulative Cost
0	0.00	0.00	0.00	0.00	0.00
1	14.00	8.00	1.60	1.60	1.60
2	4.00	4.00	0.80	0.80	2.40
3	4.00	4.00	0.80	0.80	3.20
4	4.00	4.00			
Total	26.00			3.20	

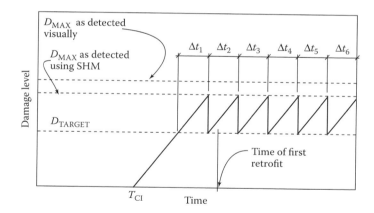

FIGURE 3.27 SHM effects on corrosion damage for constant rehabilitation schedule.

3.8.3.3 SHM Effects on Corrosion LCCA

One of the advantages of SHM is that it offers an accurate estimation of the condition of the monitored system. If an SHM is used to monitor corrosion in the above example, there is a potential for cost saving. For example, if the SHM system enables the official to detect D_{MAX} earlier than a visual monitoring system such that, in a constant retrofit schedule, the time between retrofits is 3.33 years instead of 4 years (Figure 3.27), the LCC of corrosion would be $11.11 for a nonlinear damage progression situation (see Table 3.39). The cost saving is about 13% compared with a visual monitoring situation. Such cost saving can justify the cost of the SHM project.

TABLE 3.39
Corrosion LCCA—Constant 4-Year Rehabilitation Schedule Using SHM and Nonlinear Damage Behavior

Period	Elapsed Time	Delta-T (years)	Area (m²)	Retrofit Cost	Cumulative Cost
0	0.00	0.00	0.00	0.00	0.00
1	9.33	3.33	2.22	2.22	2.22
2	3.33	3.33	2.22	2.22	4.44
3	3.33	3.33	2.22	2.22	6.67
4	3.33	3.33	2.22	2.22	8.89
5	3.33	3.33	2.22	2.22	11.11
6	3.33				
Total	26.00			11.11	

3.9   APPENDIX: CHLORIDE DIFFUSION AND CORROSION INITIATION

Assuming that the concrete is homogeneous and noncracked material, Fick's second law of diffusion can be used; it states that at any given time t

$$\frac{\partial c(x,t)}{\partial t} = D\frac{\partial^2 c(x,t)}{\partial x^2} \tag{3.22}$$

with

x	= Distance from concrete edge
$c(x, t)$	= Chloride concentration at x
D	= Diffusion coefficient.

For a semiinfinite concrete solid, the solution of Equation 3.22 is

$$c(x,t) = c_S\left[1 - erf\left(\frac{x}{\sqrt{4Dt}}\right)\right] \tag{3.23}$$

with erf ( ) representing the error function. For the sake of simplicity, we assume that initially, $t = 0$, the initial chloride concentration at depth $x = X$ is $c(X, 0) = 0$. We also assume that the corrosion will start when the chloride concentration reaches a critical value, $c(X, T_{CI}) = c_{cr}$. Now the time to corrosion initiation T_{CI} can be computed from Equation 3.23 as

$$T_{CI} = \frac{X^2}{4D}\left[erf^{-1}\left(1 - \frac{c_{cr}}{c_s}\right)\right]^{-2} \tag{3.24}$$

The function erf^{-1} ( ) is the inverse of the error function. A more accurate form of solution (3.23) can be obtained for realistic boundary conditions. See, for example, Carslaw and Jaeger (1947).

REFERENCES

ACI (2008). *Building Code Requirements for Structural Concrete and Commentary*. American Concrete Institute, Farmington Hills, MI.

Agrawal, A., Kawaguchi, A., and Qian, G. (2009a). "Bridge Element Deterioration Rates: Phase I Report," New York State Department of Transportation, Project # C-01-51, Albany, NY.

Agrawal, A., Yi, Z., Alampalli, S., Ettouney, M., King, L., Hui, K., and Patel, M. (2009b). "Remote Corrosion Monitoring Systems for Highway Bridges," In *Practice Periodical on Structural Design and Construction*, American Society of Civil Engineers, vol. 14, no. 4.

Alampalli, S. (2005) "Effectiveness of FRP Materials with Alternative Concrete Removal Strategies for Reinforced Concrete Bridge Column Wrapping." *International Journal of Materials and Product Technology*, Inderscience Publishers, Vol. 23, No. 3/4, pp. 338-347.

Bentur, A., Diamond, S. and Berke, N. (1997). *Steel Corrosion in Concrete: Fundamentals and Civil Engineering Practice*, Chapman & Hall, London, UK..

Broomfield, J. (1998). *Corrosion of Steel in Concrete*, E & FN Spon, New York, NY.

Broomfield, J., Langford, P., and McAnoy, R. (1987). "Cathodic Protection for Reinforced Concrete: its application to buildings and marine structures," *Corrosion of Metals in Concrete, Proceedings of Corrosion/87 Symposium*, NACE, Houston, TX.

Carslaw, H. and Jaeger, J. (1947). *Operational Methods in Applied Mathematics*, Oxford University Press.

Crank, J. (1983). *The Mathematics of Diffusion*, Clarendon Press, Oxford, UK.

Ettouney, M. and Alampalli, S. (2012). *Infrastructure Health in Civil Engineering: Theory and Components*, CRC Press, Boca Raton, FL.

Ettouney, M. M., Benaroya, H., and Wright, J. (1989). "Boundary Element Methods in Probabilistic Structural Analysis (PBEM)," *Applied Mathematics Modeling*, 13(7), pp. 432-441

Fazio, R. (1996). *Flexural Behavior of Corroded Reinforced Concrete Beams*, McGill University, Montreal, Canada.

Fuchs, P., Washer, G., and Chase, S. (2000). "Large-Scale Structural Monitoring Using a Scanning Laser Displacement Measurement Instrument," *Structural Materials Technology: an NDT Conference*, ASNT, Atlantic City, NJ.

Ghandehari, M. and Vimer, C. (2002). "Fiber Optic pH Monitoring for Civil Infrastructure," *Proceedings of 1st International Workshop on Structural Health Monitoring of Innovative Civil Engineering Structures*, ISIS Canada Corporation, Manitoba, Canada.

Goto, Y. (1971). "Cracks formed in concrete around deformed tension bars," *American Concrete Institute (ACI) Journal*, 68(1), 244-251.

Jahn, T. and Mehlhorn, G. (1998). "System Identification of Damage in Reinforced Concrete Structures by Measured Modal Test Data," 2nd International PhD Symposium in Civil Engineering, Budapest.

Lewis, A. and Weinmann, T. (2004). "Advances in Bridge Materials and Structural Health Monitoring Systems with Case Studies," *Proceedings, NDE Conference on Civil Engineering*, ASNT, Buffalo, NY.

Liu, Y. and Weyers, R. (1998). "Modeling of the Time to Corrosion Cracking in Chloride Contaminated Reinforced Concrete Structures." *ACI Materials Journal*, 95, 675-681

Micheal, A. P., Hamilton, H. R., Green, S., and Boyd, A. J. (2004). "Long Term Monitoring of the FRP Repair and Corrosion of Steel Reinforcement at the University Boulevard Bridge in Jacksonville Florida," *Proceedings, NDE Conference on Civil Engineering*, ASNT, Buffalo, NY.

Ong, K. and Grimes, C. (2002). "Monitoring of Stress, Temperature, and Chlorine in Concrete Structures with Passive, Wireless Harmonic Sensors," *Proceedings, NDE Conference on Civil Engineering*, ASNT, Cincinnati, OH.

Riahi, M. and Khosrowzadeh, B. (2005). "Development of Acoustic Emission AE Method for the Diagnosis of Corrosion in Steel Pipes," *Proceedings of the 2005 ASNT Fall Conference*, Columbus, OH.

So, H., Millard, S., and Law, D. (2006). "Environmental Influences on Corrosion Rate Measurements of Steel in Reinforce Concrete," *NDE Conference on Civil Engineering*, ASNT, St. Louis, MO.

Sohanghpurwala, A. A. (2006). "Manual on Service Life of Corrosion-Damaged Reinforced Concrete Bridge Superstructure Elements," National Cooperative Highway Research Board (NCHRP), Report No 558, Washington, DC.

Sprinkel, M., Weyers, R., and Sellars, A. (1991). "Rapid Techniques for the Repair and Protection of Bridge Decks," Transportation Research Record 1304, Highway and Maintenance Operations and Research, Washington, DC.

Thoft-Christensen, P. (2001). "What Happens with Reinforced Concrete Structures when the Reinforcement Corrodes?" *Proceedings of the 2nd International Workshop on "Life-Cycle Cost Analysis and Design of Civil Infrastructure Systems,"* Ube, Yamaguchi, Japan.

Thompson, N. and Lankard, D. (1995). "Improved Concrete for Corrosion Resistance," Report No. FHWA-RD-96-207, Federal Highway Administration, Washington, D.C.

Thompson, N., Lawson, K. M., Lankard, D., and Virmani, Y. (1996). "Effect of Concrete Mix Components on Corrosion of Steel in Concrete," Corrosion – 96, Denver, CO.

Trejo, D. (2002). *Evaluation of the Critical Chloride Threshold and Corrosion Rate for Different Steel Reinforcement Types.* Texas A&M University, College Station, TX.

Virmani, Y. and Clemena, G. (1998). "Corrosion Protection—Concrete Bridges," Report No. FHWA-RD-98-088, Federal Highway Administration, Washington, DC.

Vorster, M., Merrigan, J., Lewis, R., and Weyers, R. (1992). Techniques for Concrete removal and Bar Cleaning on Bridge Rehabilitation Projects, SHRP-S-336, National Research Council, Washington, DC.

Watters, D. G., Jayaweera, P., Bahr, A. J. and Huestis, D. L. (2001). "Design and Performance of Wireless Sensors for Structural Health Monitoring," Review of Progress in Quantitative Nondestructive Evaluation, Brunswick, ME, July 29-August 3, 2001, Vol. 21A, pp.969–976, D. O. Thompson and D. E. Chimenti, Eds. (*Am. Inst. Phys.,* Melville, NY, 2001).

Sohanghpurwala, A. A. (2006), "Manual on Service Life of Corrosion-Damaged Reinforced Concrete Bridge Superstructure Elements," National Cooperative Highway Research Board (NCHRP), Report No 558, Washington, DC.

Wiese, S., Kowalsky, W., Wichern, J., and Grahn, W. (2000). "Fiberoptical Sensors for on-line Monitoring of Moisture in Concrete Structures," *Proceedings of 2nd International Workshop on Structural Health Monitoring,* Stanford University, Stanford, CA.

Yunovich, M., Thompson, N., and Virmani, Y. (2005). "Corrosion protection system for construction and rehabilitation of reinforced concrete bridges," *International Journal of Materials and Product Technology,* 23(3–4), 269-285.

4 Prestressed Concrete Bridges

4.1 INTRODUCTION

Prestressed (or Posttensioned) concrete (PSC) bridges have been in use since the middle of the twentieth century. In the United States alone, the number of PSC bridges is almost 20% of all the bridges built between 1950 and 1990 (Yunovich and Thompson 2003). The popularity of PSC systems is attributed to several factors:

- Lower overall construction costs
- More efficient detailing due to lower reinforcement congestions
- Lighter weights, leading to longer bridge spans (see Figure 4.1)

These advantages can be traced directly to the efficient use of the material properties of concrete and steel. Prestressing utilizes the high compressive strength of concrete, while reducing or eliminating the use of the weaker tensile concrete strength. Use of high-strength steel in PSC systems also enables a more efficient system performance. The concept of PSC systems is fairly simple: apply an initial compressive strain field, $\varepsilon_{\text{Initial}}$, on the system before major loading demands affect the system. When these loading demands are applied later during the lifespan of the system, the net strain demands, ε_{Net}, would be mostly compressive, since any tensile strain demand that might be generated by the loading demands would be greatly offset by the initial compressive strains. The concept is shown in Equation 4.1, with the strain demand resulting from the loading being $\varepsilon_{\text{Load}}$. The initial compressive strains are generated by a set of high-strength steel tendons.

$$\varepsilon_{\text{Net}} = \varepsilon_{\text{Load}} + \varepsilon_{\text{Initial}} \tag{4.1}$$

The PSC systems can be categorized in general terms as shown in Figure 4.2. The two basic categories, pretensioned (PRT) and posttensioned (PST), differ in the sequence of applying the initial strain on the system. In PRT, the initial strains are applied before the concrete is cast. In PST, the initial strains are applied after the concrete is cast, and it attained a predetermined strength. PST systems have several subcategories as follows:

- The tendons are inserted in the duct and covered before or after the casting of concrete.
- The tendons are either bonded to the ducts, using the grouting system, or not bonded.
- The tendons are either internal or external (Figure 4.3).

For a detailed discussion of all these categories, their construction techniques and their advantages and disadvantages, the reader is referred to Collins and Mitchel (1991).

For our immediate purpose, we note that PSC bridges are quite different from conventional reinforced concrete systems in several important ways, such as

- Initial stain field
- High-strength wires
- Additional mechanical details, especially for PST systems, such as grout, ducts, anchors, and joints

FIGURE 4.1   Zilwuakee Bridge: a box girder PSC system in Michigan.

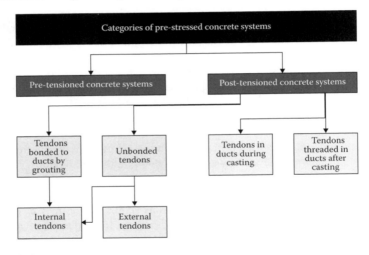

FIGURE 4.2   Categories of PSC systems.

The unique effects of these factors—and how they interact together with the parent concrete material—on the structural health of PSC systems need to be investigated. In the rest of this section, we discuss the challenges of PSC systems to structural health, define some problems that need attention, and then finally offer an overview of the rest of this chapter.

4.1.1   CHALLENGES

Some of the challenges of structural health monitoring (SHM) of PSC systems are

- Field testing of damaged tendons is very costly.
- Analytical tools for damaged (or partially damaged) structures are not well developed, especially when compared with the analytical tools for new structures.
- Ductile, or loss of ductile, behavior of damaged (or partially damaged) structures is not well understood.
- Development of PSC-specific sensors. We note that Matthys et al. (2002) used fiber optic sensors (FOS) for monitoring PSC behavior. More such specific sensors are needed.

FIGURE 4.3 External tendons inside a box girder. (From Hartle, R. et al., Bridge inspector's reference manual, Federal Highway Administration, FHWA, Report No, FHWA NHI 03–001, McLean, VA, 2002, with permission from National Highway Institute.)

4.1.2 PROBLEM DEFINITION

SHM/structural health in civil engineering (SHCE) for PSC systems can be defined as the use of different SHM techniques and components to accomplish one or more of the following:

- Determine existing strength/resilience
- Estimate safety
- Estimate existence and extent of damage of tendons
- Reach a cost-saving decision as to what, if any, repair levels are needed

4.1.3 THIS CHAPTER

To lay the ground for discussions of SHM/SHCE of PSC systems, we begin this chapter by briefly discussing different components of PSC systems. We then present different damage modes and their sources that can affect PSC systems. Structural identification (STRID) specific issues that pertain to PSC systems are discussed next. We point out the limitations of the current STRID methods and the special needs of PSC systems. We then explore some PSC damage detection strategies. Decision making examples of PSC systems follow. In recognition of the several SHM projects under way now, some detailed SHM case studies are presented. This chapter concludes by introducing a PSC-specific life cycle analysis (LCA) method with pertinent practical examples.

4.2 ANATOMY OF PSC BRIDGES

To ensure the healthy and long performance of PSC systems, we have to understand the issues that differentiate these systems from conventional concrete systems. This section presents an overview of different components of PSC systems. Specific construction techniques used for building PSC bridges are offered. Finally, we present PSC-specific SHM issues.

4.2.1 COMPONENTS OF PSC BRIDGES

Figure 4.4 shows the general components of PRT and PST systems. We discuss these components and their structural health issues next. For a detailed discussion of each of the components, see Collins and Mitchell (1991).

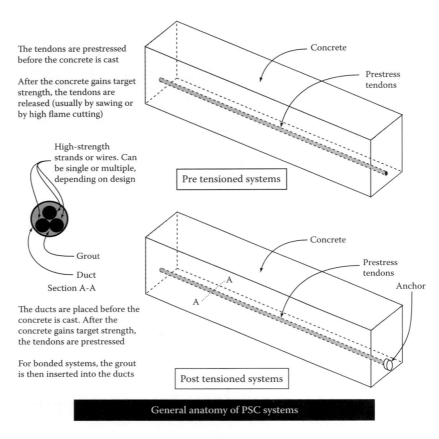

The tendons are prestressed
before the concrete is cast

After the concrete gains target
strength, the tendons are
released (usually by sawing or
by high flame cutting)

High-strength
strands or wires. Can
be single or multiple,
depending on design

Concrete

Prestress
tendons

Pre tensioned systems

Grout

Duct

Section A-A

The ducts are placed before the
concrete is cast. After the
concrete gains target strength,
the tendons are prestressed

For bonded systems, the grout
is then inserted into the ducts

Concrete

Prestress
tendons

Anchor

A

A

Post tensioned systems

General anatomy of PSC systems

FIGURE 4.4   General anatomy of PSC systems.

4.2.1.1   Tendons

High-strength steel used in PSC systems includes wires, strands, bars, or any combination of these. Any such combination is usually referred to as tendons. A wire has a strength of about 250 ksi and a diameter in the range of 0.5 to 0.7 in. A strand is a combination of wires. The most popular strand configuration is the seven-wire strand. Tendons can include a single strand or multiple strands. Straight high-strength steel bars are also used in PSC systems.

Health issues of tendons are

- Loss of strength or ductility due to aging (corrosion, embrittlement, etc.)
- Loss of pretension force

4.2.1.2   Concrete

Concrete of high compressive strengths are used in PSC systems. The strength ranges from 5 to 10 ksi or even higher. Recent advances have seen the development of high-performance concrete of compressive strengths that can be as high as 40 ksi. Health issues of concrete include

- Concrete deterioration
- Corrosion of mild steel reinforcement bonded directly to the concrete
- Creep or shrinkage
- Corrosion, stress corrosion, and hydrogen embrittlement of high-strength steel bonded directly to the concrete in PRT systems

4.2.1.3 Ducts and Grouts

For PST systems, the tendons are placed within metallic or nonmetallic ducts. For bonded PST systems, after the post-tensioning procedure, a grout is poured to fill in the empty volume between the duct and the internal tendons. For unbonded PST systems, the volume between the ducts and the tendons is not filled.

The duct system, bonded or unbonded, is the source of major deterioration problems of PST systems, which are as follows:

- For unbonded systems, the empty void between the duct and the tendons might have moisture that might cause corrosion, stress corrosion, or embrittlement of the tendons
- For bonded systems, the grout might include voids with moisture in them. Again the moisture in those voids might cause corrosion, stress corrosion, or embrittlement of the tendons

Because of this, monitoring the state of the ducts and their interior is important in SHM of PST systems.

4.2.1.4 Anchors

In PRT systems, the interaction forces between the concrete and the tendons are transferred along the whole length of the tendons; no special mechanical devices are needed to ensure such transfer. In PST systems, there is a need for anchors to transfer the interaction forces between the concrete and the tendons (see Figure 4.4). These anchors are located at the two ends of the tendons, which are usually located at the ends of the structural components or at the ends of the construction segments.

Anchors in good state are essential for the adequate performance of PST systems. Some structural health issues of the anchors are

- Loss (relaxation) of tendon forces
- Corrosion
- Cracks due to overstressing

4.2.1.5 Others

In addition to the above PSC components, we note that there are other components that are used in the overall structural assembly. These include

- Diaphragms: Diaphragms help in resisting lateral loads and providing overall stability to the system.
- Deck: Such a deck can be used as a composite deck to reduce the overall weight of the structure.

Ensuring the health of the components is essential for the performance of the overall system, since they play a role in the structural adequacy of the bridge.

4.2.2 Construction Techniques

4.2.2.1 Balanced Cantilever

A balanced cantilever method of construction places a segment of either cast-in-place or precast concrete alternatively at opposite sides of the piers (see Figure 4.5). The process continues until the segments from the opposite piers meet and get connected.

4.2.2.2 Span by Span

In this method, the bridge piers are built first. The spans are then placed on the piers. The spans can be either cast in place or built off-site and then lifted and placed on appropriate piers (see Figure 4.6).

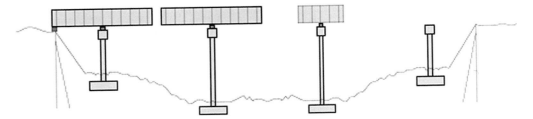

FIGURE 4.5   Balanced cantilever construction method. (From FHartle, R. et al., Bridge inspector's reference manual, Federal Highway Administration, FHWA, Report No, FHWA NHI 03–001, McLean, VA, 2002, with permission from National Highway Institute.)

FIGURE 4.6   Span-by-span construction method. (From Hartle, R. et al., Bridge inspector's reference manual, Federal Highway Administration, FHWA, Report No, FHWA NHI 03–001, McLean, VA, 2002, with permission from National Highway Institute.)

A temporary structure is needed to support the span if it is cast in place. The process continues until all the spans are erected.

4.2.2.3   Progressive Placing

This method adds bridge segments progressively, as shown in Figure 4.7. The negative moment on the superstructure can be fairly high; this would result in potential large displacement at the front end of the construction segment.

4.2.2.4   Incremental Launching

The segments are added at one end of the bridge, and then the whole system is moved over the piers as shown in Figure 4.8. Temporary piers may be needed in this type of construction.

4.2.2.5   SHM and Different Construction Techniques

We just explored different innovative and cost-effective construction techniques. However, we should remember one of the points of Chapter 3 by Ettouney and Alampalli (2012): the construction phase is analogous to the birth of human beings. It embodies two major concerns:

1. A failure during or before the construction is completed or immediately after it is completed.
2. An unnoticed construction defect that occurs during construction. The effects of such a defect can have negative implications on structural health and performance in the long run.

FIGURE 4.7 Progressive placement construction method. (From Hartle, R. et al., Bridge inspector's reference manual, Federal Highway Administration, FHWA, Report No, FHWA NHI 03–001, McLean, VA, 2002, with permission from National Highway Institute.)

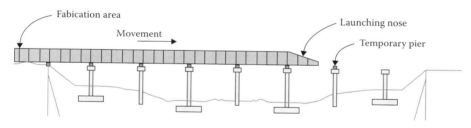

FIGURE 4.8 Incremental launching construction method. (From Hartle, R. et al., Bridge inspector's reference manual, Federal Highway Administration, FHWA, Report No, FHWA NHI 03–001, McLean, VA, 2002, with permission from National Highway Institute.)

Of course, construction quality assurance (QA)/quality control (QC) procedures have striven to lessen the impacts of these two possibilities; this is a subject beyond the scope of this chapter. For more details on construction QA/QC, see Willenbrock (1980).

What concerns us immediately is how to use SHM/SHCE techniques to guard against these two possibilities. Note that many concepts in the following discussion apply to construction of PSC systems as well as several other types of systems.

4.2.2.6 Failure during Construction and SHM

We categorize construction failures as ductile or brittle, as detailed below.

4.2.2.6.1 DUCTILE Failure

Ductile construction failures usually occur with large displacements and/or large strains. They also give indications or warning signs, such as large deflections or rotations of the tip of the double cantilever type construction or large deflection of midspan, and so on. Ductile construction failures are caused by several factors such as

- Loss of stiffness: Examples of stiffness loss are larger settlement of soil under a column footing or loss of pretension force in tendons during construction.

- Gradually applied larger forces unaccounted for in design. Such unexpected forces can be high wind or floods.

This type of construction failure can be monitored by placing displacement, or strain sensors at expected large displacement or strain areas, respectively. Continued correlation with design displacements or strains will ensure adequacy of the construction process in real, or near real, time. Figure 4.9 illustrates this concept.

4.2.2.6.2   Brittle Failure

Brittle construction failure is a dangerous event. It occurs suddenly with little or no warning signs. The causes of brittle failure can be

- Loss of strength: This can happen when the strength of the system is lower than the design strength, for example, if the concrete strength is less than the design strength.
- Loss of global stability: This occurs in low redundancy, fracture critical systems.
- Suddenly applied larger forces unaccounted for in design. This can happened if an earthquake suddenly occurs during construction. Also, an accidental construction crane impact can cause a brittle failure.

This type of failure can be monitored by a combination of the following methods:

- Monitoring both *in situ* capacity and demands. *In situ* strength of construction materials (capacity), such as concrete strength, forces in tendons, and bearing capacity of soils (or pile strengths) need to be carefully monitored. Loads (demands)—including both self-weight weight and environmental demands such as temperature, live loads, and vibration demands—need to be monitored. Real-time comparisons between those demands and capacities need to be performed. We stress the fact that measuring strains or displacements and comparing them with design values cannot give any warning about brittle-type failure: only *in situ* capacity and demands can give such a warning. Figure 4.10 illustrates this concept.
- In certain situations, detecting brittle failure can be done accurately by vibration monitoring. For example, if it is determined that brittle failure (loss of stability) might occur if a

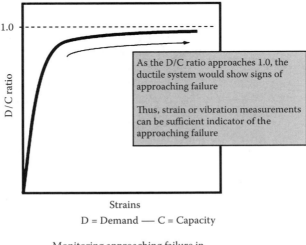

D = Demand —— C = Capacity

Monitoring approaching failure in
ductile systems

FIGURE 4.9   Concepts of monitoring ductile system failure.

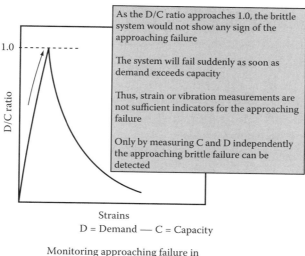

Strains

D = Demand — C = Capacity

Monitoring approaching failure in
brittle systems

FIGURE 4.10   Concepts of monitoring brittle system failure.

particular component in the system fails, vibration monitoring of such a component can detect imminent brittle failure. Failure of low internally redundant trusses is an example of such a situation. Other examples of brittle failure monitoring are given by Ettouney and Alampalli (2012) in Chapters 1 and 8.

- Careful site management that would reduce the potential of any accidentally applied forces, such as crane impact.

4.2.2.7   Construction Defects and SHM

Detecting construction defects that might have an undesired long-term effect can be done by one of three methods:

- Enforcing construction QA/QC
- Global monitoring, such as (1) vibration monitoring, (2) temperature monitoring, and (3) displacement monitoring. Continued global monitoring during construction and validating the results against design values might indicate any construction anomaly that needs to be corrected
- Local monitoring, such as (1) level of forces in tendons, (2) grout voids in bonded PST, or (3) moisture content in unbonded PST. Vigilant visual inspection as detailed by Hartle et al. (2002) is required.

4.2.3   PSC-Specific SHM

One of the most important decisions regarding SHM projects is "what to monitor?" We try to address this question about PSC systems. We approach the answer by addressing each component of the PSC system. For each component, we discuss potential damage and then offer some suggestions on how to monitor such damage. As a rule, SHM planning and strategies must follow the good practices of visual inspection, as detailed, for example, by Hartle et al. (2002).

Areas near supports: These areas are susceptible to higher stresses than the rest of the system. As such, they are subjected to cracking, spalling, and/or corrosion of internal steel (see Figure 4.11). In addition, any imbalance of reaction forces from the bearings can cause cracks that would originate from the bearing area. These signs of damage can be observed visually. Placing strain monitors can reveal excessive strain locations.

Shear zones: Shear cracks and overstress can be monitored visually or by using adequate nondestructive testing (NDT) techniques, such as strain sensors.

Tension zones: Tension zones occur at midspans or above bearing locations for continuous beams. Damage in these locations include cracking, spalling, and delaminations. The cracks can be vertical or longitudinal. Monitoring the damage can be done visually or by using strain or displacement sensors. In addition, it is helpful to place accelerometers (or any other vibration-measuring device) to correlate with strain measurements.

Diaphragms: Damage in diaphragms (see Figure 4.12) can result from relative movements of PSC beams restrained by the diaphragms. A suitable monitoring strategy is to place displacement sensors at appropriate locations on the diaphragms to capture any such relative motion.

FIGURE 4.11   Damage near supports. (From Hartle, R. et al., Bridge inspector's reference manual, Federal Highway Administration, FHWA, Report No, FHWA NHI 03–001, McLean, VA, 2002, with permission from National Highway Institute.)

FIGURE 4.12   End diaphragm in an I-beam PSC system. (From Hartle, R. et al., Bridge inspector's reference manual, Federal Highway Administration, FHWA, Report No, FHWA NHI 03–001, McLean, VA, 2002, with permission from National Highway Institute.)

Drainage areas: Moisture is the main cause of corrosion in PSC systems. So, proper drainage is essential to reduce accumulation of moisture and also corrosion activities. Each PSC system has a drainage mechanism designed to ensure proper drainage. For example:

- Adjacent box beam may develop a drainage problem if the shear keys are damaged, thus causing a leakage between the different box girders (see Figure 4.13).
- The drainage holes in box beams may get clogged.
- Drain holes in different parts of the bridge may be overrun or get clogged.
- Water may seep near the ends of beams or joints (see Figure 4.14).

In many situations, drainage problems can be detected visually. However, hidden areas, such as areas between box beams or inside sealed volumes, can be difficult to detect visually. A moisture sensor system can detect problems long before they damage the structure.

FIGURE 4.13 Leaking between adjacent box girders. (From Hartle, R. et al., Bridge inspector's reference manual, Federal Highway Administration, FHWA, Report No, FHWA NHI 03–001, McLean, VA, 2002, with permission from National Highway Institute.)

FIGURE 4.14 Leaking at end of beam. (From Hartle, R. et al., Bridge inspector's reference manual, Federal Highway Administration, FHWA, Report No, FHWA NHI 03–001, McLean, VA, 2002, with permission from National Highway Institute.)

Traffic in over passes: Damage due to traffic impacting the bottom of PSC beams are common (see Figure 4.15). The impact can cause spalling of concrete cover and perhaps expose or damage several tendons. Visual inspection can reveal these. However, the extent of the impact damage may not be detectable visually. A NDT techniques such as impact-echo (IE), can reveal the extent of damage due to such an impact.

Past repairs: It is a good practice to monitor the condition of past repairs. Visual monitoring may be adequate for simple repairs. However, it may be a good practice to have a more advanced monitoring practice for complex repairs. Such a practice includes intermittent NDT inspection. In more involved rehabilitation projects, a permanent monitoring system may be required.

Shear keys: Shear keys are used in several instances to connect precast PSC systems such as T-, double T- (Figure 4.16), or adjacent box beams (Figure 4.17). Shear keys can be damaged if over-stressed. Placing strain monitors or differential deflection measurement gages at, or near a shear key, can reveal early signs of overstressing.

FIGURE 4.15   Traffic impact damage. (From Hartle, R. et al., Bridge inspector's reference manual, Federal Highway Administration, FHWA, Report No, FHWA NHI 03–001, McLean, VA, 2002, with permission from National Highway Institute.)

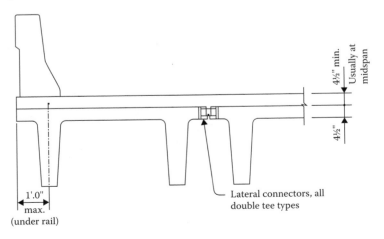

FIGURE 4.16   Shear keys for double T-girders. (From Hartle, R. et al., Bridge inspector's reference manual, Federal Highway Administration, FHWA, Report No, FHWA NHI 03–001, McLean, VA, 2002, with permission from National Highway Institute.)

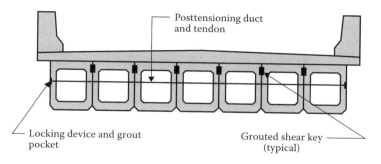

FIGURE 4.17 Shear keys for adjacent box girders. (From Hartle, R. et al., Bridge inspector's reference manual, Federal Highway Administration, FHWA, Report No, FHWA NHI 03–001, McLean, VA, 2002, with permission from National Highway Institute.)

Anchorages: Higher stresses and different material geometries are the main characteristics of the anchorage areas in PSC systems. Visual detection of minor developing damage in the anchorage area may be possible. For embedded anchorages, or for hidden damage, such as internal cracks, a more sensitive detection scheme is needed. Such schemes include intermittent NDT detection (such as IE or electromagnetic [EM] testing). Internal strain sensing may reveal the development of excessive strains. Carefully planned SHM sensing can also detect relaxation of forces in the tendons.

Ducts, grouts, and tendons: Since ducts and grouts are hidden from view, special SHM and NDT methods are needed. Several techniques of damage identification of ducts, grouts, and tendons are discussed later in this chapter.

4.3 DAMAGE TO PSC BRIDGES

4.3.1 GENERAL

There are several causes of damage to PSC systems. Some of these are specific to PSC tendons, and others are generic factors that can also affect conventional reinforced concrete systems. This section addresses the different causes of damage to PSC systems. Whenever possible, we explore the relationships between SHM/SHCE issues and the damage situation under discussion.

4.3.2 CORROSION

Generally, the effects of corrosion in PST or prestressed systems are of more concern than in conventional reinforced concrete systems. This is due to the higher possibility of loss of section or the loss of ductile behavior. In addition, there are several corrosion types that can affect high-strength wires. They are discussed next.

4.3.2.1 Conventional Corrosion

Conventional corrosion (discussed in Chapter 3) affects either PST or prestressed systems similar to the way it affects conventional mild steels in conventional reinforced concrete. In addition, high-strength wires are susceptible to two or more corrosion-type damage: stress corrosion and hydrogen embrittlement. As before, sources of conventional corrosion can be sea water, deicing (chloride) salts, or even freshwater. Penetration through concrete, sheathing, grout plugs, and end of anchors can result in conventional corrosion (see Chapter 3 for more details).

4.3.2.2 Stress Corrosion

Stress corrosion occurs in high-strength wires when the wire is subjected to a corrosive environment. The interactions of the corrosive environment with the high-strength alloy and its coatings can result in pitting and loss of ductility. Pitting will induce cracking, which will result in loss of area. As such, stress corrosion will reduce both the strength and ductility of the wires. Eventually, the wires might break in a brittle fashion at lower stresses than the original designs. Rogowsky and Robson (1998) found that in some cases even though the pitting is only 0%–5% of the cross section, cracks and eventual wire break can result. More recently, Betti et al. (2000) showed that, under controlled conditions, when exposed to pH 3.0 chloride solution, high-strength wires experienced a reduction of ultimate strains of about 50%, from about 6% ultimate strains to about 3% ultimate strains.

4.3.2.3   Hydrogen Embrittlement

Hydrogen embrittlement is a damage process that occurs in high-strength steel tendons, Mahmoud (2003a and 2003b). Mahmoud argued that, with a threshold level of hydrogen concentration, it can degrade the fracture resistance of the high-strength wires used in posttensioning of bridges (also in prestressing of suspension cables). Embrittlement was observed to cause failures of wires at low static loads Those failures were observed to be brittle, even though the failed wires were observed to exhibit ductile properties. More recently, Yamaoka et al. (1988) studied the effects of galvanization of high-strength wires on their susceptibility to hydrogen embrittlement. They reported that zinc coating on galvanized wires interacts with the alkaline cement and produces hydrogen. On the other hand, they reported that galvanization might protect the wires from hydrogen embrittlement.

The loss of ductility as a result of embrittlement is a serious matter. Without accounting for this ductility loss, the factor of safety for the system on hand is obviously overstated. Under certain conditions, this can lead to a brittle failure of the system (see Chapter 3 by Ettouney and Alampalli 2012).

As of the writing of this chapter, the Federal Highway Administration (FHWA) is sponsoring a major research, to study, among other things, hydrogen embrittlement in high-strength wires (see Betti et al. 2000). The research also aims at investigating different monitoring techniques for such a phenomenon.

4.3.3   Loss of Tendon Force

PSC systems owe their superior performance over conventional reinforced concrete systems to only one factor: prestressing tendon force. Hence, it is logical to ensure that the prestressing tendon force is kept at the design level throughout the lifespan of the bridge. Obviously, the level of the tendon force will not increase on its own. However, there are many causes for the loss (sometimes referred to as relaxation) of prestressing force during the lifespan of the bridge. We note that such a relaxation of tendon force can result from direct or indirect sources. In this section, we explore those sources and then offer some SHM-related observations:

4.3.3.1   Direct Sources

Some of the direct causes of loss of tendon forces are

Cracking: Cracking of PSC systems, at any location, can result in loss of tendon forces. Specifically, any cracking near anchors of PST systems, or corrosion-related cracking in PRT systems, can result in an effective reduction of tendon forces, that is, the force transmitted to the concrete component of the PSC system.

Degradation of mechanical properties: Generally, any loss of stiffness or strength of the overall PSC system can result in a reduction of effective tendon forces. An example is a foundation settlement in continuous PSC girders.

Corrosion: Presence of corrosion is an indication of loss of tendon force, as discussed above.

Impact: Impact in the vicinity of tendons or anchor can result in loss of tendon forces.

Creep of concrete: Creep is the deformation of concrete under constant load. It is a time-dependent property of concrete. The effects of creep should be accommodated in any design of PSC systems (Collins and Mitchell 1991).

Relaxation of steel: Similar to concrete creep, relaxation of steel is a property of steel which is defined as the time-dependent steel deformation under constant loading. Steel relaxation also results in loss of tendon force over a passage of time (see Collins and Mitchell 1991).

Shrinkage of concrete: Shrinkage of concrete is the reduction of concrete volume due to the presence of moisture. Concrete shrinkage cause loss of tendon force over the lifespan of the PSC system (Collins and Mitchell 1991).

4.3.3.2 Indirect Sources

Indirect causes of loss of tendon force are a bit more difficult to quantify. Two obvious causes are large deflections and large vibration amplitudes. Large deflections can result from excessive overload that can produce nonlinear bridge behavior. Such nonlinearity can cause a permanent deformation. One potential consequence of such permanent deformation is an effective loss of tendon force.

Large vibration amplitude can be caused by strong earthquake motions or strong wind (for long-span PSC systems). Large vibration amplitude will essentially cause reversal of stress distributions. Stress reversals, if not accommodated properly in the design and detailing of the system, can cause cracking in undesired locations, producing some loss of tendon force.

4.3.3.3 SHM and Loss of Tendon Force

Accurate evaluation of tendon force has not been given much attention in the SHM/NDT community. We observe that accurately detecting the magnitude of tendon forces offers an advantage to measuring corrosion. This is because it can be related directly to the actual capacity of the PSC system. Note that measuring corrosion offers a partial insight into the system capacity. Recall that

$$F = A(x)\sigma(x) \tag{4.2}$$

where F, $A(x)$, and $\sigma(x)$ are the total tendon force, the tendon cross-sectional area, and the average axial tendon stresses, respectively. The distance along the tendon is x. Note that Equation 4.2 implies a constant force along the tendon under consideration. This is a reasonable assumption. However, relaxing such an assumption by assuming a distance dependency of F should not change the basic premise of the current discussion. We expand Equation 4.2 as

$$F = \frac{E A(x)}{\varepsilon(x)} \tag{4.3}$$

The modulus of elasticity is E, and the axial strains are $\varepsilon(x)$.

In many situations, applying monitoring observations to Equations 4.3 can provide direct estimation of the state of health of the PSC system. As of the writing of this chapter, we are not aware of research efforts that have approached PSC monitoring in this manner.

4.3.4 DURABILITY

Much research and many publications have been devoted to the durability of concrete. A summary of important factors that influence concrete durability, their consequence, and mitigating measures is shown in Table 4.1. The table is based on a summary by Salas et al. (2004). For detailed discussions, see Schokker et al. (1999), West (1999), and West et al. (1999).

We note that the durability issues in Table 4.1 are not limited to PST systems; they are equally applicable to conventional concrete systems.

4.3.5 POSTTENSIONING DETAILS

Details of construction of PST systems can also be severely damaged, specifically

Anchorage details: The location and details of anchorage of the tendons can affect the durability of the anchorages themselves, and of the concrete in that region and ultimately of the tendons, if the anchorage details allow moisture to penetrate into the ducts. For an in-depth discussion of this durability issue and some protection measures of the anchorages, see West (1999).

Table 4.1
Durability Issues in Concrete Systems

Durability Issue	Cause	Consequence	Mitigation Measures
Sulfate attack	Sulfate interacts with C3A; in concrete, a chemical process that produces ettringite	Since ettringite has a much larger volume than its constituents, this chemical process will cause concrete cracking	Use cement with low C3A; use adequate additive in the concrete mix
Freezing and thawing	Expansion and contraction of water during the freezing-thawing cycles can cause concrete cracking		Use air-entraining additives to eliminate or reduce the problem
Alkali-aggregate reaction	Alkali interacts with aggregates in concrete, a chemical process that produces alkali-silica or alkali-carbonate	Alkali-silica or alkali-carbonate can expand and thus cause concrete cracking	Use aggregates that are more resistant to alkali
Carbonation	A chemical process in which, in the presence of moisture, carbon dioxide will penetrate, and then interact with calcium hydroxide in the concrete, producing calcium carbonate	Calcium carbonate will accelerate corrosion process	Use concrete with low permeability (low water-cement ratio, compaction, adequate curing, etc.); use sealers
Cracking (due to loading)	As concrete cracks owing to loading on the structure, moisture will penetrate and cause local corrosion		Ensure adequate crack-limiting designs and detailing

Segmental construction joint details: Poorly detailed joints in segmental construction were reported to cause major damage to these types of systems (see Woodward 2001 and Salas et al. 2004). This poor behavior can occur for either internal or external posttensioning. Therefore, discontinuous ducts were banned in the United Kingdom (Woodward 2001).

4.3.6 IMPACT

Impact or collision of vehicles can cause damage to tendons directly. It can also cause spalling of protective concrete or damage to protective ducts and/or grout in PTS. The damage from impact must be detected and then corrected as soon as possible.

Detecting impact occurrence can be done by one of two methods:

- **Visual inspection:** This is the conventional practice. It has the following disadvantages: (1) it is not immediate; some time will elapse between the impact incident and the visual inspection, and (2) it may not reveal the extent of impact severity. The method has the advantage of being simple and economical.
- **Vibration monitoring:** A continuous vibration sensor (for example, accelerometers or acoustic emission [AE] sensors) at appropriate location can also reveal impact events as they occur. This method has the advantage of real-time monitoring and accuracy. The disadvantage is that it may not be economical. Also, if not validated carefully, it can result in false reporting. We note that the efficiency of this type of monitoring can be improved by combining it with other monitoring goals such as global performance monitoring.

4.3.7 SIGNS OF DAMAGE

In many situations, it is easier to monitor the effects of loss of prestress efficiency (which can result from many sources, such as corrosion, loss of area, failure of tendons, or relaxation of prestressed tendons). For example, loss of prestress efficiency, will result in a redistribution of stresses in the PSC component. Such a redistribution of stresses can be monitored easily by observing the state of strains inside the concrete volume. This approach requires placing strain monitors in locations where the redistributed stress paths are expected to occur, such as diagonal tension locations near supports.

TABLE 4.2
Signs of Damage and Potential Monitoring Techniques

Sign and Reason of Damage	Monitoring Technique
Excessive vertical sagging (deflections) indicates excessive overload or loss of prestress forces	Monitor vertical displacements
Horizontal deflections indicate possibility of asymmetric internal stresses that can result from nonuniform prestress forces or tendon failure	Monitor horizontal displacements
Longitudinal cracks in the wearing surface may indicate failure in nearby shear keys	Place strain monitors on shear keys. Investigate overall performance of system to see if it is performing as designed. Carry out corrections as needed
Rust stains near drainage holes. This indicates potential hidden corrosion in tendons/reinforcement nearby	Place internal or external strain monitors to detect cracks. Also, perform corrosion-rate testing as needed

Since lateral strains (normal to the direction of the tendons) might occur due to the corrosion-caused expansion of tendon volume, placing lateral strain monitors might be advisable in certain situations. Table 4.2 shows additional signs of damage and potential monitoring techniques.

4.3.8 Deterioration of Systems: A Generic Damage?

Salas et al. (2004) argued that there is a parallel between conventional design for loading (such as gravity, earthquakes, and wind) and design for durability. The parallels are shown in Table 4.3. In addition, Table 4.4 shows the role of SHM components at every step of the durability design.

4.4 STRUCTURAL IDENTIFICATION

4.4.1 Governing Equations of PT-PS Systems

To establish some principles for STRID methods for posttensioned (PT)-prestressed (PS) systems, we will present some of the basic governing equations. We start by recalling that prestress or post-tension is introduced into the structural component as an initial strain. The stress field in a general PT-PS system can then be described as (Zienkiewicz 1971).

$$\{\sigma\} = [D](\{\varepsilon\} - \{\varepsilon_0\}) + \{\sigma_0\} \tag{4.4}$$

In Equation 4.4, the stress and strain vectors are $\{\sigma\}$ and $\{\varepsilon\}$, respectively. The matrix $[D]$ is the general elasticity matrix. Initial stress and strain vectors are $\{\sigma_0\}$ and $\{\varepsilon_0\}$, respectively. Initial stresses result from sources such as residual stresses. Initial strains result from thermal, creep, shrinkage, as well as prestressing and posttensioning forces. Assuming a linear behavior, the equilibrium equations of a general PT-PS system can be developed from Equation 4.4:

$$[K]\{U\} + \{F\}\varepsilon_0 + \{F\}\sigma_0 = \{F\} \tag{4.5}$$

We recognize the conventional terms $[K]$, $\{U\}$, and $\{F\}$ as the stiffness matrix, the displacement vector, and the applied forces vector, respectively. Vectors $\{F\}\varepsilon_0$ and $\{F\}\sigma_0$ represent the effects of initial strains and initial stresses, respectively. Equation 4.5 forms the basis of conventional PT-PS analysis and design procedures. It should also form the basis of STRID procedures for PT-PS systems, as discussed next.

4.4.2 Conventional Structural Modeling of PT-PS Systems

Conventional modeling of PT-PS systems involves solving a system of equations similar to Equation 4.5. To ensure accurate results, some rules need to be followed while generating the

TABLE 4.3
Analogy between Design Processes of Loading and Durability Hazards

Step	Description	Structural Loading Hazard	Durability Hazard
1	Type of hazard	Gravity loads, earthquakes, wind, bomb blast, collision, floods, etc.	Corrosion, freeze-thaw, creep, etc.
2	Define hazard intensity	Load amplitude, earthquake maximum acceleration, flood levels, etc. For dynamic loads, wave forms should be defined	Define potential of attack for a specific structure and location. Range of temperature variation, chloride content, sulfur content, etc.
3	Analysis for structural response	Conventional structural analysis for the given loading	Analysis of the structural components for the postulated attack level and severity. Some analysis can be qualitative. Statistical analysis may be used in situations where uncertainties are large
4	Material choices for optimal behavior	Use materials suitable for the load demands on hand	Use materials suitable for the durability demands on hand
5	Acceptance limits	Identify appropriate acceptance limits (maximum stresses, displacements, accelerations, strains, etc.)	Identify appropriate acceptance limits (maximum chloride content, minimum concrete cover, location of expansion joints, etc.)
6	Details	Design detailing to minimize potential damage (sizes, connections, foundations, etc.)	Design detailing to minimize potential damage (exposure to salt, exposure to humidity, connections, crack control, etc.)
7	Design iterations	Iterate on designs to optimize performance and costs	Iterate on designs to optimize performance and costs
8	Cost–benefit implications	Cost–benefit considerations and life cycle costs should be an integral part of designs	Cost–benefit considerations and life cycle costs should be an integral part of designs

numerical model. For example, Fanning (2001) recommended explicit modeling of reinforcing tendons rather than smearing concrete and steel. The model should also be capable of simulating crack behavior (Faherty 1972). Depending on the level of loading, nonlinear concrete and steel behavior need to be considered (see Kachlakev et al. 2001). Perhaps the most important requirement is that the analysis should also be capable of including a measure of initial strains.

Generally, the finite-element analyses of PT-PS systems are performed using the following steps:

- Build a model of the concrete and steel components of the system using an adequate analysis computer code
- Apply initial strains within the prestressed/PST tendons
- Apply self-weight as needed. Pay special attention to construction sequences
- Apply other design loads as appropriate

4.4.3   STRID MODELING OF PT-PS SYSTEMS

As was discussed in Chapter 6 by Ettouney and Alampalli (2012), there are three general STRID methods. We are concerned here with the applicability of those methods to PT-PS systems. Specifically, we observe that PT-PS systems are different from all other types of conventional structural systems in one major aspect: they do have initial strains. Such initial strains are their main functional parameter; it makes sense to expect that any STRID method must be capable of incorporating and accounting for such a parameter. In addition, STRID methods must also account for

TABLE 4.4
Roles of SHM Components during Design for Durability

Step	Description	SHM Role
1	Type of hazard	Sensing plays a major role in this step. (corrosion extent, humidity, strains, etc.)
2	Define hazard intensity	A priori decisions as to the type of expected deterioration attack might be needed. The designer, for example, needs to determine if the goal is to monitor hydrogen embitterment, rust, chloride penetration, sulfur, etc. Admittedly, this can be too demanding a task. Virtual sensing paradigm can help in monitoring more than one attack by monitoring the virtual environment for several attacks at once (see Chapter 5 by Ettouney and Alampalli 2012)
3	Analysis for structural response	Analyses of durability attacks are mostly empirical as of the writing of this chapter. Use of statistical analysis is also desired owing to the high uncertainties in durability parameters and the structural response to them. Because of this, the use of monitoring to evaluate structural response directly (rust, or corrosion rates, for example) can result in accurate estimates for the structural response. Also, decision making tools (Chapter 8 by Ettouney and Alampalli 2012) can help in accommodating the highly uncertain nature of structural response to durability hazard attacks
4	Material choices for optimal behavior	Many new materials are used to improve structural durability. For example, FRP materials (see Chapters 7 and 8) or corrosion-inhibiting materials (Chapter 3) are used. Short-term and long-term performances of the new materials need to be monitored and evaluated closely. Several examples of monitoring and evaluation of new materials are given in Chapters 3, 7, and 8
5	Acceptance limits	Acceptance limits of durability-related hazards are mostly qualitative and are based on statistical measurements that might not be suited for the particular structure and environment on hand. Because of this, a well-designed monitoring and evaluation system can help in providing an adequate decision making process that is both accurate and cost conscious
6	Details	Complex structural components, especially connections, seats, and/or bearings, are the most susceptible to durability hazard attacks. In many situations, it is difficult to observe physically the damage in these complex systems. Well-designed monitoring and evaluation systems can help in providing an adequate decision making process that is both accurate and cost conscious
7	Design iterations	Long-term or intermittent monitoring might not be of value during design iterations
8	Cost–benefit implications	Considerations of costs and benefits computed using decision making processes (Chapter 8 by Ettouney and Alampalli 2012), and LCA (Chapter 10 of this volume) are related to the whole issue of durability designs. Both decision making and LCA are basic components of SHM

- PT/PR force in tendons
- Details of intermediate geometry between the concrete and the tendons (sheathing, ducts, grout, etc.)
- Anchor details at ends of tendons

In the light of this, we discuss each of the methods next.

4.4.3.1 Modal Identification

The use of dynamic modal identification methods, by default, will succeed in producing accurate modal parameters, which are, by default, global parameters. Relating the modal parameters to the specific tendon parameters, for example, posttensioning force, complex interaction mechanisms, or state of cracks, is extremely difficult. More importantly, it is not mathematically clear how modal identification methods can relate to the initial strains problem in general. Because of this, using modal identification techniques for STRID of PT-PS systems must be done with clear goals and objectives *before* starting any project of this kind. Specifically, we offer the following observations:

Modal damping: Using modal damping computed as a result of a modal identification process in a PT-PS system is very valuable. It would represent the actual damping in the system, thus producing accurate results during analysis or design procedures.

Mode shapes and natural frequencies: Using identified mode shapes as a part of the analysis or design process should be done carefully. Note that the identified mode shapes, [Φ] and the natural frequencies, $\langle \omega \rangle$, are related to the matrix [K] in the sense

$$[\Phi]^T \langle \omega \rangle [\Phi] = [K] \tag{4.6}$$

As such, they do not have information within them regarding the all-important initial strains (prestress or posttension). Because of this, they can result in inaccurate design results. To understand this point further, consider Figure 4.18. It shows schematically the generic force-displacement behavior of a PT-PS system. Initially, at point "A," the system is experiencing a camber, Δ, (negative displacement) due to the initial tendon force. As the external force increases, the displacement increases until it reaches zero displacement at point "B." The first crack in the system is at point "C." The tendons would yield at point "D," then the whole system fails at "E." The stiffness matrix [K] represents the slope of line A-B-C. Thus, the identified mode shapes, in conjunction with the identified frequencies, can only represent the slope of line A-B-C. They do not include any information within them regarding the camber Δ.

Let us now assume that at the time of the STRID experiment, there has been a change in the state of the system such that it is represented by the dashed line A'-B'-C'-D'-E' in Figure 4.18. Such a change in state can result from loss of tendon prestress, creep, shrinkage, and so on. Such changes will not produce a significant, if any, change in initial stiffness; thus the measured modes, shapes, and natural frequencies should remain essentially the same. However, those changes would result in a shift in the initial line from A-B-C to A'-B'-C'. Reduction of strength would result as shown. This means that even though the capacity of the system has decreased, the STRID modal results have not changed. This can result in a false sense of safety in the design professional.

We note that in an effort to identify effects of damage on natural frequency of PSC systems, Khalil et al. (2002) performed static and dynamic tests. The authors detected a 5% frequency shift (10% stiffness loss) in a bridge with damaged tendons. They needed a baseline bridge to compare a pristine bridge with the damaged bridge. Also, Zhou, Wegner, and Sparling (2004) reviewed several vibration-based methods to detect damage in PSC systems.

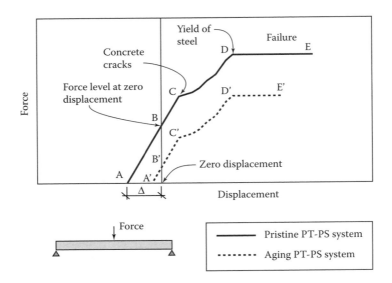

FIGURE 4.18   Behavior of pristine and aging PT-PS systems.

4.4.3.2 Parameter Identification

Using direct methods (identifying parameters needed for FE method, stiffness/mass/damping matrices) would require an accurate description of the PT/PR tendons and of the complex interaction mechanism with the concrete through the sheathing, ducts, anchors, and so on. More importantly, we recall that the parameter identification techniques are essentially minimization processes, mostly in the least square sense (see Ettouney and Alampalli 2012, Chapter 6). As such, it is essential to include all parameters on the left-hand side of Equation 4.5 in the minimization processes. We note that only the first term, the stiffness matrix term, is currently used in most parameter identification methods. Including only stiffness matrices, without the initial strains there is a good possibility that the results would converge to the wrong solution.

4.4.3.3 Neural Networks

Since neural networks methods do not include any formal definitions of equilibrium of the system on hand, the initial strain term of Equation 4.5 could not be included in any solution. Of course, the very lack of formalism is itself one of the limiting factors of the neural network method.

4.4.4 Load Testing

Perhaps the most popular form of STRID method for PT-PR systems is also the simplest: static identification method. In this type of STRID, the bridge is loaded by known weights (trucks) as a live load. The dead load of the bridge is estimated. The vertical deflections are then measured, δ, usually at bridge midspan. A finite-element model is generated and validated against the testing weights and displacements. This process usually requires simple adjustments of the numerical model of the bridge to produce adequate force-displacement relations that are similar to the test.

The main disadvantage of this method is that the modifications in the numerical model are usually arbitrary, and even though the analysis versus test results may be similar, the numerical model itself may not be accurate. The simplicity of the method is the main advantage of this approach.

4.5 DAMAGE DETECTION

4.5.1 PSC-Specific DMID Attributes

One of the general themes of this chapter has been that efficient damage identification needs to be consistent with all the attributes of the system. Even a casual look at the PSC system reveals that it is a compound of two subsystems: the conventional concrete (with its conventional steel rebars, if any) subsystem and the high-strength tendons subsystem with all its necessary mechanical details (grout, ducts, anchors, etc.). The concrete subsystem is a three-dimensional system, while the tendons subsystem is essentially a one-dimensional system. DMID methods in PSC systems can also be subdivided into three-dimensional or one-dimensional systems. Figure 4.19 shows the concepts of one-dimensional and three-dimensional detection. Generally, DIMID methods that account for this special attribute of PSC systems would be more efficient in achieving its primary goal than the more generic DMID methods. In the rest of this section, we discuss some of the DMID methods that have been used to detect damage to PSC systems. Also, Beard, et al (2003) used ultrasonic guided waves to measure attenuation of waves in bonded tendons. Such attenuation is caused by leakage into surrounding material as well as material losses (damping). Another study of the use of NDT methods for DMID in PST system was offered by Ali and Maddocks (2003).

Streicher et al. (2006) performed a study that accounted for the PSC-specific attributes. The authors used radar, ultrasonic echo, and IE to scan a box girder bridge to identify reinforcement bars and tendon grout ducts condition. The results showed that radar and ultrasonic echo complement each other. Another study that looked at PSC-specific DMID was performed by Holst et al. (2004). The authors explored three aspects of PT/PS monitoring: measurements of local force, using

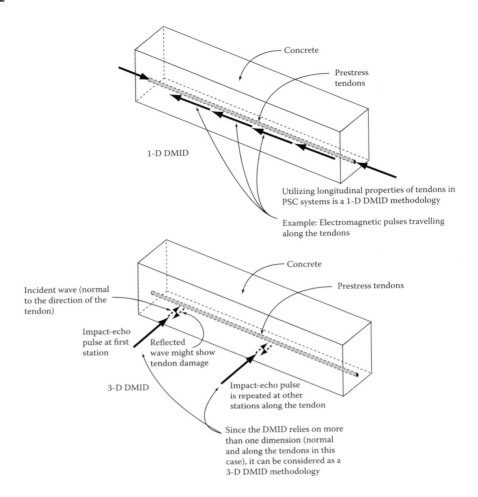

FIGURE 4.19   One-dimensional versus three-dimensional DMID methodology in PSC systems.

magnetoelastic techniques, corrosion in tendons using reflectometry, and fracture in tendons. All tests were performed under realistic and ambient conditions. Also, Beard, et al (2003) used ultrasonic guided waves to measure attenuation of waves in bonded tendons. Such attenuation is caused by leakage into surrounding material as well as material losses (damping). Another study of the use of NDT methods for DMID in PST system was offered by Ali and Maddocks (2003).

4.5.1.1   Use of Dispersion Properties of Tendons

Continuing the same logic, we note that for efficient detection of damage in grouted PST tendons, it would be efficient to accommodate the following properties:

- Axial pretension force in the tendons
- The impedance mismatch generated from the radial pressures exerted by the hardened grout on the tendons
- The increased attenuation of ultrasonic waves due to the radial pressures

The last two issues were mentioned by Kroopf et al. (2005). They discussed the potential for using guided ultrasonic waves in detecting deformation, corrosion, and fracture of thin wires. Generating guided ultrasonic waves in thin wires (tendons) can be done by using magnetostrictive effects. The generated guided waves are dispersive, that is, their group velocity is frequency dependent (see Rose 1999). For example, longitudinal guided wave in a wire with a diameter a will have an nth wave mode group velocity

$$C_n = -\frac{2 i k_n \alpha_n J_n (\alpha_n a)}{(\beta_n - k_n^2) J_1 (\beta_n a)} A_n \tag{4.7}$$

with

$$\beta_n^2 = \frac{\omega^2}{c_T^2} - k_n^2 \tag{4.8}$$

$$\alpha_n^2 = \frac{\omega^2}{c_L^2} - k_n^2 \tag{4.9}$$

$$c_T^2 = \frac{G}{\rho} \tag{4.10}$$

$$c_L^2 = \frac{E}{\rho} \tag{4.11}$$

Note that $J_n(x)$ is Bessel function of order n for the variable x. The modulus of elasticity, shear modulus, and mass density are E, G, and ρ, respectively. The driving frequency is ω, and the nth wave number is k_n. The use of the group velocity of a generated guided wave in a pitch-catch and one-dimensional time-of-flight techniques can reveal damage and their locations along the tendon.

4.5.1.2 Detection of Prestressing Force

Another PSC-specific DMID was an attempt to estimate tendon forces. Riad and Fehling (2004) tried to determine the effective tensile force in external prestressing cables on the basis of vibration measurements. Vibration method in these cases is simple, speedy, inexpensive, but not satisfactorily accurate. Since there are no accurate methods to determine free vibration lengths, the authors developed an algorithm to obtain the actual vibration length. In this case, the test mode shapes were found and were used to calibrate the finite-element model, considering the fact that its bending rigidity is corrected to determine the actual cable length.

4.5.2 DMID METHODS

4.5.2.1 Acoustic Emission

AE is stress waves that are generated in the range of 15 kHz. to 1.0 MHz It results from slow or rapid change in structural properties. When such occurrences take place, an acoustic wave propagates away from the source of the structural change. A listening AE transducer can then detect the event. One of the easiest and most beneficial uses of AE is for detecting the occurrence of damage and locating it (see Ettouney and Alampalli 2012, Chapter 7, for more details on AE).

The use of FOS and AE in detecting damage of tendons was demonstrated by Duke (2002). The experiment used two types of FOS. A Bragg-grating strain sensor was used to measure strains along the tendons. The strains were measured at 1.0-cm intervals. As such, a good strain resolution of the strands was observed. As observed earlier, strain measurements will give an indication of structural response to loadings, that is, the status of the material at any given moment on the strain spectrum.

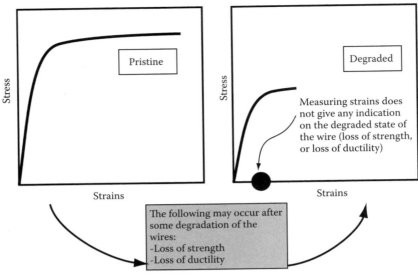

Effectiveness of measuring strains of damaged tendons

FIGURE 4.20   Effectiveness of strain measurements.

Knowledge of strain alone does not give any information about changes in the properties of tendons, such as loss of ductile behavior, or loss of strength due to stress corrosion, or hydrogen embitterment. Figure 4.20 illustrates this concept.

Duke (2002) was able to detect actual damage to tendons by using another type of FOS: an AE sensor. The AE sensor used Fabry-Perot interferometer to detect tendon breaking successfully (failure limit state; see Table 4.5).

4.5.2.2 Electromagnetic Detection

EM damage detection in high-strength tendons is based on the concept that it is the opposite of the AE concept. EM signals, usually compound pulses, are generated through one end of the tendons (the method is thus an active method as compared to the passive AE method). By observing the reflections of the signals, damage locations and sizes can be estimated (see Figure 4.21). Since the signals travel along the tendons, this method can be considered a one-dimensional DMID method. It has the advantage of detecting damage in intermediate states, rather than waiting for the wires to fail. This method can also be used for both bonded and unbounded PSC systems. For example, Rahman and Pernica (1998) used the method to detect corrosion in unbonded PSC systems. They reported that their experiment showed some sensitivity to environments such as contact of ducts/tendons with non-PST steel. The experiment showed a high degree of inaccuracy, especially for low loss of area (less than 6%).

Wang et al. (2004) studied the application of EM sensor on cable force measurements for large bridges. The most important characteristic of this technique is that the magneto elastic characterization of the field material to be measured can be done in the lab. Thus, one can use a smaller size (such as a small wire) in the laboratory for calibration and then use it for stress monitoring of large similar material in the field. Thus, it may be useful for PSC and cable/suspension bridges.

The authors tried it on cables manufactured for QiangJiang Bridge in China with calibration done at the cable-manufacturing facility. The results indicated that the temperature will influence stress measurements, but the influence is the same for a similar type of material with different dimensions. Calibration from a single rod can be used to predict the calibration of EM sensor for cables that are composed of the rods.

TABLE 4.5
NDT and SHM Methods for Corrosion in PT/PR Concrete

Method	Locates Voids?	Suitable for Ducted Tendons?	Detect Corrosion?	Quantifies Tendon Losses?	Suitability for SHM
AE	No, in AE passive mode. Can be designed for active pulsing which might detect voids	Yes	Yes (see Ettouney and Alampalli 2012, Chapter 7)	Yes, especially at failed limit state	Yes, can be automated, especially when detecting failed limit state
Penetrating radiation/ Computed tomography	Yes	Yes	Yes	Yes (if large)	Yes, with adequate safety measures
Electromagnetic (penetrating radar)	Yes	Yes	Yes	Varied results	Yes
Moisture level measurement	NA	Only unbonded ducts	Only qualitatively	No, only qualitative	Not recommended owing to the manual effort involved and the limited information this method might generate
Static magnetic fields	No	No	Yes	No, only qualitative	Labor intensive
Linear polarization	No	No	Yes	No, only qualitative	Labor intensive
Electrical resistance	No	NA	Yes	Yes	Labor intensive
Surface potential survey	No	No	Yes	No, only qualitative	Yes, with some manual effort
Impact-echo (ultrasonic)	Yes	Yes	Yes, with appropriate calibration	Varied degrees of success	Yes, with some manual effort

Source: Source: Based on Ali, M.G. and Maddocks, A.R. Evaluation of corrosion of prestressing steel in concrete using nondestructive techniques, Concrete in the 3rd Millennium, Proceedings of the 21st Biennial Conference of the Concrete Institute of Australia, Brisbane, Australia, 2003. With permission.

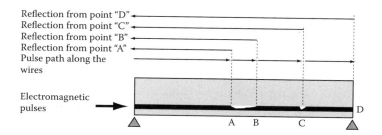

By recording all the reflections and comparing them with the initial signals, the location and sizes of damages in the wires can be detected

FIGURE 4.21 Electromagnetic DMID in PSC high-strength wires.

4.5.2.3 Moisture Level

This method is based on the fact that the presence of moisture in unbonded ducts is one of the main causes of corrosion in tendons. Thus, measuring moisture in the airspace between the ducts and the tendons can indicate potential corrosion damage. Rahman and Pernica (1998) used dry air pumps for collecting the exiting air. By measuring the moisture content of the exiting air, some estimation of corrosion potential in the tendons can be made.

The disadvantages of this method are

- Applies only for unbonded PST systems
- Gives only qualitative indications of corrosion
- Cannot predict accurate locations of the damage
- Can underpredict damage, since moisture, and hence damage, might have existed in the past

Betti et al. (2000) attempted to correlate environmental conditions with corrosion of high-strength wires of suspension bridges. Among environmental conditions considered by the researchers is the moisture level between the wires. Although the results of the research still do not apply to bonded PST or PRT systems, it is of interest to see how moisture levels can be correlated objectively to corrosion in high-strength wires.

4.5.3 OTHER METHODS

Ali and Maddocks (2003) presented a survey of different NDT methods for detecting corrosion damage in prestressed tendons. Table 4.5 is a modified version of the table they presented in their study. Some changes in the entries of the table reflect the opinion of the authors, as noted. A new column in the table also addresses the suitability of the NDT method in an SHM environment is added. In what follows, we briefly discuss some of the methods of Table 4.5.

4.5.3.1 Penetrating Radiation

Penetrating radiation, such as X-ray or gamma ray, can reveal the presence of voids as well as other damage in the tendons. The method is used in a three-dimensional mode since the waves do not follow the direction of the tendons. See Ettouney and Alampalli (2012, Chapter 7, for details of this method. Low resolution of resulting images can limit the objective evaluation of loss of area of the tendons. A practical limit to the detectable thickness is 0.6 m. Among the limitations of the method are (1) appropriate angle of incident beam, since damaged tendons can be shadowed by pristine tendons, (2) access to both sides of the test object is needed, and (3) safety concerns can limit the use of this method. This method is applicable to all forms of PSC systems. Computed tomography was also used to detect damage in PSC systems (see Buyukozturk 1998 and Martz et al. 1993).

In a comparative study by Saleh et al. (2002), the capability of the radiographic method using high energy x-ray linear accelerator to detect grout voids and broken strands in PST concrete bridges, and the feasibility of using real-time imaging technology in data collection were investigated. The tests were conducted as part of an autopsy of the posttensioning in the Fort Lauderdale Airport Interchange Ramp D Bridge, which is a curved, continuous, balanced cantilever, concrete segmental box girder superstructure. The radiographic images obtained included several features inside the concrete slabs, including reinforcement bars, posttensioning ducts, grout voids, wire cuts, and missing duct wall.

Some defects were introduced before the radiographic testing in certain locations. This was not available to investigators prior to radiographic testing and analysis. Some endoscopic inspections and core drilling were also performed to get an idea of the defects before the radiographic testing. Table 4.6 below summarizes this.

The results indicate that it is possible to detect defects in posttensioning tendons in segmental bridges. These can be voids, broken wires, foreign objects etc. Detecting concrete and grout voids is

TABLE 4.6
Results of Testing

Location ID	Induced Defects by DMJM + Haris	Defects Detected by Endoscopy	Defects Detected by Radiography	Exposure Time in Minutes
89R1	Wire cut, grout void	Grout void	Wire cut, grout void	5
88-L1	Wires cut	Grout void	Grout void	6.5
88-13C	None	Grout void	Grout void	1.5
88-13D	None	Grout void	Broken duct, grout void	18
87-11A	None	NA	Grout void, wire coil	3.3
86-L3	Wire cut, grout void	NA	Wires cut	4.6
86-L9	Wire cut, grout void	Grout void	Wire cut, grout void	2.5
85-5A	None	None	Film rejected	4.5
79-L9	None	Grout void	Wire cut, missing section of rebars	2.2
79-5B	None	None	Grout voids	4.5
79-13A	None	None	Break in conduit wall, concrete void	3
77-11B	None	NA	Concrete void	8.3

Source: Saleh, H., Goni, J., and Washer, G. Radiographic Inspection of Posttensioning Tendons Using High Energy X-Ray Machine, Proceedings, NDE Conference on Civil Engineering, ASNT, Cincinnati, OH, 2002. Reprinted from ASNT Publication.

easier than detecting broken wires. Film interpretation requires a thorough knowledge of the structural design of the structure being inspected.

4.5.3.2 Ground or Surface Penetrating Radar

EM pulses are used in this method. The pulses are generated at the higher end of the EM wave range (~1.0 MHz range). The pulse source is located at the surface of the concrete, near where the tendon damage is assumed to be located. The pulse penetrates the concrete and is reflected when it encounters the steel tendons, the voids, or any other sudden change in materials. The scattered waves should reveal those changes in material states. We note that this mode of DMID is a three-dimensional mode where the pulse source should be moved along the concrete surface to follow the approximate direction of the tendons. The maximum depth at which this method can detect tendon states is reported to be about 1.0 m (see Flohrer and Bernhardt 1992). The presence of metallic ducts makes DMID less accurate. Hillemeier (1989) reported that the use of nonmetallic ducts would render DMID using this method impossible.

4.5.3.3 Ultrasound

Washer and Fuchs (2004) performed a laboratory study using electromagnetic acoustic transducers (EMAT) to detect issues with steel tendons embedded in concrete. The EMAT sensors were placed on strands before the stressing operations. One of them acted as transmitter and two as receivers. The distance between the transducers is fixed, and this changes when the strand is tensioned. Time-delays in the detection of the ultrasonic pulse resulting from both the acoustoelastic effect and the strain effect. The results show that ultrasonic pulses have the potential to detect load in unbonded prestressing strands. This is done based on the linear behavior between pulse delay and strand loading. But, it was found that the EMAT sensor is not capable of detecting the ultrasonic pulses once the concrete is placed around the strand.

In another experiment, Fisk et al. (2002) used sonic/ultrasonic testing to identify voids in tendon ducts, so that repairs can be done to prevent reduction in long-term durability. The authors felt that this is broader than IE as the concrete condition can be determined and energy source adjusted to overcome high attenuation and still provide a wide-frequency band for high resolution. It is capable of defining frequencies for larger elements and at the same time providing relative concrete strength estimated from the transmission velocity values provided.

The tests described here were performed on bridge segments made available from the Central Artery Tunnel project in Boston.

The grout conditions defined were

- Fully grouted duct
- Slightly voided grout with air entrapment
- Voiding substantially higher than above or larger and with less than 1 m of longitudinal extent; the void may be dry or water filled
- Similar to above with greater longitudinal extent; the duct may be dry or water filled

If water presence is noted, these ducts require remedial action to prevent future corrosion and thus reduced durability.

During the testing, resonant frequencies can emanate from the duct dimensions (diameter, length, circumference, and air column presence), duct cover, wall, floor, and ceiling.

Several configurations were tried, and it was decided that placing the sensor on the wall and energizing the wall is the best test configuration. By careful analysis of the dominant frequencies in the spectrum, all configurations detected voiding, which was verified by drilling. Determining the size of the void was successful only to a limited degree.

4.5.3.4 Impact-Echo

The IE method (see Ettouney and Alampalli 2012, Chapter 7) can be used in a three-dimensional form to detect tendon damage. It has been used successfully in detecting voids and damage in conventional concrete structures (see Cheng and Sansalone 1993; Lin and Sansalone 1992 Carino and Sansalone 1992).

Colla (2002) examined the use of IE in determining the location of the tendons and evaluating them in a PST concrete beam. The tests were conducted on a dismantled beam section, which was autopsied after the testing to determine the reliability of the IE results. As usual, the relation is $d = v/(2f)$, where d is depth of reflector, v is the velocity, and f is the peak frequency in the spectrum. It is reported in this paper that due to wave reflection between concrete surface and a metal interface, such as a fully grouted posttensioning duct with metal tendons, the relation is modified to $d = v/(4f)$.

In this paper, a scanning version (two-dimensional) of IE was used. This allows two-dimensional and three-dimensional visualization. When collected in discrete steps, along the measurement lines, data can be plotted as two-dimensional images, such as frequency series and impact-echogram. By plotting frequency with testing unit on the concrete surface, a depth versus horizontal location of beam can be plotted. This can give the position of tendon. By using a depth slice of the area under investigation, the position of ducts in vertical plane can be shown. The results showed that it was successful in calculating the thickness of the concrete beam and the duct locations.

Zheng and Ng (2006) used the IE method to detect voids in PST U-beams of an existing bridge. Radar was first used for longitudinally locating the tendon ducts on the side of the girders. Then, IE technique was used to measure the thickness (inversely proportional to the measured frequency). If there was a higher frequency than that corresponding to the solid web thickness, it was presumed that there is a void in the tendon duct (see Figure 4.22). This was confirmed by careful scoring and videoscope to confirm the test results and to refill the voids. The IE test was repeated on the repaired beam to confirm that the repairs worked well. The governing equation of the experiment is

$$f = \frac{C_p}{2h} \qquad\qquad (4.12)$$

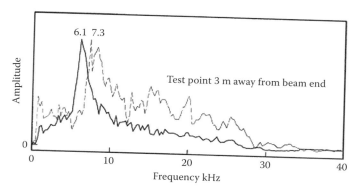

FIGURE 4.22 Impact-echo spectra from a test point 3 m away from beam end. (Reprinted from ASNT Publication.)

The dominant spectral frequency is f, the expected p-wave speed in the material is C_p and the thickness is h.

4.5.3.5 Static Magnetic Fields

Detecting damaged tendons in PRT systems can be performed using the magnetic field method (see Ettouney and Alampalli 2012, Chapter 7). By introducing a magnetic field to the test object and measuring disturbances to the magnetic field, flaws and damage can be detected (see Ghorbanpoor and Shew 1989). Note that this method cannot detect damage to tendons or voids within ducts in PST systems.

4.5.3.6 Multiple Method Testing

Krause et al. (2004) used multiple monitoring methods for DMID. A PST concrete unicellular box bridge in Germany was investigated using impulse radar, IE, and ultrasonic echo methods, using an automated scanner system to locate and assess the tendon ducts at select areas. Techniques were used from both sides of the decks. All three techniques provided information about the horizontal location of tendons and the thickness of the construction element. Impulse radar gives the depth of reinforcement and tendons.

Impulse radar and ultrasonic echo both give accurate thickness of bridge deck (to 1%), but they have to be calibrated at several locations (relative permeability for radar; wave velocity for ultrasonic and IE). Destructive tests showed that deck thickness, concrete cover, and lateral location of tendon ducts can be measured accurately.

Another multiple method experiment was reported by Cao and Davis (2004). The author used an IE and ground penetrating radar (GPR) to detect voids in PST tendon ducts. An intrusive borescope examination was also done, and it is discussed in this chapter. One issue is that all of them are not used in the same case study, that is, on similar types of structure, and make the comparison difficult.

Based on IE and borescope testing of precast posttendons I-girders simply supported by the piers, it was found that IE is about 80% effective in predicting both complete and partial voiding in tendon ducts. IE is sensitive to the geometry of the tested element. It is a very useful tool in void detection when the structural member behaves like a plate and there is only one tendon present under the test point.

Sounding, GPR, and borescope were used for external PSD ducts. Most voids identified by sounding were not found by GPR test. These voids were confirmed by borescope tests as very thin layered delaminations between the top of the polyethylene duct and the grout, caused probably by grout shrinkage at the time of construction. Partial and complete voids identified by sounding were found by GPR. Thus, proper use of GPR was found to detect significant size (larger than 22 mm) voids inside the external PST cables with polyethylene pipes. At the same time, radar can show the continuity of a void along with start and finish of the void. The main disadvantage of the GPR is that it will not work on any metallic base material such as galvanized steel sheeting. But identification of the tendon location can help before the intrusive method can be carried out.

In all cases, personnel experience is very critical for the success of these NDT methods.

4.5.3.7 Time Domain Reflectometry

Chajes et al. (2002) explored the possible use of TDR for detecting voids, and detecting and monitoring corrosion damage in embedded steel strands through lab testing. An electric pulse is sent, and the echoes returning from the device under test are observed. Discontinuity causes reflection, with time of reflection and property of the reflection giving the spatial location and the type of discontinuity, respectively. This is well understood in transmission lines. Here, the method is based on the assumption that any physical damage to the steel strand will change in radius, while voids in the surrounding grout will change the dielectric constant. As a result, corrosion and voids cause impedance change, and it can be detected with TDR.

Small-scale lab tests were conducted with corrosion being simulated electrochemically. Data from control sample and corroded sample were compared by observing the differential. TDR-monitoring wires installed on strands of a newly constructed beam for further monitoring.

We note that TDR has some limitation in DMID of tendons. First, corrosion at multiple locations along the tendon might be difficult to identify. Baseline data are also needed.

4.5.3.8 Other Methods

Other DMID methods based on electrical properties of material have been used. Rizzo et al. (2004) performed tests on the efficiency of guided acoustic waves in detecting damage of tendons. Among these methods is the linear polarization method (see Fontana 1986). Another simple method is based on measuring the electrical resistance of the tendons and inferring the loss of area due to corrosion from the measurements (see Bapu et al. 1988). Half-cell method (also called surface potential) has also been used to estimate corrosion rate in PSC systems (see Escalante 1990).

4.6 DECISION MAKING

4.6.1 COST OF FAILURE UNCERTAINTY

One of the basic issues in decision making processes is the cost implications of different decisions. As an example, the cost of bridge failure was discussed by Ettouney and Alampalli (2011, Chapter 8). The discussion was based on a simple, yet accurate, expression of the cost of bridge failure, C_{Failure}, measured in dollars, which was developed by Stein and Sedmera (2006) as

$$C_{\text{Failure}} = C_1 eWL + \frac{TDAd}{100}\left\{\left[C_2\left(\frac{100}{T}-1\right)+C_3\right]+\frac{1}{S}\left[C_4 O\left(\frac{100}{T}-1\right)+C_5\right]\right\}+C_6 X \qquad (4.13)$$

where
    C_1 = Unit rebuilding costs (in \$).
    e = Cost multiplier (in \$).
    W = Bridge width, can be obtained from National Bridge Inventory (NBI), field 52 (in ft).
    L = Bridge length, can be obtained from NBI, field 49 (in ft).
    C_2 = Cost of running cars.
    C_3 = Cost of running trucks.

D = Detour length, can be obtained from NBI, field 19 (miles).

A = Average daily traffic (ADT) on bridge can be obtained from NBI, field 29.

d = Duration of detour (in days).

C_4 = Value of time per passenger.

O = Average occupancy rate per car.

T = Average daily truck traffic (ADTT) on bridge, can be obtained from NBI, field 109 (note that it is a % of ADT).

C_5 = Value of time per truck.

S = Average detour speed (in mph).

C_6 = Cost per life lost.

X = Number of deaths from failure.

Typical values of the parameters of Equation 4.13 can be found in Stein and Sedmera (2006) or Ettouney and Alampalli (2012).

We can use that expression in an example of risk evaluation of a PST bridge SHM project. On inspecting the components of the cost, it becomes clear that most of them are fairly uncertain. They are usually *averages* of some statistical observations. Being defined as an average makes it reasonable to assume that those averages can be used as random variables instead. This leads us to the obvious conclusion that the cost of failure itself is a function of random variables, that is, it is a random variable itself. Traditionally, of course, $C_{Failure}$ has been treated as a deterministic value. This is due to the simplicity of the deterministic versus probabilistic methods. We will investigate the effects of acknowledging the uncertainties in $C_{Failure}$ next. To simplify the probabilistic evaluation process, we make use of the closed-form expression of $C_{Failure}$ by using a Taylor series approach for handling the functions of random variables. The other approach, the Monte Carlo approach, can also be used. However, we prefer the Taylor series in this situation due to the simplicity of the analytical expression of $C_{Failure}$. Such an expression contains several variables, and any of those can be assumed to be random. For the sake of simplicity, we will assume that only four of the variables are random, namely, ADT, A; average detour speed, S; duration of detour, d; and ADTT, T. The other variables are assumed to be deterministic. Following the Taylor series method, we need partial derivatives of $C_{Failure}$ with respect to all of these random variables, which can simply be evaluated as

$$\frac{\partial C_{Failure}}{\partial A} = Dd\, A_0 \tag{4.14}$$

$$\frac{\partial C_{Failure}}{\partial T} = \frac{DAd}{100}\left\{ (C_3 - C_2) + \frac{1}{S}(C_5 - C_4) \right\} \tag{4.15}$$

$$\frac{\partial C_{Failure}}{\partial d} = DA\, A_0 \tag{4.16}$$

$$\frac{\partial C_{Failure}}{\partial S} = -DAd\, \frac{A_2}{S^2} \tag{4.17}$$

$$\frac{\partial^2 C_{Failure}}{\partial A^2} = 0 \tag{4.18}$$

$$\frac{\partial^2 C_{Failure}}{\partial T^2} = 0 \tag{4.19}$$

$$\frac{\partial^2 C_{\text{Failure}}}{\partial d^2} = 0 \tag{4.20}$$

$$\frac{\partial^2 C_{\text{Failure}}}{\partial S^2} = 2\, D\, A\, d\, \frac{A_2}{S^3} \tag{4.21}$$

with

$$A_0 = \left\{ A_1 + \frac{1}{S} A_2 \right\} \tag{4.22}$$

$$A_1 = \left[C_2 \left(1 - \frac{T}{100} \right) + C_3 \frac{T}{100} \right] \tag{4.23}$$

$$A_2 = \left[C_4 O \left(1 - \frac{T}{100} \right) + C_5 \frac{T}{100} \right] \tag{4.24}$$

We need to assume that the probability distributions of the four random variables A, S, d, and T are known. This information can be obtained by one of two methods: by engineering judgment of the user or through an SHM effort that collects the statistical data for these variables. On the basis of this, it is assumed that their expected values (means) and variances are known and are defined as $\bar{A}, \bar{S}, \bar{d}$, and $\bar{T}$ for the expected values (means) and V_A, V_S, V_d, and V_T for the variances, respectively. Next, we expand the Taylor series about ε_A, ε_S, ε_d, and ε_T such that

$$\varepsilon_A = A - \bar{A} \tag{4.25}$$

$$\varepsilon_S = S - \bar{S} \tag{4.26}$$

$$\varepsilon_d = d - \bar{d} \tag{4.27}$$

$$\varepsilon_T = T - \bar{T} \tag{4.28}$$

From the definitions of expected values we have, $E(\varepsilon_A) = 0$, $E(\varepsilon_S) = 0$, $E(\varepsilon_d) = 0$, and $E(\varepsilon_T) = 0$. Also, from Equations 4.25 through 4.28, we can express the variances as

$$V_A = E(A^2) = E(\varepsilon_A^2) \tag{4.29}$$

$$V_S = E(S^2) = E(\varepsilon_S^2) \tag{4.30}$$

$$V_d = E(d^2) = E(\varepsilon_d^2) \tag{4.31}$$

$$V_T = E(T^2) = E(\varepsilon_T^2) \tag{4.32}$$

Applying the Taylor series method of Ettouney and Alampalli (2012, Chapter 8) to this problem, and limiting the order of the series to $O(\varepsilon_i^2)$, the expected value (mean) of the cost of failure can be expressed as

$$E\left(C_{\text{Failure}} \right) = \bar{C}_{\text{Failure}} + \frac{1}{2} \left(V_A \cdot \frac{\partial^2 \bar{C}_{\text{Failure}}}{\partial A^2} + V_S \cdot \frac{\partial^2 \bar{C}_{\text{Failure}}}{\partial S^2} + V_d \cdot \frac{\partial^2 \bar{C}_{\text{Failure}}}{\partial d^2} + V_T \cdot \frac{\partial^2 \bar{C}_{\text{Failure}}}{\partial T^2} \right) \tag{4.33}$$

And the expected mean square of the cost of failure is

$$E\left(C_{\text{Failure}}^2\right) = \bar{C}_{\text{Failure}}^2 + \left[V_A \cdot \left(\bar{C}_{\text{Failure}} \cdot \frac{\partial^2 \bar{C}_{\text{Failure}}}{\partial A^2} + \left(\frac{\partial \bar{C}_{\text{Failure}}}{\partial A}\right)^2\right)\right] +$$

$$\left[V_S \cdot \left(\bar{C}_{\text{Failure}} \cdot \frac{\partial^2 \bar{C}_{\text{Failure}}}{\partial S^2} + \left(\frac{\partial \bar{C}_{\text{Failure}}}{\partial S}\right)^2\right)\right] +$$

$$\left[V_d \cdot \left(\bar{C}_{\text{Failure}} \cdot \frac{\partial^2 \bar{C}_{\text{Failure}}}{\partial d^2} + \left(\frac{\partial \bar{C}_{\text{Failure}}}{\partial d}\right)^2\right)\right] + \qquad (4.34)$$

$$\left[V_T \cdot \left(\bar{C}_{\text{Failure}} \cdot \frac{\partial^2 \bar{C}_{\text{Failure}}}{\partial T^2} + \left(\frac{\partial \bar{C}_{\text{Failure}}}{\partial T}\right)^2\right)\right]$$

The over bar in 4.33 and 4.34 indicates that the function is evaluated at the mean of the random variables. The mean, variance, and the standard deviation of C_{Failure} can now be evaluated as

$$\mu_{C_{\text{Failure}}} = E\left(C_{\text{Failure}}\right) \qquad (4.35)$$

$$V_{C_{\text{Failure}}} = E\left(C_{\text{Failure}}^2\right) - E^2\left(C_{\text{Failure}}\right) \qquad (4.36)$$

$$\sigma_{C_{\text{Failure}}} = \sqrt{V_{C_{\text{Failure}}}} \qquad (4.37)$$

$$COV = \frac{\sigma_{C_{\text{Failure}}}}{\mu_{C_{\text{Failure}}}} \qquad (4.38)$$

To illustrate the applications of the uncertainty of cost of failure, consider a simple case of a bridge deck rating. The gross weight W is assumed to be 36 tons. Using analytical techniques, the analyst computed C, D, and L as shown in Table 4.7. The rest of the parameters of Equation 4.13 are also shown in Table 4.7. Table 4.7 shows the resulting deterministic deck rating, R.

Recognizing that the computed values of A, S, d, and T are based on many uncertain factors, the analyst decides to perform probabilistic analysis of the deck rating. To start with, the analyst decided to use the computed values of A, S, d, and T in Table 4.7 as the means (expected values). The analyst also made a reasonable estimate for the coefficient of variation (COV) of A, S, d, and T. The standard deviation and the variances of each parameter were then computed using 4.37 and 4.38, as shown in Table 4.8.

Using Equations 4.14 through Equation 4.34, the expected value and the expected mean square for the inventory and operational deck rating can be computed. Using 4.35 through 4.38, the rest of the statistical properties of the inventory and operational deck rating can be computed. The statistical results are shown in Table 4.9.

Some interesting observations can be made on the resulting statistics. First, note that the expected value (mean) of C_{Failure} is almost equal to the deterministic value of C_{Failure} (in Table 4.7). This is due to the effects of the variance terms in Equation 4.33. Additionally, note that the COV in Table 4.9 is higher than all the COV of the input random variables in Table 4.8, which is expected. We note that at a value of 0.247, the COV is a bit high, even though the component COV are in the range of 0.1 to 0.2, which is in the usual range of engineering uncertainties. The relatively higher value of the COV of C_{Failure} leads us to an important bridge management conclusion: there is a relatively high degree of uncertainty in estimating the cost of bridge failure. Care should be taken in basing important decisions when it comes to costs of bridge failure. There are several ways to reduce the

TABLE 4.7
Cost of Failure Example Parameters

Parameter	Description	Magnitude	Units
C_1	Unit-rebuilding costs	100	$
e	Cost multiplier	1.5	Scalar
W	Bridge width	50	Feet
L	Bridge length	300	Feet
C_2	Cost of running cars	0.5	$
C_3	Cost of running trucks	1.8	$
D	Detour length	20	Miles
A	Average daily traffic on bridge	3000	Car+Trucks
d	Duration of detour	365	Days
C_4	Value of time per passenger	7.61	$/hour
O	Average occupancy rate per car	1.63	Scalar
T	Average daily truck traffic on bridge	1200	Trucks
C_5	Value of time per truck	20	$
S	Average detour speed	30	Mph
C_6	Number of deaths from failure	2	Scalar
X	Cost per life lost	500,000	$
$C_{Failure}$	Cost of failure	431.43	$ millions

TABLE 4.8
Statistical Properties of Input Parameters

Parameter	Mean	COV	Variance
A	3000	0.1	90,000
S	30	0.15	20.25
d	365	0.2	5329
T	1200	0.1	14,400

TABLE 4.9
Statistical Parameters of Cost of Failure

Statistical Parameter	Magnitude
$E(C_{Failure})$ — $ millions	v.43
$E(C^2_{Failure})$ — ($ millions)^2	197457.0
Variance ($ millions)^2	11322.37
Standard deviation—$ millions	106.4
COV	0.247

uncertainties. One obvious method is to reduce the uncertainties in one or more of the random variables A, S, d, or T.

To produce a probabilistic statement that can be used for decision making, the analyst would need more information than that given in the statistical parameters of Table 4.9. In fact, the probability distribution function (PDF) of $C_{Failure}$ is needed to make the probabilistic statement. There is no simple way, short of a full Monte Carlo simulation analysis, to obtain the needed PDF. Instead, the analyst might choose to make a reasonable assumption about the PDF. One possible assumption is that the PDF is simply a normally distributed function. Such an assumption seems reasonable and simple.

4.7 CASE STUDIES

4.7.1 OVERVIEW

4.7.1.1 Impact-Echo in Full-Scale Monitoring

In a study by Tinkey and Olson (2008) and Tinkey, Miller, and Olson (2008), the ability of IE to detect voids in ducted tendons was established. The IE scanning method for rapid QC/QA of structural concrete components can detect and capture the image of internal anomalies that cannot be observed by visual inspection of the structural components. The tested bridge is shown in Figure 4.23. The IE concept as applied to the handheld equipment is shown in Figures 4.24 and 4.25. The testing process is shown in Figure 4.26.

FIGURE 4.23 I-390 Bridge over the Genesee River, Rochester, NY. (Courtesy of Olson Engineering, Inc.)

Olson Instruments, Inc. handheld
test head for impact-echo tests

Receiver

Impact

Flaw

Reflection from concrete/flaw
interface

Reflection from backside of
test member

*Reflection from backside occurs at a lower frequency than that
from the shallower concrete/flaw interface

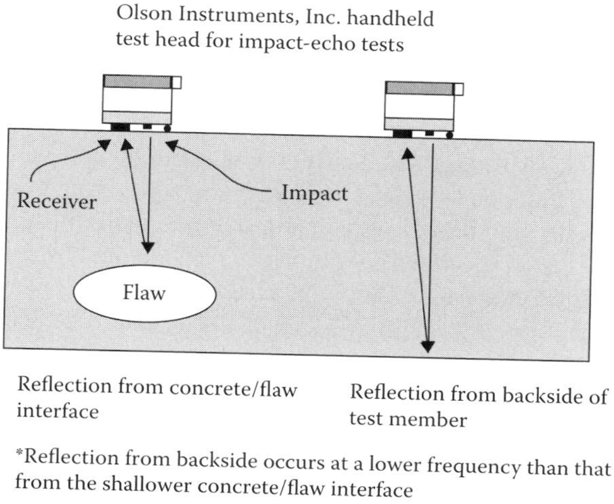

FIGURE 4.24 Impact-echo concepts. (Courtesy of Olson Engineering, Inc.)

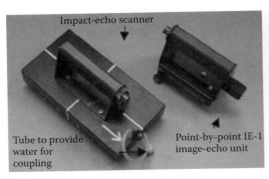
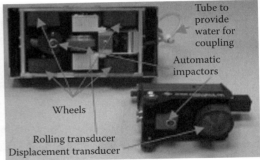

FIGURE 4.25   Impact-echo scanner unit and impact-echo point-by-point unit.

FIGURE 4.26   Impact-echo field experiment. (Courtesy of Olson Engineering, Inc.)

The IE rolling scanner (Figure 4.24) was used for rapid IE testing to either locate delaminations, honeycombing, or cracks parallel to the test surface, or measure the thickness of concrete structures with only one-sided access. The scanner (Figure 4.25) incorporates a rolling transducer assembly with multiple sensors, attached underneath the test unit with an optocoupler to track the distance. For a rough concrete surface, water was suggested as a couplant to improve displacement transducer contact conditions. Typical scanning time reported for a 13-ft line (about 160 test points) was 60 s. Data analysis and visualization were achieved by using an IE scanning software that includes raw data, digitally filtered using a Butterworth filter with a band-pass range of 1 to 20 kHz, and automatic and manual picks of dominant frequency performed on each spectrum to calculate thickness at each test point, based on the selected dominant frequency. A three-dimensional grey scale or color plot of tested specimen condition can be generated by combining the calculated IE thicknesses from each scanning line. Based on case histories—to determine integrity of concrete transfer wall, to detect grout voids in PST bridge ducts, and to detect internal cracks within concrete pavement exposed to the Alkali Silica Reaction (ASR) problem—it was concluded that IE test can detect internal voids/honeycombs, grout voids in PST bridge ducts, and internal microcracking in concrete structures. Typical results are shown in Figures 4.27 through 4.30.

The authors of the study concluded that, for a wall thickness of 30.0 cm or less, the depth of concrete cover should be less than 19.0 cm for the IE to correctly locate internal voids. They also observed that IE can be used successfully, under certain dimensional restrictions, for either steel or plastic tendons.

FIGURE 4.27 Typical impact-echo scan on box girder with duct voids. Courtesy of Olson Engineering, Inc.

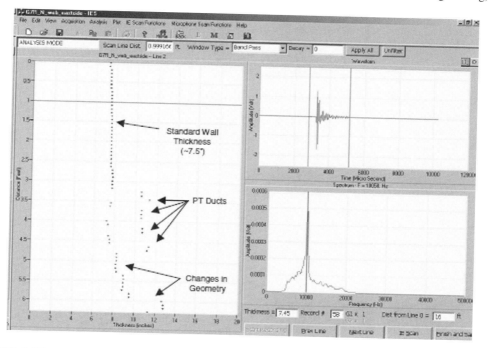

FIGURE 4.28 Impact-echo for main girder wall scan—distance (ft) versus echo depth (in). Courtesy of Olson Engineering, Inc.

In a separate study of IE and hammer tests, Limaye (2008) used IE to detect cold joints and delaminations in bridge decks, the condition of the grout in posttensioning ducts, and testing of repair quality of pressure injected cracks. Test results were acquired in terms of frequencies (Hz) for each tested location. Concrete members without any internal defects were characterized by one primary peak, which represented the thickness of the member. If the frequency obtained did not match the expected frequency, then a possible anomaly was suspected in the tested area.

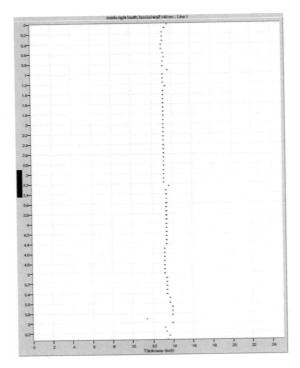

FIGURE 4.29 Impact-echo for bottom deck scan—distance (ft) versus echo depth (in), Courtesy of Olson Engineering, Inc.

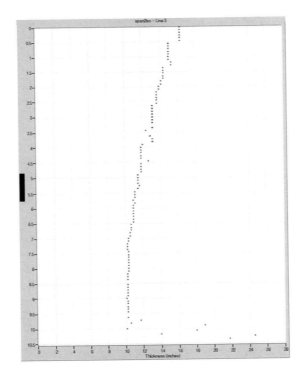

FIGURE 4.30 Impact-echo results showing actual damaged ducts. (Courtesy of Olson Engineering, Inc.)

FIGURE 4.31   Interior view of the girder area. (Reprinted from ASNT Publication.)

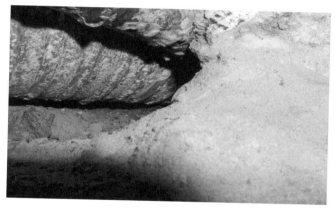

FIGURE 4.32   Debonded duct. (Reprinted from ASNT Publication.)

Box girders of a bridge were tested at three primary locations: at both ends of each girder and in middle of the girder (Figure 4.31). Testing consisted of impacting the girder surface near the sensor placed initially above the first embedded duct. Then the sensor was moved 2 in below the prior reading, and the procedure was repeated until approximately 2 in below the last duct was encountered. Most of the test results were consistent and repeatable except in the areas where cracks and delaminations were on the surface of the concrete. It was found that test results were difficult to interpret because the conditions were different at many locations, resulting in a wide frequency variation. Encountered conditions included variation in girder wall thickness and concrete cover over the ducts, concrete cover less than 3 in, debonded ducts (Figure 4.32), and duct joints wrapped with duct tape. It was difficult to distinguish between an unbonded duct and an ungrouted duct as seen in because if the duct is unbonded, then the stress wave cannot travel through the duct. Also, it was not possible to detect a partially grouted (75%) duct when there was a wide variation in frequencies caused by other conditions.

IE testing and hammer sounding were used to obtain the extent of the bottom slab below the posttensioning ducts in the closure areas. Locations of the potential anomalies were verified by drilling and removing the cores for verification.

It was reported that the IE test results over an ungrouted duct in the concrete were similar to the test results obtained over a void in the concrete because stress waves cannot pass through the duct. If there are overlapping anomalies, only the first anomaly was detected. Therefore, if there is a delamination below an ungrouted duct, detection is very difficult. Other limiting factors reported were stacked ducts, closely spaced ducts, voids created by congestion of the reinforcing steel, cold joints, and tight cracks.

4.7.2 CASE STUDY: MONITORING VOIDS IN BONDED PTS SYSTEMS

In recognition of the role that voids, or the lack of them, play in the deterioration of tendons in bonded systems, Venugopalan (2008) performed tests on a bonded PTS box girder bridge (Figure 4.33) . There were several visible signs of damage: efflorescence, cracks, and spalling (see Figure 4.34). Also, previous reports indicated the presence of voids of different sizes. At the start of the project, it was clear that there were corrosion problems. The project aimed at quantifying the level of corrosion damage.

FIGURE 4.33   Box girder PST bridge. (Courtesy of Siva Venugopalan.)

FIGURE 4.34   Signs of damage. (Courtesy of Siva Venugopalan.)

The investigator chose to quantify corrosion damage using the half-cell method. The steps of the investigations were as follows:

1. Choose adequate number of locations to be investigated along the bridge
2. Expose the tendons
3. Visually document the condition of the wires/strands. This included observing different degrees of rusting and the quality of grout
4. Measure corrosion rates of strands. This step needs to be performed while the strands are still covered by grout
5. Measure alkalinity of grout
6. Test grout samples for chloride

Some findings of the project were

1. Voids in grout of different sizes were observed (see Figure 4.35).
2. Moisture levels were recorded to vary from 10% to 36%; the latter is high enough to indicate an active corrosion process.
3. Chloride content of grout was below the limit of 0.08% set for new PST systems.
4. Corrosion rates were tested (see Figure 4.36). In the sample of 30 locations, 23% showed a high corrosion rate.
5. Grout samples testing indicated high possibility of bleeding, which can explain presence of voids in the grout.

The tests showed the importance of (1) quality of grout in reducing corrosion of tendons (a proactive measure), and (2) validity of using corrosion rate to detect corrosion process in PTS systems (a reactive measure).

4.7.3 HIGH-STRENGTH SUSPENSION CABLES

General problem description: Betti et al. (2000) observed that in-depth visual inspections of the cable systems in the suspension bridges in the New York area have shown that there are many broken wires inside the cables and at the anchorages, showing brittle fractures and, in some cases, significant section loss. Due to the high safety factors used during designs, these were not attributed

FIGURE 4.35 Observed voids in grout. (Courtesy of Siva Venugopalan.)

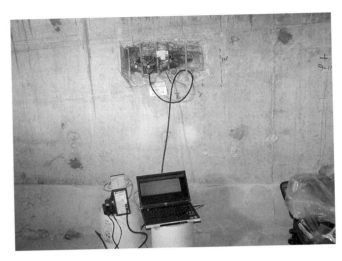

FIGURE 4.36 NDT testing of corrosion rate. (Courtesy of Siva Venugopalan.)

to overstress, and thus they indicate a higher deterioration rate of the cable strengths than from inferred section loss. This suggests that pitting and/or cracking effects may be present, whether induced by corrosion or hydrogen embrittlement or both. Hydrogen concentration analysis suggests that hydrogen does indeed absorb into the corroded wire. Ultimate strain measurements indicate that the corrosion leads to increased embrittlement of both galvanized and ungalvanized wires.

The authors concluded that currently there are no commercially available nondestructive evaluation (NDE) techniques for *in situ* evaluation of suspension bridge cables that are capable of detecting not only broken wires but also progressive corrosion of wires. In some cases, the size and configuration of such structures limit the success of NDE techniques (i.e., too large high-power radiography systems, and limited penetration depth for EM systems). AE techniques have the ability of continuously monitoring cable health by detecting wire breaks. In addition, new AE techniques can be used to monitor corrosion in local regions close to the sensors. However, AE techniques cannot provide information about prior damage. There are other NDE techniques that show promise for detecting the condition of suspension bridge cables. Such methods include neutron radiography and electromagnetically generated ultrasonic and impulse response techniques. These techniques show good promise of detecting not only wire breaks but also reduction in cross section due to corrosion and seem to be able to overcome the limitation of the penetration depth (i.e., cross-talk between wires in ultrasonic techniques).

Full-scale laboratory testing: To try to address the above problem, an FHWA-sponsored study by Columbia University and others investigating an integrated methodology that uses state-of-the-art sensing capabilities and NDT direct and indirect technologies to assess the cable condition. The study was reported by Betti et al. (2008). Several technologies were selected based on an evaluation of available technologies and applicability to large suspension bridge cables. A 20-in-diameter, 20-ft-long cable mock-up (Figure 4.37a) subjected to 1500 kips and fully instrumented, has been built at Columbia University to re-create conditions as close as possible to real operating conditions, and it will be tested inside an accelerated corrosion chamber using the selected technologies. The cable mock-up is made by 73, 127-wire hexagonal strands for a total of more than 9000, 0.196-in-diameter steel wires. Of the 73 hexagonal strands, 7 are 35 ft long and are subjected to tension load of 1100 kips, while the remaining 66 strands are 20 ft long. This cable specimen will be placed in a loading frame, properly designed for this particular action, and some of the strands will be subjected to a load so as to induce stresses up to 100 ksi (to include the effects induced by stress-corrosion cracking). The total length of the experimental setup is over 35 ft. An environmental chamber will be built around it, and the cable will be subjected to harsh environmental conditions.

Twenty-five long wires were prenotched at three locations along their length and inserted into various strands to test and calibrate the AE system placed on the cable when these wires are failed later. In addition, two strands were precorroded before being placed inside the cable to generate inside conditions that are not uniform from the beginning of the experiment so that, when direct sensing methods are tested for assessing the initial cable condition, a verification of their accuracy will be done from the beginning. Several types of sensors were strategically installed in the cross section of the cable to monitor cable and environmental conditions. Once the construction of the corrosion chamber is complete, this cable mock-up will be tested for 6 months in a mildly corrosive environment. After the test is complete, the cable mock-up will be dismantled, and it will be possible to validate the sensor readings with the "real" cable conditions. Another full scale experiment for testing degradation of suspension cables will be performed at the City College of New York by Mahmoud (2011) (see Figure 4.37b). The objectives include validation of deterioration models of suspension cables under realistic load and environmental conditions.

Acoustic emission: Another study of damage in high-strength cables was reported by Drissi-Habti, Gaillet, and Tessier (2008). In that study, AE was reviewed for detecting wire breaks and corrosion in bridge cables. When an individual wire strand of the wire cable fails, the failure event generates AE. Acoustic and pressure waves are, therefore, propagating in both directions along the wire cable. A piezoelectric transducer located on the wire strand cable or on a cable band surrounding the wire strand cable detects vibrations in the cable. Multiple transducers are positioned at predetermined locations all along the structure being monitored to obtain data from several locations. This can also be useful in detecting the location of the failure (Figure 4.38).

The system used in this project is called CASC, and it was developed in LCPC in the 1970s. Its architecture is based on Digital Signal Processing (signal treatment processor) and on software that enables the use of wireless communicating sensors and a supervision monitor. For a given AE event, each smart sensor transmits the associated wave outlook and the detection time. From this, failure location is localized and displayed on the screen, in real time. As needed, bridge supervisors can be informed through alarms delivered as cell phone messages.

(a)

(b)

FIGURE 4.37 Cable mock-up and loading frame. (a) Columbia University experiment. (Courtesy of Dr. Raimondo Betti.) and (b) Bridge Technology Consulting experiment. (Courtesy of Dr. Khaled Mahmoud.)

The results showed that AE is a useful technique suitable for detecting flaws and corrosion in metallic structures, especially cables. At the same time, AE signals coming from the sensor networks on bridge cables require skill and experience in interpreting them accurately.

Ultrasonic and radiography are considered by some researchers as capable of detecting cable failures. The high attenuation of the ultrasonic pulses was found to be limiting when inspecting the anchorages. It also required skilled technicians to accurately interpret the signals. Radiography was also investigated, and user safety concerns were an issue. The main problem with radiography was that the stack of multiple materials on the cable made the interpretation of images difficult. The process was also considered tedious and costly. Hence, considering the limitations of each of these methods, the data fusion methodology, which combines two or more techniques to get reliable detection and monitoring of damage on cables, should be considered.

EM device testing: EM method has been described by Hall (2008) as the only proven practical way to efficiently inspect wire ropes of lift bridges in addition to the visual inspection method.

A magnet circuit is created around and through a specimen of wire rope, as depicted in Figure 4.39. The magnets oriented in the ends of the sensor are arranged so that the magnetic current is directed from a positive to a negative pole. The flux bar carries the current from one pole to another; the

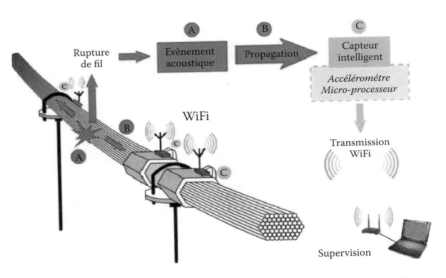

FIGURE 4.38   CASC system and the associated WiFi Setup. (Reprinted from ASNT Publication.)

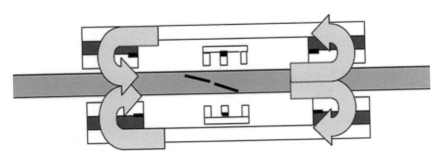

FIGURE 4.39   Sensor head diagram. (Reprinted from ASNT Publication.)

magnetic current, following the path of least resistance, is directed through the wire rope. The flux bars must be designed so that the total cross-sectional area of the bars is equal to or greater than the total cross-sectional area of the wire rope or strand under test.

Loss of metallic area (LMA) and local faults in the rope can be detected. Once the unit is placed on a rope, the hall sensors at the ends of the sensor head sense the magnetic current within the rope and convert it to voltage. Voltage is then converted to percent of LMA or weight of the rope in pounds per foot. The rope is moved through the head, or the head is moved over the rope. The LMA is relative from the best part to the worst part of the rope. The original cross-sectional area of the rope can be entered into the test data, giving metallic area content as new. The data is recorded digitally and placed in a strip chart presentation, displaying the LMA in percent from the original. When there is a pit or broken wire, flux leakage occurs. The Hall sensor in the local fault cage picks up the leakage, converts the magnetic current to voltage, and then the data is displayed in the computer program in strip chart form. A digital encoder fitted inside a tachometer wheel records the distance or location in the rope body and wire rope or sensor head speed during the test. The data is interpreted by the inspector, anomalies and defects are noted, and the rope is either retired or left in service. Figure 4.40 shows the unit, and Figure 4.41 shows typical measurement.

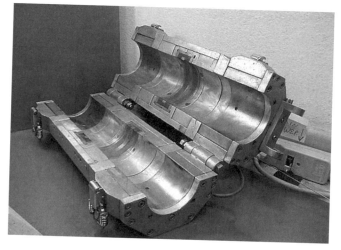

FIGURE 4.40   Sensor head equipment. (Reprinted from ASNT Publication.)

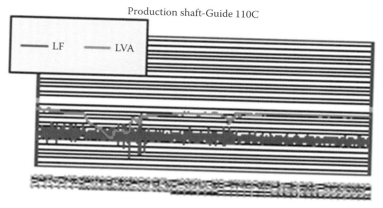

FIGURE 4.41   Typical measurement. (Reprinted from ASNT Publication.)

4.7.4 THERMAL IMAGING

Dupuis et al. (2008) inspected lab specimens using three different methods (Table 4.10), which were similar to inspection procedures implemented in field inspections of concrete bridges Method 1 involved placing the specimen on the test frame and heating from underneath while taking thermal images of the unheated surface from above. In Method 2, the heater was suspended above the specimen, heated for a period of time, and then removed so that thermal images of the heated surface could be obtained. Method 3 involved exposing the specimen to direct sunlight for a relatively long duration and then placing it on the test stand for thermal imaging of the heated surface. The temperature gradient between two sides of the specimen caused heat energy to propagate through the specimen, which is essential for acquiring thermal images where embedded flaws are detected and seen as surface temperature differences.

From Table 4.10, it is apparent that Method 1 was the most productive inspection method used. This method utilized active heating through the specimen thickness. The tests also showed that only the simulated voids located between the steel tendons and the infrared camera were detected. It was

TABLE 4.10
Concrete Specimen Inspection Summary

Specimen	Test Method	Inspected Specimen Face	What Was Detected?		
			Rebar	PT Ducts	Simulated Voids
8a	Method 3	Top	X		
	Method 1	Top	X	X	X
	Method 1	Bottom	X	X	
8b	Method 1	Top	X	X	X
	Method 1	Top	X	X	X
	Method 1	Top	X	X	X
	Method 3	Top	X	X	X
	Method 1	Bottom	X	X	
	Method 2	Top	X		
8c	Method 3	Top	X	X	X
	Method 1	Top	X	X	X
	Method 1	Bottom	X	X	
8d	Method 1	Top	X	X	
	Method 1	Bottom	X	X	
	Method 2	Top			
12a	Method 3	Top	X		
	Method 1	Top	X	X	
	Method 2	Top			
12b	Method 3	Top	X		
	Method 1	Top	X		
12c	Method 1	Top	X	X	X
	Method 2	Top			
12d	Method 1	Top	X	X	
	Method 1	Top	X	X	
12e	Method 1	Top	X	X	
12f	Method 1	Top	X	X	
	Method 2	Top			

Source: Reprinted from ASNT Publication.

also observed that, when the simulated voids were visible, they were located within plastic PT ducts. None of the simulated voids in steel ducts was detected during lab inspections.

4.8 LCA OF PT-PS SYSTEMS

4.8.1 GENERAL

This section will consider cost sources that are specific to PT-PS systems. We will argue that there are three cost categories that can affect the retrofits and life cycles of PT-PS systems. They are abnormal hazards, deterioration/durability, and substandard detailing. Simple cost models are proposed for the three categories. We then present a general example to show how the life cycle cost (LCC) model can be executed. The example shows how retrofit intervals can affect LCC in an objective manner. The section concludes with a discussion of the effects of uncertainties on the proposed LCC cost model.

4.8.2 LIFE CYCLE COST ANALYSIS (LCCA)

4.8.2.1 Cost of Abnormal Hazards

Costs of abnormal hazards (earthquakes, scour, wind, etc.) on systems are discussed in several places in this chapter. The treatment of such costs in PT-PS systems is fairly similar to that in other systems. For example, for a given abnormal hazard, the cost per unit time (traditionally a year) can be expressed as

$$\bar{C}_H = \int_h C_h \, p_h \, dh \tag{4.39}$$

The cost of retrofit on the system due to a hazard level h is C_h. The probability of occurrence of hazard level h during the unit time is p_h. The integral in Equation 4.39 is over all realistic hazard space. The total life cycle cost of an abnormal hazard is

$$C_H = \sum \bar{C}_H \tag{4.40}$$

The sum in Equation 4.40 is over the number of unit times within the lifespan of the system under consideration. Note that the equation does not include discount rates, for the sake of simplicity. Finally, the equation can be repeated for as many abnormal hazards as are appropriate.

4.8.2.2 Cost of Deterioration

We recall that conventional corrosion is one potential element of deterioration of those systems. Also, remember that we introduced a simple method of computing LCC for corrosion in conventional concrete systems in Chapter 3. We note that all elements that contribute to corrosion share several attributes:

- They all affect the system on hand in a slow, temporal manner.
- The damage (deterioration) would increase until it reaches a threshold, D_{MAX}, where a rehabilitation measure is performed.
- After such rehabilitation, the damage level drops into an acceptable target level, D_{TARGET}.

The above process, illustrated in Chapter 3, would continue for as long as the bridge is in service.

We propose to generalize this conventional corrosion model to all other sources of deterioration. To do that, we need first to quantify damage measures as a function of time for all sources of deterioration.

Conventional corrosion: Recall that we used an area to represent time-dependent damage measure in Chapter 3. That choice can be still adequate for PT-PS systems. However, since most of the corrosion damage of concern in these systems are the tendon damage, we propose to use the length of the damaged tendons as the damage measure such that

$$D_1(t) = A_1 t^{n_1} \tag{4.41}$$

where
  $D_1(t)$ = Length of damaged tendons due to conventional corrosion that needs retrofit after time, t
  A_1 = Appropriate constant
  n_1 = Appropriate constant
  Stress corrosion: A damaged length seems to be adequate for stress corrosion damage measure such that

$$D_2(t) = A_2 t^{n_2} \tag{4.42}$$

where
  $D_2(t)$ = Length of damaged tendons due to stress corrosion that needs retrofit after time, t
  A_2 = Appropriate constant
  n_2 = Appropriate constant
  Hydrogen embrittlement: Similar to earlier corrosion sources, hydrogen embrittlement damage can be modeled using a damaged length such that

$$D_3(t) = A_3 t^{n_3} \tag{4.43}$$

where
  $D_3(t)$ = Length of damaged tendons due to hydrogen embrittlement that needs retrofit after time, t
  A_3 = Appropriate constant
  n_3 = Appropriate constant
  Loss of tendon force: Damage due to loss of tendon force can be reasonably related to the amount of loss in force, that is, it has the unit of force. The relationships of force losses as a function of time were discussed earlier in this chapter. For the sake of simplicity, we propose to use the form

$$D_4(t) = A_4 t^{n_4} \tag{4.44}$$

where
  $D_4(t)$ = Loss of tendon force at time, t
  A_4 = Appropriate constant
  n_4 = Appropriate constant

Deterioration due to exposure: Damage due to exposure to element can be related to the exposed (damaged) area. It has, thus, the unit of area. It can be expressed as

$$D_5(t) = A_5 t^{n_5} \tag{4.45}$$

where
  $D_5(t)$ = Loss of tendon force at time, t
  A_5 = Appropriate constant
  n_5 = Appropriate constant

Durability issues: Damage due to different durability issues, such as sulfate attacks, can also be related to the exposed (damaged) area. It has, thus, the unit of area. It can be expressed as

$$D_6(t) = A_6 t^{n_6} \tag{4.46}$$

FIGURE 3.13 Geosynthetic reinforced soil abutment. (Courtesy of CRC Press.)

FIGURE 3.14 Curved bridge girder test setup. (Courtesy of CRC Press.)

FIGURE 3.15 Component specimen after failure showing top flange buckling. (Courtesy of CRC Press.)

Such a system has considerable value for SHM of structures, as measurements can be done with relative ease, and it also has implications for continuous SHM. The potential use of such a system for detecting reinforced concrete corrosion was discussed by Ettouney and Alampalli (2012, Chapter 5). The cost and line-of-sight requirements of these systems need to be resolved.

where

$D_6(t)$ = Loss of tendon force at time, t

A_i = Appropriate constant

n_i = Appropriate constant

The choice of the forms of relations Equations 4.41 through 4.46 is fairly subjective for the purpose of this chapter. Adequate studies are needed to choose more accurate relationships. Clearly, these studies should rely on both short-term and long-term monitoring of PT-PS systems.

We can now formalize the cost of durability between retrofits in a PT-PS system as

$$\overline{C}_{\text{DURABILITY}} = \sum_{i=1}^{i=6}\left[C_{0i} + C_i\, D_i(t)\right] \tag{4.47}$$

In Equation 4.47, the cost of retrofitting unit damage $D_i(t)$ is C_i. The initial cost of retrofitting is C_{0i}. Figure 4.42 shows a schematic view of the cost for a given issue (i). Also, note that the time measure t is not an absolute time: it is the time measured from the latest retrofit.

Finally, we can estimate the total life cycle cost of durability in a PT-PS system as

$$\tag{4.48}$$

The sum in Equation 4.48 is over the number of retrofits within the lifespan of the system under consideration. Note that, for the sake of simplicity, the equation does not include discount rates.

4.8.2.3 Cost of Substandard Details

While discussing durability (damage) effects of substandard details earlier, we observed that such damaging effects would increase as time passes. We propose a simple cost model for substandard details between retrofits as:

$$\overline{C}_{\text{DETAIL}} = \sum_{i=7}^{i=8}\left[C_{0i} + C_i D_i(t)\right] \tag{4.49}$$

In Equation 4.49, the cost of retrofitting $D_i(t)$ is C_i. Descriptions of damage due to substandard details are shown in Table 4.11.

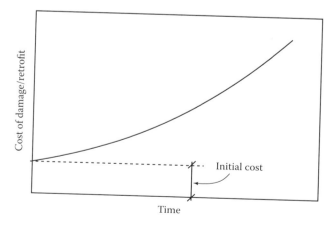

FIGURE 4.42 Cost model for deterioration.

TABLE 4.11
Details of Costs of Substandard Details

i	Source of Damage
7	Substandard anchoring details
8	Substandard segmental joint details

Damage in Equation 4.49 is measured by the deterioration of the detailing itself, as well as the effects of the deterioration on the surrounding areas of the structure. Thus, it is not easy to describe the damage in an objective fashion. It more appropriate to describe it in a subjective fashion. One possible way is to use a conventional rating range, $R_i(t)$, say from 1 to 7, to describe the damage. Thus,

$$1 \le R_i(t) \le 7 \tag{4.50}$$

with $i = 7, 8$. The detail is in a new condition when $R_i(t) = 7$. The detail is in failed condition when $R_i(t) = 1$.

Expressions of $R_i(t)$ are not known as of the writing of this chapter. However, a general expression can be

$$R_i(t) = B_i - A_i\, t^{n_i} \tag{4.51}$$

where
    B_i = Appropriate constant
    A_i = Appropriate constant
    n_i = Appropriate constant

The choice of constants B_i, A_i and n_i needs to satisfy Equation 4.50. The constants can be evaluated using monitoring results for the deterioration of substandard details in future research projects. Again, note that the time measure t is not an absolute time: it is the time measured from the latest retrofit. Finally, a damage measure $D_i(t)$ can be related to $R_i(t)$ such that

$$D_i(t) = 7.0 - R_i(t) \tag{4.52}$$

We can now estimate the total life cycle cost of substandard details in a PT-PS system as

$$C_{\text{DETAIL}} = \sum \bar{C}_{\text{DETAIL}} \tag{4.53}$$

The sum in Equation 4.53 is over the number of retrofits within the lifespan of the system under consideration. Note that the equation does not include discount rates for the sake of simplicity.

4.8.2.4 Total Costs

Total LCCA of PT-PS systems is the sum of all sources of costs as

$$\text{LCC}_{\text{PT-PS}} = C_H + C_{\text{DURABILITY}} + C_{\text{DETAIL}} \tag{4.54}$$

Note that Equation 4.54 includes only costs of retrofits and maintenance. There are numerous other cost sources, such as management, inspection, initial construction, and decommissioning.

4.8.3 EXAMPLE

We consider now a practical example to show the use of the above LCCA method. In this example, we consider only the durability and detail costs. Examples of abnormal hazard costs are shown in Chapters 1, 2, and 11. Table 4.12 shows the different parameters of the problem. The units of these parameters are as described previously. Applying Equations 4.41 through 4.52, the damage estimates as a function of time (for up to 60 years) are shown in Table 4.13. The costs as a function of time can now be computed, using Equations 4.47 and 4.49. The results are shown in Table 4.14.

The results in Table 4.14 show the total cumulative LCCA if no retrofits are done at any given year, up to 60 years. Let us assume that the thresholds of accepted damage were reached after 30 years; in such a situation, two retrofits will be needed: one after 30 years and the other after 60 years. The total LCC costs will be twice the LCC after 30 years, in Table 4.14. Similarly, if the bridge is retrofitted at 20 years interval (three total retrofits), the total LCC costs will be three times the LCC after 20 years, in Table 4.14. Also, if the bridge is retrofitted at 15 years interval (four total retrofits), the total LCC costs will be four times the LCC after 15 years, in Table 4.14. The total LCC costs for 30, 20, and 15 years retrofit intervals are shown in Figures 4.43 and 4.44 and Table 4.15. Clearly, the LCC is a function of the retrofit intervals; as such a careful evaluation of all pertinent cost parameters is needed for optimal cost and performance bridge operations.

This procedure can be used to manage and optimize LCC for PT-PS bridges. The process can be improved by using adequate discount rates. Also, improving the prescribed models for LCC by

TABLE 4.12
LCC Example Parameters

Damage Source	i	B_i	A_i	n_i	C_i	C_{0i}
Conventional corrosion	1	NA	0.015	2	10	70
Stress corrosion	2	NA	0.1	1.5	12	30
Hydrogen embrittlement	3	NA	0.1	1.5	12	30
Loss of tendon force	4	NA	1	0.75	15	40
Exposure	5	NA	2.5	1.2	1.5	10
Durability issues	6	NA	2	1.3	1.2	10
Anchoring details	7	6.5	0.3	0.75	20	20
Segmental joint details	8	6.7	0.3	0.75	25	30

TABLE 4.13
Damage as a Function of Time

Source of Damage	Time (Years)											
	5	10	15	20	25	30	35	40	45	50	55	60
Conventional corrosion	0.38	1.50	3.38	6.00	9.38	13.50	18.38	24.00	30.38	37.50	45.38	54.00
Stress corrosion	1.12	3.16	5.81	8.94	12.50	16.43	20.71	25.30	30.19	35.36	40.79	46.48
Hydrogen embrittlement	1.12	3.16	5.81	8.94	12.50	16.43	20.71	25.30	30.19	35.36	40.79	46.48
Loss of tendon force	3.34	5.62	7.62	9.46	11.18	12.82	14.39	15.91	17.37	18.80	20.20	21.56
Exposure	17.25	39.62	64.45	91.03	118.98	148.08	178.16	209.13	240.88	273.34	306.46	340.19
Durability issues	16.21	39.91	67.60	98.26	131.33	166.45	203.38	241.94	281.97	323.36	366.02	409.85
Anchoring details	0.50	1.19	1.79	2.34	2.85	3.35	3.82	4.27	4.71	5.14	5.56	5.97
Segmental joint details	0.30	0.99	1.59	2.14	2.65	3.15	3.62	4.07	4.51	4.94	5.36	5.77

TABLE 4.14
Cost of Retrofit of Different Damage as a Function of Time

Source of Damage	Time (Years)											
	5	10	15	20	25	30	35	40	45	50	55	60
Conventional corrosion	73.75	85.00	103.75	130.00	163.75	205.00	253.75	310.00	373.75	445.00	523.75	610.00
Stress corrosion	43.42	67.95	99.71	137.33	180.00	227.18	278.48	333.58	392.24	454.26	519.47	587.71
Hydrogen embrittlement	43.42	67.95	99.71	137.33	180.00	227.18	278.48	333.58	392.24	454.26	519.47	587.71
Loss of tendon force	90.16	124.35	154.33	181.86	207.71	232.28	255.85	278.58	300.62	322.05	342.94	363.37
Exposure	35.87	69.43	106.68	146.54	188.47	232.11	277.25	323.69	371.32	420.01	469.69	520.28
Durability issues	29.45	57.89	91.12	127.91	167.59	209.74	254.06	300.33	348.37	398.04	449.22	501.82
Anchoring details	30.06	43.74	55.73	66.74	77.08	86.91	96.34	105.43	114.25	122.82	131.18	139.35
Segmental joint details	37.58	54.68	69.66	83.43	96.35	108.64	120.42	131.79	142.81	153.52	163.97	174.19
Total LCC	384	571	781	1011	1261	1529	1815	2117	2436	2770	3120	3484

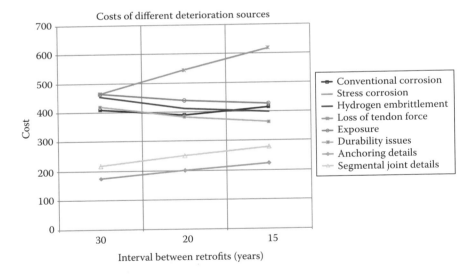

FIGURE 4.43   Costs of different deterioration sources.

incorporating an integrated SHM monitoring system will add immensely to the accuracy of the results.

4.8.4 CONSIDERATIONS OF UNCERTAINTIES

Close inspection of Equation 4.54 would reveal an inconsistency. The expression of cost of abnormal hazards C_H includes some measure of hazard uncertainty as expressed in the probability

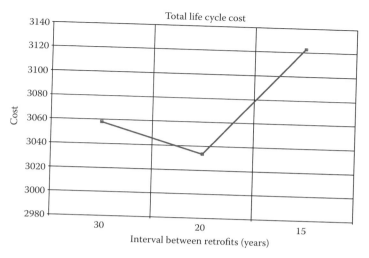

FIGURE 4.44 LCC as a function of retrofit intervals.

TABLE 4.15
LCC of Different Damage Sources and Retrofit Intervals

Source of Damage	30 Years	20 Years	15 Years
Conventional corrosion	410	390	415
Stress corrosion	454	412	399
Hydrogen embrittlement	454	412	399
Loss of tendon force	465	546	617
Exposure	464	440	427
Durability issues	419	384	364
Anchoring details	174	200	223
Segmental joint details	217	250	279
Total LCC	3058	3033	3123

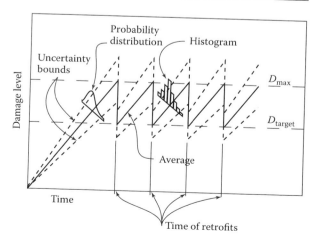

FIGURE 4.45 Uncertainty and time-dependent damage.

term p_h. The costs of durability or substandard details do not have any uncertainly expressions. Clearly, uncertainties need to be considered in any LCCA for accurate decision making. To objectively accommodate some uncertainty measure in cost estimations, let us consider the generalized Figure 4.45. The generalized figure includes uncertainty bounds as time progresses. The uncertainty bounds reflect probability distributions or histograms. We need to know, a priori, either those probability distributions or histograms for uncertainty considerations. These can be estimated by monitoring similar systems. The LCC can then be computed using a stochastic simulation technique such as the Monte Carlo method.

REFERENCES

AASHTO. (1994). *Manual for Conditions Evaluation of Bridges*, American Association of State Highway and Transportation Officials, Washington, DC.

AASHTO. (1996). "Standard Specifications for Highway Bridges," 16th edition including 1997 and 1998 Interim Specifications, American Association of State Highway and Transportation Officials, Washington, DC.

Ali, M. G. and Maddocks, A. R. (2003). "Evaluation of Corrosion of Prestressing Steel in Concrete Using Nondestructive Techniques," *Concrete in the 3rd Millennium*, Proceedings of the 21st Biennial Conference of the Concrete Institute of Australia, Brisbane, Australia.

Bapu, S., Nayak, R., Rajagopalan, N., Srinivasan, C., Rengaswamy, S., and Iyer, M. (1988). "Monitoring of corrosion of prestressing steel cables in prestressed concrete bridges," *Transactions of the SAEST*, 23(2–3), 235-248.

Beard, M., Lowe, M., and Cawley, P. (2003). "Ultrasonic guided waves for inspection of grouted tendons and bolts," *ASCE, Journal of Materials in Civil Engineering,* 15(3), 212-218.

Betti, R., Khazem, D., Carlos, M., Gostaudas, R., and Aktan, E. (2008). "Monitoring Corrosion Inside Main Cables of Suspension Bridges," *NDE/NDT for Highwaya and Bridges: Structural Materials Technology (SMT)*, ASNT, Oakland, CA.

Betti, R., Smyth, A. W., Testa, R. B., Duby, R., and West, A. C. (2000). "Deterioration of the Mechanical Properties of Wires in Suspension Bridge Cables," *Structural Materials Technology: an NDT Conference*, ASNT, Atlantic City, NJ.

Buyukozturk, B. (1998). "Imaging of concrete structures," *NDT & E International*, 31(4), 233-243.

Cao, H. and Davis, A. (2004). "NDT for the detection of Voids in Post-Tensioned Tendon Ducts," *Structural Materials Technology;* an NDT Conference, ASNT, Buffalo, NY.

Carino, N. J. and Sansalone, M. (1992). "Detection of voids in grouted ducts using the impact echo method," *ACI Materials Journal*, 89(3), 296-303.

Chajes, M., Hunsperger, R., Liu, W., Li, J., and Kunz, E. (2002). "Nondestructive Evaluation of Pre- and Post-tensioning Strands Using Time Domain Reflecotmetry," *Proceedings, NDE Conference on Civil Engineering*, ASNT, Cincinnati, OH.

Cheng, C. and Sansalone, M. (1993). "Effects of impact-echo signals caused by steel reinforcing bars and voids around bars," *ACT Materials Journal*, 90(4), 421-434.

Colla, C. (2002). "Scanning Impact-Echo NDE of Post-Tensioning Ducts in Concrete Bridge Beam," *Proceedings, NDE Conference on Civil Engineering*, ASNT, Cincinnati, OH.

Collins, M. and Mitchell, D. (1991). *Prestressed Concrete Structures*, Prentice-Hall, Englewood Cliffs, NJ.

Drissi-Habti, M., Gaillet, L., and Tessier, T. (2008). "Acoustic Emission Technique as a Tool for Structural Health Defects Monitoring of Bridge Cables," *NDE/NDT for Highways and Bridges: Structural Materials Technology (SMT)*, ASNT, Oakland, CA.

Duke, J. (2002). "Health Monitoring Of Post Tension Tendons," Virginia Transportation Research Council Report VTRC 03-CR, Charlottesville, VA.

Dupuis, K., Olsen, K., Musgrove, R., Pearson, E., and Pollock, D. (2008). "Thermal Imaging of Concrete Members to Locate Grouted Post-Tensioning Ducts and Identify Simulated Voids," *NDE/NDT for Highways and Bridges: Structural Materials Technology (SMT)*, ASNT, Oakland, CA.

Escalante, E. (1990). "Effectiveness of potential measurements for estimating corrosion of steel in concrete," *Corrosion of Reinforcement in Concrete*, Elsevier Applied Science, New York, NY.

Ettouney, M. and Alampalli, S. (2012). *Infrastructure Health in Civil Engineering: Theory and Components*, CRC Press, Boca Raton, FL.

Faherty, K. F. (1972). *An Analysis of a Reinforced and a Prestressed Concrete Beam by Finite Element Method*, Doctorate's Thesis, University of Iowa, Iowa City.

Fanning, P. (2001). "Nonlinear Models of Reinforced and Post-tensioned Concrete Beams," *Electronic Journal of Structural Engineering*, University College Dublin, Earlsfort Terrace, Dublin 2, Ireland.

Fisk, P., Holt, R., Sargent, D., and El-Beik, A. (2002) "Detection and Evaluation of Voids in Tendon Ducts by Sonic Resonance and Tomography Techniques," Proceedings, NDE Conference on Civil Engineering, ASNT, Cincinnati, OH.

Flohrer, C. and Bernhardt, B. (1992). "Detection of prestressed steel tendons behind reinforcement bars, detection of voids in concrete structures—A suitable application for radar systems," International Conference on NDT in Civil Engineering, Vol 1.

Fontana, M. (1986). *Corrosion Engineering*, McGraw-Hill, New York, NY.

Ghorbanpoor, A. and Shew, T. (1989). "Detection of flaws in bars and cables in concrete bridge structures," *Transportation Research Record*, (1211).

Hall, D. (2008). "Electromagnetic Inspection of Wire Ropes - Vertical Lift Bridges," *NDE/NDT for Highways and Bridges: Structural Materials Technology (SMT)*, ASNT, Oakland, CA.

Hartle, R., Ryan, T., Mann, J., Danovich, L., Sosko, W., and Bouscher, J. (2002). "Bridge Inspector's Reference Manual," *Federal Highway Administration, FHWA, Report No, FHWA NHI 03–001, Vol 1*, McLean, VA.

Hillemeier, B. (1989). "New Methods in the Rehabilitation of Prestressed Concrete Structures," IABSE Symposium, Vol 57, Lisbon.

Holst, A., Hariri, K., Wichmann, H., and Budelmann, H. (2004). "Innovative Non-Destructive Techniques for the Monitoring of Prestressed Concrete Structures," *Proceedings of 2nd International Workshop on Structural Health Monitoring of Innovative Civil Engineering Structures*, ISIS Canada Corporation, Manitoba, Canada.

Kachlakev, D. I., Miller, T., Yim, S., Chansawat, K., and Potisuk, T. (2001). "Finite Element Modeling of Reinforced Concrete Structures Strengthened With FRP Laminates," California Polytechnic State University, San Luis Obispo, CA and Oregon State University, Corvallis, OR, Oregon Department of Transportation.

Khalil, A., Wipf, T., Greimann, L., and Russo, F. (2002). "Static and Dynamic Testing on a Damaged Prestressed Concrete Bridge," *Proceedings of 1st International Workshop on Structural Health Monitoring of Innovative Civil Engineering Structures*, ISIS Canada Corporation, Manitoba, Canada.

Krause, M., Streicher, D., Wiggenhauser, H., Haardt, P., and Holst, R. (2004). "Non-Destructive Testing of a Post Tensioned Concrete Bridge Applying an Automated Measuring System," *Proceedings, NDE Conference on Civil Engineering*, ASNT, Buffalo, NY.

Kroopf, M., Pedrick, M., and Tittmann, B. (2005). "Remote Sensing Using Ultrasonic Guided Waves in Thin Wires," *Proceedings of the 2005 ASNT Fall Conference*, Columbus, OH.

Limaye, H. (2008). "Experience of Testing Posttensioned Bridges with Impact-Echo Technique," *NDE/NDT for Highways and Bridges: Structural Materials Technology (SMT)*, ASNT, Oakland, CA.

Lin, Y. and Sansalone, M. (1992). "Detecting flaws in concrete beams and columns using the impact-echo method," *ACI Materials Journal*, 89(4), 394-405.

Mahmoud, K. (2003a). "Hydrogen Embrittlement of Suspension Bridge Cable Wires," In *System-Based Vision for Strategic and Creative Design*, Bontempi, F. Ed, Swets & Zeitlinger, Lisse.

Mahmoud, K. (2003b). "An Absorbing Process," *Bridge Design and Engineering*, No. 31, London, UK.

Mahmoud, K. (2011) Personal communications, Bridge Technology Consulting, New York, NY.

Martz, H., Schneberk, D., Roberson, G., and Monteiro, P. (1993). "Computerized tomography analysis of reinforced concrete," *ACI Materials Journal*, 90(3), 259-264.

Matthys, S., Taerwe, L., moerman, W., Waele, W., Vantomme, J., Degrieck, J., Sol, H., Roeck, G., and Jacobs, S. (2002). "Dynamic Monitoring of a Prestressed Concrete Girder by Means of Fiber Optic Bragg Grating Sensors," *Proceedings of 1st International Workshop on Structural Health Monitoring of Innovative Civil Engineering Structures*, ISIS Canada Corporation, Manitoba, Canada.

Olson Engineering, Inc. (2008), "Nondestructive Testing And Evaluation Demonstration Investigation Impact Echo Scanning Results For I-390 Over Genesee River Rochester, NY," Report to NYS-DOT, Albany NY.

Rahman, A. and Pernica, G. (1998). "Assessing New Techniques for Evaluating Posttensioned Buildings," Construction Technology Updates, No 19, Institute for Research in Construction, National Research Council of Canada, Ottawa, Canada.

Riad, K. H. and Fehling, E. (2004). "Determination of the Effective Tensile Force in External Prestressing Cables on the Basis of Vibration Measurements," *Proceedings, NDE Conference on Civil Engineering*, ASNT, Buffalo, NY.

Rizzo, P., de Scalea, F., and Marzani, A. (2004). "Guided Acoustic Waves for Health monitoring of Tendons and Cables," *Proceedings of 2nd International Workshop on Structural Health Monitoring of Innovative Civil Engineering Structures*, ISIS Canada Corporation, Manitoba, Canada.

Rogowsky, D. and Robson, N. (1998). "Influence of Corrosion and Cracking on Tendon Strength and Ductility," *Workshop on Environment-Assisted Cracking of Unbonded Posttensioned Tendons*, Calgary, Alberta, Canada.

Rose, J. (1999). *Ultrasonic Waves in Solid Media*, Cambridge University Press, Cambridge, UK.

Salas, A, Schokker, A., West, J., Breen, J., and Kreger, M. (2004). "Conclusions, Recommendations and Design Guidelines for Corrosion Protection of Posttensioned Bridges," Research Report FHWA/TX-04/0–1405-9, Texas Department of Transportations and Federal Highway Administration, Austin, TX.

Saleh, H., Goni, J., and Washer, G. (2002). "Radiographic Inspection of Posttensioning Tendons Using High Energy X-Ray Machine," *Proceedings, NDE Conference on Civil Engineering*, ASNT, Cincinnati, OH.

Schokker, A., Koester, B., Breen, J., and Kreger, M. (1999). "Development of High Performance Grouts for Bonded Posttensioned Structures," Research Report 0–1405-2, Center for Transportation Research, Bureau of Engineering Research, The University of Texas at Austin, TX.

Stein, S. and Sedmera, K. (2006). "Risk-Based Management Guidelines for Scour at Bridges with Unknown Foundations," National Cooperative Highway Research Program, NCHRP Project 24–25, Washington, DC.

Streicher, D., Algernon, D., Wostmann, J., Behrens, M., and Wiggenhauser, H. (2006). "Automated NDE of Posttensioned Concrete Bridge using Radar, Ultrasonic Echo and Impact Echo," *NDE Conference on Civil Engineering*, ASNT, St. Louis, MO.

Tinkey, Y., Miller, P., and Olson, L. (2008). "Internal Void Imaging Using Impact Echo Scanning," *NDE/NDT for Highways and Bridges: Structural Materials Technology (SMT)*, ASNT, Oakland, CA.

Tinkey, Y. and Olson, L. (2008). In *Transportation Research Record: Journal of the Transportation Research Board, No.* 2070, Transportation Research Board of the National Academies, Washington, DC.

Venugopalan, S. (2008). "Corrosion Evaluation of Post-Tensioned Tendons in a Box Girder Bridge," 10th International Bridge Management Conference, Buffalo, NY.

Wang, M.L., Zhao, Y, and Wang, G. (2004) "Applications of Magnetoelastic Sensors to Force Measurements in Large Bridge Cables," Proceedings, NDE Conference on Civil Engineering, ASNT, Buffalo, NY.

Washer, G. and Fuchs, P. (2004). "Ultrasonic Health Monitoring of Prestressing Tendons," *Proceedings, NDE Conference on Civil Engineering*, ASNT, Buffalo, NY.

West, S. (1999). "Durability Design of Post-Tensioned Bridge Substructures," Ph.D. Dissertation, The University of Texas at Austin, TX.

West, S., Larosche, C., Koester, B., Breen, J., and Kreger, M. (1999). "State-of-the-Art Report about Durability of Posttensioned Bridge Substructures," Research Report 0–1405-1, Center for Transportation Research, Bureau of Engineering Research, The University of Texas at Austin, TX.

Willenbrock, J. (1980). "Construction QA/QC Systems: Comparative Analysis," *ASCE Journal of Construction Division*, 106(3), 371-387.

Woodward, R. (2001). "Durability of Post-Tensioned Tendons on Road Bridges in the UK." Durability of Posttensioning tendons. fib-IABSE Technical Report, Bulletin 15, Ghent (Belgium).

Yamaoka, Y., Hideyoshi, T., and Kurauchi, M. (1988). "Effect of galvanizing on hydrogen embrittlement of prestressing wire," *Journal of the Prestressed Concrete Institute*, Chicago, IL.

Yunovich, M. and Thompson, N. (2003). "Corrosion of Highway Bridges: Economic Impact and Control Methodologies," *Concrete International*, American Concrete Institute, ACI, January.

Zheng, Y. and Ng, K. (2006). "Detection of Voids in Tendon Ducts of Posttensioned U-Beams for an Existing Bridge," *NDE Conference on Civil Engineering*, ASNT, St. Louis, MO.

Zhou, Z., Wegner, L., and Sparling, B. (2004). "Vibration-Based Detection of Multiple Damage States on a Prestressed Concrete Girder," *Proceedings of 2nd International Workshop on Structural Health Monitoring of Innovative Civil Engineering Structures*, ISIS Canada Corporation, Manitoba, Canada.

Zienkiewicz, O. (1971). *The Finite Element Method in Engineering Science*, McGraw-Hill, New York, NY.

5 Fatigue

5.1 INTRODUCTION

5.1.1 GENERAL

Fatigue is the process that describes how damage accumulates as a stress/strain cycle within metals. As the number of stress/strain cycles increase, the cumulative damage increases. Under certain conditions, the cumulative damage will lead to metal failure.

The initial damage due to stress/strain cycles will occur as cracks. As the number of cycles increase, the crack will grow. Depending on the stress/strain magnitude and the type of material, the crack growth either will be arrested or will continue until the material fails. There are three modes (types) of crack formations (Barsom and Rolfe 1999):

- Mode I: Tearing mode where two sides of the crack pull away from each other.
- Mode II: Sliding mode, where the two sides of the crack slide against each other in a symmetric fashion.
- Mode III: Shearing mode, where the two sides of the crack slide against each other in an asymmetric fashion.

Most of fatigue cracks in civil infrastructures are of type I. Factors affecting fatigue life include loading, design errors, manufacturing, construction or assembly defects, material defects, poor quality assurance (QA) or quality control (QC), residual stresses, and environmental conditions such as temperature or humidity. Another important aspect to keep in mind is that fatigue is highly uncertain. Any efforts regarding fatigue effects or mitigation measures need to account for this uncertainty to some degree.

One of the important fatigue failure properties is that the threshold of the fatigue failure stresses is a function of the stress level. As the stress level is reduced, the number of cycles that can occur before failure is increased. At the limit, there is a stress value below which no failure will occur no matter how many stress cycles are applied to the material. Such a relationship is known as S-N curves. This phenomenon is known as high-cycle fatigue. If a cyclic-loading condition causes plastic behavior in the metal, then the number of cycles before fatigue failure is reduced; this phenomenon is known as low-cycle fatigue.

Fatigue is considered a serious hazard in civil infrastructures. There are many aspects to fatigue, some of which are shown in Figure 5.1.

5.1.2 OVERVIEW OF FATIGUE DETECTION METHODS

Effects of fatigue hazard are detected usually by one of two approaches: direct or indirect, as shown in Figure 5.2. Direct fatigue detection is estimated by direct measurement of fatigue effects, such as cracks. Different detection methods of nondestructive testing (NDT)/structural health monitoring (SHM) can be used. Indirect methods measure parameters that affect fatigue and then estimate fatigue effects based on those parameters. This approach is defined as virtual sensing paradigm (VSP). VSP can either detect fatigue damage or estimate fatigue remaining life. All these methods are discussed in detail in Section 5.4.

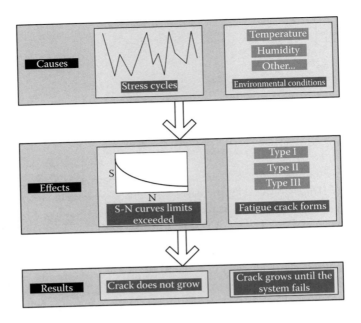

FIGURE 5.1 Fatigue causes and effects.

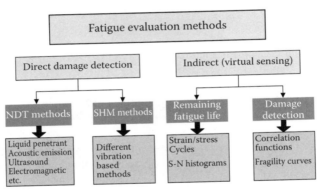

FIGURE 5.2 Fatigue detection methods.

5.1.3 This Chapter

This chapter will discuss different NDT techniques that can detect fatigue damage. We offer a similar discussion about SHM methods. VSP is then introduced. We note that VSP is applicable to many other hazard effects, not only fatigue. We present a step-by-step guide to monitoring fatigue remaining life in bridges. This chapter concludes by presenting two fatigue-monitoring case studies. Figure 5.3 shows the composition of this chapter.

5.2 NDT TREATMENT OF FATIGUE

5.2.1 Overview of Methods

NDT methods have been used in fatigue-related problems in many ways. The first step in choosing an adequate NDT method for a fatigue-related activity is to determine whether such an activity is to estimate remaining fatigue life or whether fatigue damage has already occurred. Table 5.1 shows the utility of different NDT methods as they apply to the temporal sequencing of the fatigue problem.

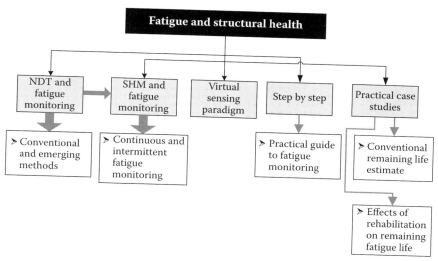

FIGURE 5.3 Outline of fatigue chapter.

TABLE 5.1
Temporal Utility of NDT Methods for Fatigue Hazard

NDT Method	Before Event	After Event
Leak testing	NA	Can be used in a virtual manner to detect fatigue cracks in fluid or pressure vessels. NA for most civil infrastructures, including bridges
Liquid penetrant testing	NA	Can be used to detect surface cracks
Penetrating radiation	NA	
Electromagnetic	NA	
Acoustic emission	Can be used to measure stress signals, then detect remaining fatigue life	
Ultrasonic	Can be used to measure stress signals and then to detect remaining fatigue life in a virtual manner	
Visual/manual	NA	Only larger size surface cracks can be observed
Thermography	NA	Can be used to detect surface cracks

Based on Table 5.1 and the objectives of the activity, several NDT methods can be discarded. Also, we note that most NDT methods are applicable only after the fatigue damage has occurred. Acoustic emission (AE) and ultrasonic methods can be used before and after the event since both methods rely on observing some form of stress (or strain) wave forms, either actively or passively. These wave forms can be analyzed to evaluate stress cycles, which can virtually produce an accurate estimate of the remaining fatigue life.

5.2.1.1 Identification, Utility, and Prioritization

We explored the concepts of identification adequacy, utility, and prioritization for different NDT and SHM methods for damage detection. It is of interest to discuss these concepts as they relate to the use of different NDT methods in fatigue problems. Tables 5.2 and 5.3 show the identification adequacy and utility of different NDT methods, respectively. Note that the utilities of NDT methods presented are specifically for bridge structures. These utilities might vary for other types of civil infrastructure or other types of engineering systems (mechanical, electrical, etc.).

TABLE 5.2
Identification Adequacy of NDT Methods for Fatigue

NDT Method	Scale (Size) of Crack	Failure Modes
Leak testing	Crack size can be detected virtually by relating to the pressure of the leaking fluid	Full depth of plate
Liquid penetrant testing	Minimum 1.0 µm cracks	Surface
Penetrating radiation	Changes of density of about 2%. Microwave can detect cracks of order of 0.25 µm. Computed tomography can detect cracks to 0.1 mm	Surface, interior
Electromagnetic	Cracks to 0.03 mm	Surface, interior
Acoustic emission	Varies	Surface, interior
Ultrasonic	Cracks to 0.01 mm.	Surface, interior
Visual/manual	Varies	Surface
Thermography	Varies	Surface, interior

TABLE 5.3
Utility of NDT Methods for Fatigue (as Applied to Bridges)

NDT Method	Size of Equipment	Simplicity	Low Environmental Effects?	Maintenance Needs	*In situ* versus Laboratory Settings	Cost
Leak testing	M	M	M	L	M	M
Penetrating radiation	L	L	L	L	L	L
Electromagnetic	M	M	M	M	M	M
Acoustic emission	H	H	H	M	H	H
Ultrasonic	H	H	H	M	H	H
Visual/manual	H	H	H	L	H	M
Thermography	M	M	L	L	M	L

H, high utility; L, low utility; M, medium utility; NA, not applicable.
Maintenance needs include needs of continuous testing.
In situ versus laboratory-setting utility evaluates if the NDT method can easily be used for an SHM *in situ* investigation.
Low-cost utility indicates higher costs and reverse is true.

For a specific fatigue investigation, the professional should evaluate NDT methods using Tables 5.2 and 5.3. The prioritization of NDT methods can then be investigated quantitatively by following the method of Ettouney and Alampalli (2012, Chapter 7).

We note that Tables 5.2 and 5.3 deal with fatigue effects only, that is, detection of fatigue cracks *after* they occur. Virtual estimation of remaining fatigue life by using NDT methods is discussed in Sections 5.4 and 5.7.

5.3 SHM TREATMENT OF FATIGUE

5.3.1 OVERVIEW

Many authors have explored fatigue behavior in bridges, for example, Tarris et al. (2002) discussed remote monitoring of plate girder bridges. The investigation included identifying sources of fatigue cracking in plate girders and effects of retrofit. Jalinoos et al. (2008) presented the Federal Highway

Administration steel-testing program. It uses three laboratory sensors for active defects: (1) electro-chemical fatigue sensor, (2) Eddy current, and (3) AE sensors. This section explores several SHM fatigue-related aspects.

5.3.2 Factors Affecting Fatigue

It is important to understand the way different factors can affect fatigue damage formation, both before and after the damage occurs. Some of the important factors are (see Table 5.4):

Surface condition: Surface condition is an important factor in fatigue formation. In any SHM experiment, surface conditions, both before and after crack formation, must be included. In virtual fatigue sensing, correlations between surface conditions (grounded, polished, corroded, machined, etc.) must be well documented.

Crack size: Crack size can affect the method of detection. It also can affect the number of sensors, sensor locations, and the type of sensors. Temporal and special resolution of the experiment is directly affected by the size of cracks that can be detected experimentally.

Type and magnitude of loading: Type of loading and magnitude of the loading are directly correlated to fatigue damage and remaining fatigue life. Therefore, the type and magnitude of loading are essential parts in any SHM experiment, virtual or direct.

Temperature: Correlation between temperature and fatigue damage is well established (Barsom and Rolfe 1999). Measurement of temperature can be used to detect fatigue damage in a virtual experiment; this is particularly true when the fatigue formation temperature correlation is well established.

Residual stresses: It is well known that residual stresses can have a major effect on fatigue damage formation (Barsom and Rolfe 1999). Placing sensors near areas that are suspected to have higher residual stresses is needed.

Metal corrosion: Corrosion of metals can lead to ductile degradation of engineering properties and can eventually cause fatigue fracture. Therefore, the state of corrosion/rust in the structure must be studied carefully. Since corrosion of metals can develop in hidden areas, between steel components, for example, care in designing SHM experiment is needed.

Other factors: Other factors, such as fretting and humidity, can affect fatigue formation. These need to be considered whenever fatigue effects are studied.

TABLE 5.4
Importance of Factors Affecting Fatigue

Factor	Importance before Crack Formation	Importance before Crack Formation
Surface condition	H	H
Crack size	NA	H
Load type (cyclic, random, sudden, etc.)	H	M
Load magnitude	H	M
Temperature	NA	M
Residual stresses	H	H
Corrosion	M	L

H, high; L, low; M, medium; NA, not applicable.

Uncertainties: Correlating a discrete number of sensing locations to a fairly uncertain event such as fatigue damage will require good statistical analysis. Since some fatigue failures can be catastrophic, a conservative analysis is warranted in those situations.

5.3.3 Stress-Cycle Count

Correlating measured stress or strain signals to a number of cycles is an integral part of estimating the remaining fatigue life of systems. Many procedures are available. The most popular method is the time domain rainflow method of cycle counting. Also, for stationary random signals, rainflow cycle counting can be performed in the frequency domain. We summarize both methods next.

5.3.3.1 Rainflow: Time Domain

The rainflow method is a simple method for counting equivalent stress cycles in a general stress wave form. The wave form can be general, with no restrictions on its properties. This makes the rainflow method ideal for estimating fatigue remaining life in SHM projects. A simple rainflow algorithm was developed by Downing and Socie (1982), and the method was developed into an ASTM standard E 1049–85 (1985). The method envisions a random stress time history, as shown in Figure 5.4, which is turned 90° (Figure 5.5). A series of rain drops originates from the positive peaks 1, 5, and 6 and slides on top of the lines representing the time history as if it were the inclined roofs of a building. Thus, the flow of the rain drops on positive peaks can be decomposed into a series of paths, and each path will constitute a single half stress cycle. The range between the origin of the path and the path termination point is the magnitude of the half stress cycle. The origin, path, and termination of each of the series of the rain drops are governed by the following simple rules:

1. The half cycle (path) will terminate if it reaches the end of the time history.
2. The half cycle will terminate if it merges with another path that originated at another positive peak.
3. It meets another positive peak with higher magnitude.

The process or counting half cycles and the corresponding stress ranges will continue until the end of the time history is reached. A similar process is repeated for the negative peaks of the time history, with similar rules. It should be expected that for a long enough time history, most of the positive half cycles will have their exact negative half cycles, thus producing a count for full cycles of the time history of interest.

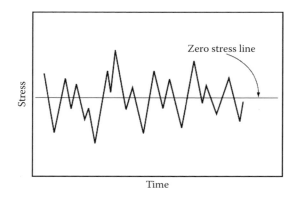

FIGURE 5.4 Typical general stress time history.

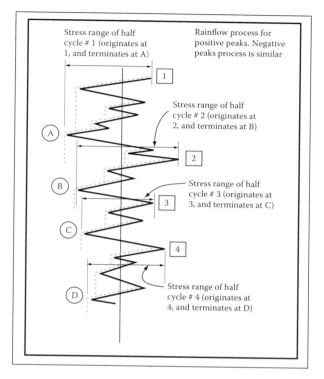

Stress range of half cycle # 1 (originates at 1, and terminates at A)

Rainflow process for positive peaks. Negative peaks process is similar

Stress range of half cycle # 2 (originates at 2, and terminates at B)

Stress range of half cycle # 3 (originates at 3, and terminates at C)

Stress range of half cycle # 4 (originates at 4, and terminates at D)

FIGURE 5.5   Rainflow process for positive peaks.

As an example of applying the rainflow process to stress time history of Figure 5.4, consider Figure 5.5. The first half cycle of the positive peaks originates at point #1 and terminates at point A. Three additional stress half cycles are counted. The stress ranges in the four half cycles are shown. A similar process for the negative peaks is performed, and the total number of equivalent complete stress cycles is counted, with their corresponding stress ranges.

5.3.3.2   Rainflow: Frequency Domain

If the measured time signal is stationary and Gaussian, rainflow cycle counting can be performed in the frequency domain. These conditions usually occur during random motions of systems on hand. For example, wind-induced vibrations can generate time signals that satisfy the rainflow frequency domain analysis. Note that nonstationary signals, such as those produced by individual truck movement on highway bridges, are not suited for the frequency domain method. We present the method developed by Bishop and Sherratt (1989) for the evaluation of rainflow method using frequency domain. Consider the one-sided power spectral density (PSD) of a measured such a stress signal $G(f)$, where f is the frequency scale. The nth moment of the PSD is expressed as

$$m_n = \int_0^\infty f^n G(f)\mathrm{d}f \tag{5.1}$$

If can be shown that the expected number of zero crossings per second of the signal is

$$E[0] = \sqrt{\frac{m_2}{m_0}} \tag{5.2}$$

Similarly, the expected number of peaks per second and the expected irregularity factors of the signal are, respectively,

$$E[P] = \sqrt{\frac{m_4}{m_2}} \tag{5.3}$$

and

$$\gamma = \sqrt{\frac{E[0]}{E[P]}} = \sqrt{\frac{m_2^2}{m_0 \, m_4}} \tag{5.4}$$

We can now compute the total number of stress range cycles N_i for the whole length of measured time history T as

$$N_i = T \, E[P] \tag{5.5}$$

More importantly, the number of cycles n_i for a given stress range S_i is

$$n_i = N_i \, p(S_i) \mathrm{d}S \tag{5.6}$$

Where $p(S_i)$ is the probability density function of the occurrence of the stress range S_i.

There have been several expressions for $p(S_i)$. Perhaps the most general is the expression by Dirlik (1985), which developed the following:

$$p(S) = \frac{D_1 / Q e^{-Z/Q} + (D_2 / R^2 + D_3 Z) e^{-Z^2/2}}{2\sqrt{m_0}} \tag{5.7}$$

All variables of Equation 5.7 are functions of the moments of the PSD as computed by Equation (5.2). They are detailed in Section 5.9.

5.3.4 NDT Applications

5.3.4.1 Merrimac Free Ferry

The Merrimac Ferry carries for free Wisconsin SR 113 traffic across the south end of Lake Wisconsin on the Wisconsin River in Merrimac, WI (Figure 5.6). The 80-ft-long, three-lane ferry has been in operation since 1963, and recent inspections showed fatigue cracks at several locations in the hull. Strain gauges and a remote monitoring system were used to monitor live traffic and wind-loading effects.

Prine and Socie (2000) describe some of the issues faced during the installation: water issues due to condensation, security and safety of the system, issues with using power from an unregulated source (which can cause noise in data), high temperatures, and the use of wired sensors spread across the boat, requiring multiple computers and battery sources. Data was collected using wireless transmission from the ferry.

Strain gauge data (Figure 5.7) is recorded in both rainflow counting mode (for fatigue analysis) and short-burst time histories (triggered by preset threshold strain to check rainflow data reliability). Test data was also collected using known weight truck driven on and off at preset paths, and the strains observed in some welds indicated the reasons for cracking due to stress concentrations. Data was measured for nearly a year.

Based on the test data, the load posting was reduced for safety reasons.

FIGURE 5.6   The Merrimac Free ferry. (Courtesy of CRC Press.)

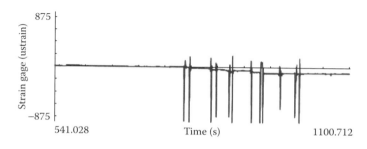

FIGURE 5.7   Strain gauge results. (Courtesy of CRC Press.)

5.3.4.2   Welded Railway Bridges

Details of the influence of welding residual stresses on the fatigue life of structural components are important, in cases such as the use of the results of fatigue testing of relatively small welded specimens without high-tensile residual stresses, analysis of the effect of such factors as overloading, spectra loading, and application of the improvement treatments. An advanced ultrasonic method with a new portable device was used by Kudryavtsev et al. (2000) for the residual stress measurement in the zones of welded elements of the railway bridge span.

The ultrasonic method of stress measurement is based on the acoustic elasticity effect, according to which the velocity of elastic wave propagation in solids is dependent on the mechanical stresses. The acoustic relationships are the theoretical grounds for the development of acoustic methods of stress measurement and in investigations of physical-mechanical properties of materials by means of ultrasound. For the measurement of residual stresses in a material with known mechanical properties, the propagation velocities of the longitudinal ultrasonic wave and shear waves of orthogonal polarization are determined. The mechanical properties are represented by the proportionality coefficients, which can be calculated or determined experimentally under the uni- or bi-axial loading of a sample of the considered material. The change of the acoustic wave velocity in structural materials under mechanical stresses amounts to tenths of a percent, and hence high resolution, reliable, and fully computerized equipment and software are required.

The measurement showed that the tensile residual stresses near the welds reach 200–240 MPa. These stresses are one of the factors leading to the origination and propagation of the fatigue cracks in welded elements. The application of the improvement treatment caused a significant redistribution of residual stresses and a 45% increase in the limit stress range of the welded element.

5.3.4.3 Fatigue and Bridge Rating

The State Road Viaduct is a 3232-m-long dual structure and I-95 in Philadelphia. It consists of noncomposite, cover-plated, multibeam continuous units integrally framed to steel box girder or parallel twin plate girder pier caps, and noncomposite multigirder simple and continuous units supported on conventional reinforced concrete pier bents. The analysis conducted by Lai and Ressler (2000) indicated that the fatigue cracking theoretically will start to develop at those locations because their fatigue lives have been exhausted. However, field inspections detected no fatigue cracks. So, load testing was conducted to obtain actual stress levels for fatigue life determination (Figure 5.8). The strain test data also proved to be very valuable in understanding the in-service structure.

Follow-up testing showed that the analytical fatigue life estimate procedures yielded a very conservative fatigue life estimate for the viaduct. Other conclusions from the data are

1. The structure was designed as a noncomposite one, but it exhibits stronger composite action.
2. The measured stresses are a lot lower than the analytical stresses and were attributed to unintended composite action.
3. The measured stress range cycles are much less than the projected truck traffic counts on the viaduct.

5.3.4.4 Long-Range Ultrasonic Fatigue Detection

Ultrasonic Lamb waves were experimented by Woodward and McGarvie (2000) to detect known fatigue cracks in I-40 steel bridge girders in New Mexico. Lamb waves have the capacity for propagating long distances, allowing the inspection to take place from the abutment or from some other easily accessible location. Lamb waves have the capacity for propagating over long distances in thin plates such as the web of a steel bridge girder. Fatigue cracks in steel beams have been detected at ranges of over 100 ft. The generation of Lamb waves can be accomplished using conventional ultrasonic flaw detectors and transducers. However, results are limited to short ranges with limited crack-detecting abilities when using conventional equipment.

Linear ultrasonic methods refer to the case where a Lamb wave is propagated down to girder's web. If a crack or other discontinuity is present, part of the wave is reflected back to the transducer. This reflected wave (echo) contains information on the location and severity of the discontinuity.

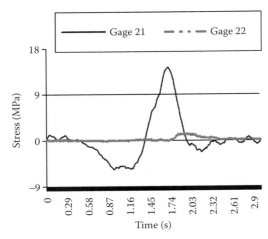

FIGURE 5.8 Typical measured stress. (Courtesy of CRC Press.)

FIGURE 5.9   Fatigue crack—I-40, Edith Street Bridge. (Courtesy of CRC Press.)

Nonlinear methods use truck traffic or another external energy source to open and close the crack while it is being isonified by a Lamb wave. The interaction of the Lamb wave and the opening and closing of the crack gives additional information as to the character of the discontinuity. Both methods are capable of crack detection and health monitoring of existing cracks.

The I-40 Edith Street Bridge was initially inspected visually and by using a dye penetrant as needed to determine the size and geometry of the fatigue cracks (see Figure 5.9). Eleven beams were selected for further study. Nine contained fatigue cracks, while two uncracked beams were used as controls.

The area of the reflected wave (echo) received from the diaphragm connection was analyzed, using linear methods to determine if a crack was present. In general, the area of the echo increased if a crack was present when compared to the control beam. Of interest was that one of the control beams also exhibited an increase in area of the echo. A further detailed conventional anglebeam ultrasonic inspection of the welded area of the beam indicated a crack. Thus, the Lamb wave method found a crack which was not found using conventional visual inspection methods, further verifying the potential of the Lamb wave method.

5.3.4.5   Remaining Fatigue Life Using Load Tests

The design fatigue life of a bridge component is based on the stress spectrum that the component experiences and the fatigue durability. Changes in traffic patterns, volume, and any degradation of structural components can influence the fatigue life of the bridge. A fatigue life evaluation reflecting the actual conditions has value for bridge owners. Procedures are outlined in the AASHTO (1990) to estimate the remaining fatigue life of bridges using the measured strain data under actual vehicular traffic. Lund and Alampalli (2004a and 2004b) presented an actual case study describing the testing of the Patroon Island Bridge (Figure 5.10). The bridge consists of 10 spans. Spans 3 through 9 are considered the main spans and consist of steel trusses and concrete decks. Spans 1, 2, and 10 are considered approach spans and consist of plate girders. The overall bridge length is 1795 ft. Strain data from critical structural members were used to estimate the remaining fatigue life of selected bridge components. The results indicate that most of the identified critical details have an infinite remaining safe fatigue life, and others have a substantial fatigue life. Cracked floor beams were not addressed in this analysis, but have been recommended for retrofitting or replacement. This application is described in detail in Section 5.7.

5.3.4.6   Condition Assessment of Concrete Bridge T-Girders

Impact echo testing (IET) was used by Gassman and Zein (2004) to detect change in stiffness of a T-beam cut from an old bridge (Figure 5.11), which was in good condition based on visual

FIGURE 5.10 Patroon Island Bridge. (Courtesy of New York State Department of Transportation.)

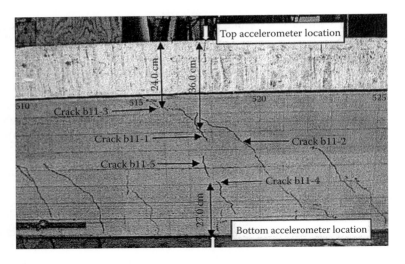

FIGURE 5.11 Typical cracks. (Reprinted from ASNT Publication.)

inspection. IET was used to assess initial condition, and then it was subjected to fatigue loading (8.5–81.2 kip load cycle on the girder) of 20k, 100k, 500k, 1000k, 1500k, and 2000k cycles at 2-Hz rate (Figure 5.12). The results showed that the P-wave velocity can be correlated to change in stiffness and radial distance to the cracks. The authors note that testing from top of the deck results in clearer peaks than testing from the bottom of the girder due to the reduced effect of side boundaries on the response. Figure 5.13 shows the final results of the experiment.

5.3.4.7 Paint Sensors

Zhang (2004) presented a piezoelectric paint sensor for monitoring fatigue. Piezoelectric paint was fabricated using three components: piezoelectric ceramic particles, a polymer binder to facilitate suspension of filler during the application and to bind the filler together after curing, and chemical additives to enhance the paint mixing, deposition, and curing properties. Corona polling method was used to induce piezoelectricity. In this method, large DC potential is applied to a

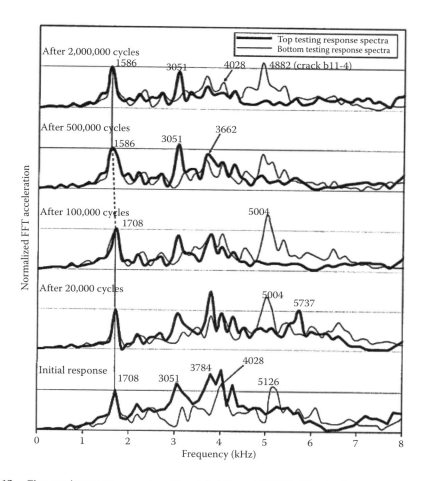

FIGURE 5.12 Changes in response spectra. (Reprinted from ASNT Publication.)

No.	NDE testing: Crack frequency (Hz)	NDE testing: Crack distance (cm)	Visual: Crack distance (cm)
b11	4028 (T)	36	36
	5615 (T)	24	28
	5126 (B)	27	28
b12	1953 (T)	74	71.5
	5615 (T)	24	25.5
	3417 (T)	42	46.5
	5004 (B)	29	27
b13	5859 (T)	25	31
	4028 (T)	36	39.7
	4638 (B)	31	29

No.	NDE testing: Crack frequency (Hz)	NDE testing: Crack distance (cm)	Visual: Crack distance (cm)
b14	5859 (T)	25	22
	2319 (B)	63	63
	3173 (T)	45	46
b17	5126 (T)	30	29
	5859 (T)	27	28
	1953 (T)	79	76
b18	4028 (T)	39	42.5
	2319 (T)	67	67
b19	3662 (T)	37	37
	3662 (B)	37	37

(T) Detected from top testing.
(B) Detected from bottom testing.

FIGURE 5.13 Final crack distances. (Reprinted from ASNT Publication.)

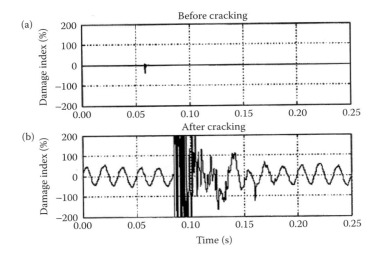

FIGURE 5.14 Damage index time history of piezoelectric sensor (a) before cracking and (b) after cracking. (Reprinted from ASNT Publication.)

set of needles (corona points) which act as field intensifiers, causing an ionization of surrounding molecules. One surface is electroded and grounded. The charge from the needle is then sprayed onto the electrode surface, causing an electric field.

The lab experiment included an aluminum sheet covered with this paint clamped to a heavy steel block. An impact hammer was used to excite free vibrations of the aluminum sheet with output voltage measured using a digital signal analyzer. Results indicated that this sensor can measure the vibration reliably with good repeatability (Figure 5.14).

SHM can be done using a piezoelectric sensor with a multiple-electrode configuration. If there is a crack crossing the sensor, then the voltage signal in two channels (assuming two-electrode configuration) will be the same; if not, the channels will give different voltage signals. The advantages are (1) it is a self-powered sensor, (2) it can be directly deposited onto the surface and hence can be fabricated to suit any surface configuration, (3) it can be optimized for the type of material under investigation, (4) due to ease of processing, complex sensor patterns can be utilized, and (5) the paint can be used potentially for replacing the conventional paint as smart paint.

5.3.4.8 Impact-Echo Testing of Fatigue

Impact echo method is a technique that generates elastic stress waves into a test structure and uses the monitoring of the frequency of reflections to detect internal damage. Zein and Gassman (2006) used impact echo to measure the quality of concrete and the extent of damage in concrete specimens by measuring the reflected P-waves. This study used impact echo method to test four T-beam girders from a decommissioned bridge, two intact and two retrofitted with fiber reinforced polymer (FRP) laminates to enhance their flexural strength. Both static and cyclic load testing was done. Based on the testing, it was found that the carbon fiber reinforced polymers (CFRP) structures had higher stiffness than the intact/control specimens as measured by the lower P-wave velocity (see Figure 5.15). On the basis of the study, the authors concluded that measuring P-wave velocities periodically during the life cycle of a concrete bridge member can be used to effective bending stiffness and maximum deflection, both of which can be used to estimate the remaining service life of the structural member.

5.3.4.9 Active Crack Growth Detection

Miceli, Moshier, and Hadad (2008) describe the use of Electrochemical Fatigue Sensor (EFS) sensors to detect active crack growth. The EFS system is an NDT method that has potential to detect active crack growth, either of known cracks or in areas that are susceptible to fatigue cracking.

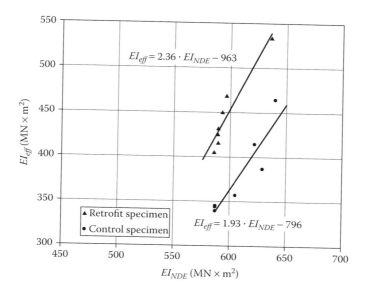

FIGURE 5.15 Correlation between computed and measured bending stiffness. (Reprinted from ASNT Publication.)

The system includes two sensors, a reference sensor and a crack sensor, a potentiostat that applies constant polarizing voltage between the structure and the sensors, a ground, and data collection and analysis software. During testing, the inspection areas encompassed by the sensors are anodically polarized to create a passive film on the areas of interest. This polarizing voltage produces a DC base current in the electrochemical cell. As the structure is exposed to cyclic stresses, the current flowing within the cell fluctuates in a complex relation to the variations in the mechanical stress. This results in an AC current superimposed on the base DC current. During cyclic loading, the fatigue process causes microplasticity and strain localization on a very fine scale. The interaction of the cyclic slip and the passivating process (due to the applied polarizing voltage) causes temporary and repeated changes to the passive layers. These disruptions, including both dissolution and repassivating processes, give rise to transient currents (Figure 5.16 through Figure 5.19). Depending on the properties of the materials, the loading conditions, and the activity of the cracks under inspection, this transient current provides information on the status of fatigue damage at that location.

Crack sensor is placed at the tip of the crack, and the reference sensor is placed away from the crack but in the area subjected to the same stress field. In general, crack growth is indicated when the ratio of the crack sensor output to the reference sensor output in both the frequency and time domains is at least 2.0. This has been termed the "energy ratio." Current output for the crack measurement sensor in the range of 1.5 to 1.9 times that of the reference indicates that microplasticity may be occurring at that location and that the area is at an elevated risk for future crack growth. Those areas should be kept under observation. Output below 1.5 generally indicates that little to no crack growth is taking place. These are general and simplified guidelines for quickly determining the crack activity.

This system has been applied in several cases, including in New York. Retrofits were designed and installed at four locations in Patroon Island Bridge in New York, where distortion-induced cracking occurred in the floor beam webs, at the connection with the gusset plates. The designed retrofit required softening the connection by removing a section of the floor beam web in the shape of a large teardrop (Figure 5.20). Two locations on each of the four retrofit locations were inspected with the EFS system to determine if active fatigue cracks were present and if the four retrofits had been successful in reducing the likelihood of future fatigue cracking.

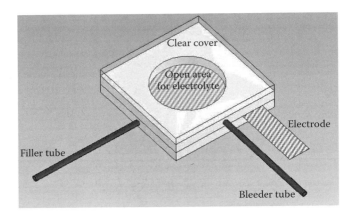

FIGURE 5.16 Schematics of the EFS sensor. (Courtesy of Marybeth Miceli.)

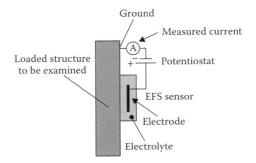

FIGURE 5.17 Schematics of the EFS sensor in use. (Courtesy of Marybeth Miceli.)

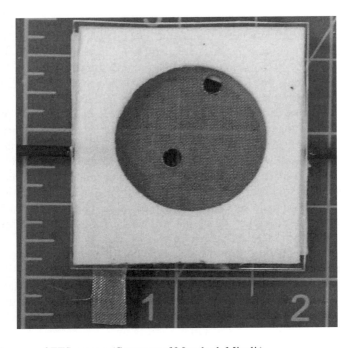

FIGURE 5.18 Close up of EFS sensor. (Courtesy of Marybeth Miceli.)

FIGURE 5.19 EFS sensor in wireless assembly. (Courtesy of Marybeth Miceli.)

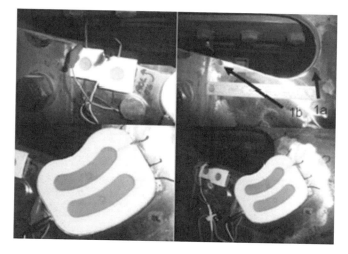

FIGURE 5.20 Installed EFS sensor. (Courtesy of Marybeth Miceli.)

5.4 VIRTUAL SENSING PARADIGM

5.4.1 FATIGUE AND GENERAL SENSING

We showed earlier that the fatigue problem can be subdivided into two phases: before damage and after damage. Before the damage occurs, we are interested in knowing if fatigue damage might occur and, if so, when. After the damage occurs, we are interested in knowing if it really occurred, where, and how extensive it was. In both phases, the fatigue damage is treated indirectly (except, of course in direct NDT damage detection methods). We call this indirect damage handling "virtual sensing," which is the subject of this section. Since VSP has a wider use than just fatigue damage in the SHM field, we first discuss the general need for VSP. We then introduce the theoretical basis of VSP. A simple step-by-step guide to the general use of VSP is discussed next. We then explore some examples of VSP use, both in fatigue damage detection and in other situations of SHM field.

5.4.2   GENERAL NEED FOR VIRTUAL SENSING

Damage identification methods used for structural systems can be classified into direct and inferred classes (Figure 5.21). Direct damage identification methods detect the damage directly. Inference methods do not detect or measure damage directly, but identify it by measuring some parameters that would give us a means to identify the damage. Vibration-based structural identification is a good example of the inferred class of damage identification. The logic of this approach is to measure the dynamic responses of the structure, identify a set of structural properties (such as modal properties) using the dynamic response, and then correlate the damage to the identified changes in the structural properties. The logic is fairly simple, but the vibration-based approach has proved to be very limited in the infrastructure area in relating the damage to the measured vibration metrics. Thus, both the direct and inferred methods have limitations, especially when applied to general SHM projects (Table 5.5).

This section discusses a technique known as VSP that takes advantage of both classes of damage identification while eliminating the limitations of both. VSP has its roots in the aerospace field. It aims at establishing a monitoring system to detect small fatigue cracks as they develop and grow. It tries to generalize the main logic behind inferred damage detection while retaining the advantages of the direct detection method. It identifies most or all of the parameters that can cause damage and relates the parameters to the expected properties of damage (existence, location, size, and severity) quantitatively. By measuring the parameters, quantitative relationships can be used to identify the damage. Three widely different examples are presented; they show the general nature and the advantages of using VSP in identifying damage.

5.4.3   THEORETICAL BACKGROUND

We can generalize the VSP logic to develop more robust damage identification techniques. We start first by expressing the inferred damage identification logic as follows:

$$D = f(x_i) \tag{5.8}$$

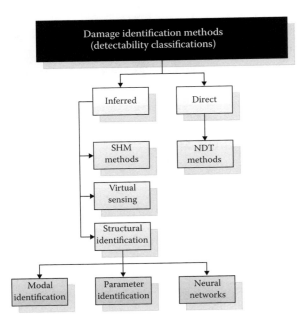

FIGURE 5.21   Damage identification detectability classifications.

TABLE 5.5
Advantages and Disadvantages of Direct and
Inferred Damage Identification Techniques

Issue	Direct	Inferred
Accuracy	Relatively accurate	Limited accuracy
Computational demands	Limited	Can be demanding
Range	Local	Global
Applicability to SHM	Varies	Applicable

$$S_j = g(y_k) \tag{5.9}$$

$$x_i = h(S_j) \tag{5.10}$$

From Equations 5.8 through 5.10, one can obtain

$$D = q(y_k) \tag{5.11}$$

where

y_k = Measured parameters
D = Damage of interest
x_i = Factors or parameters that identify the damage
S_j = Intermediate tools that can be used to infer damage, D, from measurements y_k, through the function h.

The above equations are fairly general. But, in the following examples, their use is examined in specific situations.

5.4.4 Step-by-Step Guide to VSP

Practical utilization of VSP can follow these steps:

1. Define goals of VSP. For example, define the type of damage D that needs to be identified and the area where such damage is located. Note that such an area can be local or global. In VSP, there should be no limitation on the size of the damage detection area.
2. Define parameters of VSP, x_i, y_k, and S_j. The accuracy of VSP depends on the completeness of the parameter definition. Note that the measured parameters y_k need to be easily measured in order for the VSP to be cost effective.
3. Define formal relationships between parameters, that is, $f(x_i)$, $g(y_k)$, and $h(S_j)$. These formal relationships can be analytical, empirical, or numerical.
4. Start SHM experiment and apply the relationship of (4) to identify D.

5.4.5 Case Studies

5.4.5.1 Vibration of Structures

Vibration measurements have been used as a means of structural identification or, in some cases, structural damage identification. The input in the process is a measured set of structural motions $U(t)$. The output of the process can be structural modal properties—a set of mode shapes Φ and modal frequencies ω. In some cases, the output can be physical structural properties, such as structural stiffness or damage in structures (see Doebling et al. 1996 and Alampalli 2000). Figure 5.22 shows this process in a systematic fashion. In all cases, the structural parameters or the damage are not measured directly; rather, they are identified by using analytical techniques relating vibration to physical parameter or damage. Thus, this process can be described as a virtual sensing technique.

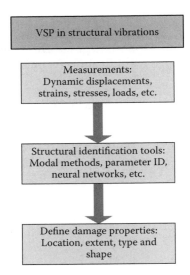

FIGURE 5.22   Virtual sensing paradigm (VSP) and structural vibrations techniques.

Using vibration sensing to identify damage via structural identification can be easily demonstrated by Equations 5.8 through 5.11. From a set of structural displacement measurements x_i, some important parameters y_k, of a given structure S_j, where $j = 1, 2$ can be identified. For $k = 1$, the structure is in a pristine state, whereas for $k = 2$, the structure is in a damaged state. For modal identification methods y_k represents modal information, with k as a counter for different identified modes, which are usually a few of the low natural structural modes. For parameter identification methods, y_k can represent material properties, spring constants, or finite-element properties. In this case, k represents a count for the number of parameters that are being identified.

It should be noted that there are severe limitations to using vibration of structures as a virtual sensing of structural damage. Alampalli and Ettouney (2007) presented a guide for the accurate use of vibration techniques to virtually detect structural damage. They mentioned four factors that would affect the use of structural vibrations: error scale, dynamic scale, spatial scale, and geometric modeling. It was shown that for a successful virtual sensing process all the four factors should be accommodated. Without accommodating the demands of these factors, the virtually sensed damage might not be accurate.

5.4.5.2   Acoustic Emissions

The AE method is an NDT technique that senses stress waves resulting from material damage, such as fatigue cracks Bassim, M. 1987. Traditionally, AE sensing needed several sensors to determine the damage location. In addition, it was not easy to identify the size and shape of the damage, a task that was mostly qualitative and required great personal expertise. A novel virtual sensing technique by Vahaviolos et al. (2006) used the AE method to identify damage location, size, and orientation in a fairly objective manner. The technique is based on the following steps:

1. Use an AE sensor to record the AE time signal from the damage source.
2. Use wavelet transform (WT) (Hubbard 1998) to generate the time frequency spectrum of the time signal.
3. Generate dispersion curves for the type of problem and material under consideration. These dispersion curves can be generated analytically or numerically (Rose 1999).
4. Compare the WT of the recorded time signal in Step 2 with the generated dispersion curves of Step 3, as shown in Figure 5.23.

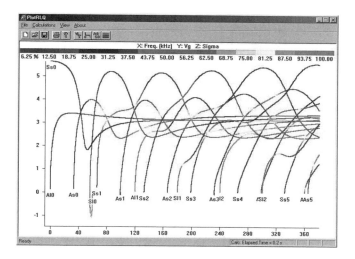

FIGURE 5.23 Dispersion curves used for acoustic emission damage detection. (Courtesy of Physical Acoustics Corporation.)

5. From the properties of the dispersion curves, it is possible to compute the distance between the source (damage) and the AE sensor.
6. Using a previously generated database of damage size and orientation, and correlating this database with the WT spectrum, it is possible to estimate the source (damage) size and orientation.

This method is a good example of the virtual sensing technique since it uses precomputed information in the form of dispersion curves and damage properties database to correlate with real-time recordings away from the damage. It is simple and can be generalized with minimal human interaction.

5.4.5.3 Deterioration of High-Strength Wires

High-strength wires are used for load transfer in suspension bridges, cable-stayed bridges, posttensioned and prestressed buildings, and bridge components. Monitoring their deterioration is, thus, an important goal for stakeholders of those important structures. It has been argued that the deterioration of high-strength wires would result in several failure modes, such as stress corrosion cracking, brittle fracture, and fatigue. Each of these modes has one or more causes, as shown in Table 5.6 (Mahmoud 2003).

Direct monitoring of deterioration due to different causes shown in Table 5.2 is not an easy task due to the harsh environment in which high-strength wires are usually found. For example, they are usually encased inside reinforced concrete girders in buildings or bridges. In the case of suspension bridges, there are hundreds of high-strength wires that are bundled together inside a metal casing that forms the main suspension cables of the bridge (see Figures 5.24 and 5.25). Due to the difficulties in direct monitoring, virtual sensing techniques have been suggested (Mahmoud 2003). Virtual monitoring in these situations can be done by identifying all major environmental and mechanical parameters that can contribute to all the failure causes in Table 5.6. Some of the environmental conditions that are easier to monitor are temperature, humidity, pH levels, presence of corrosion promoters such as sulfides, hydrogen concentration, among other factors. Mechanical conditions include loading levels, stress cycles, and material properties.

As soon as all main parameters are identified, a neural network approach (or any other regression-type method) can be used to correlate these parameters with the deterioration of the high-strength wires in a controlled laboratory setting. The computed neural network can then be used in a real-life setting by measuring the parameters, which is an easy task, and forecasting the deterioration of the real-life structure. Figure 5.26 shows the steps in this approach.

TABLE 5.6
Failure Modes and Causes of
Deterioration in High-Strength Wires

Failure Mode	Cause
Brittle fracture	Local corrosion attack (pitting)
	Hydrogen embrittlement
Stress corrosion	Anodic stress cracking
	Hydrogen-induced stress corrosion
Fatigue	Corrosion fatigue
	Fretting fatigue

FIGURE 5.24   Photo showing suspension bridge cables.

FIGURE 5.25   Damages in high-strength wires. (Courtesy of William Moreau.)

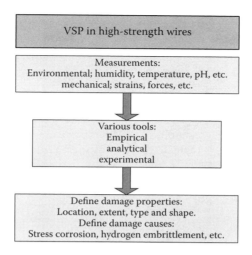

FIGURE 5.26 Virtual sensing paradigm (VSP) and high-strength wires.

TABLE 5.7
VSP Relations to the Three Case Studies

Parameter	Structural Vibrations	High-Strength Wires	Acoustic Emissions	Remaining Fatigue Life
D	General degradation in mechanical properties, such as cracks or loss of stiffness	Stress corrosion or hydrogen embrittlement	Small cracks in plates or shells	Time to fatigue damage occurrence
y_k	Dynamic displacements and/or strains	Environmental (humidity, pH levels, temperature, etc.), and mechanical conditions (stresses, strains)	Acoustic emission time signals	Stress ranges, number of cycles
X_i	Damage existence, size, extent, shape, etc.			
S_j	Structural identification methods		Wavelet transforms, dispersion relationships, etc.	Empirical and experimental tools
Fatigue Related?	Yes. Care is needed for scale issues	Yes	Yes	Yes

5.4.5.4 Remaining Fatigue Life

Remaining fatigue life problem tries to predict the time at which fatigue damage might occur. The remaining fatigue life formulation was developed empirically as a function of several parameters, including the number of cycles and the corresponding stress ranges (the formulation will be presented in more detail in Section 5.7). The number of cycle-stress range relationships can be estimated either analytically or experimentally (by performing an SHM project). Thus, remaining fatigue life can be considered a practical example of VSP. Table 5.7 relates the above case studies with the formal relations of Equations 5.8 through 5.11.

5.5 STEP-BY-STEP APPROACH FOR REMAINING FATIGUE LIFE

Strain measurements to evaluate the fatigue remaining life of bridge components have been used by several researchers (see, for example, Zhou 2006; Lund and Alampalli 2004a and 2004b; Alampalli and Kunin 2001). We summarize the essential steps usually followed in this technique. Similar steps were also described by Zhou (2006).

1. Identify critical bridge components that might be susceptible to fatigue. This can be accomplished by structural analysis of the bridge, or by personal judgment, or both. While analyzing the bridge, accurate as-built dead load estimates must be used. Truck and traffic loads must be realistic, yet conservative. Impact coefficients as recommended by appropriate specifications must be used. Stress reversal should be identified. Different bridge components that have higher stress ranges are then designated as fatigue susceptible.
2. Perform an infinite fatigue life check on each of the components identified in step #1. For components that do not meet the infinite remaining fatigue life, proceed to the next step.
3. Estimate remaining fatigue life (see the example of Patroon Island Bridge in Section 5.7). Based on the resulting remaining fatigue life, perform a decision making analysis to determine the best course of action. One potential decision making technique is bridge life cycle analysis (BLCA), which is described later. Another decision making technique is the value of additional information technique described in Chapter 8 in relation to bridge load testing.
4. If it is decided that more accurate information is warranted, an SHM experiment can be performed. It aims at evaluating remaining fatigue life, using strain measurements at the locations identified in step #3.
5. Identify the optimum number of strain gauges to be used in the experiment. The locations of the gauges should be consistent with the analysis results of step #1. Also, the gauge locations should reflect as-built details, as well as the judgment of the professional engineer on site. In choosing the experimental setup, the sampling rate of the setup needs to be considered carefully. Recall that the goal of the measurements is to estimate both the strain amplitude as well as the equivalent strain cycles. Since the strain cycles are a direct function of the dynamic properties of the global as well as the local bridge components, it is essential that the gauge-sampling rate be adequate to capture this fact. An inadequate sampling rate can result in coarse results that will produce smaller number of strain cycles—an unsafe result.
6. Start the strain measurements under real traffic conditions. Measurements should be long enough to account for different traffic patterns and any reasonable dynamic effects that might result from traffic. A good practice is to perform the measurements as more than a one-time setting. This will establish both extra redundancy and realism of results.
7. Calibrate the strain gauge results. Calibration can be established by relating known weight of trucks with measured strains by different gauges.
8. Using the time histories of step #6, compute the equivalent strain cycles. A popular method for computing equivalent cycles from time histories is the rainflow technique. From the cycles, form the strain histograms.
9. Use the histograms to compute the infinite fatigue life at different locations.
10. For locations where infinite fatigue life is not obtained, compute remaining fatigue life. A decision making process is needed, as described in step #3.

Figure 5.27 illustrates the structural health in civil engineering (SHCE) remaining fatigue life steps. Note that the process includes all four components of SHCE. Decision making processes are present in and intertwined with every step of the project. This is essential to ensure a meaningful and successful outcome.

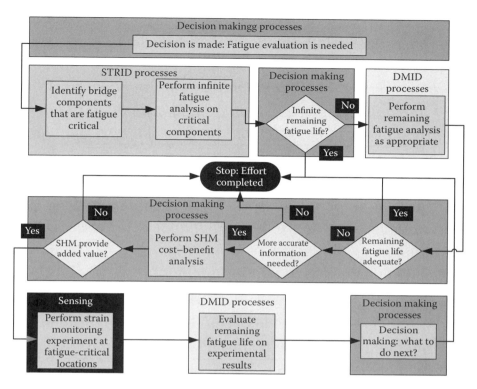

FIGURE 5.27   SHM process in determining remaining fatigue life.

5.6   DYNAMIC AND FATIGUE ANALYSIS OF A TRUSS BRIDGE WITH FIBER-REINFORCED POLYMER DECK

Bridge owners are constantly looking for new materials, construction practices, and new designs to build bridges with long service lives and low life cycle costs. Since higher dead loads reduce the live-load capacity, lightweight materials are very attractive. At the same time, since the major cause of deterioration of most bridge components is corrosion, materials with noncorrosive properties are also of great interest. As noted in several other case studies in this book, FRP materials have been gaining popularity among bridge owners during the last decade due to their noncorrosive properties and light weight compared to conventional materials such as concrete and steel. One of the applications of interest is to replace heavy concrete bridge decks with lighter composite decks to extend their service life by increasing their live-load capacity. In these situations, it is very important to study the system response to make sure that the deck replacement does not change the secondary or nondominant failure mechanisms to become controlling mechanisms. This case study describes one such situation where fatigue response could be an issue.

The New York State Department of Transportation (NYSDOT), as part of its initiative to develop effective rehabilitation methods, experimented with FRP decks to replace deteriorated bridge decks. The first FRP deck used on a state highway bridge is in New York, and it was built to improve the live-load capacity of a 60-year-old truss over Bentley Creek in Wellsburg, NY. The bridge has a length of 42.7 m and a width of 7.3 m, curb to curb, with 27 degree skewed supports. The floor system is made up of steel transverse floor beams at 4.27-m center-to-center spacing with longitudinal steel stringers. Figures 5.28 and 5.29 illustrate the plan and elevation views of Bentley Creek Bridge, respectively. Figure 5.29 also shows a detailed section of floor beam with FRP deck.

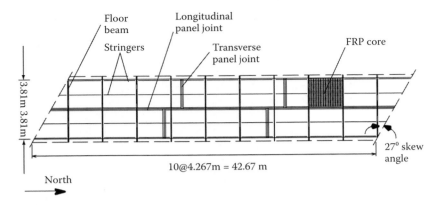

FIGURE 5.28 Bentley Creek Bridge plan view. (Courtesy of New York State Department of Transportation.)

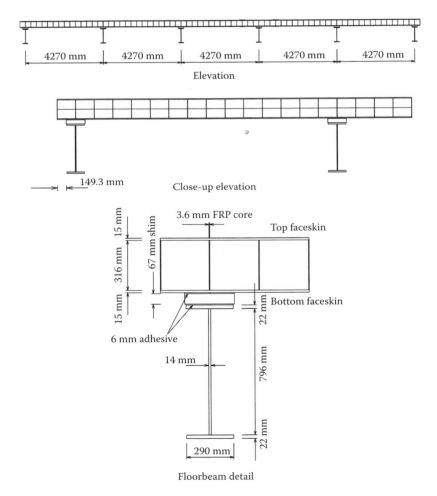

FIGURE 5.29 FRP deck geometry of Bentley Creek Bridge. (Courtesy of New York State Department of Transportation.)

A static load test and detailed analysis was conducted to verify the design assumptions made and to study possible failure mechanisms that the deck may be subjected to during its service life (see Aref and Chiewanichakorn 2002; Alampalli and Kunin 2001; Alampalli and Kunin 2003). Thermal behavior of this bridge was also studied through temporal-thermal stress simulations to determine the effect of accidental fires on the deck (Alnahhal et al. 2006). An important aspect that may be overlooked in such situations is the effect of the rehabilitation process on the dynamic response of the structure and fatigue life of the bridge. By reducing the self-weight of the deck, dynamic characteristics of the structure would be altered. Hence, a study was initiated by NYSDOT to investigate the dynamic fatigue response of this truss bridge with FRP deck. This was accomplished by comparing the dynamic responses of the FRP deck system with a generic reinforced concrete deck system after completing the rehabilitation process by subjecting these systems to three-dimensional finite-element analysis (FEA) dynamic fatigue simulations (Aref and Chiewanichakorn 2002; Chiewanichakorn et al. 2007). Predicted load-induced fatigue lives were determined based on numerical results and fatigue resistance formulae provided by AASHTO-LRFD specifications (AASHTO 2004).

Three-dimensional FEA was used to perform dynamic fatigue simulations of two deck systems, that is, FRP and concrete decks. The entire bridge, including the steel trusses, was modeled and analyzed using a general purpose FEA package ABAQUS. The FEA model was verified using the data from load test results conducted by the NYSDOT (see Figure 5.30), immediately after the rehabilitation, to verify some of the design assumptions used (see Alampalli and Kunin 2001, 2003).

The dynamic characteristics of FRP and concrete decks were obtained from finite-element models by performing eigenvalue/frequency analyses. The first five fundamental frequencies of the bridge were extracted and used to determine Rayleigh damping parameters. In dynamic fatigue simulation, a nominal AASHTO fatigue truck was used (AASHTO 2004). This truck has a constant spacing of 9 m between the 145 kN axles. The dynamic load allowance of 15% was imposed on the design truck for fatigue and fracture limit state evaluation. Lane load was not considered in this study. By assuming a maximum allowable speed on Bentley Creek Bridge of 50 mph, it would take

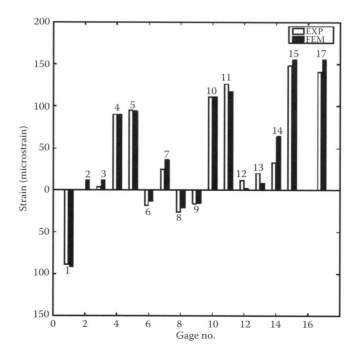

FIGURE 5.30 Finite element model verification results. (Courtesy of New York State Department of Transportation.)

approximately 2 s for a truck to travel across the bridge. A moving truck was simulated at eight discrete time instants (0.25-s time interval). An impact time duration at each time instant was assumed to be one-thousandth of a total traveling time, that is, 0.002 s (see Figure 5.31).

Dynamics analyses were conducted on both FRP and concrete decks. Two main components, that is, trusses and floor beams, were expected to be potentially vulnerable to fatigue damage and failure. Dynamic time-history analyses of Bentley Creek Bridge subjected to a moving fatigue truck were determined, considering the total time duration of 10 s with 2 s of truck traveling time and 8 s of free structural vibration that was considered to be sufficient for predicting its fatigue life based on dynamic responses. For FRP deck system, Figures 5.32 and 5.33 illustrate dynamic responses of four different members on the west truss of both deck systems.

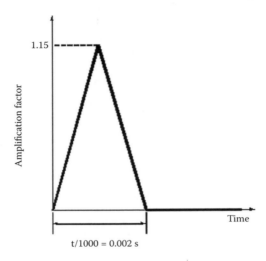

FIGURE 5.31   Loading protocol. (Courtesy of New York State Department of Transportation.)

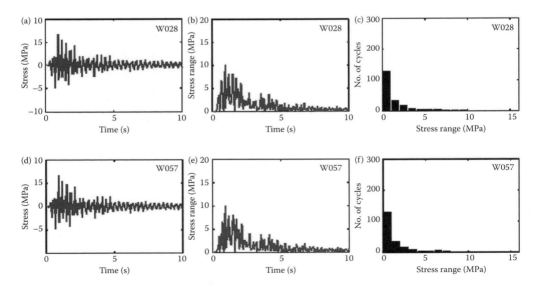

FIGURE 5.32   Response history of critical members on west truss of FRP deck (a) member #1, time history, (b) member #1, stress range history, (c) member #1, stress range histogram, (d) member #2, time history, (e) member #2, stress range history, and (f) member #2, stress range histogram. (Courtesy of New York State Department of Transportation.)

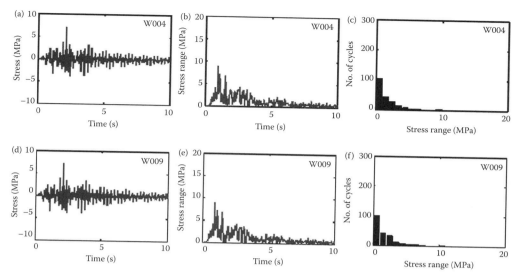

FIGURE 5.33 Response history of critical members on west truss of concrete deck (a) member #3, time history, (b) member #3, stress range history, (c) member #3, stress range histogram, (d) member #4, time history, (e) member #4, stress range history, and (f) member #4, stress range histogram. (Courtesy of New York State Department of Transportation.)

Fatigue life of each deck system was determined using AASHTO-LRFD fatigue resistance formulae (see Aref and Chiewanichakorn 2002; AASHTO 2004; Chiewanichakorn, Aref, and Alampalli 2007).

$$(\Delta F)_n = \left(\frac{A}{N}\right)^{1/3} \geq \frac{1}{2}(\Delta F)_{TH} \tag{5.12}$$

and

$$N = (365)(75)(n(ADTT)_{SL}) \tag{5.13}$$

where

A = Constant (MPa³)
n = Number of stress range cycles per truck passage
$(ADTT)_{SL}$ = Single-lane Average Daily Truck Traffic (ADTT) as specified in Article 3.6.1.4 in AASHTO-LRFD
$(\Delta F)_{TH}$ = Constant amplitude fatigue threshold (MPa)

In this study, a number "75" in Equation 5.12 was replaced by an unknown variable "y" representing a fatigue life in years as the code considers the design life to be 75 years in this equation. Equations 5.8 and 5.9 can be rearranged into

$$y = \frac{A}{n(365)(ADTT)_{SL}(\Delta F)_n^3} \tag{5.14}$$

When

$$(\Delta F)_n \geq \frac{1}{2}(\Delta F)_{TH} \tag{5.15}$$

And

$$(ADTT)_{SL} = p \times (ADTT) \tag{5.16}$$

TABLE 5.8
Predicted Fatigue Life (75-year ADTT Equivalent to Infinite Life in AASHTO)

| | Fatigue Life (years) | | | |
| | East Truss | | West Truss | |
Location	FRP	Concrete	FRP	Concrete
1	Infinite life	23, (E113)	63, (W028)	31, (W004)
2	Infinite life	20, (E128)	63, (W057)	23, (W009)
3	Infinite life	Infinite life	Infinite life	31, (W105)

Source: Courtesy of New York State Department of Transportation.

where $(ADTT)_{SL}$ is the number of trucks per day in one direction averaged over the design life, (ADTT) is the number of trucks per day in a single lane averaged over the design life, and p is the fraction of truck traffic in a single lane—taken as 1.0 in this study.

For the given ADTTs of 275 and 384 in years 2002 and 2020, respectively, the results show that trusses and steel floor beams would be able to sustain an infinite number of cycles without introducing any fatigue fracture or failure. Most importantly, the results (see Table 5.8) showed great improvement in fatigue life of the bridge after the replacement of concrete deck with FRP deck. The fatigue life of FRP deck system almost doubles when compared with the concrete deck system. Numerical results also indicated that the stress range induced on floor beams lies within an infinite life regime. Hence, they would be functioning well without any load-induced fatigue failures.

5.7 ESTIMATING FATIGUE LIFE OF BRIDGE COMPONENTS USING MEASURED STRAINS: PRACTICAL APPLICATION

5.7.1 INTRODUCTION

The design fatigue life of a bridge is generally based on the truck traffic data, prevalent specifications, and the fatigue durability of the components in the bridge. When the theoretical remaining fatigue life has been exhausted but the bridge is functioning well, engineers are left with the difficult choice of whether or not it should be rehabilitated/replaced. Thus, an evaluation based on actual conditions can benefit bridge owners and play a major role in making cost-effective optimal bridge management decisions. This is especially true when the bridges are a part of busy interchanges.

The remaining safe fatigue life of bridge components can be evaluated on the basis of the procedures outlined in the AASHTO (1990) and strain data from critical structural members. The procedure, in general, involves the following steps:

1. Problem identification: Critical details requiring estimation of remaining fatigue life should be identified and appropriate instrumentation plan developed to capture high in-service tensile strains that these details are subjected to.
2. Data acquisition: Based on the analysis of traffic patterns and count data, representative data collection periods should be determined. Then, installation of strain gauges at appropriate locations should be accomplished and data collected.
3. Data analysis: From the data collected, strain cycles should be extracted. This is normally accomplished using rainflow algorithm, as per the ASTM Specifications (2004) and Downing and Socie (1982), and categorizing into a manageable number of strain bins.

4. Calculation of the effective stress: Effective stress representing the stress spectrum should be estimated. This is normally accomplished using Miner's rule (AASHTO 1990) for each of the critical details.

5. Estimation of remaining fatigue life: If a detail has low stresses compared to the fatigue strength or if the stress is dominantly compression, then the detail will likely have an infinite fatigue life, and the code (AASHTO 1990) quantifies these ideas. If the factored effective stress is less than the limiting stress range, and the factored tension portion of the stress range is less than half the compression portion of the stress range, then the remaining fatigue life is infinite. Otherwise, the finite remaining life estimate should be calculated on the basis of the detail type, the estimated lifetime average daily truck volume in the outer lane, the number of stress cycles per truck passage, the reliability factor, and the bridge age. A case study is given below, which uses the above concept. The study is based on NYSDOT (2002).

5.7.2 PATROON ISLAND BRIDGE STRUCTURE

The Patroon Island Bridge (Figures 5.34 and 5.35) is part of a major interchange in Albany, NY, and carries I-90 over the Hudson River. The bridge consists of 10 spans, 1795 ft long, and has been in

FIGURE 5.34   Patroon Island Bridge. (Courtesy of New York State Department of Transportation.)

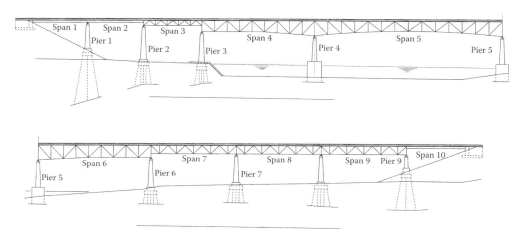

FIGURE 5.35   Elevation view of Patroon Island Bridge. (Courtesy of New York State Department of Transportation.)

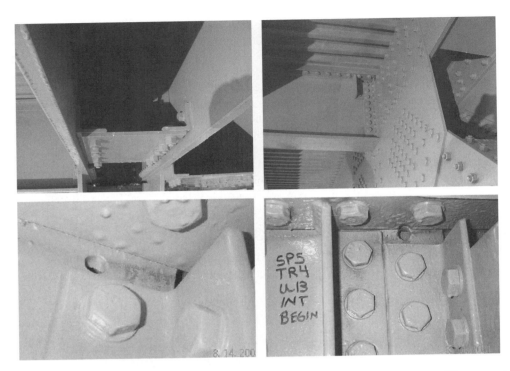

FIGURE 5.36 Typical details requiring evaluation. (Courtesy of New York State Department of Transportation.)

service since 1968 with an average daily traffic of 70,787 in 1998. Spans 3 through 9 are considered the main spans and consist of steel deck trusses and concrete decks. Spans 1, 2, and 10 are considered approach spans and consist of plate girders.

The bridge and two ramp structures have a documented history of cracks in the connection welds and floor beams. These include the webs of floor beams, sole plates between the floor beams and girders, truss to bearing welds, and transverse web stiffeners (see Figure 5.36). The cracking between the truss and the bearing is not expected to be fatigue related since compression is expected here. Many of the problems have been related to the floor beam webs in spans 4, 5, and 6. The floor beam web is attached to the truss using an angle. A top strap plate creates continuity between the top flanges of the floor beam on each side of the truss but is not attached to the truss. Cracking occurred in the floor beam webs between the truss connection angle and the top flange of the floor beam. The cracking was attributed to the deformation of the truss in the longitudinal direction with respect to the slab, which caused large out-of-plane bending stresses in the floor beam webs. The truss is exerting an out-of-plane load on the web of the floor beam, but the floor beams are constrained by the stringers connected to the slab. The slab is very stiff along the length of the bridge when compared to the top truss chords in the bridge, and the web of the floor beam must transmit the shear between the truss and the slab. Similar damage in floor beams has been observed in the literature (see Fisher 1977; Yen et al. 1989).

5.7.3 Detail Identification

Bridge details, current status, and structural problems were evaluated. Based on this evaluation, fatigue critical details on the Patroon Island Bridge were identified. An instrumentation plan was then developed to collect strain data near critical details. The strain transducers were generally located on the bottom flanges of the beams in positive moment regions, on the top flanges of the negative moment regions, and on the main load-carrying plates of the truss members, to capture peak tensile stresses. Gauge locations can be found in Tables 5.9 and 5.10.

TABLE 5.9
Infinite Fatigue Life Check

Span	Member	Critical location	Detail	Gage	Stress Factor	$R_s S_r$ (ksi)	Remaining Safe Life
2 BF	FB, 24WF84	Bolted connection with stringer	B	1	1	2.30	Infinite
2 BF	Stringer, 21WF55	Shear studs, midspan	C	2	1	1.73	Infinite
2 BF	Stringer, 21WF55	Shear studs, end	C	2	1.86	3.21	Infinite
2 BF	Girder	Sole plate to girder fillet weld (L > 4 in), point B	E′	3	1.25	1.80	Finite
2 BF	Girder	Toe of transverse stiffener, bottom, center of span 2	C	3	1.4	2.02	Infinite
2 Main	FB, 24WF84	Midspan, bolted connection with stringer	B	4	1	1.64	Infinite
2 Main	Stringer, 21WF55	Shear studs, midspan	C	5	1	1.43	Infinite
2 Main	Stringer, 21WF55	Shear studs	C	5	1.86	2.66	Infinite
2 Main	Girder	Midspan of span 2, transverse toe stiffener	C	6	1	1.78	Infinite
2 Main	Girder	Sole plate to girder fillet weld (L > 4 in), point B	E′	6	0.9	1.60	Finite
3	Truss L4-L6 (T)	Longitudinal fillet weld	B	7	1	1.72	Infinite
4	Truss U0-L1 (T)	Bolted diaphragms and intermediate stays	B	8	1	1.53	Infinite
4	Truss L3-L4 (T)	Welded intermediate stay plates	E	9	1	1.83	Finite
4	Truss U4-U5 (C)	Diaphragm welds	C	10	1	0.53	Infinite
4	Truss U8-U9 (T)	Bolted diaphragms and intermediate stays	B	11	1	0.42	Infinite
5	Truss U9-L10 (T)	Bolted diaphragms and intermediate stays	B	12	1	1.24	Infinite
5	Truss U13-L14 (T)	Bolted diaphragms and intermediate stays	B	13	1	1.41	Infinite
5	Truss U14-U15 (©)	Diaphragm welds	C	14	1	0.49	Infinite
5	Truss L14-L15 (T)	Bolted diaphragms and intermediate stays	B	15	1	0.72	Infinite
8	Truss L19-U20 (T)	Longitudinal fillet weld between top flange and web	B	20	1	1.61	Infinite
9	Truss U20-U21 (T)	Welded intermediate stays	E	21	1	1.00	Infinite
9	Truss L25-L26 (T)	Welded intermediate stays	E	22	1	1.74	Finite
9	Truss L29-U30 (T) 12WF92	Rolled member with bolted end connection	B	23	1	1.70	Infinite
10	FB, 30WF99	Bolted connection with stringers	B	24	2	0.93	Infinite
10	Stringer, 21WF55	Shear studs, midspan	C	25	1	1.44	Infinite
10	Stringer, 21W55	Shear studs, end	C	25	1.86	2.68	Infinite
10	Girder	Midspan, toe of transverse stiffener	C	26	1	1.12	Infinite

BF, ramp from 787 northbound to I-90 eastbound; FB, floor beam; ML, main line of bridge; R_s, reliability factor; S_r, effective stress in ksi.

TABLE 5.10
Remaining Safe Fatigue Life for Critical Details

Span	Member	Location	R_sS_r (ksi)	Detail	K	Span (ft)	Y_f (y)
2 BF	Girder	Sole plate to girder fillet weld (L > 4 in), point B	1.80	E′	1.1	110	27
2 Main	Girder	Sole plate to girder fillet weld (L > 4 in), point B	1.60	E′	1.1	110	53
4	Truss L3-L4 (T)	Welded intermediate stay plates	1.83	E	2.9	225	122
9	Truss L25-L26 (T)	Welded intermediate stay plates	1.74	E	2.9	170	146

BF, ramp from 787 northbound to I-90 eastbound.

Horizontal and vertical gauges were installed on the webs of several floor beams near their respective connections to truss members in span 5. These floor beams were expected to have large out-of-plane stresses induced by a relative displacement in the direction of traffic, between the upper truss nodes and the slab. The gauges were expected to provide the data necessary to calculate and compare the effective in-plane stresses in the floor beam webs for reinforced and unreinforced connections, and they were not intended to predict fatigue life. After reviewing the lower-than-expected strains, it became apparent that the gauges were too far outside of the floor beam's web gap, between the truss connection angle and the floor beam's top flange. The procedure outlined in Guide Specifications for Fatigue Evaluation (AASHTO 1990) applies only to uncracked members subjected to primary stresses. Hence, the fatigue life of the floor beams in spans 4, 5, and 6 was not estimated. Repair or replacement of the floor beams in spans 4, 5, and 6 was recommended.

5.7.4 FIELD TESTING AND ANALYSIS

Once the fatigue critical locations were identified, strain gauges were installed, and data was collected. Traffic count data revealed that Monday to Thursday can be considered normal traffic days, and hence all data was collected during this part of the week. Forty-eight continuous hours of strain data was collected during three instrumentation phases due to cable limitations and data storage capabilities.

Strain transducers from Bridge Diagnostics, Inc. were used in the testing (BDI 2001). These transducers have a full Wheatstone bridge, with four active 350-ohm gauges embedded in aluminum housing, and an operating temperature range of −50°C to 1200°C. They were attached to the bridge components with threaded mounting tabs using quick-setting adhesive. A typical strain time history is shown in Figure 5.37. A limited number of conventional foil gauges were used in the field testing to verify the strain data from the BDI gauges. Type T (copper/constantan) thermocouples were used to measure the temperature fluctuations, to determine if the thermal expansion and contraction would affect the fatigue life prediction. However, the strain data was processed in 10-minute sets, so that the temperature fluctuation in each set was negligible. These thermocouples have the ability to measure temperatures ranging from −270°C to +400°C.

5.7.4.1 Analysis

All the time histories were reviewed carefully to eliminate instrument-related errors. In some cases, unreasonable spikes with magnitudes of 100 times the expected strain were noticed in the data. These were attributed to sporadic radio signals. If a small portion of the data was affected by these spikes, then that portion of the data was neglected in the analysis. After the strain/time history data was collected, turning points (peaks and valleys) were extracted from the strain/time history.

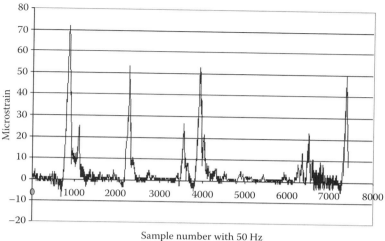

FIGURE 5.37 Typical strain time history. (Courtesy of New York State Department of Transportation.)

Once the peaks and valleys were extracted, the rainflow algorithm was used to extract strain cycles from the reorganized turning point data. The large number of low-magnitude strains will reduce the calculated effective stress, so the low-amplitude cycles will be removed if they are below a threshold level. Figure 5.38 shows a typical histogram with an amplitude filter. The procedures in specifications (AASHTO 1990) do not provide guidance on using a strain cycle amplitude filter; however, it has been determined that very low stresses do not cause fatigue damage. A study conducted by Dr. Fisher for the American Institute of Steel Construction (Fisher 1977) indicates that a truck with a gross vehicle weight less than 20 kip has a very low probability of causing fatigue damage, hence the strain ranges corresponding to a 20-kip truck were determined with a load test and then used as a strain cycle threshold. The strain ranges below the strain threshold were then ignored for the calculation of effective stress. Low-amplitude filtering gives a higher and more conservative estimate of the effective stress.

Figure 5.39 shows the strain time history for a rolling truck test. The strain ranges were extracted from this data by scaling them linearly to represent a 20-kip vehicle. Fatigue testing was done in more than one phase due to operational limitations, but rolling truck tests were done only during one phase of the testing. Hence, a correlation was found between the threshold strain and the effective stress, calculated without amplitude filtering, as shown in Figure 5.40. The regression equation from Figure 5.40 and the effective strain, calculated without amplitude filtering, were used to estimate the strain caused by a 20-kip truck.

The effective stress was calculated using the strain histograms and Miner's rule.

$$S_r = \sum_{i=1}^{n} (f_i S_{ri}^3)^{1/3} \tag{5.17}$$

where S_r = effective stress amplitude, f_i = fraction of stress ranges within interval i, and S_{ri} = average stress range for interval i, and n = number of histogram bins.

The effective stress range can be calculated from the effective strain amplitude ε_{ri} and elastic modulus E:

$$S_{ri} = E\varepsilon_{ri} \tag{5.18}$$

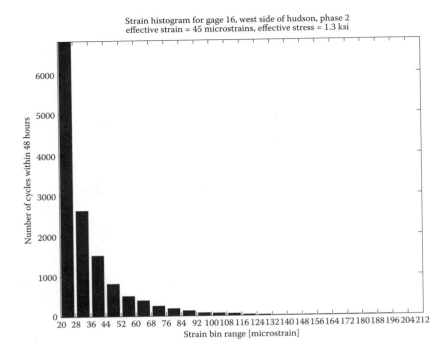

FIGURE 5.38 Typical strain histogram with a 20 microstrain threshold. (Courtesy of New York State Department of Transportation.)

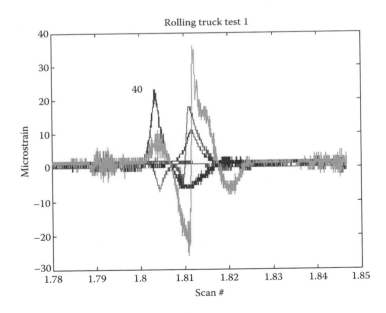

FIGURE 5.39 Typical strain data from loaded trucks. (Courtesy of New York State Department of Transportation.)

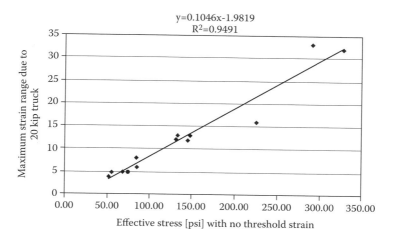

y=0.1046x-1.9819
R²=0.9491

FIGURE 5.40 Effective stress correlated to test data for obtaining threshold strain. (Courtesy of New York State Department of Transportation.)

and

$$S_r = E\left(\sum_{i=1}^{n}(f_i\,\varepsilon_{ri}^3)^{1/3}\right)$$ (5.19)

The effective stress will then be factored as described in specifications (AASHTO 1990) to account for the procedures for determining the stress and redundancy of the member in question. Hence, the term "factored effective stress" is used in this chapter to show the difference. Interpolation based on structural analysis was required to estimate the effective stress at locations that did not have instrumentation.

5.7.5 FATIGUE LIFE EVALUATION

Once the effective stress is determined, it is then compared with the limiting stress range for infinite fatigue life S_{FL}. If the limiting stress range S_{FL} is greater than the factored effective stress, then the corresponding fatigue detail has an infinite remaining life. Otherwise, the remaining fatigue life is calculated. The limiting stress ranges from AASHTO (1990) are about a third of the constant amplitude fatigue thresholds found in AASHTO (1999). A detail's limiting stress range is dependent on its fatigue durability. The detail categories, limiting stress range S_{FL} and detail constant, K are given in specification (AASHTO 1990). All structural members considered in this analysis were taken as nonredundant, except for the stringers. Most details that were checked showed an infinite fatigue life, so no further calculations were necessary for these details (see Table 5.9).

Four of the details have a factored effective stress greater than the limiting stress range, SFL, and the stresses exceeded tension limits, so the fatigue life is finite. The following equation was used to estimate the remaining safe fatigue life (AASHTO 1990):

$$Y_f = \frac{f\,K\times10^6}{T_a\,C(R_s\,S_r)^3} - a$$ (5.20)

where Y_f = remaining fatigue life in years, $f = 1.0$ for calculating remaining safe life, K = detail constant found in AASHTO (1990), T_a = estimated lifetime average daily truck volume in the outer lane, C = stress cycles per truck passage, R_s = reliability factor associated with calculation of the stress range, S_r = effective stress, and a = present age of the bridge in years = 30 years in 1998. R_s includes the effects of redundancy and procedures used (i.e., field data or an alternative) for stress range. Note that the field strain data was not used to estimate the value of C. $C = 1$ was assumed in this study for all girders and trusses (AASHTO 1990). The stringers and floor beams tested had an infinite life, and hence C value was not required for these members.

The present average daily truck volume in the outer lane is estimated as given below (Guide 1990):

$$T = (\text{ADT}) F_T F_L = 2230 \tag{5.21}$$

where

  ADT = Average daily traffic volume in 1998
  F_T = 0.07 (fraction of trucks excluding panel, pickup, and other 2-axle/4-wheel trucks)
  F_L = 0.45 (fraction of trucks in the outer lane)

For an annual traffic volume growth rate of 4.5% and a bridge 30 years old (age of Patroon Island Bridge in 1998), the ratio of the lifetime average daily truck volume and the present average daily truck volume T_a/T will be 1.345 (AASHTO 1990; Moses et al. 1987). Hence, the estimated lifetime average daily truck volume will be

$$T_a = T\left(\frac{T_a}{T}\right) \tag{5.22}$$

The remaining safe fatigue life in the weld between a girder and a sole plate for a floor beam was found to be 27 years (see Table 5.10). The remaining safe fatigue life for tensile truss members L3-L4 and L25-L26 was found to be 122 and 146 years, respectively. Neither of these two truss member has been retrofitted with bolted stays or diaphragms to replace the welded connections.

5.7.6 CONCLUSIONS

Estimates of remaining fatigue life of bridges are very useful for bridge owners to make cost-effective bridge management decisions. This case study gives an example using Patroon Island Bridge in New York, where the remaining fatigue life of some bridge components was estimated on the basis of actual traffic conditions by continuous monitoring of the bridge during normal traffic loads. The fatigue life procedure used here has inability to estimate the fatigue life of damaged members.

5.8 BLCA AND FATIGUE

5.8.1 OVERVIEW

Fatigue is one of those slow hazards that affect most of constructed facilities. It is also one of those hazards that require the continuous vigilance of the owners. It requires continuous inspection, maintenance, and rehabilitation. The decision maker is also usually interested in estimating the economic implications of different decisions being taken on fatigue hazard. One technique that can be used is BLCA as applied to fatigue. In Chapter 10, we argue that BLCA includes three components: cost life cycle, benefit life cycle, and lifespan estimation. We explore the three components as they apply to fatigue hazard.

5.8.2 BRIDGE LIFE CYCLE COST ANALYSIS

General fatigue cost methods rely on the simple relation

$$C_{\text{FATIGUE}} = C_{\text{Event}} \, p_f \tag{5.23}$$

With C_{EVENT} is the cost of fatigue failure and p_f is probability of failure. In the literature, Equation 5.23 uses failure as the formation of fatigue crack in a given component (see Fu et al. 2003). The probability of fatigue failure for steel members is estimated by

$$p_f = \frac{((g(x_1,\mu,\sigma) - g(x_2,\mu,\sigma))}{(1 - g(x_2,\mu,\sigma))} \qquad (5.24)$$

where

x_1 = Total estimated lifespan
x_2 = Current age of the bridge
μ = Mean fatigue remaining life
σ = Standard deviation of remaining life

Note that

$$x_1 = BLS + x_2 \qquad (5.25)$$

Bridge lifespan BLS is discussed later.

The function $g(a, b, c)$ is the cumulative truncated lognormal distribution of random variable a, with a mean of b and standard deviation c (Benjamin and Cornell 1970).

An empirical form, AASHTO 1990, for estimating remaining fatigue life of the component is

$$Y = \frac{f K \times 10^6}{\alpha TC (R_s S_r)^3} \qquad (5.26)$$

A different version of Equation 5.26 was developed by Fu et al. (2003) as

$$Y = \frac{LOG\left(\left(\frac{(f K \times 10^6)}{(TC(R_s S_r)^3)}\right)(u(1+u)^{x_2}) + 1\right)}{LOG(1+u)} \qquad (5.27)$$

where

K = Factor depending on type of bridge component
f = 1, for safe life evaluation
f = 2, for mean life evaluation
C = Number of truck passage cycles
R_s = Reliability factor
S_r = Stress range that corresponds to number of cycles, C
T = Annual daily truck volume
T_a = Average lifetime daily truck volume in the outer lane
u = Annual traffic growth rate

Thus,

$$\mu = Y|_{f=2} \qquad (5.28)$$

The standard deviation σ can be evaluated as

$$\sigma = \mu\left(-\beta + \sqrt{\beta^2 + 2\ell n\left(\frac{\mu}{\mu_1}\right)}\right) \qquad (5.29)$$

The natural logarithm is ℓn, while β is a reliability index that is defined as

$\beta = 2$, for redundant systems

$\beta = 3$, for nonredundant systems

Finally,

$$\mu_1 = Y\big|_{f=1} \tag{5.30}$$

Using Equations 5.24 through 5.30 completely defines p_f. Cost of fatigue failure C_{EVENT} is more difficult to estimate. However, Fu et al. (2003) suggested that the cost of component repair, replacement, or a combination of both might be used as an estimate for C_{EVENT}.

5.8.2.1 Other Models

Cost of fatigue failures for reinforced concrete bridge decks can be estimated in similar steps as before (Fu et al. 2003).

5.8.2.2 SHM Role

SHM can play a major role in evaluating p_f. Monitoring and improving the estimate of parameters such as C, S_r, T, T_a, and/or u can enhance accuracy of the process. Examples of using SHM to improve estimate of C and S_r are given in Sections 5.8.3 and 5.8.4. Examples of the value of improving estimate of T and T_a are given in Chapter 8.

5.8.3 Bridge Life Benefit Analysis

Benefits of mitigation efforts, such as maintenance, repair, or replacement of components, can be estimated by computing the probability of failure after the mitigation effort. p_{fm} The governing equations are the same as Equations 5.24 through 5.30. The mitigation effort benefit can then be computed as

$$B_{FATIGUE} = C_{Event}\, p_{fm} \tag{5.31}$$

It is expected that $p_{fm} < p_f$.

Equations 5.23 and 5.31 have been used to compute the impact of mitigation effort by Fu et al. (2003) as

$$C_{Total} = C_{EVENT}\,(p_{fm} - p_f) \tag{5.32}$$

The role of SHM in computing mitigation benefits is as important as it is in computing costs. Improvements in accuracy in estimations of C, S_r, T, T_a, and/or u can be achieved by careful planning of SHM projects.

5.8.4 Bridge Lifespan Analysis

Computations of infinite fatigue life conditions and the actual remaining fatigue life are well known and have been discussed elsewhere in this chapter. We note that those computations produce values that are local, that is, these values are pertinent only to specific locations. A computed infinite fatigue life at a particular location does not have any significance for any other location in the bridge. Similarly, a remaining fatigue life of Y at a location does not mean that the lifespan of the bridge is Y. Let us explore this subject in more detail.

Consider a remaining fatigue life at N_f components to be Y_i, with $i = 2 \dots N_f$, It is reasonable to state that as $N_f \rightarrow \infty$ we get

$$\text{Min}(Y_i) \leq \text{BLS}\big|_{Fatigue} \leq \text{Max}(Y_i) \tag{5.33}$$

TABLE 5.11
Fatigue Bridge Lifespan Estimates

Case	BLS $\vert_{\text{Fatigue}}$	Comment
Low or no redundancy with the critical load paths are well known	$\text{BLS}\vert_{\text{Fatigue}} = \text{Min}(Y_i)$	Ensure that some estimates of Y_i cover the critical load paths
Accurate and high resolution stress analysis of the bridge is available	$\text{Min}(Y_i) \leq \text{BLS}\vert_{\text{Fatigue}} \leq \text{Max}(Y_i)$	Estimate Y_i at many locations in the analytical model

The fatigue-based lifespan estimate is $\text{BLS}\vert_{\text{Fatigue}}$. This is the lifespan of the bridge if the only factor affecting it is fatigue hazard. Functions Min( ) and Max( ) indicate the minimum and maximum values of the vector Y_i, respectively. Equation 5.33 is based on the fact that the occurrence of the first fatigue failure, at $\text{Min}(Y_i)$, does not necessarily mean that the lifespan of the bridge is reached. Also, it is reasonable to state that after numerous fatigue failures the bridge lifespan will be reached; this explains the upper limit in Equation 5.33.

For the more practical case of finite N_f, we can only state that

$$\text{Min}(Y_i) \leq \text{BLS}\vert_{\text{Fatigue}} \tag{5.34}$$

Obviously, the usefulness of Equations 5.33 and 5.34 is fairly limited. It is more desirable to estimate $\text{BLS}\vert_{\text{Fatigue}}$ directly from Y_i in the form

$$\text{BLS}\vert_{\text{Fatigue}} = f(Y_i) \tag{5.35}$$

Unfortunately, function $f(\ )$ cannot be easily determined. We can make practical use of Equations 5.33 through 5.35 as in Table 5.11.

For the general case, evaluating $f(\ )$ will depend on the analyst's judgment.

5.9 DIRLIK RAINFLOW EMPIRICAL SOLUTION

For a known moment m_n of a single sided PSD $G(f)$ of a random signal, the probability density function of the occurrence of the stress range S_i, $p(S_i)$ can be modeled as

$$p(S) = \frac{D_1 / Q e^{-Z/Q} + \left(D_2 / R^2 + D_3 Z\right) e^{-Z^2/2}}{2\sqrt{m_0}} \tag{5.36}$$

with

$$Z = \frac{S}{2\sqrt{m_0}} \tag{5.37}$$

$$D_1 = \frac{2(X_m - \gamma^2)}{1 + \gamma^2} \tag{5.38}$$

$$D_2 = \frac{1 - \gamma - D_1 + D_1^2}{1 - R} \tag{5.39}$$

$$D_3 = 1 - D_1 - D_2 \tag{5.40}$$

$$R = \frac{\gamma - X_m - D_1^2}{1 - \gamma - D_1 + D_1^2} \qquad (5.41)$$

$$Q = \frac{5(\gamma - D_1 - D_2 R)}{4D_1} \qquad (5.42)$$

$$X_m = \frac{m_1}{m_0} \sqrt{\frac{m_2}{m_4}} \qquad (5.43)$$

And finally

$$\gamma = \sqrt{\frac{m_2^2}{m_0 \, m_4}} \qquad (5.44)$$

REFERENCES

AASHTO. (1990). "Guide Specifications for Fatigue Evaluation of Existing Steel Bridges." American Association of State Highway and Transportation Officials, Washington, DC.

AASHTO. (1999). "Standard Specification for Highway Bridges." Sixteenth Edition with 1999 Interim Revisions, American Association of State Highway and Transportation Officials, Washington, DC.

AASHTO. (2004). *LRFD Bridge Design Specifications. American Association of State Highway and Transportation Officials (AASHTO)*, 3rd Edition, Washington, DC.

Alampalli, S. (2000). "Effects of testing, analysis, damage, and environment on modal parameters," *Journal of Mechanical Systems and Signal Processing*, 14(1), 63–74.

Alampalli, S. and Ettouney, M. (2007). "Structural Identification, Damage Identification and Structural Health Monitoring," *Proceedings of SPIE*, Vol. 6531, San Diego, CA.

Alampalli, S. and Kunin, J. (2001). "Load Testing of an FRP Bridge Deck on a Truss Bridge." Special Report 137, Transportation Research and Development Bureau, NYSDOT, Albany, New York, NY.

Alampalli, S. and Kunin, J. (2003). "Load testing of an FRP bridge deck on a truss bridge." *Journal of Applied Composite Materials*, Kluwer Academic Publishers, 10(2).

Alnahhal, W-I., Chiewanichakorn, M., Aref, A. J., and Alampalli, A. (2006). "Temporal thermal behavior and damage simulations of frp deck." *Journal of Bridge Engineering, ASCE*, 11(4), 452–464.

Aref, A. and Chiewanichakorn, M. (2002). The Analytical Study of Fiber Reinforced Polymer Deck on an Old Truss Bridge. Report submitted to New York State Department of Transportation, Transportation Research and Development Bureau, and Transportation Infrastructure Research Consortium.

ASTM. (2004). "Standard Practices for Cycle Counting in Fatigue Analysis." E1049–85, American Society for Testing and Materials, West Conshohocken, PA.

Barsom, J. and Rolfe, S. (1999). *Fracture and Fatigue Control in Structures*, an ASTM publication No. MNL41, West Conshohocken, PA.

Bassim, M. (1987), "Macroscopic Origins of Acoustic Emission," in *Nondestructive Testing Handbook, Volume 5: Acoustic Emission*, Miller, R, Techincal Editor, ASNT, Columbus, OH, pp. 45–62.

BDI. (2001). "BDI Strain Transducer Specifications." Bridge Diagnostics, Inc., Boulder, CO.

Benjamin, J. and Cornell, A. (1970). *Probability, Statistics, and Decision for Civil Engineers*, McGraw-Hill, New York, NY.

Bishop, N, and Sherratt, F. (1989). "Signal processing for fatigue in both the time and frequency domains," *IMA Conference on Mathematics in the Automotive Industry*, University of Warwick, UK.

Chiewanichakorn, M., Aref, A. J., and Alampalli, S. (2007). "Dynamic and fatigue response of a truss bridge with fiber reinforced polymer deck." *International Journal of Fatigue, Elsevier Science*, 29(8), 1475–1489.

Dirlik, T. (1985). "Application of computers in Fatigue Analysis," PhD Thesis, University of Warwick, UK.

Doebling, S. W., Farrar, C. R., Prime, M. B., and Shevitz, D. W. (1996). "Damage Identification and Health Monitoring of Structural and Mechanical Systems from Changes in Their Vibration Characteristics: A Literature Review," Los Alamos National Laboratory Report, LA- 13070-MS.

Downing, S. D. and Socie, D. F. (1982). "Simple rainflow counting algorithms.' *International Journal of Fatigue*, 4(1), January, 31–40.

Ettouney, M. and Alampalli, S. (2012). *Infrastructure Health in Civil Engineering: Theory and Components*, CRC Press, Boca Raton, FL.

Fisher, J. (1977). "Bridge Fatigue Guide: Design and Detail," American Institute of Steel Construction, New York, NY.

Fu, G., Feng, W., Dekelbab, W., Moses, F., Cohen, H., Mertz, D., and Thompson, P. (2003). "Effect of Truck Weight on Bridge Network Costs," Transportation Research Board Report No. 495, Washington, D.C.

Gassman, S. and Zein, A. (2004). "Condition Assessment of Concrete Bridge T-Shaped Girders under Fatigue Loading," *Proceedings, NDE Conference on Civil Engineering*, ASNT, Buffalo, NY.

Hubbard, B. (1998). *The World According to Wavelets*, A. K. Peters, Natick, MA.

Jalinoos, F., Rezai, A., Moore, M, and Fuchs, P. (2008). "FHWA Steel Bridge Testing Program," *NDE/NDT for Highways and Bridges: Structural Materials Technology (SMT)*, ASNT, Oakland, CA.

Kudryavtsev, Y. F., Kleiman J. I., and Gustcha, O.I. (2000). "Ultrasonic Measurement of Residual Stresses in Welded Railway Bridge," *Structural Materials Technology: an NDT Conference*, ASNT, Atlantic City, NJ.

Lai, L. L-Y. and Ressler, P. R. (2000). "NDT and NDE on an I-95 Viaduct," *Structural Materials Technology: an NDT Conference*, ASNT, Atlantic City, NJ.

Lund, R. and Alampalli, S. (2004a). "Estimating Fatigue Life of Patroon Island Bridge Using Strain Measurements," New York State Department of Transportation Special Report 142, Albany, NY.

Lund, R. and Alampalli, S. (2004b). "Estimating Remaining Fatigue Life of Bridge Components: A Case Study," *Proceedings, NDE Conference on Civil Engineering*, ASNT, Buffalo, NY.

Mahmoud, K. M. (2003). "Hydrogen Embrittlement of Suspension Bridge Cable Wires," In *System-Based Vision for Strategic and Creative Design*, Bontempi, A. Ed, Swets & Zeitlinger, Lisse.

Miceli, M., Moshier, M., and Hadad, C. (2008). "Fatigue Crack Growth Activity Determination During Inspections on Highway Bridges in Three States," *NDE/NDT for Highway and Bridges: Structural Materials Technology (SMT)*, ASNT, Oakland, CA.

Moses, F., Schilling, C. G., and Raju, K. S. (1987). "Fatigue Evaluation Procedures for Steel Bridges," Report 299, National Cooperative Highway Research Program, Transportation Research Board, Washington, D.C.

NYSDOT. (2002). "Structural Integrity Evaluation of Patroon Island Bridge." New York State Department of Transportation, Albany, NY.

Prine, D. W. and Socie, D. (2000). "Continuous Remote Monitoring of the Merrimac Free Ferry," *Structural Materials Technology: an NDT Conference*, ASNT, Atlantic City, NJ.

Rose, J. L. (1999). *Ultrasonic Waves in Solid Media*, Cambridge University Press, London, UK.

Tarris, D., Greimann, L., Phares, B., Wood, D., and Wipf, T. (2002). "Remote Continuous Monitoring of a Plate Girder Bridge," *Proceedings of 1st International Workshop on Structural Health Monitoring of Innovative Civil Engineering Structures*, ISIS Canada Corporation, Manitoba, Canada.

Vahaviolos, S., Wang, D., and Carlos, M. (2006). "Advanced Acoustic Emission for On-stream Inspection of Petrochemical Vessels," ECNDT 2006, We.3.6.5, 2006.

Woodward, C. and McGarvie, A. (2000). "Long Range Ultrasonic Fatigue Crack Detection," *Structural Materials Technology: an NDT Conference*, ASNT, Atlantic City, NJ.

Yen, B. T., Huang, T., Lai, L-Y., and Fisher, J. (1989). "Manual for Inspecting Bridges for Fatigue Damage Conditions." Report No. FHWA-PA-89–022, Lehigh University, Bethlehem, PA.

Zein, A. S. and Gassman, S. L. (2006). "Use of Apparent P-Wave Velocity Measurements to Estimate the Bending Stiffness of a Concrete Bridge Girder," *NDE Conference on Civil Engineering*, ASNT, St. Louis, MO.

Zhang, Y. (2004). "Piezoelectric Paint Sensor for Fatigue Crack Monitoring," *Proceedings, NDE Conference on Civil Engineering*, ASNT, Buffalo, NY.

Zhou, Y. E. (2006). "Assessment of Bridge Remaining Fatigue Life through Field Strain Measurement," *ASCE Journal of Bridge Engineering*, Vol. 11, No 6, pp. 737–744.

6 Fiber-Reinforced Polymer Bridge Decks

6.1 INTRODUCTION

Although fiber-reinforced polymer (FRP) materials have been accepted by several civil engineers in the recent past (Burgueno et al. 2001; Chajes et al. 2001; Alampalli et al. 2002; Alampalli and Kunin 2003; Shekar et al. 2003; Turner et al. 2003), the methods of applying them are still in their infancy (ASCE/CERF 2001; Aboutaha 2004). Not much data on their service life, durability, maintainability, reparability, and life cycle costs are yet available. So, several of the earlier applications were designed very conservatively (Alampalli et al. 2002; Aref and Chiewanichakorn 2002; Alampalli and Kunin 2003; Chiewanichakorn et al. 2003a, 2003b). More data are still needed for proper design, construction, maintenance, and optimization of sections to make them more cost-effective. So, there is a strong need for structural health monitoring (SHM) to document the in-service performance of bridges retrofitted with FRP materials under service loads. The following sections explore two main applications of FRPs in bridge structures. The basic use of each application is described along with a discussion on how structural health in civil engineering (SHCE) can be used to maximize the benefits of the use of FRP in each application (Alampalli and Ettouney 2006).

6.1.1 THIS CHAPTER

We first discuss the general aspects of FRP bridge decks. Material properties, different types of bridge decks, and manufacturing techniques are first presented. General structural health issues and guidelines for the design of the decks are then discussed. Some implications and techniques of inspection, monitoring, and retrofitting are offered whenever appropriate. The next section looks in depth at the structural health components for FRP bridge decks. Conventional and new techniques in monitoring, Structural Identification (STRID), and Damage Identification (DMID) are discussed, and knowledge gaps, as of the writing of this chapter, are presented. The new-material status of FRP bridge decks gives the monitoring of their behavior an even greater importance. Some examples of analysis of monitoring results and the needed decision making processes are discussed next. Several case studies of FRP bridge decks from different viewpoints are presented. Finally, the life cycle analysis (LCA) methodology suitable for new materials in civil infrastructures is offered; it is applicable to LCA of any new material such as the emerging high-performance concrete, high-performance steel, or nanomaterials.

6.2 THE ADVENT OF FRP BRIDGE DECKS

6.2.1 GENERAL

6.2.1.1 History of FRP Bridge Decks

FRP materials are relatively new to bridge-engineering applications. However, several engineers are accepting these materials due to the advantages they offer when compared to conventional materials. Some of the advantages include increased corrosion resistance and less weight. In several cases, where superstructures are in good condition but are restricted to legal loads, replacing the concrete deck with FRP deck offers higher live-load capacity with relatively minimal repairs to the super-structure. This is very attractive to bridge owners as they can extend the life of the superstructure instead of replacing it entirely.

At the same time, the recent trend of minimizing public inconvenience due to closure of a bridge and staying away from the repairs as long as possible (get-out-quick-and-stay-out philosophy) also encourages innovative materials and construction techniques. As most FRP decks are prefabricated and are light weight, the installation time and bridge closure time are significantly less than for those decks made of conventional materials. Use of heavy equipment is also minimized or is not needed. Additionally, thanks to their expected long-term durability and low life cycle costs resulting from their longevity and less maintenance due to absence of corrosion, FRP decks may see good growth in the near future.

FRP materials are engineerable, and their degradation mechanisms are very different from those of steel and concrete. Most bridge engineers using these materials are unfamiliar with them. Most of the durability information is obtained on the basis of the tests done in other industries for different applications. So, studying the durability of these decks comprehensively is very important to make sure that they do really offer the expected long-term durability.

6.2.2 FRP MATERIAL PROPERTIES

FRP materials used in civil infrastructures are composed of two main ingredients: resins and the matrix (sometimes called reinforcements or fibers). Resin is the glue that transfers internal forces to and from the matrix. The matrix is the load-carrying component. It can be oriented in almost any direction within the resin. It is made of numerous types of materials, including fiber glass or carbon fibers. In addition to the fibers and resins, there can also be fillers that are added to the resins during the manufacturing process to reduce cost. There can also be additives that are used to enhance the mechanical and durability properties of the composite, such as increasing its fire resistance.

Mechanical properties of FRP used in civil infrastructures vary greatly. These values depend of the manufacturers, manufacturing processes, and of course the base ingredients in the composite. Table 6.1 shows the mechanical properties of two FRP materials used in building a bridge deck (Aref, Chiewanichakorn, Alnahhal 2004).

Note that the ultimate strengths for tension, compression, and shear vary greatly, with the materials being strongest in tension and weakest in shear. The weight density of FRP materials is in the range of 0.05–0.065 lb/in^3. For comparison purposes, if we assume that a typical design yield steel stress is 33 ksi, a typical steel modulus of elasticity is 29,000 ksi, and a typical steel weight density is 0.28 lb/in^3, then we can define a relative modulus and strength densities of FRP to steel as

$$E_\rho = \frac{E_{FRP}}{\rho_{FRP}} \frac{\rho_{STEEL}}{E_{STEEL}} \qquad (6.1)$$

$$\sigma_\rho = \frac{\sigma_{FRP}}{\rho_{FRP}} \frac{\rho_{STEEL}}{\sigma_{STEEL}} \qquad (6.2)$$

TABLE 6.1
Mechanical Properties of Some Typical FRP Materials

Property	FRP Material 1	FRP Material 2
Modulus of elasticity (ksi)	2680	4310
Shear modulus (ksi)	850	900
Ultimate tensile strength (ksi)	45	90
Ultimate compressive strength (ksi)	32	69
Ultimate shear strength (ksi)	16.5	17.6

Source: Courtesy of Dr. Amjad Aref.

The relative modulus and strength densities of FRP to steel are shown in Table 6.2.

Table 6.2 shows the strongest attribute of FRP materials in civil infrastructures. They do exhibit very high strength-density ratios compared to steel and hence to concrete. On the other hand, they are much more flexible than steel and concrete, due to their relatively low modulus to density ratios.

We note that one of the advantages of FRP materials is their versatility. There is a vast class of FRP which exhibit high stiffness. For example, Jones (1999) reported some FRP material properties that are shown in Table 6.3. We use Equations 6.1 and 6.2 to produce the relative densities in Table 6.4. The table shows that when the advantage of light weight is included, there can be a great strength or stiffness advantage for FRP materials. Such an advantage is direction dependent; this fact needs to be taken into consideration. Table 6.4 also shows that the discrepancy in thermal expansion ratios between steel and FRP might be a cause for concern when combining these two materials; again, this fact needs to be considered during the design process.

6.2.3 ANATOMY OF AN **FRP** BRIDGE DECK

6.2.3.1 Sandwich Geometry

6.2.3.1.1 General

FRP bridge decks have many geometric similarities, yet there are also numerous geometric differences between them. First, it is recognized that FRP constructs are mostly thin plating, with the thickness of the construct being much smaller than the other two dimensions. FRP deck geometry tries to account for that and maximize the structural efficiency of the FRP deck components. As such, the top and bottom surfaces are used to resist stresses in axial tension/compression, a core is

TABLE 6.2
Relative Stiffness (Modulus) and Strength Ratios of FRP Materials of Table 6.1

Property	FRP Material 1	FRP Material 2
Modulus of elasticity (ksi)	0.43	0.69
Ultimate tensile strength (ksi)	6.36	12.73
Ultimate compressive strength (ksi)	4.53	9.76

TABLE 6.3
Material Properties of High-Stiffness and High-Strength FRP

Material	Stiffness	Strength	Thermal Expansion $(10^{-6}$ in/in/F)	Weight Density (lb/in³)
Steel	$E = 30,000$ ksi; $v = 0.3$	$\sigma_{max} = 30$ ksi	$\alpha = 6.5$	0.282
High-modulus graphite-epoxy (GY-70-HYE1534)	$E_1 = 42,000$ ksi; $E_2 = 1000$ ksi; $v_{12} = 0.25$; $G_{12} = 700$ ksi	$X_t = 90$ ksi; $X_c = 90$ ksi; $Y_t = 2$ ksi; $Y_c = 28$ ksi; $S = 4$ ksi	$\sigma_1 = -0.58$; $\sigma_2 = 16.5$	0.061
High-strength graphite-epoxy (AS-3501)	$E_1 = 18,500$ ksi; $E_2 = 1600$ ksi; $v_{12} = 0.25$; $G_{12} = 650$ ksi	$X_t = 169$ ksi; $X_c = 162$ ksi; $Y_t = 6$ ksi; $Y_c = 25$ ksi; $S = 7$ ksi	$\sigma_1 = 0.25$; $\sigma_2 = 15.2$	0.055

Source: Courtesy CRC Press.

TABLE 6.4
Stiffness and Strength Density of FRP

Material	Stiffness Density	Strength Density	Thermal Expansion Ratio
High-modulus graphite-epoxy (GY-70-HYE1534)	$E_1 = 6.47$; $E_2 = 0.15$; $G_{12} = 0.11$	$X_t = 13.87$; $X_c = 13.87$; $Y_t = 0.31$; $Y_c = 4.31$; $S = 0.62$	$\sigma_1 = -0.09$; $\sigma_2 = 2.54$
High-strength graphite-epoxy (AS-3501)	$E_1 = 3.16$; $E_2 = 0.27$; $G_{12} = 0.11$	$X_t = 28.88$; $X_c = 27.69$; $Y_t = 1.03$; $Y_c = 4.27$; $S = 1.20$	$\sigma_1 = 0.04$; $\sigma_2 = 2.34$

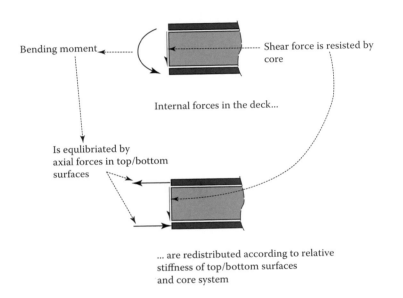

FIGURE 6.1   Structural principles of FRP Deck.

used to transmit shear between them. This basic construct tries to prevent or reduce any bending stresses in the otherwise thin-walled FRP plating. Figure 6.1 shows the basic structural principles that govern this type of geometry.

In what follows, we review the basic geometric components of a typical FRP bridge deck, as shown in Figure 6.2. In doing so, we also point to the health issues of each of the components.

6.2.3.1.2   Top and Bottom Surfaces

The main function of the top and bottom surfaces in an FRP deck is to resist top and bottom axial stresses so as to form the overall flexural resistance of the global deck. For the surfaces to perform this primary function, the stresses and strains within them must be within an acceptable limit. For appropriate performance, the two surfaces must work as a unit through the core. This is achieved by the shear stresses between the surfaces and the core. Two more functions are needed from the top and bottom surfaces. The top surface will have to transmit the direct traffic loading in bearing. The bottom surface must transmit the reactions safely to the supports of the deck.

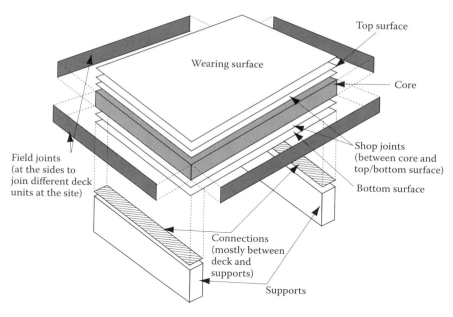

FIGURE 6.2 Anatomy of FRP deck.

To follow the structural health of the top and bottom surfaces, all of the above-mentioned functions must be monitored. For example, some or all of the following might be observed in any SHM for top/bottom surfaces:

- Axial stresses/strains at appropriate locations. Such locations can be the middle of the deck. The number and orientation of the sensors must account for the orientation of the fibers within the material.
- Bearing pressures from traffic on the top surface.
- Behavior of the joint system used to join the surfaces and the core. This is especially needed near the supports of the deck, where the shearing stresses in the deck are expected to be large.
- The expected complex stress and strain behavior near the supports.

6.2.3.1.3 Core Construct

The core of a typical FRP deck has the function of ensuring that the top and bottom surfaces will act in unison. As such, it transmits the required shear between the two surfaces, as shown in Figure 6.1. In addition, the core has to transmit the direct stresses from the FRP supports to the top surface and the direct traffic load from top to bottom surfaces. All these functions require that the core must resist shearing stresses as well as normal stresses (mostly compression). The geometry of FRP cores varies greatly. Honeycomb cores have been used for several FRP bridge decks (see Figures 6.3 and 6.4).

Based on the above, a SHM of the core of an FRP deck might include

- Monitoring the normal and shear stresses (or strains) near the supports
- Monitoring the behavior of the joint system used to join the surfaces and the core

6.2.3.2 Adhesively Bonded Pultruded Geometry

The adhesively bonded pultruded concept utilizes the two-dimensional geometry of pultruded manufacturing process in constructing the bridge deck. First, a series of two-dimensional pultruded

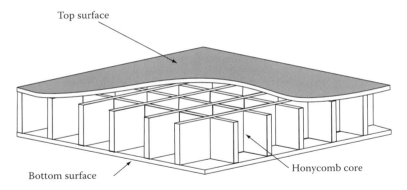

FIGURE 6.3 Honeycomb core.

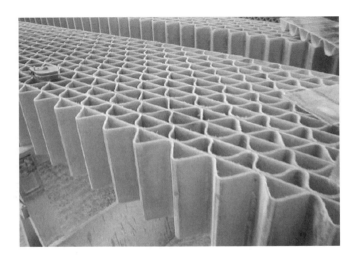

FIGURE 6.4 Typical honeycomb core. (Courtesy of Jerome O'Connor.)

geometries are created with the desired length in the third direction. These components are then joined together in the lateral direction of the deck using appropriate adhesives, thus forming the whole deck assembly. Figure 6.5 shows the concept. These classes of decks also have top and bottom surfaces, as well as a shear core. The shear core is a set of longitudinal plates designed for efficient deflections and strength (both vertical and lateral). See Liu (2007) for a detailed discussion on the different geometrical arrangements of this class of bridge decks. One of the main performance issues is of course the strength and durability of the unit-to-unit adhesives. The adhesives are usually required to be stronger and more durable than the parent FRP material, as shown by Liu (2007).

6.2.3.3 Hybrid Decks

6.2.3.3.1 Concepts

Table 6.1 shows that FRP materials are stronger in tension than in compression. Because of this, general flexural behavior, which is the main mode of behavior of bridge decks, is not optimal. To optimize flexural behavior, hybrid geometries of FRP decks were introduced by many professionals. The hybrid concept is fairly similar to the conventional steel-concrete composite designs. It utilizes the high capacity of concrete in resisting compression as an enhancement to the relatively weaker compression capacity of FRP. Several hybrid FRP geometries are discussed below.

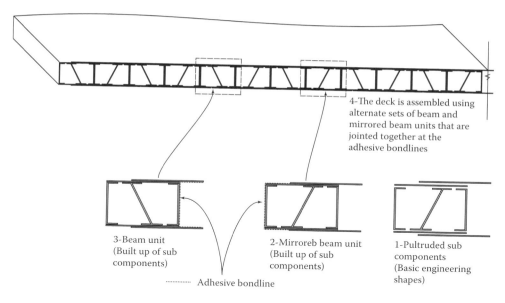

FIGURE 6.5 Components of adhesively pultruded FRP bridge deck.

FIGURE 6.6 Aref hybrid FRP bridge deck. (Courtesy of Dr. Amjad Aref.)

6.2.3.3.2 Aref Hybrid System

The Aref hybrid system was devised at University at Buffalo (see Figure 6.6.) It introduces a small volume of concrete at the compression side (top side) of an FRP pultruded system. The additional concrete was shown to enhance the flexural behavior of the system.

6.2.3.3.3 Wagner CFT Hybrid System

The Wagner CFT hybrid system was used effectively as a deck superstructure bridge in Erie County, NY. The bridge relies on high-strength concrete as the deck/compression component. The webs are made of pultruded FRP elements. The tensile bottom is made of glass fiber reinforced polymers (GFRP) plating as shown in Figure 6.7. The system is efficient on several fronts. The compression concrete can provide durable wearing surface. The FRP webs and tensile bottom plating are highly

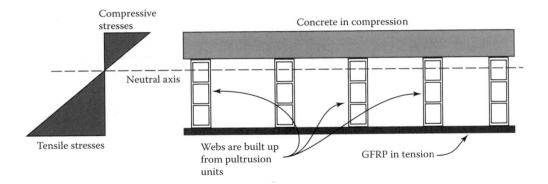

FIGURE 6.7   Wagner CFT hybrid system.

efficient in resisting tension forces. Similar hybrid concepts were used elsewhere (see, for example, McDad et al. 2004).

6.2.3.4   Joints and Connections

6.2.3.4.1   General

There are three types of joints in the FRP deck system:

- Shop joints: These are joints constructed in the manufacturing plant.
- Field joints: These joints connect different FRP panels and components; they are erected on the site.
- Connections between FRP deck system and the supporting elements. The supporting elements can be stringers, cross girders, bridge main girders, or bridge abutments.

We discuss each system next.

6.2.3.4.2   Shop Joints

Shop joints are the joints between the core and the top and bottom surfaces. They are usually adhesive-type joints. The adhesives are administered in the shop during the manufacturing process of the FRP deck. The main function of these joints is to ensure that the top and bottom surfaces function in an integrated manner. As such, the joints will have to transmit the needed shear stresses between the top/bottom surface and the core. In addition, they will need to transmit any existing normal stresses from top (direct traffic loads) or bottom (support reactions) surfaces to the core. Some possible monitoring activities for shop joints are as follows:

- State of stresses in the adhesive (that includes any possible shear lag between the adhesive and the parent material)
- Manufacturing quality of the joint

The quality of the joint is of special interest for FRP bridge decks. This is due to the relatively recent use of FRP as bridge decks, resulting in a limited body of knowledge. Secondly, it is recognized that these joints are going to be *inside* the deck after it is manufactured. As such, it is going to be impossible to visually inspect the state of adhesion in the joint. Special methods must be used to accommodate these problems, as discussed in Sections 6.4 and 6.5.

6.2.3.4.3   Field Joints

Another type of joints within the FRP deck skeleton is the field joints, that is, the joints that will be constructed in the field. Consider the situation when the bridge deck is so large that it cannot be manufactured and then transported as a single unit. In such a situation, the deck is composed of two or more units. These smaller units are manufactured individually and then transported to the site. An example of these types of joints is the shear keys. The shear key joint assumes that the shear key will transfer the shear between the two sides of the joint, whereas the flexure might be transferred by either mechanical fasteners or adhesives. In the site, shear keys are used to connect the different FRP units to form an integrated bridge deck. Another simpler example of field joints is when field-applied adhesives are used to transfer both shear and flexure. Figure 6.8 shows typical load path demands on an all-adhesive field joint. Figure 6.9 shows typical load path demands for mechanical shear key, with adhesives that transfer flexure.

6.2.3.4.4 Connections

Connections are used to connect the FRP deck to other components of the bridge system. An example would be the connecting of a deck to the supporting steel of concrete girders, shown in Figure 6.10, or connecting the deck to supporting abutments. The details of the connections vary greatly, depending on the type, magnitude, and frequency of the loading, the magnitude of the reactions, and the relative motions (creep or temperature effects, for example).

There are several types of potential connections:

- Stud shear connectors
- S-Clips (Figure 6.11)
- Bolted connection (Figures 6.12 and 6.13)

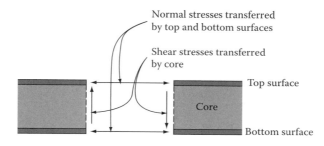

FIGURE 6.8 Load path in field joints (all-adhesive).

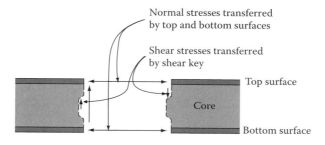

FIGURE 6.9 Load path in field joints (mechanical shear key).

FIGURE 6.10 Connecting FRP bridge deck to steel supports. (Courtesy of New York State Department of Transportation.)

FIGURE 6.11 S-connector. (Courtesy of New York State Department of Transportation.)

- Adhesives (Figure 6.14)
- Pop rivets
- Stirrups—concrete beams (Figure 6.48)

6.2.3.5 Wearing Surface

There are three functions of wearing surfaces on an FRP bridge deck. The first is to act as a load path from the traffic and pedestrian load sources into the supporting structural deck. In addition, it protects the structural deck from various factors (weather, traffic, etc.). Finally, it should provide adequate texture for the movement of traffic and pedestrians. To accomplish the first function, the wearing surface should have adequate bearing strength. To attain the second function, the wearing surface should be durable. In all cases, the wearing surface and the FRP structure below it should have adequate (durable and strong) adhesion between them. Because of the differences in performance demands between the FRP structure and the wearing surface, different materials are used for each. Figure 6.15 shows the basic construct of an FRP deck wearing surface.

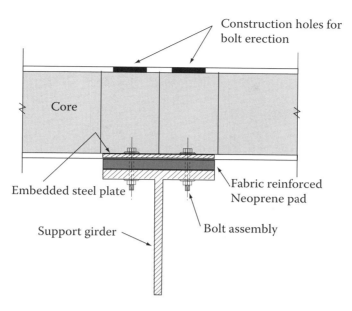

FIGURE 6.12 Bolted connection of an FRP deck with a supporting steel girder.

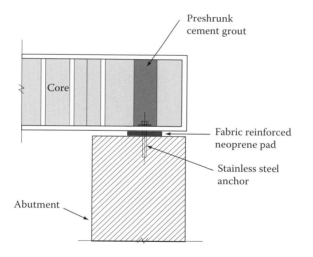

FIGURE 6.13 Connection of an FRP deck with an end abutment. (Courtesy of Dr. Aref.)

Three material types can be used as wearing surface: polymer based, asphalt concrete, or concrete. Aref and Chiewanichakorn (2002) studied several aspects of deck wearing surface behavior.

6.2.4 MANUFACTURING TECHNIQUES AND PROBLEMS

6.2.4.1 Manufacturing Techniques

There are three manufacturing techniques of FRP bridge decks: (1) pultrusion (Figure 6.16), (2) vacuum-assisted resin transfer molding (VARTM) (Figure 6.17), and (3) open/closed mold hand lay-up (Figures 6.18 and 6.19). The manufacturing techniques are described in detail in MDA (2008). There are, of course, several differences between the FRP products of the three manufacturing processes. Table 6.5 shows a comparison of different merits of the three manufacturing methods.

FIGURE 6.14   Adhesive connection. (Courtesy of Jerome O'Connor.)

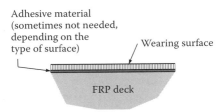

FIGURE 6.15   Typical wearing surface.

FIGURE 6.16   FRP bridge deck manufacturing: pultrusion. (Courtesy of Jerome O'Connor.)

Table 6.6 shows a quantitative comparison of two of the manufacturing techniques offered by Bakis et al. (2002).

6.2.4.2   Quality Control/Quality Assurance Issues

Quality control (QC) and quality assurance (QA) are an integral part of the health of FRP decks. QC measures and techniques of manufacturing, transportation, and erection are needed to ensure

FIGURE 6.17 FRP bridge deck manufacturing: VARTM. (Courtesy of Jerome O'Connor.)

FIGURE 6.18 FRP bridge deck manufacturing: open mold—1. (Courtesy of Jerome O'Connor.)

an acceptable delivered product. Similarly, QA measures and techniques are needed to ensure that the delivered product meets the specifications. The subjects of QC and QA are beyond the scope of this chapter. However, we will only discuss briefly the interrelationships between structural health techniques and the QC/QA of an FRP bridge deck.

QC of manufacturing: Besterfield (2001) identified three basic approaches for total quality management of any product or process: (1) principles and practices, (2) quantitative tools and techniques, and (3) nonquantitative tools and techniques. The principles and practices include management methods at the manufacturing facility, close interaction with the clients, as well as setting performance measures. It is the latter two points that are of interest when considering the health of FRP decks. The client (bridge officials or consultants) must be involved in setting needed performance measures for the FRP manufacturer. These performance measures can be acceptable tolerances in dimensions, average acceptable defects in shop joints,* or acceptable material properties.

---

* An example on this topic is presented in Section 6.4.

FIGURE 6.19   FRP bridge deck manufacturing: open mold—2. (Courtesy of Jerome O'Connor.)

TABLE 6.5
Comparison of Different FRP Decks Manufacturing Methods

	Feature				
Manufacturing Method	Ability to get Custom Sizes	Adherence to Dimensional Tolerances	Cost Attractiveness	Ability to Incorporate Special Features	Overall Quality
Pultruded	L	H	L	L	H
VARTM	H	L	H	H	M
Open mold	H	M	H	M	M

H, high; M, medium; L, low.
Source: Courtesy of Jerome O'Connor.

Quantitative QC tools and techniques include sampling techniques and reliability analysis. Sampling techniques include sample size for defect inspection, as well as the acceptance statistics (sometimes measured as a function of standard deviations). Reliability analysis includes the evaluation of how the product (FRP deck) would perform during the specified lifespan. Nonquantitative QC tools and techniques include adherence to ISO* 9000 and ISO 14000. ISO 9000 is a standard for quality management systems, while ISO 14000 is a standard for environmental management systems. For an in-depth description of QC tools, the reader is referred to Besterfield (2001).

From the above discussion, it is clear that the tools for structural health (experiments, monitoring, and decision making) can be used for QC manufacturing efforts. Some of these tools are included in Table 6.7.

QC of transportation: Even the highest quality products can be exposed to accidental damage during transportation. To ensure high-quality transportation, adherence to tightly specific standards is needed. It is advisable that a responsible official ensures application of an appropriate QA during transportation.

---

* ISO is an acronym for International Organization for Standards.

TABLE 6.6
Comparison between FRP Manufacturing Methods

	Deck System	Depth (mm)	kg/m²[a]	Dollars/ m²	Deflection[b] (reported)	Deflection[c] (normalized)
Sandwich construction	Hardcore composites	152–710	98–112	570–1184	L/785[d]	L/1120
	KSCI	127–610	76[e]	700	L/1300[f]	L/1300
Adhesively bonded pultrusions	DuraSpan	194	90	700–807	L/450[g]	L/340
	Superdeck	203	107	807	L/530[h]	L/530
	EZSpan	229	98	861–1076	L/950[i]	L/950
	Strongwell	120–203	112	700[j]	L/605[k]	L/325

[a] Without wearing surface.
[b] Assumes plate action.
[c] Normalized to HS20IM for a 2.4-m center-to-center span between supporting girders.
[d] HS20+IM loading of a 203-mm-deep section at a center-to-center span between girders of 2.7 m.
[e] For a 203-mm-deep deck targeted for RC bridge deck replacements.
[f] HS20+IM loading of a 203-mm-deep deck at a center-to center span between girders of 2.4 m.
[g] HS20+IM loading of a 203-mm-deep deck at a center-to-center span between girders of 2.2 m.
[h] HS20+IM loading at a center-to-center span between girders of 2.4 m.
[i] HS20+IM loading at a center-to-center span between girders of 2.4 m.
[j] For a 171-mm-deep deck with a wearing surface under experimental fabrication processes.
[k] HS20+IM loading of a 171-mm-deep section at a center-to-center span between girders of 2 m.
Source: Bakis, C. et al., *J. Compos. Construct.*, 6, 2002. With permission from ASCE.

TABLE 6.7
QA and Structural Health Techniques

Manufacturer QC Issue	Structural Health Issue	How to Do It?
Interaction/communications	Planning and manufacturing stages	Built in to the contract
Performance measures	Decision making techniques	Utilize past experiences and life cycle cost analysis
Sampling issues	Decision making, measurements, structural modeling, and damage identification	Utilize all techniques to evaluate both sampling size and acceptance limits. Pay special attention to joints and connections
FRP deck reliability	Decision making, measurements, structural modeling, and damage identification	Utilize reliability techniques and life cycle cost analysis techniques to ensure a reliable product (Product will perform as designed for the designed lifespan)
Adherence to ISO 9000 and ISO 14000	Adherence is part of project management (decision making)	Built-in to the contract

6.2.4.3 Proof Testing: A Case Study

Nondestructive testing (NDT) and SHM techniques were used in a QA proof testing for a new FRP project (McDad et al. 2004). The FRP deck is a hybrid FRP-reinforced concrete deck. The FRP beams were shown to exhibit a factor of safety of about 3. The desired composite FRP-reinforced concrete action was established by designing a special connection. The QA of the FRP beams was established by testing a 4-beam sample out of the total of 24 beams required for the project. The proof testing was performed using acoustic emission (AE). ASME (1998) regulations for AE testing were generally followed. The emission monitoring was done during the reloading of the specimens, not during the initial loading. After calibrating each of the AE sensors, each of the four beams was reloaded in steps. Three of the four beams were found to exhibit adequate performance. The fourth specimen failed the test. Further investigation showed suspected delaminations in that specimen. During the test, precautions to reduce wind noise effects on AE sensing were taken. Continued SHM using AE is reportedly continuing.

6.2.5 ISSUES RELATED TO STRUCTURAL HEALTH

6.2.5.1 General

We can put the structural health issues of FRP bridge deck into structural and performance categories (Figure 6.20). Speaking of structures, there are local and global structural issues. Structural health considerations vary depending of the category and type of issue.

6.2.5.2 Structural Issues: Local

6.2.5.2.1 Delaminations

One of the main local failure modes in an FRP deck is the delaminations between the surface plating and the webs (in an adhesively pultruded system) or the core (in a sandwich system). Potential mitigating schemes were offered by Hassan et al. (2003). By utilizing a three-dimensional woven fiber methodology into the manufacturing of the pultruded FRP deck, increased resistances to potential delaminations were achieved. For sandwich panels, the authors suggested the use of three-dimensional fiber insertions across the foam core. These insertions increased shear resistance and reduced the potential of delaminations.

We note that delaminations, especially localized delaminations, constitute one of the most recurring damage in an FRP bridge deck. Therefore, monitoring for delaminations should be an integral part of any long-term SHM project. Unfortunately, conventional strain or vibration monitoring cannot reveal such damage due to the localized nature of the damage. AE, ultrasonic, or thermography methodologies can be of help in executing such projects.

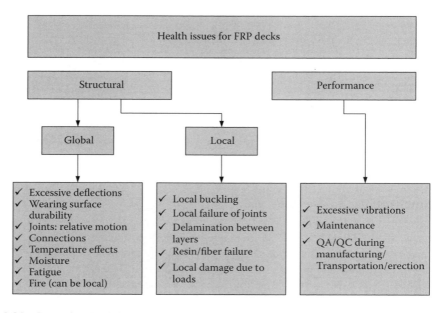

FIGURE 6.20   Issues in FRP bridge deck.

6.2.5.2.2 Debonding

Almost all FRP bridge decks are made of subcomponents attached together by either mechanical or adhesives means. One form of the localized failure of these joining or connecting is debonding. Smith, Hassan, and Rizkalla (2004) and Liu (2007) both studied debonding issues between FRP components that are adhered together. Among factors that affect debonding are the quality of adhesives, surface preparations, direct impact, and environmental effects (such as pH level and high temperature).

Quality of adhesives and surface preparations are both manufacturing and erection issues. They should be addressed during the manufacturing/construction processes. Karbhari et al. (2005) observed that impacting FRP components can cause hidden damage below the impact surface. As such, detecting impact damage, especially for top surface, can be a challenging task. Thermography or shearography offers reasonable solutions: they can cover a large surface remotely with reasonable detection capabilities. Environmental effects can be monitored either by directly monitoring deterioration behavior (see Alampalli and Kunin 2003) or by monitoring the actual environmental factors, such as pH and temperature levels. In the latter case, a virtual sensing paradigm can be used to detect potential damage from those monitored environmental effects.

6.2.5.2.3 Local Damage due to Loads

This type of damage includes top surface damage that results from snow plows or wind or flood-borne debris. Aref et al. (2004) concluded that snow plow damage is not a failure-producing event. However, he recommended that such damage should be retrofitted immediately. Similarly, after wind storms or major flooding, it is reasonable to check for debris-caused damage and retrofit as appropriate.

6.2.5.2.4 Other Local Effects

There are many other local damage such as (1) local failures of core or surfaces (plate or web buckling, local joint failures), (2) resin versus fiber failures, or (3) minor fire damage. The localized damage and others can be monitored using any of the schemes discussed in this chapter.

6.2.5.3 Structural Issues: Global

6.2.5.3.1 Excessive Deflections

Due to the relative flexibility of FRP decks, Federal Highway Administration (FHWA 2008) recommended a maximum deflection-to-span ratio limit of 1/800. Generally, to meet such a limit, the deck needs to be well within the elastic limit. Because of this, if unusually excessive deflections are observed during the normal operations of the deck, they would indicate major damage. They can be an indication that some major structural components within the deck are behaving nonlinearly. Nonlinear behavior is not an allowed design condition (as of the writing of this chapter): it might be an indication of an imminent failure. Swift action is needed in these situations. Fortunately, monitoring for excessive deflections is an easy task: simple displacement transducers in critical deck locations can produce adequate results.

6.2.5.3.2 Wearing Surface Failures

Wearing surfaces on FRP bridge decks are perhaps the component in the deck assembly to experience the fastest rates of deterioration. This rapid deterioration was observed in several situations. The signs of deterioration include cracking of surface, loss of adhesion, and spalling of wearing surface material. Additional indirect effects of wearing surface deterioration include damage of the underlying deck itself.

Visual inspection of the wearing surface remains the most effective means of tracking its health. Unfortunately, visual inspection cannot help in quantifying any problem with the adhesion between the wearing surface and the underlying deck. In addition, any indirect damage in the deck under the wearing surface is impossible to verify visually.

In an automatic health-monitoring project for wearing surfaces, some of the possible activities are shown in Table 6.8.

Since the causes of damage or deterioration of the wearing surfaces are mostly nature or traffic related, it is very difficult to design a realistic laboratory experiment that accounts for most of the causes in Table 6.8. What would add to the difficulties is that many of these causes interact and affect each other. For example, higher temperature would affect the properties of the adhesion between the wearing surface and the deck below. However, in recognition of the relatively high costs of wearing surface replacement, on-site monitoring of wearing surfaces might be a cost-effective measure in the long run.

6.2.5.3.3 Field Joints: Relative Motion of Joints Causes Failure of Joint

Field joints, as was discussed earlier, are needed when the FRP deck is too large to be manufactured and transported as a single unit. Field joints are used to attach smaller units into an integrated FRP deck. Figures 6.8 and 6.9 indicate that shear and flexure must be transferred by the field joint construct. In addition to transferring the shear and flexure, a field joint must ensure compatibility in deformation (strains) between the two sides of the joint. Figure 6.21 shows how the high deformation (strains) demands of the all-adhesive field joints can affect both top and bottom surfaces and the core material.

We can summarize that if a field joint provides incompatible deformations (strains), the potential effects can be (1) damaged wearing surface, (2) damaged top/bottom surface, and/or (3) damaged core.

In a health-monitoring project for field joints, Table 6.9 shows some potential measures that can help in monitoring and identifying the damage and its source.

6.2.5.3.4 Creep

Creep is defined as the time-dependent inelastic deformation of material (Mclintock and Argon 1966). This can be applicable also to FRP materials. Deformations under constant loads and over

TABLE 6.8
Health Monitoring of Wearing Surface

Problem	Possible causes	Experiment	Sensor Attributes
Cracking/ spalling	Inadequate durability properties	Measure strains, applied loads, temperature, construction defects	Visual inspection, Acoustic emission, Self-sensing materials
	High temperature		Thermographs
	Inadequate strength		Strains, load cells
	Low construction quality		GRP
Loss of adhesion	High stresses (shear/normal) and inadequate adhesion properties	Strains	Strains
	High temperature	Measure temperature	Thermographs
Damage to the underlying deck	Nonuniform stress distribution between wearing surface and underlying deck might cause excessive stresses beyond the design range	Vibration monitoring, especially with higher modal behavior load tests	Strains, velocities

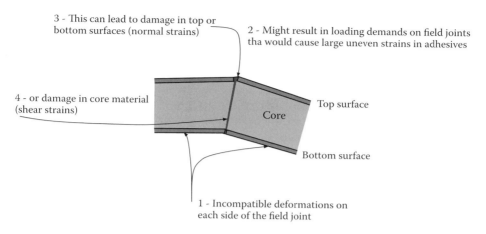

3 - This can lead to damage in top or bottom surfaces (normal strains)

2 - Might result in loading demands on field joints tha would cause large uneven strains in adhesives

4 - or damage in core material (shear strains)

Core

Top surface

Bottom surface

1 - Incompatible deformations on each side of the field joint

FIGURE 6.21   Incompatible deformation (strain) effects on all-adhesive field joints.

TABLE 6.9
Health Monitoring of an FRP Field Joint

Problem	Possible causes	Experiment
Damage of top/bottom surfaces at joint	Inadequate surface-to-surface adhesion (top/bottom)	Visual inspection might be sufficient to detect damage
Large joint rotation		Use inclometer or strain sensors. Large joint rotation usually occurs as a result of abnormal behavior elsewhere. Their presence might indicate a major safety issue
Wearing surface damage		Visual inspection is usually used to observe potential surface damage. Thermography can be used to observe potential of large surface area
Core damage	Inadequate shear key design or inadequate adhesion of core	GPR or impact-echo procedures can be used in this situation
Large vertical slippage at joint		Strain sensors might be adequate to detect potential damage. Also, localized NDT techniques can be used
Wearing surface damage		The damage would be confined to an area that is near the shear key or the core problem region. Localized NDT methods, such as ultrasound, acoustic emission, or GPR might be used to detect damage

time are not desirable for engineering systems, definitely not for bridge decks. Thus, creep can be subdivided into three stages, according to Mclintock and Argon (1966): transient, steady state, and tertiary stages. In the transient stage, the creep strain rate is proportional to $t^{-2/3}$. During the steady-state creep, the creep strain rate becomes constant. The rapid increase of creep strains in the tertiary stage is due to reductions of sections of the component on hand.

Creep strains are influenced by temperature, type of FRP, and geometric properties of the construct. For example, near joints, where stresses tend to be higher, creep strains are expected to be higher. For discontinuous fibers, it is also expected that creep effects will be high. Perhaps the single most important factor that affects creep behavior is temperature effects, both direct applied temperature effects (temperature cycles), and/or residual thermal effects. Hull and Clyne (1996) discussed the direct relationship between temperature cycles and creep.

Due to the criticality of creep strains in the performance of FRP bridge decks, any structural health experiments should consider creep strains. The following are some considerations for monitoring creep strains in a typical FRP deck:

1. Install sensors near joints that are expected to have higher stress levels.
2. The core of a typical FRP deck can be sealed and hence it can entrap heat. A judicial choice of spots with most expected high temperatures must be made, perhaps with the aid of thermal analysis. The creep measurements should be at those locations.
3. Since temperature is the single most important cause for creep, the inclusion of pairing sensors, for both strain and temperature measures, is always recommended.
4. Creep strain sensors can be either the general strain sensors or specialized creep sensors. The advantage of a conventional strain sensor is its multitasking ability and cost-effectiveness. The advantage of a specialized creep sensor is its accuracy, but it may be costlier than the conventional sensor.

6.2.5.3.5   Connections

Because of the complexities of connections of FRP bridge decks, it is not surprising that damage to these connections can occur more often than to the other components of the deck. Difficulties in analysis and design of the connections are some of the reasons for the complexities. For example, accurate analytical methods are either extremely involved or nonexistent (as of the writing of this chapter)*. Construction techniques are another reason for such complexity; the accurate torque of a bolt so as to achieve the required pressure without damaging the resin matrix is difficult. Controlled laboratory testing can be of help in evaluating the performance of a particular connection design. Laboratory testing cannot uncover defects that are due to connection manufacturing or connection construction. Also, visual inspection can reveal the damage to connections. Unfortunately, connection damage can be hidden inside the matrix, or the whole surface can be hidden away from the inspector's line of sight.

It seems that a health-monitoring program for connections in an FRP deck can help to identify potential problems during the lifespan of the deck. Table 6.10 shows how the focus of monitoring programs can change during the life of the deck.

6.2.5.3.6   Fatigue

Fatigue can have numerous definitions; in general, it concerns itself with the reduction of capacity of the subject matter under specific demands as a function of passing of time[†]. The common load-time curve (S-N) is shown in Figure 5.1. The subject of fatigue in metals is covered in depth in Chapter 5. In this section, we consider the fatigue of FRP materials as they are used in constructing bridge decks. The time dependency of FRP material response is one of the major differentiating features that sets it apart from conventional engineering materials (steel and/or reinforced concrete). Table 6.11 shows the relative importance of time-dependent demands for different engineering materials.

Several authors have investigated the fatigue effects on FRP decks by applying a large number of cyclic forces on the deck and measuring the resulting strains or displacements, searching for any sign of deterioration or damage. Alampalli and Kunin (2002) and Aref et al. (2005) studied the performance of FRP decks in the field in real-life conditions. No sign of degradation of strength/behavior was reported. It seems that the behavior of FRP decks under accelerated fatigue testing conditions is acceptable.

There are, however, many issues that are not settled as of the writing of this chapter. The two most important issues are (1) fatigue behavior at discontinuities, and (2) interaction with other

---

[*] The reader is referred to Section 6.3.4 on structural modeling.
[†] Time dependency in this section does not include any inertial effects.

TABLE 6.10
Change of Monitoring Focus during LifeSpan of Connections

Period	Focus
During construction—Before commissioning	Manufacturing/construction defects
Intermediate (up to 2–5 years)	Fatigue/creep/moisture/temperature effects
Very long term	Abnormal conditions (extreme temperature, earthquakes, accidental impacts, etc.)

TABLE 6.11
Time-Dependent Demands for Common Engineering Materials

		How Important Is the Physical Phenomenon in the Time-Dependent Behavior of the Material? (H = High, M = Medium, L = Low)	
Physical Phenomenon		Steel	Reinforced Concrete	FRP
Fatigue (cyclic loading)		M	M	M
Stress rupture		M	L	M
Creep		L	H	M–H
Environmental conditions	Temperature	L	L	M
	Moisture	M	H	L
	Alkaline	L	H	L
	UV	L	L	H
Microdamage in brittle materials		M	L	H

time-dependent loadings (Table 6.11). There is paucity of information about fatigue behavior near shop joints, field joints, and connections. Dutta et al. (2003) investigated the relationship between fatigue and temperature levels. However, there is a need for a closer look at the fatigue behavior of shop joints. In addition, the interaction of the time-dependent effects needs to be studied. We next discuss the two issues from monitoring viewpoints.

Fatigue behavior at discontinuities: When monitoring fatigue performance, there are several questions that need answers: *Where? How long? What?* The location, or locations, of fatigue performance should be near discontinuities; these are the locations where fatigue effects are expected to be the maximum. Unfortunately, the locations of discontinuities in an FRP deck are the most difficult to place sensors at (field joints, shop joints, and connections). Strains are perhaps the most obvious measure for direct fatigue effects. Near discontinuities, sensors must be small, fairly reliable (since it is difficult to maintain or replace them), and easy to attach to the system. Finally, the length of time (field monitoring) or number of stress cycles (laboratory monitoring) depends on the type of monitoring. For field monitoring, the length of time can vary from few months to several years (see Aref et al. 2005).

Interaction between time-dependent concerns: We showed that fatigue is not the only time-dependent issue that can reduce the capacity of FRP decks. Because of this, it is important to investigate the interaction between all of those issues in any monitoring project. It is possible to isolate direct fatigue effects in laboratory monitoring. Unfortunately, it is much more difficult to isolate the time-dependent issues from each other in a field-monitoring project. Alternatively, in laboratory monitoring, the different time dependencies can be accelerated and controlled. A comparison of field and laboratory testing for time-dependent issues is shown in Table 6.12.

TABLE 6.12
Monitoring Time-Dependent Concerns

Issue	Laboratory	Field
Cost	Lower	Higher
Realism	Less realism	High degree of realism
Control of conditions	Better control	No control on natural conditions
Ease of measurements	Easier	More difficult

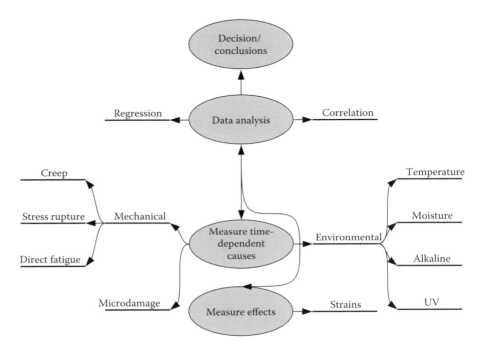

FIGURE 6.22   Time-dependent monitoring.

Whether the monitoring is done in a laboratory or in the field, it is essential to perform the concurrent measuring of many of the time-dependent issues. Figure 6.22 shows this need. The parallel measurements are then correlated and the appropriate conclusions made. This chapter includes an example of using correlation and regression analysis in a decision making effort.

6.2.5.3.7   Alkaline, Ultraviolet, and Moisture Effects
Presence of alkaline is known to degrade the properties of bare glass fibers. This is particularly true for alkaline from concrete, a situation that would occur (1) if a concrete wearing surface is used on top of the FRP deck (Figure 6.15), or (2) if the deck is supported directly over a concrete girder. We observe that the presence of the resin matrix would help in protecting the fiber from the degradation due to alkaline; however, the extent of the degradation, in the long run, is not well known. Ultraviolet (UV) is known to be harmful to most polymers. The harmful effects are generally of lesser importance in the case of bridge decks than in other FRP applications. One reason is that FRP bridge deck plating is thicker than other applications. And since the UV harmful effects are concentrated in a tiny area beneath the exposed surface, the overall effect of the UV is not as great as other applications. Another reason is that the wearing surface can protect the deck from exposure to UV

effects. However, the sides of the deck would be exposed to UV effects, and care must be taken to mitigate such harmful effects. Finally, it might come as a surprise to many that moisture can have a degrading effect on FRP material. Moisture can alter the structure of the resin itself. In addition, it will eventually affect the fiber, leading to pitting and pinching in the deck. Moisture effects on FRP material can change the stiffness by about 10% in 10–15 years.

To lessen either alkaline or UV effects, several mitigation techniques are well known. These include adequate painting from the direct UV effect or isolation from direct contact with alkaline sources. In a health-monitoring program, it is advisable to follow the correlation between cause and effect, that is, monitoring for alkaline levels and also for potential damage by the alkaline presence. The former can be achieved by using appropriate alkaline-measuring sensors. The latter can be achieved by visual inspection or by appropriate strain sensors. Similar strategies can be followed for UV presence/effect-monitoring programs. Mitigating moisture effects in FRP construction can be done by the use of adequate covering material or a more moisture-resistant resin. A monitoring program for moisture must include both cause and effect. As such, measuring moisture and strains must be included in the program.

6.2.5.3.8 Extreme Events (e.g., Fire)

Aref et al. (2004) studied analytically the behavior of an FRP deck during fire. The study included coupled thermal and mechanical analyses. They concluded that, in certain situations, fire can pose a safety concern for the deck. As such, it was recommended that immediate emergency measures and retrofit actions are taken in case of fire.

6.2.5.4 Performance Issues

6.2.5.4.1 Excessive Vibration

One of the aforementioned advantages of using FRP for bridge decks is the high strength and stiffness densities. FRP offers a fairly lightweight construct in relation to its stiffness and strength. In general, this is a great advantage. Yet, there are certain situations where the high stiffness density produces undesired results, namely, possible excessive vibrations. Shekar et al. (2004) reported large vibrations and potentially long-term harmful effects.

6.2.5.4.2 QA/QC during Manufacturing, Construction, and Acceptance

Many researchers and owners have long been aware of the importance of QA/QC issues for FRP deck. For example, Christie, R. et al. (2001) indicated the need for documentation and implementation of QA/QC procedures both during manufacturing and erection of FRP decks. Camata and Shing (2004) indicated the need for QA/QC during manufacturing, especially for joints between top/bottom surfaces and the interior core. Some reasons why QA/QC implementation is needed are that it

1. Ensures minimum degree of consistency of product
2. Helps in decision making
3. Helps in any maintenance and/or inspection procedures
4. Promotes higher degree of confidence in the design and analysis processes
5. Promotes confidence in quality of components, especially those that are hidden (inside joints between top/bottom surfaces and the core)

Perhaps the reader is wondering: "Why do QA/QC issues belong to a structural health book in the first place?" Well, there is a strong link between QA/QC procedures and structural health, especially for new and innovative subjects such as FRP bridge decks. Consider Table 6.13. It shows some potential items in a generic QA/QC plan that is specific to FRP bridge decks. It also shows how these items can be used in any structural health project.

TABLE 6.13
Relationship between QA/QC and Structural Health of FRP Decks

QA/QC Issue	Structural Health Issue
Consistent manufacturing of top/bottom-to-core joints	NDT of samples; acceptance criteria
Manufacturing and erection of field joints (all-adhesives or shear key)	Strain, load, and fatigue measurements
Transportation	Visual inspection
Material properties, material mix, and manufacturing processes	Structural identification and modeling, damage identification and modeling
Connection tolerances and details	Connection behavior: strains, fatigue, temperature effects

6.2.6 FHWA Guidelines and SHM

FHWA issued a set of guidelines (2008) for designing FRP bridge decks. We summarize in Table 6.14 some of those guidelines and potential monitoring techniques that can ensure proper deck performance.

6.3 HEALTH OF FRP BRIDGE DECK

6.3.1 GENERAL

6.3.1.1 Overview

SHCE concepts can be used to study the gaps in current knowledge and to identify the studies required further. These concepts have been routinely used to evaluate infrastructure, including FRP bridge decks (Henderson (2000); Alampalli et al. 2002; Aref and Alampalli 2001; Burgueno et al. 2001; Chajes and Shenton 2001; Reising et al. 2001; Shekar et al. 2003; Turner et al. 2003; Zhao and DeWolf 2002). There have been numerous experimental and analytical studies to investigate the behavior of FRP bridge decks, but an integrated study covering their long-term durability is not yet available. This section outlines the issues related to FRP bridge decks needing attention for future implementation of a successful SHCE program.

6.3.1.2 SHCE and FRP Decks

The concept of SHCE is an integration of SHM with decision making processes and asset management (Ettouney and Alampalli 2000, 2002). SHM involves instrumenting the bridge with sensors and measuring structural deformations and/or stresses under loads to compare those used in the design or obtained through analysis. Thus, the main components of structural monitoring include measurements and sensing, structural identification, and damage/deterioration identification. SHM can be considered the first of the three legs of the SHCE. This concept was advocated by Ettouney and Alampalli (2000, 2002) with the argument that any SHM project that does not incorporate decision making/cost–benefit ideas in all its tasks cannot be a successful project and should not be pursued. Thus, for efficient and meaningful upkeep of the structural health of our infrastructure inventory, all these issues must be well integrated and covered.

As noted above, FRP bridge decks were first introduced to bridge applications on experimental basis, in most cases to take advantage of their light weight, to replace deteriorated concrete bridge decks and increase the live load capacity of bridges. Until recently, most of the monitoring and research studies of these decks predominantly focused on bridge decks alone, with not much attention paid to the entire structure. The future use of FRP decks in bridge applications

TABLE 6.14

FRP Bridge Decks Design Guidelines and Potential Monitoring Techniques

Issue	FHWA Guide	NDT—SHM Support Potential
Long-term creep	Design strains less than 20% of ultimate	Monitor strains at critical locations
Environmental durability	Use 65% of specified material properties	Monitor deflections as well as strains; monitor critical joints and connections using appropriate NDT methods
Low modulus of elasticity	Design to be driven by deflection rather than strength	Monitor modulus of elasticity using adequate NDT method
Deflection limit	1/800 of span length	Monitor deflections
Compression failure in lamina	Use factor of safety of 5	Use adequate NDT methods (thermography, ultrasonic, acoustic emission, etc.) to detect local buckling at critical locations.
UV hazard	Use appropriate paint	Use adequate NDT method: surface coverage and large surface area (thermography, ultrasonic, shearography, etc.) to detect potential UV degradation
Unintended composite action with steel girders	Design for composite action or provide for slip mechanism	Strain monitoring for both FRP deck and supporting girder
Ambient thermal stresses	Design for 100° difference between top and bottom deck surfaces	Temperature gauges at critical locations
Bonded FRP curbs	Protect from impact by bridge rail	Visual inspection for potential impact damage

will depend on their practicality and cost-effectiveness as part of the entire bridge system. Hence, a system approach, instead of a component approach, is needed in studying these decks. Application of concepts of SHCE to FRP decks will yield better understanding of the use of FRP decks for infrastructure applications. Figure 6.20 shows the performance and structural issues related to FRP decks. SHCE concepts can be of immense use in the following areas (Alampalli and Ettouney 2006):

- Component health versus system health: FRP bridge decks are built using several components, such as
  - Top and bottom skins/plates to resist stresses in axial/flexural compressive and tensile forces, with a core to transmit shear between them (see Figure 6.2)
  - Longitudinal and transverse joints and construction joints: limitations of manufacturing, construction, and transportation dictate the size of the FRP deck segments, necessitating connections between segments (see Figures 6.23 through 6.26)
  - Connections are required to attach the deck to the superstructure and the supports (see Figure 6.27).
  - A wearing surface is added to provide a good riding surface (see Figure 6.28).

There is limited information concerning the behavior of the entire system. This, coupled with the effect of in-service conditions on the behavior of FRP decks and the type of inspection they are subjected to, makes it important to design them for implementation in a service setting, that is, the actual deck rather than the individual components. Thus, all these aspects become very important for structural health application involving an FRP deck (Ettouney and Alampalli 2000; GangaRao and Shekar 2003; Alampalli and Ettouney 2006).

FIGURE 6.23 FRP bridge deck in New York during installation. (Courtesy of New York State Department of Transportation.)

FIGURE 6.24 Sections of FRP deck delivered to construction site on a truck. (Courtesy of New York State Department of Transportation.)

6.3.2 HEALTH ISSUES OF FRP BRIDGE DECKS

6.3.2.1 General Issues

This section discusses the capacity versus demand issues and component versus system issues as they relate to FRP decks and SHCE.

6.3.2.1.1 Capacity and Demands

Ettouney and Alampalli (2002) argued that capacity/demand (CD) ratio of a structure should be the backbone of any SHCE program. Both the capacity and demand of the structure are time dependent due to the continuously changing environment. Considering that the optimum CD ratio of any performance measure is 1.0, several CD ratios may be needed to ensure a healthy structure. The SCHE program should follow all of those CDs while the program is in progress.

FIGURE 6.25 FRP superstructure delivered to site before installation. (Courtesy of New York State Department of Transportation.)

FIGURE 6.26 FRP superstructure showing shear key to connect two sections. (Courtesy of New York State Department of Transportation.)

Capacity of an FRP bridge deck can be defined in a direct or indirect manner. For example, direct capacity can be stresses (strains), deflections, or temperature. Indirect capacity can be measured as a moisture limit, creep limit, number of fatigue cycles, or UV exposure limits, and so on. Well-established capacity limits must be prepared before embarking on any SHCE program for an FRP bridge deck. Chapter 2 by Ettouney and Alampalli (2012) discusses the attributes of time-dependent capacity of a structural system.

Demands on FRP bridge decks can be traffic-type demands (weight demand) or environmental demands. Traffic (weight) demands are self-evident and can be controlled by the bridge official. Environmental demands can be general or FRP specific. Examples of general environmental demands are earthquakes or wind. Among the FRP-specific environmental demands are temperature, moisture, or UV exposure. Chapter 2 by Ettouney and Alampalli (2012) discusses the attributes of time-dependent demands of a structural system.

FIGURE 6.27   FRP deck during construction to show the connection with the girders.

FIGURE 6.28   Installation of wearing surface over an FRP deck.

To have a proper CD ratio in an SHCE application, there is a need for both *in situ* measurements and a set of structure-specific algorithms that translate *in situ* measurements to structural capacity. The demands on the structure are defined by the bridge official. FRP-specific algorithms must be established to find the final CD ratio of the deck under consideration.

6.3.2.1.2 Component Health versus System Health

FRP materials have been used in aerospace and other industries for several decades, and so their behavior under mechanical and environmental loading has been studied and documented extensively. However, FRP bridge decks are built using several components. In most cases, there are top and bottom skins/plates to resist stresses in axial/flexural compressive and tensile forces, with a core to transmit shear between them. These coupled with connections, complex support conditions, longitudinal and transverse joints intended for load transfer, construction joints, intended/unintended composite action between the super structure and the deck, bonding materials, and wearing surfaces make these systems complex. All these components together would form the FRP deck system. There is limited information about the behavior of the entire system. Considering that they are subjected to harsh in-service conditions and, largely, inspected visually, it is important to design them for implementation in a service setting, that is, the actual deck rather than the individual components. Thus, all these aspects become very important for structural health application involving an FRP deck.

At the same time, the deck versus the entire structure health and durability is of paramount value. For example, the fatigue life and strength of the FRP deck may be similar to those of a conventional concrete deck, but the effect of a lighter and more flexible FRP deck could have an effect on the life of the entire bridge. On the one hand, the lower weight/flexibility would reduce the stress levels due to the deck's own weight, thus improving the fatigue remaining life of the bridge. On the other hand, by changing the weight (mass) characteristics of the bridge, a new dynamic vibration mode will result. The impact coefficients of this new vibration mode can be either beneficial or detrimental to bridge behavior. Thus, global or system behavior should be considered and investigated thoroughly to evaluate the long-term durability of the structure before a decision on such a replacement is considered.

6.3.2.2 Material Issues

Some of the FRP-specific material issues will be briefly discussed in the section below, and SHCE concepts will be applied to study the specific service-behavioral issues that need consideration when using FRP bridge decks.

6.3.2.2.1 Fatigue

Detecting and countering fatigue effects on FRP bridge decks as part of a generic SHCE, as detailed earlier, requires monitoring/measuring fatigue effects, identifying damage while they are being formed, and deciding on appropriate mitigation/corrective measures. Several authors have studied fatigue effects in FRP composite materials (Konur and Matthews 1989). Most of those studies consider general situations. The basic S-N relationships for the FRP material are published in those studies. The main approach is to measure displacements and strains within the FRP bridge deck, under realistic conditions, and then using appropriate analytical methods to extend them to compare them with analytical results, using, for example, the finite-element (FE) method. The available S-N curves for that particular FRP material are then checked, and an overall assessment of the fatigue condition of the bridge deck is issued.

To have a reliable SHCE methodology for fatigue effects in FRP bridge decks, we need to improve the knowledge in three distinct areas. First, the fatigue effects in realistic connections of the deck need to be understood more. In particular, the S-N relationships for typical connections are needed. Secondly, the effect of fatigue in the integrated system needs more understanding, as was mentioned in a gap analysis report (ASCE 2001). Finally, a comprehensive qualitative correlation between the fatigue distress signs and the state of structure needs to be studied.

Based on the above, a reasonable SHCE program for fatigue effects can involve the following steps:

1. For a typical in-service FRP bridge deck, perform required analyses to determine the locations where fatigue effects are most likely to occur. Also, choose some locations where fatigue effects are not likely to occur, to form a baseline database. Special attention should be paid to the behavior of the connections and joints
2. Install a suitable set of sensors in the locations identified in Step 1
3. Start the measurement process under normal traffic conditions, while continuously comparing measurements with predicted analytical results obtained from Step 1
4. Maintain accurate reporting of the comparisons from Step 3

Adopting an SHCE program similar to the above would add to the knowledge base. Future programs would benefit from the results of the suggested program, and in those future programs the cost–benefit reasoning can be added.

6.3.2.2.2 Creep

Creep, in FRP bridge decks, can have an important effect on the overall health of the deck. Creep behavior in FRP materials can be as important as it is in reinforced concrete material, while being very different. One of the differences is its strong correlation with fatigue behavior and causes. Because of this, special SHCE measures must be taken for FRP bridge decks to ensure continued and adequate deck performance. There are several tests for evaluating creep behavior of composite materials. For example, ASTM D2990–95 for tensile, compressive, and flexural creep, and the creep rupture of plastics. In addition, several authors have studied the phenomenon (ASCE 2001). Most of the available literature on the subject involves generic testing, analysis, and observations.

Specific creep studies in FRP bridge decks are not available; therefore, an SHCE program for monitoring creep effects on FRP bridge deck must be instituted. An initial program should include, among other things, measurements of creep and relaxations at specific locations of the FRP deck and comparisons with baseline measurements. The effects of normal traffic, moisture, and temperature should be documented. Such a program will not yield maximum benefit without a parallel program that considers fatigue effects, as detailed earlier. Future creep effect programs should also add cost–benefit evaluations to the study.

6.3.2.2.3 Moisture

Exposure to moisture can have an adverse effect on FRP material, contrary to conventional wisdom (Fried 1967). This adverse effect can be easily mitigated through the judicial use of resin material and continued vigilance in monitoring the condition of the system under consideration. Moisture during curing of the resin is important, but moisture levels during the service life of the FRP bridge deck are also important. An effective strategy for moisture effects in FRP bridge decks can then involve the following program:

1. Establish a moisture performance set of criteria for bridge decks. Such criteria would relate cumulative moisture levels, expected service life, and type and location of the FRP under consideration. The use of available testing procedures (Johnson and Houston 1990) can be of immense help in establishing such criteria
2. For an existing or new FRP bridge deck, install appropriate moisture sensors. The number and locations of such sensors would depend on the type and complexity of the deck
3. Continuously observe the moisture levels and compare with the criteria developed in Step 1. Cost–benefit conclusions can be continually made by the bridge owner using this comparison

It should be noted that the interaction between moisture effects and other mechanical effects (such as fatigue or creep) is expected. So, an SHCE for FRP bridge decks should include as many of those effects as possible.

6.3.2.2.4 Temperature and UV

Like moisture, temperature plays an important part in the performance of the FRP bridge deck. First, a controlled curing temperature of the deck during manufacturing is of paramount importance. Second, the freezing, freeze-thaw, and temperatures above the curing temperatures can lead to significant changes in the expected performance of the FRP deck. Finally, cyclic temperature changes (both range and number of cycles) can also have an effect on performance. Because of this sensitivity to temperature, many authors and tests have studied the issue (Miyano et al. 1999; Gomez and Casto 1996). Unfortunately, no comprehensive study is available for FRP bridge decks.

Currently, there are several essential topics that need immediate understanding (ASCE 2001): bonding/debonding due to freeze-thaw conditions, softening due to higher temperatures (including during the placement of hot asphalt wearing surfaces), mutual effects of moisture and temperature, and combined effects of temperature and mechanical behavior and properties of FRP bridge decks. An SHCE of the existing or new FRP bridge decks to study these effects is sorely needed. When the results from a program such as the above are available, a comprehensive program can be established. The aim of such a program is to arrive at cost–benefit decisions when accounting for temperature in FRP bridge decks.

UV effects can have unwanted degradation effects on the mechanical properties of the deck. Because of this, numerous tests are available to inspect this phenomenon on FRP systems. For example, ASTM D 1435, ASTM D 4364, and ASTM G 151 are some of the existing test specifications for FRP material.

There are several instruments that measure the harmful UV effects (such as loss of surface gloss, chalking, or pitting). Also, there exist mitigating solutions, mainly in the form of gel coating. It should be noted that coating an FRP deck with a mitigating gel would only delay the harmful UV effects. This indicates the need for an SHCE methodology for UV observation and a cost-effective methodology for FRP bridge decks. To establish such a methodology, an acceptable UV versus time criterion is needed. This criterion will establish (similar to the moisture and temperature effects, above) a threshold for cumulative UV levels, at which a new coating might be needed for the affected areas. With such a criterion on hand, a general program can be initiated for FRP bridge decks. The program will include the installation of different UV sensors in the deck at representative/critical locations. Upon continuous monitoring, these thresholds are checked continuously; if/when a particular threshold is reached, the bridge official is notified, and a proper cost–benefit decision is made.

6.3.2.2.5   Fire

Fire on FRP bridge decks can result from accidental fuel spill on the deck, intentional malicious acts, or any other source. SHCE methods cannot prevent the fire; however, it can be of paramount importance in assessing the impact of the fire on the structure and helping decision makers with the follow-up.

When an FRP deck is subjected to a fire, the intense heat can adversely affect the mechanical properties of the FRP material to the extent that a structural collapse might ensue, either immediately or later. It is important to note that there are several ASTM-testing specifications that are available to test fire effects on FRP material. Some of these tests are ASTM E84–98, ASTM E162–98, and ASTM E136–96a. Currently, the only means of making such an assessment, in case of the few already-built FRP decks, is visual inspection. Some NDT methods are available, but there is no standardized approach to evaluate this important issue.

In view of the above, it seems that an FRP bridge deck-specific SHCE program is needed. Such a program would use available sensing techniques and, in case of fire, the sensors would evaluate the structural response (global or local) to asses the damage level and thus help the bridge owner reach an appropriate decision on cost-effective repair.

6.3.2.3   Design

Most FRP decks available in the market are proprietary systems and are engineered on the basis of the specifications given by individual owners. Since they are engineerable, they can be optimized to suit various loading conditions to maximize their use while minimizing life cycle costs. This also makes it hard to write prescriptive specifications, as their short-term and long-term properties vary considerably. At present, no unified specifications exist to guide bridge owners in design and analysis of these decks. In several cases, the failure mechanisms are not understood well and require special attention. Distribution factors, impact factors, stress fields and concentrations, allowance for creep, allowable strain limits, and so on, are still under investigation, and not much knowledge base is yet available. So, most FRP decks are obtained through performance-type specifications, where performance is defined by structural deformations and stresses under predetermined loads set by the individual owner.

Most of the decks built to date are designed using FE analyses. The available postconstruction load test data are also fragmented and are not coordinated so as to be very useful. Most of the test data are on the performance of the deck alone and do not concentrate on the entire bridge system, as this is relatively complicated and expensive. The test and analysis data also account for performance immediately after the construction, and very little data are available on the long-term performance of the decks. Due to lack of such durability data, specifications do not systematically account for manufacturing, transportation, and construction methods, the deterioration rates expected in the life-time, maintenance, compatibility with existing structure, and inspection issues.

Lack of this durability data also makes it hard to optimize the designs for life cycle costs without valid information on required maintenance and inspection costs. One such example is matching the coefficient of thermal expansion of the deck to that of the wearing surface using fibers in different orientations. There have been several failures of wearing surfaces, making the cost of these decks significantly higher than the original costs. Thus, a well-defined and coordinated SHCE program is required and should be designed to monitor an FRP bridge deck from the manufacturing phase through several cycles of its in-service life. The bridge system also should be monitored for system performance. This will lead to efficient, durable, optimized FRP deck designs.

6.3.2.4 Manufacturing

FRP decks are shop fabricated, that is, manufactured in facilities away from the construction site. So, they are compared with precast concrete or steel units on the quality of the final product. However, to date, not all FRP bridge deck manufacturing processes have been standardized to the same level as those for precast reinforced concrete or steel units. In steel or precast concrete plants, a variety of nondestructive and destructive technologies are available for QA of materials, manufacturing process, and the final product. In case of FRP decks, manufacturers use different fabrication techniques, manufacturing processes and standards, besides different QC approaches. Tolerances vary considerably from pultrusion to wet lay-up methods, as the manual operations vary significantly. No standard approaches or testing methods are available for acceptance of the final product. Since the durability of the decks depends on the curing and assembly during manufacturing, these operations should be well monitored. So, for a successful SHCE program, manufacturing techniques and processes should be monitored carefully; a QA/QC system must be in place. For example, Anderegg et al. (2002) offered a study relating QA/QC issues with long-term SHM for Carbon Fiber Reinforced Polymers (CFRP) cables. The study is based on statistical analysis that can be generalized for use in FRP decks.

Since QC is an important aspect of any structural component used in bridges or any other type of infrastructure, there is an obvious interest in developing methods for improving QC manufacturing techniques. Fuchs (2008) described the development of a laser system for steel bridge fabrication for QC and eliminating shop assembly. The system is envisioned to produce a complete permanent record where the measurements would be traceable and certifiable to replace current data taken by hand and recorded in tables on paper. For cases where fabricated girders are erected by another party, this permanent record can help protect from situations where the erector or other contractors are at fault and try to blame the problem on fabrication error. If a fabrication error occurs, the system can identify it before painting and before the girder leaves the shop. Figure 6.29 illustrates the concepts of the QC system.

Turning back to the specific case of bridge deck manufacturing, generally, there are two ways to achieve QA/QC systems, depending on whether the deck is new or existing. For new decks, a

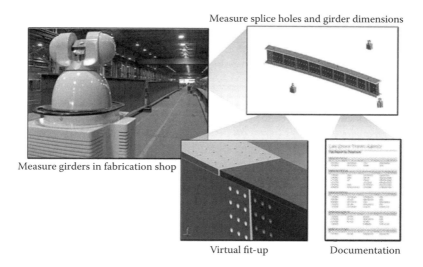

FIGURE 6.29 System concept for laser measurements of steel bridge girders. (Reprinted from ASNT Publication.)

standardized record of all pertinent manufacturing issues that might affect any SHCE program must be prepared before, during, and after the manufacturing process. Such a standardized report would be invaluable while preparing any SHCE program for the FRP deck. For existing FRP decks that do not have a standardized report, such a report should be prepared later according to the best ability of the bridge owner or the contractor.

6.3.2.5 Transportation

Transporting manufactured FRP decks from the plant site to the construction site should be considered an integral part of the SHCE of FRP decks. Note that some transportation-related damage to the structure may not be visible, and there are no nondestructive methods to detect most invisible damage suffered by the deck during transit. Temperature, moisture, and impact conditions may also have a direct effect on the performance and longevity of FRP materials. There are no data available on these issues to prepare specifications for transportation, storage, and handling. Hence, instrumenting and monitoring decks during transportation may shed some light on possible degradation/damage inflicted during transportation and reveal types of QA methods for the acceptance of a deck at the site. To facilitate the integration of transportation QA and any future SHCE for FRP decks, a standardized transportation report should be developed.

6.3.2.6 Erection

One of the advantages of FRP decks is that they may require light cranes and handling equipment and the fast construction/erection process. This phase starts with lifting the FRP deck from the vehicle that transported it to laying, aligning, and connecting it to the appropriate bridge components. It ends with any testing required for acceptance of the structure. The handling of the deck and the environment (temperature, etc.) during the erection process may have a significant influence on its durability and behavior. So, this should be carefully considered and made an integral part of any successful SHCE program. Environmental conditions for placement should be defined appropriately, as needed. Temperature, moisture, humidity, and lifting/laying down procedures should be considered during the design and manufacturing processes and well documented. When the deck is placed on the superstructure, its compatibility and alignment with existing systems should be measured carefully and well documented. Any unexpected problems, and the impromptu solutions that might have an effect on future performance and strength, should be reported in time. This reporting, in a standardized fashion, will assist in developing a standardized erection procedure for the future. Nondestructive test methods should be developed or specified to make sure the deck was not damaged during the erection process. These data should be stored, as this baseline information may be needed to develop future inspection and maintenance procedures.

6.3.2.7 Inspection

Currently, the FRP decks are inspected visually by most owners. Several owners depend on the judgment of the inspector/engineer based on visual observation for signs of deterioration or damage and limited NDT such as coin tapping or hammering routinely used on concrete decks. The qualifications of the inspectors vary from limited bridge experience to being professional engineers. Successful visual inspection relies on the experience of the inspector. Building such an experience is a long-term process, and at present there is very limited experience in inspecting FRP decks.

Some owners conduct field testing under known vehicle weights to make sure that the deck is serving as intended. The scope of these tests, in most cases, is limited by the required expertise and the costs involved. These tests, in general, also do not indicate the presence, if any, of local defects and minor deterioration/cracks, which can magnify into major problems in future. Infrared thermography, microwave, impact echo, and other advanced NDT methods are under investigation, but at present not much data are available. Automatic inspection methods, which rely on a set of sensors and measurements, are under investigation. Measuring different quantities, such as temperature, moisture, strains or displacements, can be an integral part of an SHCE program. Different algorithms that use these types of measurements to identify damage in FRP decks must be developed and improved.

6.3.2.8 Maintenance

There is very little information on the type and extent of maintenance required for FRP bridge decks. For long-term durability and life cycle costs, maintenance needs and schedules should be considered during the analysis, design, and manufacturing stages. Several of the issues identified by the owners of the FRP decks, based on their in-service experience, include failure of the wearing surface, delaminations in the face skin, warping at high temperatures due to expansion and contraction of the joints, failure of UV coatings or paint, and leakage at the joints. Due to the experimental nature of the decks so far, these issues are new to bridge owners, and there are no methods to detect premature failures such as the above. The major concern is the lack of a knowledge base on remedial measures, if any, of the damage that happens. Some of the above may not affect the immediate use of the structure, but others such as mechanical damage due to traffic or environment, snow plow operations, fire, and joint failure require immediate attention if they affect the load-carrying capacity of the bridge. Owners are also particularly interested in knowing the load rating of a bridge subjected to this type of damage and answers to questions such as: can one lane of the bridge be opened if the damage is in the other lane? Other issues will be the effect of any maintenance work on the durability of the entire system.

So, maintenance issues should be considered a part of an SCHE process, and possible scenarios requiring maintenance, types of maintenance needed, and any necessary remedial actions should be properly investigated and documented. Installation of passive sensors at appropriate locations during fabrication/manufacturing to detect moisture or delaminations is one way to account for some of these to initiate proper maintenance procedures in time. Fiber optic sensors (FOS) can be installed, and strain/stress data can be collected cost-effectively during routine inspections, and maintenance cycles can be tied to the observed data. Since most of these decks are designed using FE analysis, many of the "what if?" scenarios (such as the effect of expected mechanical or joint damage on load rating and traffic patterns) can be investigated such that the required maintenance procedures can be established from the beginning.

6.3.3 Repair of FRP Bridge Decks: A Case Study

FRP bridge decks are one possible alternative to conventional reinforced concrete decks. FRP materials are engineerable, and so deck designs and configurations vary significantly. This, combined with the relatively less experience with them, makes it hard to establish a standard repair method for possible damage in-service. There are also very few documented in-service problems/damage and associated repair methods.

The bridge on South Broad Street over Dyke Creek in Wellsville, NY, was built in 1974 as a steel multigirder with an open grate deck. The open grate deck was replaced in 2000 with an 8- in-thick honeycombed FRP deck. The FRP deck was originally surfaced with an acrylic-modified cementitious overlay, but was later replaced with an asphalt overlay due to cracking and delaminations of the overlay (see Figure 6.30).

The 2004 bridge inspection revealed two large delaminated areas, one in each span. An attempt was made to pump resin into the deck at the affected areas by drilling holes in the top skin, but this was not successful. In 2006, a research project was initiated to design a viable and structurally sound repair method, using materials compatible with the surrounding deck material to extend the life of the FRP deck, which could also be used for future FRP bridge decks of similar design and configuration. The repair method was developed working with An-Cor, a company in North Tonawanda, NY.

The asphalt overlay, the top skin, and the wet cells in the proximity of the two delaminated areas were removed leaving the bottom skin intact. Two plies made of alternating layers of the fiberglass mat and a resin and catalyst mixture were placed on the intact bottom skin. An-Cor's prefabricated cells (see Figure 6.31) were set in place on a layer of pouring mix composed of milled fiberglass and a resin/catalyst mixture. The cells were then grouted into place. Expandable foam was placed in each of the cells leaving an approximately ¾-in-gap at the top of each cell. Grout and pouring mix was placed on top of the foam to fill the remaining space. After everything cooled and dried, additional plies of the fiberglass mats and the resin/catalyst mixture were placed on top of the cells. Once the repair of the deck was complete, hot patch was applied for the wearing surface.

Some of the challenges associated with the repair are (1) The patch area has to be protected during and after the repair to prevent moisture/water intrusion and it was made sure that the vehicles on the bridge did not drive over the repair area causing harm to the vehicles and/or the bridge. A steel plate was used to cover the repair area and sealed with a perimeter of cold patch, but it was found that this posed a risk if the plate moved due to vehicle pounding; (2) The repair could be completed

FIGURE 6.30 South Board Street Bridge—Wellsville, NY.

FIGURE 6.31   FRP core-filled detail.

only in temperatures above 40°F, while temperatures too high may need prolonged cooling time for the resin to accommodate exothermic reaction; (3) The repair also could not take place during any type of precipitation, to prevent moisture trapping in the deck; and (4) If all of the wet cells are not removed, this can result in further cracking and delaminations due to additional freeze-and-thaw cycles after the repair.

The inspections conducted in 2007 showed that the bridge deck and wearing surface of the two repair sites completed in 2006 appeared to be working well (Forenz, S., 2007, pers. comm.).

6.3.4   STRUCTURAL MODELING

6.3.4.1   General

In most FRP deck experiments, there is invariably an analytical phase of the experiment, where the experimental results are compared with analytical/numerical structural modeling effort. The most popular method of analysis is the FE method. We discuss some aspects of FE modeling first. During the discussion, we point to some of the important, but ignored, FE modeling issues when analyzing FRP systems. We then introduce a scale independent analysis method, which is both efficient and accurate and which can be of help in modeling FRP decks.

6.3.4.2   FRP Bridge Deck Properties through Monitoring

Since FRP bridge decks are fairly new in implementation, the as-built material properties need to be verified by on-site testing. A study by Shekar et al. (2004) aimed at investigating real-life damping in FRP bridge decks. It also aimed at evaluating limitations on accelerations for human comfort. The study instrumented and tested two of the FRP bridge decks (known as "superdeck"). The results indicate that (1) the degree of compositeness between the FRP deck and steel stringer is 100% in the case of shear studs and about 50% where mechanical fasteners and adhesive bonding were used instead of shear studs; (2) the TLDF was close to what was predicted by AASHTO; and (3) strains and deflections were within the allowable limits. Dynamic tests indicated (1) low damping of about 0.5%; (2) dynamic load factors were less than AASHTO allowable of 0.33, but were about 0.09 for one bridge and about 0.30 for other bridge; and (3) the accelerations were perceptible to bridge users and fell in the unacceptable range according to the Canadian Code. We note that there are no such limits in AASHTO. AASHTO does refer to the 1991 Canadian Code, but the authors used the 1983 Canadian Code, and the reasons were not stated.

6.3.4.3 Conventional FE

The FE technique enables the analyst to simulate a wider variety of geometry. In addition, FE technique enables the use of nonlinear materials in the model. In using a shell element, it was assumed that the cross-section of the deck remains plane. The core, top, and bottom surfaces were *smeared* together using equivalent cross-sectional properties of the deck. It was found that such an approximation produced a fairly good overall displacement match with experiments.

Another technique is to use a shell-smeared element. Such a technique, though capable of producing good displacement results, is incapable of producing accurate shear stresses (or strains) at the core-top (or bottom) surface interface. The reason is obvious: the cross section of a typical FRP deck does not remain plane, due to the mismatch of core and surface stiffness. The accurate evaluation of the shear interface is important since it is a likely location for failure or damage of the deck construct. What is needed is a more accurate FE modeling of both the core and the surfaces. Aref et al. (2004) used an explicit FE modeling for the surfaces and the core. By using solid elements to simulate the all-important components of the deck, they obtained good matching with test results and proceeded to estimate the effects of fire and snow plowing on the health of the deck.

All the above methods were based on static linear analysis of FRP decks. Sometimes, there is a need to simulate failure mechanisms of an FRP deck. In such situations, a more detailed geometric modeling is needed. In addition, a nonlinear material model must be used in combination with suitable failure criteria of different materials. A common failure criterion in composite material that has been used by researchers is the Tsai-Hill criterion (see Hill 1950, and Jones 1999, for theory, and Aref et al. 2004 for applications to FRP deck limit states applications). The criterion is attractive because of its simplicity.

Let us assume that the adhesives used in joining the top/bottom surfaces to the core are investigated; the ultimate capacity of such adhesives as well as their interaction with the top/bottom surfaces need to be loaded until failure. In such a situation, the adhesives need to be modeled separately from the core and the top/bottom surfaces (in conventional linear elastic modeling, the adhesives are assumed to be rigid, that is, the core and the surfaces are rigidly connected). In addition to explicit adhesive modeling, a suitable nonlinear material model must be used.

When the investigations require dynamic analysis, the modeling requirements can also vary. Let us consider, for example, a situation where the investigator needs to evaluate the vibration characteristics of an FRP deck. The investigator decides that the vibration amplitudes are much lower than the elastic limit of the FRP material. Because of this, a simple, semianalytical, dynamic model can be used. Also, a simple *smeared* FE model can be used. On the other hand, let us assume that the investigator desires to study an earthquake condition or a blast-loading condition, where it is expected that part of the deck might fail. In such a situation, a more detailed FE model must be used. In addition, a suitable nonlinear material model, with appropriate failure criterion, must be used. Table 6.15 summarizes the different methods used for FRP decks and the advantages of each.

TABLE 6.15
Matching Analysis Techniques with Analysis Goals

Type	Linearity	Detail	Method
Static	Linear, within elastic limits	Only global displacements needed	Semianalytical or *smeared* FE model
	Nonlinear, within elastic limits	Detailed strains in core and joints	FE: explicit modeling of core and surfaces
	Nonlinear/failure	Failure modes	FE: explicit modeling of adhesives, core, surfaces and other complex details
Dynamic	Vibrations/noise	Only global displacements needed	Semianalytical or *smeared* FE model
	Earthquakes/blast/impact	Behavior beyond elastic limits	FE: explicit modeling of adhesives, core, surfaces and other complex details

6.3.4.4 Scale Independent Element

The general three-dimensional geometry of a pultruded bridge deck can be described as

$$G(x_1, x_2, x_3) = G_1(x_1)G_{23}(x_2, x_3) \tag{6.3}$$

The x_1 is the pultruded direction, and x_2 and x_3 are the vertical and lateral directions, respectively. In most practical cases, $G(x_1, x_2, x_3)$ is independent of x_1; thus,

$$G_1(x_1) = 1 \tag{6.4}$$

and

$$G(x_1, x_2, x_3) = G_{23}(x_2, x_3) \tag{6.5}$$

The geometry described by Equation 6.5 is applicable to the scale independent element (SIE) geometry (see Chapter 6 by Ettouney and Alampalli 2012). Thus, the dynamic equation of a pultruded bridge deck can be described by an SIE element in the sense:

$$\begin{Bmatrix} P_0 \\ P_N \end{Bmatrix} = [D] \begin{Bmatrix} U_0 \\ U_N \end{Bmatrix} \tag{6.6}$$

The force vectors $\{P_0\}$ and $\{P_N\}$ are nodal forces (or moments) at the two ends of the deck, $x_1 = 0$ and $x_1 = L$, respectively. Similarly, displacement vectors $\{U_0\}$ and $\{U_N\}$ are the corresponding nodal displacements (or rotations) at the two ends of the deck, $x_1 = 0$ and $x_1 = L$, respectively. The size of the four vectors, $\{P_0\}$, $\{P_N\}$, $\{U_0\}$, and $\{U_N\}$ is M. $[D]$ is a complex symmetric dynamic stiffness matrix of dimension $2M \cdot 2M$. The discretization level M is a function of the driving frequency, Ω, as well as the overall geometry $G_{23}(x_2, x_3)$. It is independent of the overall length L. Figure 6.32 shows a typical SIE as applied to a pultruded FRP bridge deck.

SIE, similar to any other element in the general sense of FE methodology, accepts loads that are specified at its nodal points. When there are loads such as wheel loads in the middle of the deck, conventional FE can be used in combination with the SIE. Figure 6.33 shows this concept.

Overall, SIE can provide an extremely efficient computational resource, permitting the scaling of analysis to a very high frequency ranges and thus avoiding computational and numerical computations of the conventional FEs methods (see Ettouney et al. 1989, 1990, 1997).

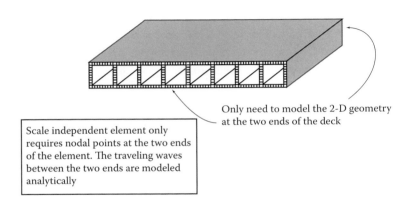

Only need to model the 2-D geometry at the two ends of the deck

Scale independent element only requires nodal points at the two ends of the element. The traveling waves between the two ends are modeled analytically

FIGURE 6.32   Basic SIE geometry for an FRP bridge deck.

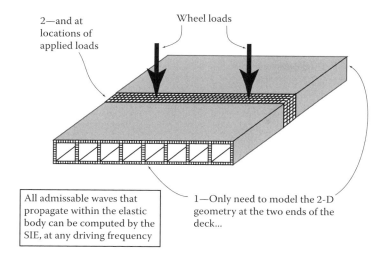

2—and at locations of applied loads

Wheel loads

All admissable waves that propagate within the elastic body can be computed by the SIE, at any driving frequency

1—Only need to model the 2-D geometry at the two ends of the deck...

FIGURE 6.33 SIE modeling of FRP bridge decks.

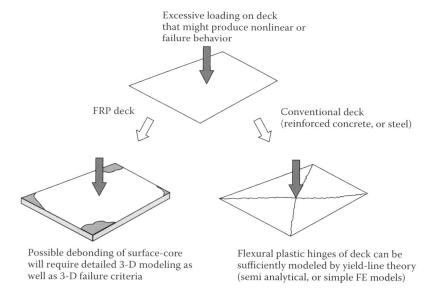

Excessive loading on deck that might produce nonlinear or failure behavior

FRP deck

Conventional deck (reinforced concrete, or steel)

Possible debonding of surface-core will require detailed 3-D modeling as well as 3-D failure criteria

Flexural plastic hinges of deck can be sufficiently modeled by yield-line theory (semi analytical, or simple FE models)

FIGURE 6.34 Comparison of modeling needs for conventional and FRP Decks.

6.3.5 DAMAGE DETECTION

6.3.5.1 General

We note that failure modes of FRP decks are, in general, different from the failure modes in conventional decks. For example, flexural failure (plastic hinge formation) is perhaps the most common failure mode in a reinforced concrete deck. A common failure mode in an FRP deck is the debonding of the core and the surfaces. This is an interesting observation that has important implications. While modeling plastic hinges in a concrete deck is a simple undertaking, modeling the FRP debonding requires much more effort. Figure 6.34 shows a schematic representation of this situation.

This leads us to an obvious but usually ignored supposition that when embarking on a structural modeling of an FRP deck, careful consideration must be given to the purpose of the analysis and the available budget. If the budget does not permit the needed level of analysis, the decision maker must consider a different route to achieve the desired goal.

6.3.5.2 Dispersion Curves

Chapter 7 by Ettouney and Alampalli (2012) discussed the use of dispersion curves in combination with AE signals to detect location, size, and shape of damage in plate-like geometry. Recall that dispersion curves relate, for a given wave mode, group wave velocities to the driving frequencies in a particular system. Dispersion curves can be analytically or numerically developed for simple geometries. Computational demands increase fast as (1) the system of interest becomes more complex, and (2) the driving frequency range increases.

SIE methodology offers an elegant and efficient approach for developing dispersion curves of pultruded FRP bridge decks. We argued earlier that because of the special geometry condition of the pultruded FRP bridge decks (Equation 6.5), the SEI methodology can be applied efficiently. Recall from Chapter 6 by Ettouney and Alampalli (2012) that for the geometry of Figure 6.35 (which is typical for pultruded FRP decks) all admissible propagating waves, at a particular driving frequency, Ω, in the construct, can be evaluated by solving the quadratic eigenvalue problem

$$[\lambda^2 S_{21} + \lambda(S_{11} + \lambda^2 S_{22}) + S_{12}]A = 0 \tag{6.7}$$

The matrices S_{21}, S_{11}, S_{22}, and S_{12} are complex matrices of order M. They are defined in Ettouney and Alampalli (2012, Chapter 6). There are $2M$ solutions for Equation 6.7. The eigensolution pairs, λ_i, and, A_i, represent the ith wave number and wave mode shape, respectively; both are functions of the driving frequency. Thus, by solving Equation 6.7 for as many driving frequencies as needed, the relationship

$$\lambda = \lambda(\Omega) \tag{6.8}$$

completely defines the dispersion curve for the ith wave mode. The complete dispersion relations are defined for the deck of interest.

6.3.5.3 Monitoring Existing Decks

Shahrooz et al. (2002) described the testing of four different types of FRP deck systems under similar and real-life conditions. The four systems were installed and monitored on the same bridge on different spans to conduct a comparative study under the same traffic and environmental conditions. The performance was monitored through field documentation, instrumented long-term continuous

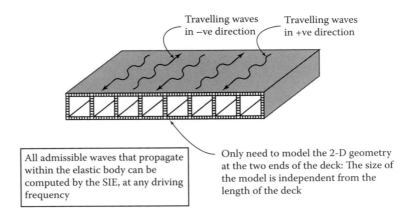

FIGURE 6.35   Dispersion model for FRP bridge decks.

monitoring with periodic recording of data, controlled static and dynamic truck load tests, and multireferenced modal tests. Reference tests were conducted shortly after the decks were installed for baseline data to be compared with future test data during the 2-year monitoring period reported in this chapter.

The panel movements relative to girders were measured and found to be very small. Absolute vertical and horizontal movements were measured and in some cases correlated to environmental loading. Girder distribution factors and impact factors were also monitored. Based on these tests, it was concluded that the girder distribution factors were generally comparable to the ones for the original RC deck. In general, impact factors were lower than AASHTO limit except for one deck system. Proper support underneath the FRP deck systems is important to prevent panel movements, and thermal characteristics of FRP panels are influenced by the panel cross-section details in addition to thermal components of the individual components of the panels.

The findings of this research have had a major influence on the understanding of the behavior of FRP bridge decks.

6.3.5.4 Fiber Optic Sensing

Baukat and Gallon (2002a) used FOSs in fabrication plant for FRP decks, and strain data was obtained during the testing. The data correlated well with the load data. There was no comparison with any other gauges. The authors hid the cable in a groove within the deck system. Also, Crocker et al. (2002) studied the use of FOS in investigating fatigue, and repeated loadings on several FRP bridge decks showed the potential for FOS use in this class of structures. Baukat and Gallon (2002b) investigated the embedding of FOS within the FRP bridge deck. They showed that a prepackaged sensor can be used by the end-user, and it will reduce cost and enable long-term monitoring.

6.3.5.5 Thermography

Duke et al. (2006) looked at the use of infrared thermography for detecting damage in FRP bridge structures. The theory behind the use of infrared thermography was used at first to show how thermographic images are affected by imperfections, geometry, and material imperfections. The advantage of this method, if successful, is that it allows for rapid, noncontact inspection of FRP structures. At the same time, it was noted that creating an appropriate thermal gradient is very difficult especially if the object has a complex geometry or material architecture.

Laboratory models showed that thermography can be used to detect resin-starved regions as well as delaminated regions (see Figure 6.36). The tests on a section of an actual deck, heated using hot air circulated through a tube section, showed the presence of disband between the top plate and the tube section, prompting redesign of the attachment. This section of the FRP deck was planned to be used for weigh-in-motion (WIM) site. It was observed that thermographic inspection during the WIM operations was complicated by the fact that truck tires passing on the deck generate heat that might interfere with damage detection (Figure 6.37).

6.3.5.6 Ultrasound and AE

Wang and Chang (2000) used a network of piezoelectric actuators to send signals throughout the structure (in this case, a composite plate). The SHM system also includes a network of piezoelectric sensors to detect the signals. This network of actuator and sensors represent an active structural health monitoring system (ASHMS). In addition to the actuators/sensors, the ASHMS system also includes a signal generator that excites the actuator by a narrow band signal. The sensor's measurements are processed using an online signal processor. Finally, the signals are interpreted using prepackaged algorithms that relate geometries and time-of-flight (TOF) input and output signals to estimate the most likely damage size in the plate. Figure 6.38 shows the schematics of the identification system and a typical ASHMS system. Note that the ASHMS is based on TOF algorithm (see Ettouney and Alampalli 2012 for the mathematical basis of TOF). Table 6.16 shows

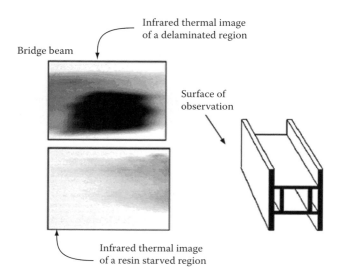

FIGURE 6.36   Infrared thermal images of a resin-starved region (top) and a delaminated region (bottom). (Reprinted from ASNT Publication.)

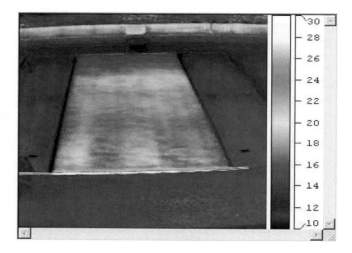

FIGURE 6.37   Infrared thermal image of an FRP deck. (Reprinted from ASNT Publication.)

the estimated and actual damage conditions for two different damage geometries. An expanded ASHMS system can be of use for real-time measurements of damage in different components of FRP bridge decks.

Acoustic emission testing was also performed on an FRP beam by McDad et al. (2004). The authors planned to use AE also as part of a health-monitoring effort after deployment.

6.3.5.7   Wavelets

The impact effects on composite structures (plates) were studied by Staszewski et al. (2000). They used wavelet analysis for their damage identification efforts. Since impact loading and the potential response to it are highly nonstationary, the use of wavelets for signal processing rather than the conventional Fourier transforms is appropriate. The advantages of the wavelet approach, which gives information about both the time and frequency contents of the analyzed data, were beneficial

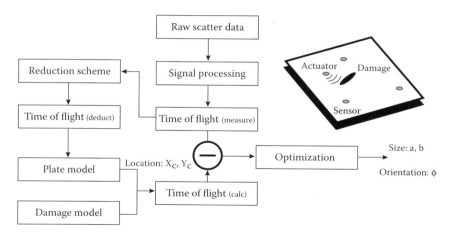

FIGURE 6.38 Logic and schematics of ASHMS system. (From Wang, C. and Chang, F. Built-in diagnostics for impact damage identification of composite structures, *Proceedings of 2nd International Workshop on Structural Health Monitoring,* Stanford University, Stanford, CA, 2000. With permission from Dr. Fu-Kuo Chang.)

TABLE 6.16
ASHMS Results

	Geometry 1			Geometry 2		
	Damage Size			Damage Size		
Test	**Length (in)**	**Width (in)**	**Angle**	**Length (in)**	**Width (in)**	**Angle**
Estimated	1.4	0.6	2°	0.9	0.5	27°
Actual	1.2	0.5	0°	0.8	0.4	32°

Source: With permission from Dr. Fu-Kuo Chang.

in the project. In addition to using wavelets as a damage identification method, the authors also used piezoceramic sensors in their experiment. The piezoceramic sensors offer the advantage of being used for sensing and for being used as actuators.

Another use of wavelets as a damage identification technique was demonstrated by Xiaorong et al. (2000). They tested a 400- × 60-mm Carbon Fiber Reinforced Polymers (CFRP) plate. The plate had a small Teflon insert to simulate the delamination. They used two sensors to measure responses and one actuator to generate signals. Upon analyzing the damaged and undamaged responses of the CFRP plate, they applied a wavelet technique on the measurements. They reported that by inspecting the wavelet spectra, they were able to detect delamination (as simulated by the Teflon insert) as well as its location and size.

Lemistre et al. (2000) offered another application of the use of wavelets to detect delamination in FRP plates. They generated Lamb waves in the test plate (700 × 700 × 2 mm) using nine piezoelectric transducers (three as emitters and six as receivers). A delamination was intentionally introduced into the plate to investigate whether the wavelet technique can detect it. By changing the excitation frequency and monitoring the times of arrival of flexural waves, they were able to produce an experimental dispersion curve. By comparing the experimental results with the theoretical dispersion diagrams of flexural waves in plates (Figure 6.39), they found that three wave modes were identified properly by the wavelet analysis. Other wave modes were not observed, probably due to the presence of delamination. The authors then used the measured time of arrivals from the wavelet spectra to locate accurately the delamination region.

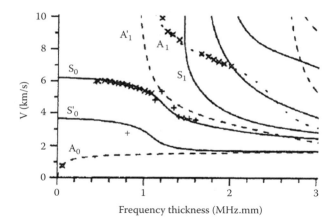

FIGURE 6.39 Theoretical and experimental dispersion curves in thin plates. (From Lemistre, M. et al., Damage Localization in Composite Plates Using Wavelet Transform Processing on Lamb Wave Signals, *Proceedings of 2nd International Workshop on Structural Health Monitoring,* Stanford University, Stanford, CA, 2000. With permission from Dr. Fu-Kuo Chang.)

6.4 DECISION MAKING AND FRP BRIDGE DECKS

6.4.1 DECISION MAKING

One of the most important features of FRP decks is that it involves new technologies, both material and manufacturing. There are many unknown factors in the whole experience. Unlike conventional engineering materials, such as reinforced concrete and steel, FRP behavior does not have a large body of experience that can help decision makers in taking decisions under uncertainties. Because of this, the use of FRP material in the field of infrastructure, where safety is of utmost concern, has been fairly different from that of other more conventional materials. Some of the questions usually asked when embarking on an infrastructure project that uses FRP materials are

- Given several FRP-manufactured components, how many samples should be tested to ensure adherence to specifications? When to reject, or to accept the components, based on the test results?
- What kind of QC should be followed in an FRP project?
- What are the procedures to be followed by the continued monitoring of an in-service FRP bridge deck?

We discuss these questions in the form of case studies next. We observe that there are several other important questions that every decision maker should be asking. Due to space limitations, we cannot address all of the possible issues. However, the rules of decision making under uncertainties are simple, and the examples in this and other sections in this chapter should be of sufficient help in many other situations.

6.4.2 DECISION MAKING CASE STUDY: RELIABILITY OF SINGLE SAMPLE (USE OF NDT METHODS)

6.4.2.1 Introduction

We discussed earlier numerous issues that can afflict the health of an FRP deck. We also note the shortage of information about the use of FRP as bridge decks. These factors make closer monitoring and inspection of FRP decks to be even more important than bridge decks that are constructed of conventional (steel/concrete) materials. Another important factor that affects the use of FRP material in bridge decks is that, to the best of our knowledge, as of the writing of this chapter, *all* bridge decks in the United States have been designed and manufactured on a case-by-case basis. This makes difficult any sampling of realistic bridge decks: any FRP deck in use is a unique entity. How can an SHM effort be done to ensure conformity to design, manufacture, or transportation specifications? Decision making techniques can help in resolving this important question.

Let us consider, for example, the quality of the joint between the core and the top (or bottom) surface. The importance of this joint was discussed in several sections earlier in this chapter. In spite of its importance, there is no monitoring experiment that directly evaluates the quality of the joint. The quality of the joint was only monitored by assessing the behavior, over time, of the displacements and/or strains, of different parts of the bridge. Yet, no direct evaluations of the joint do exist as of the writing of this chapter.

Perhaps the main reason for such a gap in knowledge is that the joints are hidden *inside* the deck; after manufacturing and erection, it is impossible to inspect the joint visually. In addition, the joint can be fairly long, and it is not realistic to inspect it all. Finally, the absence of standardized QA/QC (manufacturing, transportation, or erection) procedures makes acceptance procedures of such joints very difficult indeed.

We will suggest a hypothetical example that considers all of these points. The example will present the use of NDT and decision making techniques to reach a logical decision about the joint in question. A single-unit FRP deck is manufactured, transported, and erected on site. The deck length is 16 ft, width 8 ft, and total surface area 128 ft^2. The bridge official needs to investigate the quality of the joint between the top surface and the core.

6.4.2.2 High-Quality Manufacturing

Let us assume that the QA procedure of the manufacturer indicates that the average occurrence of single defect (say, of ¼-in length) in the joint is 0.5% for every 1 ft^2 of the surface (top or bottom) of the deck, on an average. The bridge official estimates that the defects in the joint have Poisson's probability distribution (see Ettouney and Alampalli 2012, Chapter 8). The random variable X is the exact number of ¼-in defects in a given square foot on the top/bottom surface joint with the core. The question is: what is the probability that no defect is found in any given square foot on the top (or bottom) surface of the deck? Symbolically, we need to evaluate $P(X = 0)$.

From the QA documents of the manufacturer, μ can be estimated as $\mu = 0.005$; then we can evaluate the required probability as

$$P(X = 0) = f(0) = \frac{(0.005)^0 e^{-0.005}}{0!} \tag{6.9}$$

$$f(0) = e^{-0.005} = 0.995012 \tag{6.10}$$

Now, if the official needs the probability of exactly one joint defect in any given one square foot on the surface:

$$P(X = 1) = f(1) = \frac{(0.005)^1 e^{-0.005}}{1!} \tag{6.11}$$

$$f(1) = 0.004975 \tag{6.12}$$

The above results are reassuring indeed. The probability of no joint defects, in a given 1 ft^2, is as high as 0.995012, while the probability of a single defect is a paltry 0.004975. Even more assuring. We compute from Equations 6.9 and 6.11

$$P(X > 1) = 0.00001245 \tag{6.13}$$

The probability that there is more than one joint defect in a square foot is almost nonexistent. These results clearly should give confidence to the official: the joint in question should be accepted.

6.4.2.3   Medium- to Low-Quality Manufacturing

Let us consider now a situation where the QA procedure of the manufacturer indicates that the average occurrence of a single ¼-in joint defect is 25% for every 1 ft^2 of the surface. The probabilities for this situation are

$$f\ (0) = 0.7788 \tag{6.14}$$

$$P(X > 0) = 0.2212 \tag{6.15}$$

The probability is that one or more ¼-in-joint defect in a given 1 ft^2 is 0.2212. This value is high; the official should think seriously about acceptance or rejection of the deck. More considerations are needed.

The official might take two possible decisions/actions in this situation:

1. Evaluate, structurally, the actual risk taken
2. Reject the deck

6.4.2.4   Evaluate Structural Risk

Continuing with our example, it seems that the decision maker must contend now with either assessing the situation further or rejecting the deck. To further assess the situation, he might use the somewhat not too common, but potently strong, tool of probabilistic structural analysis (probabilistic FE method/probabilistic boundary element method). Both methods were discussed by Ettouney and Alampalli (2012, Chapter 8). It accepts uncertain properties of the structure and loadings and computes the probabilistic properties of the outcome. In the current example, let us assume that the core-surface joint is found to have the probability of a ¼-in defect in any given square foot to be μ. Note that we have explored two situations, $\mu = 0.005$ and $\mu = 0.25$, so far. We also assumed that the probability distribution for the number of ¼-in defects to be a Poisson's distribution. To apply either of probabilistic FE method or probabilistic boundary element method, we need to estimate the mean and the standard deviation of Poisson's distribution as a function of μ, by using the methods of Ettouney and Alampalli (2012, Chapter 8). We can then model the bridge deck and find the required probabilistic statements, using either of the two methods as described in detail in the aforementioned reference. Rejecting the deck is the last resort. Obviously, the official must seriously consider the financial and social implications.

6.4.2.5   Concluding Remarks

In this example, we explored several decision making techniques concerning the health of core-to-surface joints. These can easily be applied to many of the other FRP bridge deck components. We had to make some hypothetical assumptions to represent a complete example. Those assumptions reflect, among other things, the fact that the use of FRP decks is still in its infancy. Most importantly, we do believe that this example portrays the immense importance of QA/QC procedures for FRP manufacturers and bridge officials. In the next section, we explore further the issue of QA/QC.

6.4.3   Lack of QC Values

The previous section was based on a simple assumption: the existence of a QC average probability value of the ¼-in defect in the joint. Let us assume that there are no QC values for possible manufacturing defects. How can the bridge official ensure that the joints in the field will function as designed? One possible way to deal with this situation is to monitor certain performance measures in the deck for any signs of overall deck performance deterioration. For example, the displacements and strains of the FRP deck in Section 6.5.1 of this chapter was monitored over a long period of time; since no measurable deterioration has been observed, it can be deduced that the joints are functioning properly.

A more direct approach would be to institute a QA procedure that is specifically designed for the joint performance. Two parameters would be needed to accomplish such a procedure. The first is an acceptance limit of the number and location of defects; the second is the establishment of a procedure for estimating the defects on the delivered product (deck).

To establish an acceptance limit of defects, a comprehensive probabilistic structural analysis of the FRP deck and the joints must be undertaken. This analysis is similar to the one described above. The purpose of the analysis is to define a quantitative acceptance limit to the probability of a defect of particular size.

The other side of the problem is to estimate the actual state of defects in the deck. This will involve two specific steps: (1) an NDT program onsite, and (2) and an analysis of the results of the program. As of the writing of this chapter, the authors know of no specific NDT effort to estimate or measure joint defects for FRP decks. Perhaps a thermography technique can be adopted. Karbhari et al. (2005) described such a technique that looks for manufacturing/erection defects for FRP wraps that are used to strengthen/stiffen existing reinforced concrete structures.

6.4.4 DECISION MAKING CASE STUDY: QC ISSUE

Many of the recently erected FRP decks have ongoing monitoring efforts (see, for example, McDad et al. 2004). These efforts aim at tracking changes in the deck behavior over time, mainly displacement and/or strains. We will present some scenarios of possible changes in the monitoring results and also some decision making analysis for the scenarios. Though these scenarios are hypothetical, we believe that the decision making analyses and methods can be beneficial in real-life situations.

6.4.5 DECISION MAKING CASE STUDY: CONTINUED MONITORING

6.4.5.1 General Overview

Let us consider a simple situation of a single-span, two-unit FRP deck. The deck spans a small creek, and it is rectangular in plan. The span of the deck is 10 ft, and the width of each unit is 15 ft, for a total width of the bridge of 30 ft. The decks are constructed from top/bottom surfaces and a honeycomb-type core. The field joint between the two units is formed by an all-adhesive construction. The deck is supported by two abutments; a simple bolted connection similar to that in Figure 6.12 is used. A bituminous wearing surface is used on top of the wearing surface. The decks are designed for maximum displacement of 80% of AASHTO allowable and maximum strains of 20% of ultimate allowable. The overall deck geometry is shown in Figure 6.40.

When the FRP deck was constructed, a set of three strain sensors were placed at different locations on the deck, as shown in Figure 6.40. When the deck was first erected, a numerical validation of the strains was performed, and the analysis and the measurements were within 10% of each other. The bridge official decided to continue the monitoring of strains to follow any deviation from these measured/computed strains. The strains are monitored quarterly by placing the design trucks on the bridge and measuring the resulting strains. Table 6.17 shows the normalized measured strains for the three sensors for 21 consecutive quarters. Figure 6.41 shows the time history of the normalized strain measurements. The normalization is done relative to an arbitrary constant such that the values of the normalized strains are of the order of unity. Since our interest in this section is the relative strain values, the normalized strains will be easier to follow without any loss of generalization.

Casual inspection of Figure 6.41 indicates no unusual behavior of the strains at the three sensor locations. Sensors 1 and 2 seem to behave similarly, while sensor 3 does exhibit some differences

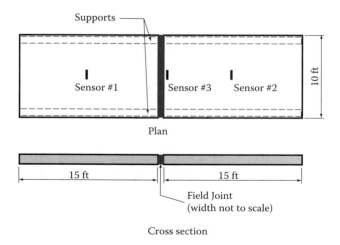

FIGURE 6.40   Continued monitoring example.

TABLE 6.17
Measured Strains on a Quarterly Basis

		Sensor Displacements (in)		
Year	Quarter	1	2	3
1	1	1.190487	0.980335	1.071438
	2	1.116527	0.952846	1.007223
	3	1.124164	0.956418	1.013923
	4	1.043715	0.923125	0.936931
2	5	1.089868	0.941215	0.981277
	6	1.061155	0.92748	0.951543
	7	1.056313	0.924447	0.94881
	8	0.994792	0.903917	0.895312
3	9	1.049016	0.925004	0.918876
	10	1.157804	0.972312	1.02797
	11	1.153628	0.97023	0.967913
	12	1.075892	0.935586	0.95453
4	13	1.079743	0.937055	0.978942
	14	1.120163	0.955028	1.074265
	15	1.062575	0.926513	0.988121
	16	1.044474	0.923421	0.979265
5	17	1.006323	0.908419	1.070144
	18	1.040459	0.915162	1.048998
	19	1.12947	0.959657	1.114823
	20	1.089656	0.941001	1.088999
6	21	1.027888	0.916975	1.205855

from the other two; however, there is no apparent cause for concern. The strain values remain within 10% of the computed numerical strains. The deck seems to be behaving in a usual manner. Based on this, it seems that the appropriate decision is to *do nothing.*

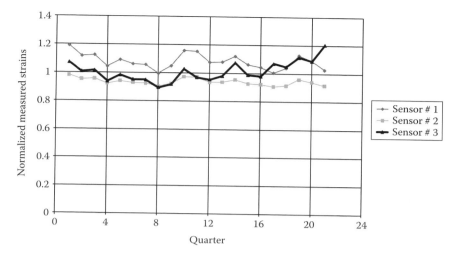

FIGURE 6.41 Time history of strains in three sensors.

TABLE 6.18
Regression Line Slope for the First 2 years (8 Quarters)

Slope of Regression Line (quarter^{-1})		
Sensor 1	**Sensor 2**	**Sensor 3**
–0.02159	–0.00888	–0.01985

6.4.5.2 When Does a Change Occur?

Upon further reflection, there is an immense amount of information in the measurements of Table 6.17 and Figure 6.41. Let us apply some of the simple processing techniques of Ettouney and Alampalli (2012, Chapter 8) to investigate the matter further. For example, let us assume that a linear regression was performed on each of the measurements. The slope of the regression line of a data set, α_2, can be computed for any measured time length. We find α_2 for each of the three strain measurement for the first 2 years of the experiment (8 quarters); the results are shown in Table 6.18.

Since the slope of the regression line is for strains per unit time (quarter), the changes in the strains are not large. We note that all the changes are with a negative sign, which indicates more or less similar performances at all sensor locations, as expected.

We now compute α_2 for the final 13 quarters of the experiment (more than 3 years). The results are shown in Table 6.19. We observe something different: while the slope of the regression lines continued on its small and negative value, the slope of the strains at sensor 3 became positive. There is a definite change in the behavior of sensor 3.

6.4.5.3 Correlation Analysis

Let us continue the data analysis of the three sensors. The coefficient of correlation matrix will be computed. Let us compute the matrix for the first 2 years (8 quarters). Table 6.20 shows the correlation matrix. We observe that it is symmetric, as expected. The diagonal terms in the matrix have values of unity. The off-diagonal terms have nearly unit values. This indicates that the three sensors in the first 2 years behave in near-perfect correlation. The deck is performing as expected and as modeled in the numerical verification effort.

TABLE 6.19
Regression Line Slope for the
Last 3 Years (13 Quarters)

Slope of Regression Line (quarter^{-1})

Sensor 1	Sensor 2	Sensor 3
-0.00491	-0.00218	0.016404

TABLE 6.20
Correlation Coefficient Matrix—First
8 Quarters

	Sensor 1	Sensor 2	Sensor 3
Sensor 1	1	0.997515	0.999412
Sensor 2	0.997515	1	0.998780
Sensor 3	0.999412	0.998780	1

TABLE 6.21
Correlation Coefficient Matrix—Final
13 Quarters

	Sensor 1	Sensor 2	Sensor 3
Sensor 1	1	0.994673	-0.10445
Sensor 2	0.994673	1	-0.08583
Sensor 3	-0.10445	-0.08583	1

The correlation coefficient matrix for the final 13 quarters (3 years) is now computed as shown in Table 6.21. It reveals that sensor 1 and sensor 2 strains are still behaving in almost perfect correlation. However, sensor 3 has limited correlation with sensors 1 and 2. This correlation result is consistent with the earlier slope of regression line results. In the final 13 quarters, sensor 3 is exhibiting a behavior not consistent with the numerical results. The data processing (regression and correlation) has revealed interesting information that casual evaluation has failed to produce.

6.4.5.4 Discussion of Regression and Correlation Results

The 5-year-long monitoring has revealed a deviation from the normal in the recordings of sensor 3. This conclusion is reached by

- Data analysis only, using regression and correlation techniques.
- The conclusion is reinforced by the aid of numerical (structural) modeling, which validated the measured data in the first 2 years, but exposed differences with the measurements during the following 3 years.

The decision maker is faced with many choices; Figure 6.42 shows a road map of the possible actions and the reasons for them. The figure shows that there are three possible actions to be taken in this situation.

Visual observation: Perhaps the first action is to visually inspect the deck for some clues as to the condition in the field. The location of sensor 3 near the field joint should give some hint that the joint

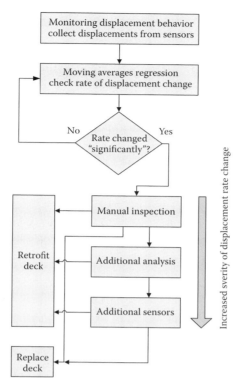

FIGURE 6.42   Decision making processes.

or some other component in its vicinity is not performing as designed. Special attention should be given to the condition of the wearing surface, the adhesive conditions in the joint, relative motion across the joint, as well as any debonding in the deck material itself.

Additional analysis: Based on the field observations, additional analysis might be needed. The additional analysis will need to explain

- Why there was a good correlation of behavior in the first 2 years?
- Why there was deviation of behavior between the sensors?

Note that the change of regression line slope indicates, in a broader sense, a change of stiffness. However, the correlation matrices behavior in Tables 6.20 and 6.21 indicate a nonlinear behavior. So, it seems that the analysis must account for changes of stiffness as well as nonlinear behavior of the deck.

Based on the above, the analysis should have the following:

- Nonlinearity of elements and adhesives should be included.
- Any observed behavior should be modeled appropriately (cracking, debonding, etc.).
- Creep effects might be included, if creep is deemed to be a possible contributing factor.

Additional sensors: Another course of action is to add more sensors or refocus the monitoring efforts. For example, if creep is suspected to be a contributing factor to the observed behavior, then temperature sensors might be added to confirm this viewpoint. If debonding is a potential factor, then adding load cell sensors as well as more strain sensors along the field joint might be considered.

6.5 CASE STUDIES

6.5.1 Rte 248—Bennetts Creek Bridge

The bridge on Route 248 over Bennetts Creek, NY, was an 860-mm-thick slab bridge built in 1926 (FHWA 2007). The single-span superstructure has a span of 7.6 m and a width of 10 m. It has a skew angle of 30%. It is located in a rural New York state road, with less than 17% truck traffic. The bridge was in a serious state of deterioration in 1998 with a condition rating of 1.97 (Figure 6.43). The traffic weight was restricted to 10 tons. As a result, long costly traffic detours were in place. A quick alternative was needed in view of the negative economic impact in the area.

As a prelude to the construction of the new bridge, the old bridge was demolished (Figure 6.44). New concrete abutments were constructed (Figures 6.45 and 6.46). The new design was based on a hollow-cell core E-glass FRP structure designed and built at the manufacturing plant (Figure 6.47). The weight of each of the two prefab panels was 7.7 metric tons. The bridge joints (Figure 6.48) were pourable silicon sealant. The wearing surface was 3/8-in epoxy polymer concrete. The design load was HS25 with a deflection limit of span/800. The bridge construction was completed in less than

FIGURE 6.43 Old superstructure over Bennetts Creek. (Courtesy of New York State Department of Transportation.)

FIGURE 6.44 Removed old superstructure. (Courtesy of New York State Department of Transportation.)

FIGURE 6.45 Construction of abutments. (Courtesy of New York State Department of Transportation.)

FIGURE 6.46 Abutments ready for new superstructure. (Courtesy of New York State Department of Transportation.)

15 months (Figures 6.49 and 6.50). When the bridge was tested (Figures 6.51 and 6.52), the measured deflections were less than the design deflection: the bridge was stiffer than what the design had predicted. The bridge was rated at 212 tons/284 tons inventory/operating loadings, respectively (Figure 6.53)—much higher than the before-construction ratings. Since the opening of the new bridge in 1998, its performance has been monitored. Figure 6.54 shows the measured strains in the period from 1998 through 2005 at different locations. The bridge performance is steady, with no signs of degradation or deterioration.

6.5.2 Bentley Creek Bridge

6.5.2.1 General

In 1999, the first FRP deck on a state highway system was installed in New York State, on State Route 367 in the village of Wellsburg, New York (Alampalli and Kunin 2003), to improve the live-load capacity of a 50-year-old truss bridge (see Figure 6.55) by replacing an old deteriorated concrete bridge.

FIGURE 6.47 New FRP deck at manufacturing plant. (Courtesy of New York State Department of Transportation.)

FIGURE 6.48 Connection details. (Courtesy of New York State Department of Transportation.)

FIGURE 6.49 Moving new FRP superstructure into position. (Courtesy of New York State Department of Transportation.)

FIGURE 6.50   Final construction steps. (Courtesy of New York State Department of Transportation.)

FIGURE 6.51   Load testing of new FRP superstructure/deck. (Courtesy of New York State Department of Transportation.)

6.5.2.1.1 Analysis and Design

Proven analysis procedures or design standards were not available at the installation time, so several assumptions were made during the design of the system. Not much data was available on the effect of fatigue, creep, temperature, in-service environment, and so on, on the deck, and these were accounted for by conservatively limiting the allowable stress in the FRP materials to 20% of their ultimate strength; and deflection was limited to span/800 (Alampalli and Kunin 2003). The sides and underside of the deck were painted to avoid UV exposure, and the effects of creep were deemed negligible due to the light weight of the FRP deck. The honeycomb deck was designed using a FE analysis by the manufacturer. The deck was designed assuming no composite action between the superstructure floor beams and the FRP deck. But, to be conservative, the connections were designed for required horizontal shear assuming full composite action, and hence the true extent of composite action was not known. Due to the experimental nature of the deck and nonexistent durability data, an in-service load testing and subsequent analyses using SHCE concepts were

FIGURE 6.52   Response measurements. (Courtesy of New York State Department of Transportation.)

FIGURE 6.53   Subsequent load tests. (Courtesy of New York State Department of Transportation.)

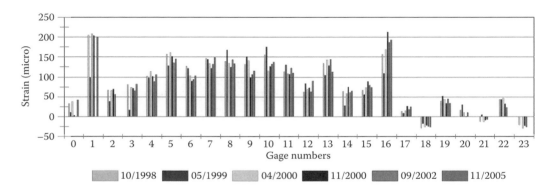

FIGURE 6.54   Long-term monitoring of superstructure/deck. (Courtesy of New York State Department of Transportation.)

FIGURE 6.55 State Route 367, NY: Bentley Creek Bridge. (Courtesy of New York State Department of Transportation.)

considered necessary to ensure safe and cost-effective use of FRP decks in future. This section will describe these efforts.

6.5.2.1.2 Testing

As a first step, the objectives of the SHCE program were defined (Alampalli and Kunin 2003). The load testing was conducted for three reasons. First, to verify two of the assumptions made during the design and construction: no composite action between the FRP deck and the beams and deck joints will transfer the loads across effectively. Second, a load rating based on actual measured data under loads was desired. The final reason was to support an FE analysis used to investigate the entire deck system and study the failure mechanisms of the structure with FRP deck. After the load testing and verification of the load test results, analysis was planned for further understanding of the deck behavior under live loads and its failure mechanisms.

Sensors and instrumentation were designed to meet the test objectives described above, while giving consideration to factors such as redundancy, resources, cost for data collection and analysis, access, total test time required, and inconvenience to bridge users due to traffic closures during the load testing. The results indicated that there was no composite action between the deck and the superstructure and verified the load ratings obtained analytically. The results showed that the deck joints were only partially effective in load transfer between different panels. The data also indicated some localized bending effects, which could play a role in the strain distribution of FRP components such as wearing surfaces.

The deck joint issues may not pose a problem to the deck itself, but it needs further investigation for long-term durability. At the same time, this may have an effect on the performance of the entire bridge system. The same issue can be taken with the local bending effects as they can have significant influence on the durability of the wearing surfaces. The decision taken based on the above load test results was to continue with the FE analysis using the load test data for calibration in order to investigate the failure mechanisms of the entire bridge system. It was also recommended that a load test should be considered in the future if the combination of the in-service loads and environmental exposure can weaken the joints. Thermal issues and probable damage scenarios were also studied to provide guidance to maintenance personnel.

6.5.2.1.3 Detailed Failure Analysis

A detailed static FE analysis was performed to investigate the failure mechanisms and thermal behavior of an FRP deck system (Aref and Chiewanichakorn 2002; Chiewanichakorn, Aref, and

Alampalli 2003a, 2003b). Three plausible types of failure mechanisms, namely, FRP deck failure, joint failure, and steel floor beam failure were considered. Samples of the FRP deck were lab tested, according to ASTM procedures, to obtain the properties for use in the analysis. The FE model of the bridge superstructure was verified by comparing the analysis with the test data. The results indicated that the buckling failure of the floor beams would dominate and govern the overall structural failure of this particular bridge (Chiewanichakorn et al. 2003b). This shows the importance of component versus system-level analysis. Fatigue performance of the deck itself was deemed satisfactory based on the fatigue tests conducted by Dutta et al. (2002).

6.5.2.1.4 *Durability and Other Issues*

During the erection (see Figures 6.56 and 6.57), parts of the face skin of a FRP panel were delaminated due to mishandling while lifting with the crane. These were repaired on-site by the manufacturer. This shows the need for standardized procedures for erection and transportation and the need for their being part of a successful SHCE program.

FIGURE 6.56 FRP bridge deck during erection. (Courtesy of New York State Department of Transportation.)

FIGURE 6.57 Placing FRP deck. (Courtesy of New York State Department of Transportation.)

Several of the FRP decks built in New York, including this FRP deck, used a thin wearing surface to take advantage of the light weight. A 10-mm polymer concrete wearing surface was used on the subject FRP deck. The wearing surface failed completely, within 3-6 months, during the first winter. This was attributed to improper preparation of the FRP deck surface. The wearing surface was replaced after proper surface preparation in the field and has been working well since the repair. During the replacement of the wearing surface, it was also found that some delaminations, as a result of the erection process, were not properly fixed. These were repaired before the wearing surface was replaced. The fact that the deck was shop-fabricated and repaired in the field by the manufacturer illustrates the need for appropriate QA/QC procedures for acceptance at the shop and after installation, as part of a successful SHCE program.

The most pressing issue on hand is therefore to address the problem of durability of wearing surfaces on FRP bridge decks and their interaction with FRP decks under traffic, thermal and environmental loads. So, a project was initiated by the NYSDOT, which involved extensive experimental investigation of wearing surface materials for GFRP bridge decks and its interaction with the GFRP decks at extreme temperatures (Aboutaha 2004). In addition, a full-scale tire testing was conducted to examine the system under simulated traffic loads at extreme temperatures. This research showed that most wearing surfaces used—the polymer concrete and polymer-modified concretes—will have poor long-term performance due to adhesion and mismatch of thermal properties. Hence, a hybrid wearing surface system made of both these concretes was proposed and will be investigated in the future. The project also gave general recommendations for construction, inspection, and maintenance of wearing surface materials for FRP bridge decks.

6.5.2.1.5   Concluding Remarks

This case study illustrated the use of SHCE concepts for the use of FRP decks. Several of the issues investigated shed light on the information and procedures needed for future projects. It also showed that issues, ranging from manufacturing to system durability, should be considered during the planning stages to take appropriate decisions during the project life and to minimize life cycle costs. This will also help to optimize future deck designs (to make sure all appropriate CD ratios are closer to unity) and to gain acceptability from designers, managers, and maintenance personnel. Until such integrated studies are done, CD ratios will continue to be much greater than unity, on the conservative side, to accommodate unknown factors.

6.5.2.2   Dynamic Effects

6.5.2.2.1   Overview

FRP decks have been used by several owners to replace old reinforced concrete bridge decks to increase their live-load capacity by taking advantage of the light weight of these decks. Their effectiveness is normally verified by load testing. The fatigue durability of FRP decks has also been studied, and the results showed that they perform very well. At the same time, the effect of deck replacement on dynamic and fatigue characteristics of the entire bridge system is not very well understood. This is important as change in the structure's mass due to deck replacement is expected to result in changes in the structure's vibration characteristics, and it can influence the fatigue life of bridge components. This case study is an example of a study that shows that system-level analysis is required before any changes are made since the component changes can negatively affect the long-term durability.

6.5.2.2.2   Modeling

A study was also conducted, using the Bentley Bridge described in the previous section (Figure 6.58), to determine the lower order natural frequencies and mode shapes of the bridge structure before and after it was retrofitted with the new lighter FRP deck. Two distinct FE models were created, using STAAD Pro FE modeling and the analysis software package (STAAD PRO 2002) to study

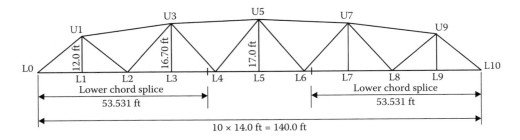

FIGURE 6.58 Truss elevation. (Courtesy of New York State Department of Transportation.)

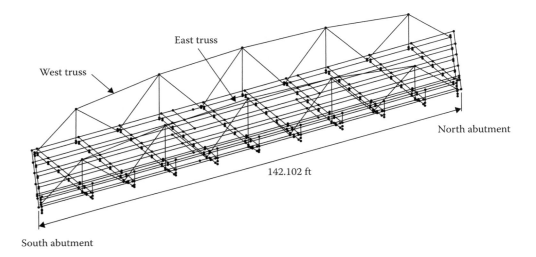

FIGURE 6.59 FRP deck bridge finite-element model. (Courtesy of New York State Department of Transportation.)

the effect of deck replacement on dynamic properties of the system. The first model represents the bridge with the original concrete deck (concrete deck model), and the second represents the current structure with the FRP deck (FRP deck model). The model is shown in Figure 6.59, and further details can be found elsewhere (Albers et al. 2007).

Validation of the FRP deck model was conducted using the field testing of the bridge after the FRP deck was installed (Alampalli and Kunin 2003). The axial stresses in the floor beam were computed at the six gauge locations on a floor beam supporting the deck in the field tests (see Figure 6.60). Appropriate changes were made to the model, based on the measured field data, to obtain good agreement, so that the model can be used for dynamic analysis (Table 6.22).

6.5.2.2.3 Vibration Analysis

Dynamic analysis was conducted to obtain the natural frequencies and mode shapes. Physical tests were also conducted using strain gauges for two selected truss members. Strains were recorded, at a rate of 500 samples/s, when the bridge was set into a free vibration mode after passage of a heavy truck. A typical time history for one of the gauges and corresponding fast Fourier transforms are shown in Figure 6.61. This figure displays the "ring down" portion of the transient response and reveals the free vibration of the span. The fundamental natural frequency was estimated from these data. This frequency (4.9 Hz) nearly coincided with those obtained using the FRP deck system FE model (4.77 Hz) and illustrates that FRP deck model is capable of predicting vibration behavior of the actual bridge structure.

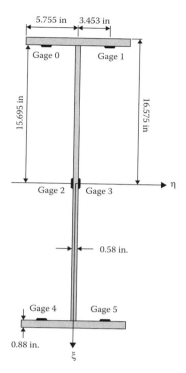

FIGURE 6.60 Instrumentation on a floor beam. (Courtesy of New York State Department of Transportation.)

TABLE 6.22
Comparison of Test and Model Data

Evaluation Method	$\|\sigma_{A\chi}\|$ (psi)	$\|\sigma_{B\xi}\|$ (psi)	$\|\sigma_{B}\eta\|$ (psi)	$\|\sigma\Sigma\|$ (psi)
Test (load case 1)	15	73	2668	2756
Model (load case 1)	0	11	2860	2871
Test (load case 2)	14.5	43.5	1276	1334
Model (load case 2)	0.0	30.0	1445	1475

Natural frequencies and mode shapes for the bridge with a concrete deck were compared to those with an FRP deck, and, as expected, frequencies of bridge with the FRP deck were consistently higher than for those with a concrete deck. Modal mass participation were used to identify the appropriate mode shapes obtained using the concrete deck model and FRP deck model. The increase in fundamental natural frequency was about 45% (3.28 vs. 4.77 Hz). Using the time histories, damping ratio was estimated to be about 1% indicating that bridges with FRP deck systems are typically lightly damped. This result is consistent with other reported FRP deck dynamic behavior (Shekar, Petro, and GangaRao 2004).

Results also indicated that fundamental bending mode shape was not strongly affected by the deck replacement. Many modes of the bridge were found to be primarily flapping and planar-bending motions of the trusses. These modes were mostly insensitive to the change in deck mass due to the deck replacement.

6.5.2.2.4 Estimation of Member Forces

Dead–load and live–load forces in eight selected/representative truss members, influenced by the deck replacement—members L5-L6, U1-U3, U5-L5, and U9-L8 on the east (E) and west trusses

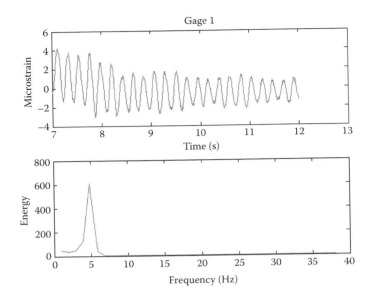

FIGURE 6.61   Strain time history and fast Fourier transform for member L5-L6 on the west truss.

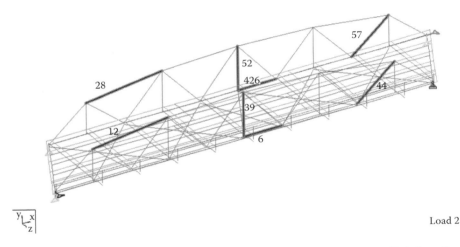

Load 2

FIGURE 6.62   Selected truss members on east and west trusses. (Courtesy of New York State Department of Transportation.)

(W)—were investigated. These members are shown in Figure 6.62, respectively, as elements 6, 12, 39, and 44 for the east truss and 426, 28, 39, and 57 for the west truss. It should also be noted that this study did not address floor beams and other elements of superstructure, which may also be critical to the behavior of the bridge (Aref and Chiewanichakorn 2002).

Dead–load forces in the selected members were compared in Figures 6.63 through 6.65, respectively, which indicated that the member forces for the FRP deck bridge, as expected, were generally lower than those for the concrete deck model. For the selected top chord member (U1-U3), this reduction was over 50%. Live–load forces in the selected members were also obtained for HS-20 truck loading at four different speeds on the east lane of the bridge. The results for the 5 mph (crawl speed) were assumed to give the static live–load forces in the members. These results show that static live–load forces

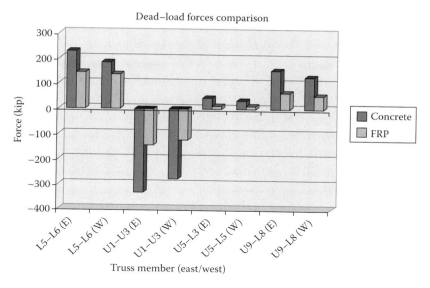

FIGURE 6.63 Effect of deck replacement on member forces resulting from Dead loads. (Courtesy of New York State Department of Transportation.)

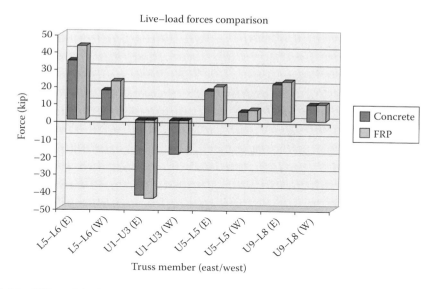

FIGURE 6.64 Effect of deck replacement on member forces resulting from Live loads. (Courtesy of New York State Department of Transportation.)

for the FRP deck bridge were generally higher than those for the concrete deck bridge. The greatest force increase was noted on members L5-L6 on the bottom chord of the east truss (about 25%).

Total forces in the selected members due to dead load and live load are also shown in Figures 6.63 and 6.64 and compared with corresponding design forces obtained from the bridge plans. Comparing the total forces for the two-deck systems, it can be concluded that the total forces are generally lower for the FRP deck system, by over 50% for some members. The total forces for both deck systems were much lower than those estimated for the members design.

FIGURE 6.65 Effect of deck replacement on member total forces. (Courtesy of New York State Department of Transportation.)

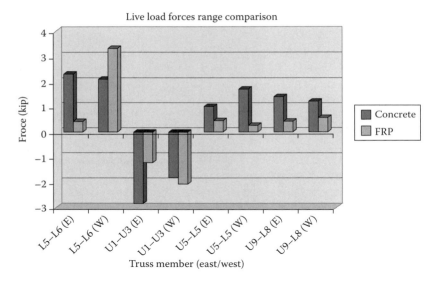

FIGURE 6.66 Effect of deck replacement on live–load force range. (Courtesy of New York State Department of Transportation.)

6.5.2.2.5 Fatigue Analysis

An attempt was made here to investigate how fatigue design of the bridge is affected by the changes in deck material. Figure 6.66 shows the force ranges for the selected members under the two-deck systems. For each member, the force range shown in Figure 6.66 represents the highest of the three force ranges obtained for the member under the three crossing speeds. The results show that the stress range for most elements is lower for the FRP deck system compared to the concrete deck system it replaced. At the same time, the stress range for at least one of the bridge members increased by over 30% due to deck replacement. Thus, even though the total forces are generally lower for the FRP deck system, it should be noted that the fatigue life depends on the live-load stress range and not on the total stress, AASHTO (1996). So, there is a possibility that some members may be more susceptible to fatigue than before. Thus, before replacing the concrete deck with a lighter FRP deck, the effect of replacement on various failure mechanisms at global and component levels should be investigated.

6.5.2.2.6 Closing Remarks

The results indicate that the fundamental frequency for the structure with the FRP deck is about 45% higher than that for the original concrete deck structure. But, fundamental bending mode shape was not strongly affected by the deck replacement. Many modes of the bridge were found to be primarily flapping and planar-bending motions of the trusses. These modes were mostly insensitive to the change in deck mass due to the deck replacement.

Eight selected members were investigated under the two-deck systems for dead load, live load, dynamic impact amplification, and fatigue stress range. For the dynamic impact amplification and fatigue investigations, member forces were obtained from the respective models for an HS-20 truck crossing the structure at various speeds. The investigation concluded that, for the FRP deck bridge, dead–load forces were lower, live-load forces were higher, and the total forces were lower than those for the concrete deck bridge. The fatigue analysis indicated that, before replacing a concrete deck with a lighter FRP deck, the effect of replacement on various failure mechanisms (such as fatigue) at global and component levels should be investigated.

6.5.2.3 Maintenance Issues

The next step of SHCE process of the new bridge deck included investigating the effects of the fire and mechanical damage, which might be faced by the FRP deck during its lifespan. These were in response to questions which arose, after the structure was in-service, regarding the reserve capacity during a high ambient temperature gradient of 60°C, due to possible damage from a burning vehicle on the structure, or possible accidental fire under the deck, or mechanical damage from snow plow operations. This would prepare the owner with possible repair mechanisms if such damage/situations occurred during the lifespan of the bridge.

6.5.2.3.1 Fire Damage

A thermal stress analysis showed that the FRP deck would have an adequate amount of reserve capacity during a high ambient temperature gradient of 60°C (Aref et al. 2004). Of more concern was the exposure of the FRP deck to fire hazard, say from traffic accidents. The FRP material degrades faster than concrete or steel when exposed to higher temperatures (see Figure 6.67). To explore this hazard, Aref et al. (2004) built a detailed FE model that is capable of coupled mechanical and thermal analysis of the FRP deck. The Tsai-Hill failure criterion of FRP laminates were used to model failure of the deck material. For plane problems, the criterion is expressed by

$$\frac{\sigma_1^2}{X_1^2} - \frac{\sigma_1\sigma_2}{X_1X_2} + \frac{\sigma_2^2}{X_2^2} + \frac{\tau^2}{S^2} = 1$$

(6.16)

With

σ_i = Normal stress in the ith direction, $i = 1, 2$
X_i = Lamina principal normal strength in the ith direction, $i = 1, 2$
τ = Shear stress
S = Lamina shear strength

The lamina fails when the Tsai-Hill criterion reaches or exceeds 1.0. Exploring several traffic geometries that might generate fire hazards, the deck was found to reach and exceed the criterion early as 440 s after the fire initiation (see Figure 6.68). It was, thus, concluded that a combination of the truck live load with the most severe thermal effect could cause a failure of the FRP deck. The FRP

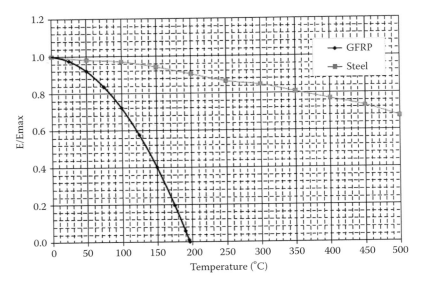

FIGURE 6.67   Relative stiffness of steel and GFRP materials. (Courtesy of Dr. Amjad Aref.)

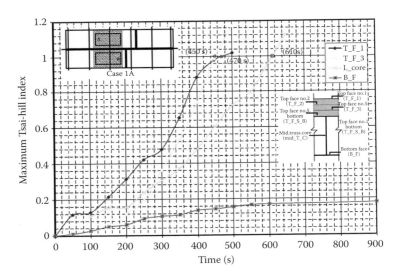

FIGURE 6.68   Elapsed time performance of Tsai-Hill criterion. (Courtesy of Dr. Amjad Aref.)

deck demonstrated lower heat resistance compared to steel and concrete decks. The researchers recommended that any FRP bridge on fire has to be cleared of people and vehicles quickly and the damaged region repaired before further use. More details can be found in the detailed report describing these studies (Aref et al. 2004).

6.5.2.3.2   Snow Plow Damage
Damage simulations showed that the FRP deck would have sufficient reserve capacity in the worst-case scenario when the 33% of the top face skin is removed by a snow-plowing process (Aref et al. 2004). Analysis also indicated that due to possible additional impacts and overloads effects, it is

recommended that the damaged portion of the deck be closed and repaired as soon as the problem is detected.

6.6 LCA FOR FRP BRIDGE DECKS

6.6.1 PRINCIPLE OF NEW MATERIALS LCA

LCA for FRP bridge decks offers a unique perspective that we have not yet seen in this chapter. The uniqueness here is the new material. We immediately observe that FRP is a fairly new material in the field of civil infrastructures. After the long dominance of conventional materials such as masonry, wood, steel, and concrete, FRP is a new material that offers many new advantages and some shortcomings. We discussed many of these in the previous sections. With those advantages and shortcomings, new models of costs and benefits, as well as some challenges of estimating lifespan, should be considered. On the basis of this, we suggest a new principle in the subject of LCA that is specific for new materials as follows:

Life-cost analysis for new construction materials should include costs of additional monitoring, costs of additional demands, benefits of new advantages, as well as accurate estimates of expected lifespan.

We call this the principle of new materials LCA or NMLCA for short. Each of the components of NMLCA is discussed below.

6.6.1.1 Additional Costs of Monitoring

The officials responsible for bridges in all examples in this chapter decided wisely to monitor the new FRP bridge decks for many years after completion of the project. This should be an essential practice whenever new materials are introduced in civil infrastructures. Given the demanding environments of bridges and the imperative safety concerns, it is logical to monitor projects that use new materials. Conventional engineering materials, such as steel or concrete, have a wide body of experience. So, even though monitoring of structures built with conventional materials is *preferable,* it is not as *essential* as with new materials.

Thus, we define the cost of monitoring per period (usually a year) as $C_{MONITOR}$. This would include all labor, equipment, data collection, and processing, as well as decision making.

6.6.1.2 Costs of Additional Demands

Costs of any new material should include two parts: generic costs and costs that are due to the newness of the material. Generic costs are those that would be incurred if any other material is used. New material–specific costs can be established by tabulating all costs due to the new material. Each of these costs can then be included in LCA in one of three forms:

- Quantify those costs in monetary form, if possible
- Estimated costs can be entered in a qualitative scale, or
- The costs that can affect lifespan; the effects of those costs should also be accounted for in the lifespan estimation of the component

As an example, we illustrate some of the new-material related costs of FRP decks in Table 6.23. The qualitative costs can be included in an LCA, using an approach described in Chapter 8.

The total costs per period of the components using the new material can now be modeled as

$$C_{NM} = C_0 + \sum_i C_{NM\_i} \tag{6.17}$$

The generic costs of a typical component are C_0.

TABLE 6.23
Costs of FRP Bridge Decks

Costs	Type	Symbol	How to include it in LCA
Wearing surface	Quantitative	C_{NM-1}	Wearing surface can be a major life cycle cost issue. Better quality surfaces at higher initial costs might prove to be an optimal life cycle cost saving solutions
Damping	Quantitative	C_{NM-2}	Low damping ratios and higher stiffness density can result in a higher vibration response (seismic and/or traffic impact effects). The cost of this issue can be easily quantified
Fire	Quantitative	C_{NM-3}	Damages (hence costs) from fire hazard can be computed using an approach similar to Aref et al. (2004). Probabilities of fire hazards can be estimated using available data and experience
Low ductility	Quantitative	C_{NM-4}	Brittle behavior, even with higher factors of safety, needs closer study. Note that monitoring brittle failures is difficult and requires additional costs
Environmental degradations	Qualitative	C_{NM-5}	Costs of repairs of environmental or durability damage need to be considered
Potential durability issues	Quantitative	C_{NM-6}	
Other?	To be determined	C_{NM-7}	Other sources of costs can be added on a case-by-case basis

TABLE 6.24
Benefits of FRP Bridge Decks

Benefit	Type	Symbol	How to include it in LCA
Light weight	Quantitative	B_{NM-1}	Movable bridges. Higher LL due to lesser dead load. Relieve older superstructures from DL demands of conventional decks. Monetary value of additional traffic can easily be computed
Speedy construction	Quantitative	B_{NM-2}	Fewer detours, less inconvenience to traffic, labor and management cost savings (working hours)
Corrosion	Qualitative	B_{NM-3}	Lack of corrosion damage can save maintenance costs when compared to other types of decks
Lighter equipment	Qualitative	B_{NM-4}	Need for heavy cranes and other large equipment is eliminated
Lifting postings	Quantitative	B_{NM-5}	When replacing older decks, higher ratings are possible. Monetary value of higher LL can easily be computed
Other?	To be determined	B_{NM-6}	Other types of benefits can be added on a case-by-case basis

6.6.1.3 Advantages of New Material (Additional Benefits)

We propose that each of the benefits of any new material should be included in the benefits side of the LCA under consideration. This can be done by tabulating all potential benefits from the new material. Each of these benefits can then be included in LCA in one of three forms:

- Quantify those benefits in monetary form, if possible.
- Estimated benefits can be entered in a qualitative scale, or
- The benefits are related to lifespan, which will be discussed in the next section.

As an example, we illustrate some of the potential benefits of FRP decks in Table 6.24.

The qualitative benefits can be included in an LCA using an approach described in Chapter 8. The total benefits per period of the components using the new material can now be modeled as

$$B_{NM} = B_0 + \sum_i B_{NM\_i}$$

$$(6.18)$$

The generic benefit of a typical component is B_0. Generic benefits are those benefits that would be incurred if any other material is used.

6.6.1.4 Expected Lifespan

The total lifespan of components using the new material, for example, an FRP bridge deck, within an infrastructure can be affected by numerous local issues. These local effects can be modeled as $\alpha_{NM-i}\Delta T_{NM-i}$. Factor $\alpha_{NM-i} = \pm 1.0$ simulates whether the ith effect increases or decreases the lifespan. Factor ΔT_{NM-i} has a time units and describes the addition or reduction of the lifespan that the ith effect has on the component on hand. Table 6.25 shows the lifespan issues for FRP bridge decks. Note that $i = 1, 2 \ldots 5$ in this case.

The total lifespan of a component using the new material can now be modeled as

$$LS_{NM} = LS_{NM\_0} + \sum_i \alpha_{NM\_i} \Delta T_{NM\_i} \tag{6.19}$$

The generic lifespan of a typical component is LS_{NM-0}.

6.6.2 LCA MODEL FOR NEW MATERIALS

The LCAs of a component that uses a new material can be estimated as

$$LCC_{NM} = \sum_{j=1}^{j=LS_{NM}} (C_{NM})_j \tag{6.20}$$

Similarly, the life cycle benefits of a component that uses a new material can be estimated as

$$LCB_{NM} = \sum_{j=1}^{j=LS_{NM}} (B_{NM})_j \tag{6.21}$$

The discount rate is dropped from Equations 6.20 and 6.21 for the sake of simplicity. The net benefit or the net costs can be easily computed and the appropriate effects of using new materials established in an unbiased manner.

TABLE 6.25
Life Span Issues of FRP Bridge Decks

Issue	Type	Symbol	How to include it in LCA
Environmental degradations	Quantitative	$-\Delta T_{NM-1}$	UV, moisture, spills, etc. might have effects on life span of decks. Testing and experience can give accurate estimates of such effects. For example, Smith, Hassan, and Rizkalla (2004) noted the significant impact of environmental conditions on longevity of bonding of adhesives in FRP applications
Potential durability issues	Quantitative	$-\Delta T_{NM-2}$	Direct impacts, delaminations, manufacturing or erection defects, heat damage, scratches, etc. can cause long-term durability problems that can reduce lifespans
Corrosion	Quantitative	$+\Delta T_{NM-3}$	Absence of corrosion can increase lifespan of the deck. Testing and experience can give accurate estimates of such effects
High safety factors	Quantitative	$+\Delta T_{NM-4}$	
Other?	To be determined	$\alpha_{NM-5}\Delta T_{NM-5}$	Other lifespan issues can be added on a case-by-case basis

REFERENCES

AASHTO. (1996). "Standard Specifications for Highway Bridges," 16th edition including 1997 and 1998 Interim Specifications, American Association of State Highway and Transportation Officials, Washington, DC.

Aboutaha, R. (2004). "Durability of Wearing Surfaces for FRP Bridge Decks," Research Report, NYSDOT, Albany, NY.

Alampalli, S. and Kunin, J. (2001). "Load Testing of an FRP Bridge Deck on a Truss Bridge," Special Report 137, New York State Department of Transportation, Albany, NY.

Alampalli, S. and Ettouney, M. M. (2006). "Long-term issues related to structural health of FRP bridge decks." *Journal of Bridge Structures: Assessment, Design and Construction*, Taylor and Francis, 2(1), 1–11.

Alampalli, S. and Kunin, J. (2002). "Rehabilitation and field testing of an FRP bridge deck on a truss bridge." *Journal of Composite Structures*, Elsevier Science, 57(1–4), 373–375.

Alampalli, S. and Kunin, J. (2003). "Load testing of an FRP bridge deck on a truss bridge." *Journal of Applied Composite Materials*, Kluwer Academic Publishers, 10(2), 85–102.

Alampalli, S., O'Connor, J., and Yanotti, A. (2000). "Design, Fabrication, Construction and Testing of an FRP Superstructure," Special Report 134, New York State Department of Transportation, Albany, NY.

Alampalli, S., O'Connor, J., and Yannotti, A. (2002). "Fiber reinforced composites for the superstructure of a short-span rural bridge." *Journal of Composite Structures, Elsevier Science*, 58(1), 21–27.

Albers, W., Hag-Elsafi, O., and Alampalli, S. (2007). "Dynamic Analysis of Bentley Creek Bridge with FRP Deck," Special Report 150, Transportation Research and Development Bureau, New York State Department of Transportation, Albany, NY.

American Society of Civil Engineers, ASCE. (2001). "Gap Analysis for Durability of Fiber- Reinforced Polymer Composites in Civil Infrastructures." ASCE/CERF report, Washington, DC.

Anderegg, P., Broennimann, R., Nellen, P. M., and Sennhauser, U. (2002). "Reliable Long-Term Health Monitoring of CFRP Cables in Bridges," *Proceedings of 1st International Workshop on Structural Health Monitoring of Innovative Civil Engineering Structures*, ISIS Canada Corporation, Manitoba, Canada.

Aref, A. J. and Alampalli, S. (2001). "Vibration characteristics of a fiber-reinforced polymer bridge superstructure." *Composite Structures*, 52(3–4), 467–474.

Aref. A. J., Alampalli, S., and He, Y. (2005). "Performance of a fiber reinforced polymer web core skew bridge superstructure: failure modes and parametric study." *Journal of Composite Structures*, Elsevier Science, 69(4), 500–509.

Aref, A. J. and Chiewanichakorn, M. (2002). "The Analytical Study of Fiber Reinforced Polymer Deck on an Old Truss Bridge", Report submitted to New York State Department of Transportation Research and Development Bureau, and Transportation Infrastructure Research Consortium. Albany, NY.

Aref, A., Chiewanichakorn, M., and Alnahhal, W. (2004). "Temporal Thermal Behavior and Damage Simulations of FRP Deck," Report submitted to New York State Department of Transportation Research and Development Bureau, and Transportation Infrastructure Research Consortium. Albany, NY.

ASCE/CERF. (2001). *Gap Analysis for Durability of Fiber Reinforced Polymer Composites in Civil Infrastructures*, Civil Engineering Research Foundation, ASCE, Reston, VA.

ASME. (1998). "Acoustic emission examination of fiber-reinforced plastic vessels," *American Society of Mechanical Engineers Boiler and Pressure Vessel Code*. New York, NY

Bakis, C., Bank, L., Brown, V., Cosenza, E., Davalos, J., Lesko, J., Machida, A., Rizkalla, S., and Triantafillou, T. (2002). "Fiber-reinforced polymer composites for construction-state-of-the-art review." *American Society of Civil Engineers, Journal of Composites for Construction*, 6(2), 73–87.

Baukat, D. and Gallon, A. (2002a). "Performance of Fiber Bragg Gratings Sensors During Testing of FRP Bridge Decks," *Proceedings, NDE Conference on Civil Engineering*, ASNT, Cincinnati, OH.

Baukat, D. and Gallon, A. (2002b). "Performance of Fiber Bragg Grating Sensors During Testing of FRP Composite Bridge Decks," *Proceedings of 1st International Workshop on Structural Health Monitoring of Innovative Civil Engineering Structures*, ISIS Canada Corporation, Manitoba, Canada.

Besterfield, D. H. (2001). *Quality Control*, 6th Edition, Prentice Hall, Upper Saddle River, NJ.

Burgueno, R., Karbhari, V. M., Seible, F. and Kolozs, R. T. (2001). "Experimental dynamic characterization of an FRP composite bridge superstructure assembly." *Composite Structures*, 54(4), 427–444.

Camata, G. and Shing, P. B. (2004). "Evaluation of GFRP Deck Panel for the O'Fallon Park Bridge," Colorado Department of Transportation, Report No. CDOT-DTD-R-2004–2, Denver, CO.

Chajes, M. J., Shenton, H. W. III, and Finch, W. W., Jr. (2001). "Performance of fiber-reinforced polymer deck on steel girder bridge." *Transportation research Record*, 1770, 105–112.

Chiewanichakorn, M., Aref, A. J., and Alampalli, S. (2003a). "Failure analysis of fiber-reinforced polymer bridge deck system." *Journal of Composites Technology and Research, ASTM*, 25(2), 119–128.

Chiewanichakorn, M., Aref, A. J., and Alampalli, S. (2003b). "Structural Behavioral Study of an FRP Deck System Using FEA." *Proceedings of the National Workshop on Innovative Applications of Finite Element Modeling in Highway Structures*, New York, NY.

Christie, R., Fagrell, B., Hiel, C., Hooks, J., Karbhari, V., Karshenas, M., Liles, P., Lopez-Anido, R., Meggers, D., Sikorsky, C., Seible, F., Seim, C., Till, R., Williams, D. and Yannotti, A. (2001). "HITEC Evaluation Plan for FRP Composite Bridge Decks," *Civil Engineering Research Foundation*, Washington, DC.

Crocker, H., Shehata, E., Mufti, A., and Stewart, D. (2002). "Development of a Smart GFRP Bridge Deck," *Proceedings of 1st International Workshop on Structural Health Monitoring of Innovative Civil Engineering Structures*, ISIS Canada Corporation, Manitoba, Canada.

Duke, J. C., Miceli, M., Horne, M. R., and Mehl, N. J. (2006). "Infrared Thermal Imaging for NDE of FRP Bridge Beams and Decks," NDE Conference on Civil Engineering, ASNT, St. Louis, MO.

Dutta, P. K., Kwon, S., and Lopez, A. R. (2002). "Fatigue Evaluation of FRP Composite Bridge Deck Systems under Extreme Temperatures." *Proceedings of International Conference on Durability of FRP Composites for Construction*. Montreal, Canada, 2, 665–675.

Dutta, P. K., Kwon, S., and Lopez-Anido, L. (2003). "Fatigue Performance Evaluation of FRP Composite Bridge Deck Prototypes Under High and Low Temperatures," Transportation Research Board annual meeting, Washington, DC.

Ettouney, M. and Alampalli, S. (2002). "Overview of Structural Health Engineering," 2002 Structures Congress and Exposition, Denver, CO, 225–226.

Ettouney, M. and Alampalli, S. (2012). *Infrastructure Health in Civil Engineering: Theory and Components*, CRC Press, Boca Raton, FL.

Ettouney, M. M. and Alampalli, S. (2000). "Engineering Structural Health." ASCE Structures Congress 2000, Philadelphia, PA.

Ettouney, M. M. Benaroya, H., and Wright, J. (1989). "Linear Dynamic Behavior of Semi Infinite Trusses," *Proceedings, Seventh VPI & SU Symposium on Dynamics and Control of Large Structures*, Blacksburg, VA.

Ettouney, M. Benaroya, H., and Wright, J. (1990). "Wave Propagation in Hyper Structures," *Proceedings, Space 90, Engineering, Construction, and Operations in Space*, Albuquerque, NM.

Ettouney, M. M., Daddazio, R., and Abboud, N. (1997) "Some Practical Applications of the use of Scale Independent Elements for Dynamic Analysis of Vibrating Systems," *Computers and Structures*. 65(3), 1997.

FHWA. (2007). "Route 248 over Bennetts Creek," http://www.fhwa.dot.gov/bridge/frp/deck248.cfm. Accessed on December 24, 2008.

FHWA. (2008). "Current Practices in FRP Composites Technology: FRP Bridge Decks and Superstructures," Federal Highway Administration, http://www.fhwa.dot.gov/bridge/frp/deckprc.cfm. Accessed on December 25, 2008.

Fried, N. (1967). "Degradation of Composite Materials: The Effect of Water on Glass Reinforced Plastics." *Proceedings of the 5th Symposium on Naval Mechanics*, Philadelphia, PA.

Fuchs, P. (2008). "Laser Instrumentation to Aid in Steel Bridge Fabrication," *NDE/NDT for Highways and Bridges: Structural Materials Technology (SMT)*, ASNT, Oakland, CA.

GangaRao, H. and Shekar, V. (2003). "Field Testing and Evaluation of Composite Bridges," Draft Final Report, WVDOT, Charleston, WV.

Gomez, J. and Casto, B. (1996). "Freez/Thaw Durability of Composite Materials." *Proceedings of the 1st International Conference on Composites in Infrastructures: Fiber Composites in Infrastructures*, Tucson, AZ.

Hassan, T., Eugin, M., and Rizkalla, S. (2003). "Innovative 3-D FRP Sandwich Panels for Bridge Decks," *Proceedings, 5th Alexandria International Conference on Structural and Geotechnical Engineering*, Alexandria, Egypt.

Henderson, M. P., Editor (2000). "Evaluation of Salem Avenue Bridge Deck Replacement: Issues Regarding the Composite Materials Systems Used," Final Report prepared for State of Ohio Department of Transportation, Ohio Department of Transportation, Columbus, Ohio.

Hill, R. (1950). *The Mathematical Theory of Plasticity*, Oxford University Press, London.

Hull, D., and Clyne, T. (1998). *An Introduction to Composite Materials*, Cambridge University Press, Cambridge, UK.

Johnson, C. F. and Houston, D. Q. (1990). "Environmental Test Methods for Liquid Molded Composites," *Proceedings of the American Society of Materials*, Materials Park, OH.

Jones, R. (1999). *Mechanics of Composite Materials*, Taylor and Francis, Philadelphia, PA.

Karbhari, V. (2001). "HITEC Evaluation Plan for FRP Composite Decks,"

Karbhari, V., Kaiser, H., Navada, R., Ghosh, K., Lee, L. (2005). "Methods for Detecting Defects in Composite Rehabilitated Concrete Structures," Submitted to Oregon Department of Transportation, and Federal Highway Administration, Report No. SPR 336.

Konur, O. and Matthews, F. L. (1989). "Effect of the properties of the constituents on the fatigue performance of composites: a review," *Composites*, 20(4), 345–361.

Kumar, P., Chandrashekhara, K. and Nanni, A. (2001). "Structural Performance of A FRP Bridge Deck," ASCE J. of Composites for Construction, 8(1), 35–47.

Lemistre, M., Gouyon, R., Kaczmarek, H., and Balageas, D. (2000). "Damage Localization in Composite Plates Using Wavelet Transform Processing on Lamb Wave Signals," *Proceedings of 2nd International Workshop on Structural Health Monitoring*, Stanford University, Stanford, CA.

Liu, Z. (2007). "Testing and Analysis of a Fiber-Reinforced Polymer (FRP) Bridge Deck," PhD. Dissertation, Civil Engineering Department, Virginia Polytechnic Institute, Blacksburg, VA.

McDad, P., Fowler, T., Medlock, R., and Ziehl, P. (2004). "Structural Health Monitoring of an Efficient Hybrid FRP/Reinforced Concrete Bridge System," *Proceedings of 2nd International Workshop on Structural Health Monitoring of Innovative Civil Engineering Structures,* ISIS Canada Corporation, Manitoba, Canada.

McLintock, F. A. and Argon, A. S. (1996). *Mechanical Behavior of Materials.* Addison-Wesley, NYC, NY.

MDA. (2008). "Composite Basics: Composite Manufacturing," Composite Growth Initiative of the American Composites Manufacturing Association, http://www.mdacomposites.org/mda/psgbridge_cb_mfg_process.html. Accessed on December 27, 2008.

Miyano, Y., Nakada, M., Yonemori, T., and Tsai, S. W. (1999). "Time and Temperature Dependence of Static, Creep and Fatigue Behavior for FRP Adhesive Joints." *Proceedings of the 12th International Conference on Composite Materials*, Paris.

NYSDOT. (1938). "Plans for Reconstruction of a Portion of the Pennsylvania State Line- Wellsburg, State Highway no. 1710," Reconstruction Contract No. FAS-RC 40-36, State of New York Department of Public Works, Division of Highways, Albany, NY.

Reising, R. M. W., Shahrooz, B. M., Hunt, V. J., Lenett, M. S., Christopher, S., Neuman, A. R., Helmicki, A. J., Miller, R. A., Kondury, S. K., and Morton, S. (2001). "Performance of five-span steel bridge with fiber-reinforced polymer composite deck panels." *Transportation Research Record*, 1770, 113–123.

Shahrooz, B. M., Reising, R., Hunt, V., Helmicki, A., and Neumann, A. R. (2002). "Testing and Monitoring of a Five-Span Bridge with Fiber Reinforced Polymer Deck Systems," *Proceedings, NDE Conference on Civil Engineering*, ASNT, Cincinnati, OH.

Shekar, V., Aluri, S., Laosiriphong, K., Petro, S., and GangaRao, H. (2004). "Field Monitoring of Two Fiber Reinforced Polymer Bridges," *Proceedings, NDE Conference on Civil Engineering*, ASNT, Buffalo, NY.

Shekar, V., Petro, S. H., and GangaRao, H. V. S. (2003). "Fiber-reinforced polymer composite bridges in West Virginia." *Transportation Research Record*, 1819(2), 378–384.

Smith, G., Hassan, T., and Rizkalla, S. (2004). "Bond Characteristics and Qualifications of Adhesives for Marine Applications And Steel Pipe Repair," *Technical Report: CFL-RD-04-01, IPS, North Carolina State University (NCSU), Constructed Facilities Lab (CFL)*, Raleigh, North Carolina.

STAAD.PRO. (2002). Users Guide Research Engineers, Intl. (REI), Yorba Linda, CA.

Staszewski, W., Biemans, C., Boller, C., and Tomlinson, G. (2000). "Impact Damage Detection in Composite Structures-Recent Advances," *Proceedings of 2nd International Workshop on Structural Health Monitoring*, Stanford University, Stanford, CA.

Turner, M. K., Harries, K. A., and Petrou, M. F. (2003). "In-situ Structural Evaluation of GFRP Bridge Deck System." Transportation Research Board 82nd Annual Meeting CD-ROM, Washington, DC.

Wang, C. and Chang, F. (2000). "Built-in Diagnostics for Impact Damage Identification of Composite Structures," *Proceedings of 2nd International Workshop on Structural Health Monitoring*, Stanford University, Stanford, CA.

Winkelman, T. J. (2002). "Fiberglass Reinforced Polymer Composite Bridge Deck Construction in Illinois," Illinois Department of Transportation, Physical Research Report No. 145, Springfield, IL.

Xiaorong, Z., Baoqi, T., and Shenfang, Y. (2000). "Study on Delamination Detection in CFRP Using Wavelet Signal Singularity Analysis," *Proceedings of 2nd International Workshop on Structural Health Monitoring*, Stanford University, Stanford, CA.

Zhao, J. and DeWolf, J. T. (2002). "Dynamic monitoring of steel girder highway bridge." *Journal of Bridge Engineering, ASCE*, 7(6), 350–356.

7 Fiber-Reinforced Polymers Wrapping

7.1 INTRODUCTION

7.1.1 OVERVIEW

The popular bridge application of fiber-reinforced polymers (FRP) is for column wrapping. Generally, the FRP-wrapping process is to attach a thin layer (or layers) of FRP material to concrete bridge components. The FRP wrapping extends the life of the structure by reducing the exposure to environmental conditions or giving additional ultimate capacity (see Figures 7.1 through 7.3). The benefits are illustrated below, based on literature:

- Wrapping FRP layers around reinforced concrete columns can result in increased column strength (shear resistance, ductility, lap splice capacity, flexural stiffness and strength, etc.) due to the extra confinement that the wrapping will induce in the column (Priestley and Seible 1991).
- If applied properly, the FRP layers can result in inhibiting corrosion in steel reinforcement of columns (Alampalli 2005), as shown in Figure 7.1.
- Strengthening and retrofitting deficient structural components can be accomplished using external application of FRP materials (see Figures 7.2 and 7.3). The FRP wraps can provide an extra tensile strength and/or an extra stiffening measure to underdesigned concrete components such as bridge decks or girders (Hag-Elsafi et al. 2004), as shown in Figure 7.2.
- The light weight of the materials makes it easy to use, and need lighter construction equipment, compared to conventional repairs, is needed.
- Cost-effective compared to conventional rehabilitation (Hag-Elsafi et al. 2002), as shown in Figure 7.3.
- The time for repairs is relatively less compared to conventional repairs, and so inconvenience to public is minimized or in some cases averted completely (Hag-Elsafi et al. 2004).
- Reduces the exposure to salt and moisture in cold climates where salt is used in winter for better traction on the road (Figure 7.1).
- Seismic rehabilitation of buildings and bridges. The seismic retrofit of columns of the FDR drive in New York City (Figures 7.4 and 7.5) is a typical example of such seismic retrofit projects.

We should mention that FRP wrapping has been used with material other than concrete, such as steel (see Chahrour and Soudki 2004), or wood (see Taheri et al. 2002). This chapter will concentrate on using FRP wrapping with concrete systems.

7.1.2 SEISMIC RETROFITS

Some of the early applications of FRP use in civil infrastructures were to retrofit seismically deficient bridges. The advantage of improving strength, stiffness, and ductility of reinforced concrete components by wrapping layers of FRP laminates around them was attractive to bridge owners. The

FIGURE 7.1   FRP wrapping to reduce exposure to road salt and inhibit corrosion. (Courtesy of New York State Department of Transportation.)

FIGURE 7.2   FRP wrapping of beams to increase flexural strength of bridge T-beams. (Courtesy of New York State Department of Transportation.)

cost-effectiveness and speed of construction of such solutions were added reasons for the popularity of using FRP laminates in seismic retrofit projects. Figures 7.4 and 7.5 show how the FDR drive in New York City was seismically retrofitted using FRP wrapping (Alberski, T., pers. comm.). Note how the retrofitted columns look almost as if they were originally built, with no sign of the FRP retrofit. A summary of some seismically retrofitted bridges in California can be found in FHWA (2008).

7.1.3   THIS CHAPTER

This chapter is concerned with different applications of FRP wrapping in civil infrastructures. First, we study some theoretical and physical attributes of the material as used in civil infrastructure applications. Recognizing the fact that the use of new materials in the highly demanding civil infrastructure field requires strict safety, durability, and cost-effectiveness, there is an

FIGURE 7.3 FRP Laminates to increase stiffness of capbeams. (Courtesy of New York State Department of Transportation.)

FIGURE 7.4 FDR, New York City. (Photo courtesy of Dr. Tadeusz Alberski.).

FIGURE 7.5 Finished FDR strengthening project. (Photo courtesy of Dr. Tadeusz Alberski.).

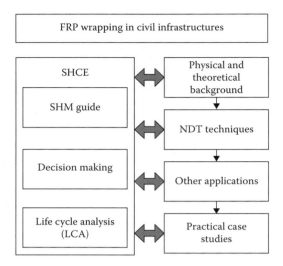

FIGURE 7.6 Organization of this chapter.

obvious need for effective performance monitoring. We explore next the use of nondestructive testing (NDT) techniques in monitoring the performance of FRP utilization. As we have seen, seismic retrofits using FRP materials were extremely successful. We discuss other equally successful applications of these relatively new materials. Different practical case studies are then introduced. We turn our attention to the more general subject of structural health monitoring (SHM) as applied to FRP use in civil infrastructures. The chapter will conclude with FRP-related decision making and life cycle analysis (LCA). Figure 7.6 shows sections of this chapter and how they relate together.

7.2 PHYSICAL AND THEORETICAL BACKGROUND

7.2.1 GENERAL

FRP wraps are thin material fabrics that have two basic components: the fabric usually referred to as fiber and the encompassing resin referred to as the matrix. The main function of the fiber is to transfer the load, while the main function of the resin is to protect the fiber and to transfer the load to and from the fiber. The mechanical properties of the fiber and the resins are shown in Tables 7.1 and 7.2.

FRP wrappings are very light, but they exhibit a high-tensile elastic modulus as well as high-tensile strength. They are also very thin (thickness of single ply ranges from 0.02 to 0.10 in), easy to handle, and easy to conform to geometries of other structural surfaces. Because of all these advantages, they have been used extensively in retrofitting civil infrastructures.

We can then reduce the FRP-wrapping subject into three components: the structural component that needs to be enhanced, the FRP-wrapping composite, and the bonding between the FRP wrapping and the structural component. By understating how the three constructs interact and how they get damaged or deteriorated, we can provide for optimum SHM strategies.

In this section, we explore the basic mechanical behavior of FRP wrappings. We then discuss the potential sources of damage when used in civil infrastructures as well as the related SHM issues. We finally try to quantify FRP-wrapping damage limit states, which is an essential ingredient for quantitative decision making processes.

TABLE 7.1
Mechanical Properties of Fiber

Fiber	Mass Density (kg/m³)	Elastic Modulus (GPa)	Tensile Strength (MPa)
E-Glass	2570–2600	69–72	3.5–3.7
Kevlar 49	1440	131	3.6–4.1
Carbon (HS)	1700–1900	160–250	1.4–4.9
Carbon (HM)	1750–2000	338–436	1.9–5.5
Carbon steel	7790	205	0.6

Source: Astrom, B., *Manufacturing of Polymer Composites*, Chapman & Hall, London, UK, 1997.
Courtesy of CRC Press.

TABLE 7.2
Mechanical Properties of Resin

Resin	Elastic Modulus (GPa)	Strength (MPa)	Cure Shrinkage (%)	Strain to Failure (%)
Polyester	3.1–4.6	50–75	5–12	1.0–6.5
Vinylester	3.1–3.3	70–81	2.1–3.5	3.0–8.0
Epoxy	2.6–3.8	60–85	1–5	1.5–8.0

Source: Astrom, B., *Manufacturing of Polymer Composites*, Chapman & Hall, London, UK, 1997. Courtesy of CRC Press.

7.2.2 BASIC MECHANICS

FRP wrappings are basically very strong and stiff tensile constructs. They can, thus, provide two modes of improving structural behavior. First, they can be used to stiffen or strengthen the tensile component of a flexural (or axial) structural member (Figure 7.7). They can also be used to enhance the confinement properties of structural members that have limited confinement (Figure 7.8). A typical use of increased flexural stiffness or strength is for reinforced concrete beams, as shown in Figure 7.9. By applying FRP wrapping at the tension side of a flexural beam, the performance of the steel rebars can be enhanced, thus optimizing the compression-rich concrete beam. They can also be used to enhance shear behavior by supplementing limited shear reinforcement near the supports of the beam, as shown in Figure 7.9. It is well known that concrete compressive strength can be increased by increasing the confinement capacity. Many existing systems have limited confinement capacity that meets modern design demands. By wrapping the FRP constructs around the reinforced-concrete columns, the confinement of the column would increase, thus increasing the compressive strength of the column, as shown in Figure 7.10.

7.2.3 SOURCES OF DAMAGE

Damage classification of FRP wrappings has been studied by several authors (Karbhari et al. 2005 and Sanyal 2007, for example). The classifications followed the steps of the FRP process from the properties of the basic components up to the operational conditions after the erection process, following Karbhari et al. (2005):

Damage of the components: such as inadequate storage of resin, manufacturing flaws of the fabric, or mishandling of the fabric.

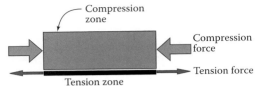

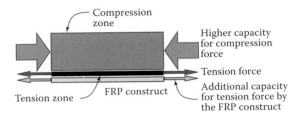

Original equilibrium diagram:
Limited flexure due to limited tension capacity, even
though there is an excess compression capacity

Equilibrium diagram with FRP construct:
Extra flexural capacity due to extra tension capacity that is supplied
by the FRP construct, thus optimizing the compression capacity of
the original structure

FIGURE 7.7   FRP wrappings as an axial tension retrofit tool.

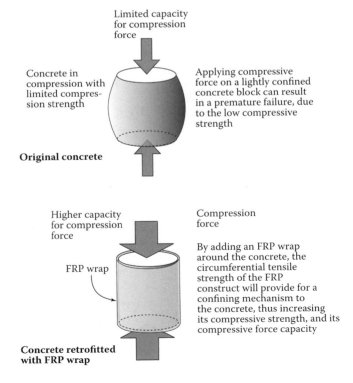

FIGURE 7.8   FRP wrappings as a confinement retrofit tool.

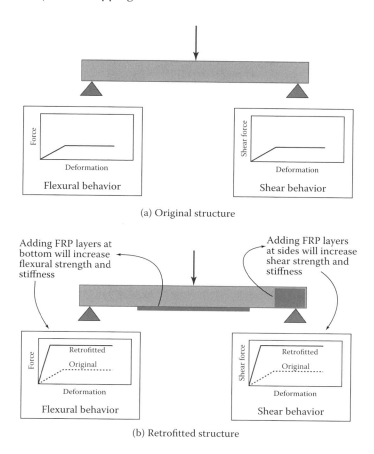

FIGURE 7.9 Use of FRP wrappings to enhance tensile behavior (a) original structure and (b) retrofitted structure.

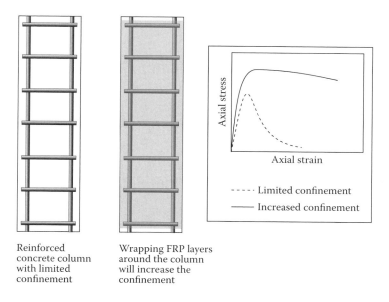

FIGURE 7.10 Use of FRP wrappings to enhance confinement.

Damage from site and material preparation: such as resin moisture absorption or inadequate primer coating. In addition, mismatch of fiber and resin, insufficient curing, knots or bumps can also cause damage.

Damage during erection process: such as handling damage, nonuniform infiltration, or air entrapment during lay-up of the plies.

Damage during operations: such as exposure to fire or aggressive environment, surface scratches or impact damage. Resulting damage can be fiber breakage, microcracking in resin, debonding, and/or delaminating.

What is obvious is that there are several sources and types of damage that can affect the FRP wrappings and degrade their intended design performance. The details of those types of damages can be found in many sources (e.g., Karbhari et al. 2005).

7.2.4 FRP-Wrapping Damage Limit States

We need now to quantify the damage properties (occurrence, size, and frequency) as related to decision-making processes (effects and mitigation). Such a relationship has not been well studied in the past. Decision makers have approached FRP wrappings damage identification (DMID)-decision making on a case-by-case basis, mostly in a qualitative manner. We try now to quantify this process or at least to lay down some quantitative principles for the process.

Fortunately, we observe that Karbhari et al. (2005) have placed FRP-wrapping damage in three basic categories according to damage size or frequency, as shown in Table 7.3

To streamline the damage frequency information, we relate an observed damage diameter, number of occurrences, and a reference area to a nondimensional damaged area, as in Table 7.4. In the following developments, we use the nondimensional damaged area as a measure of damage frequency of occurrence.

TABLE 7.3
Damage Categorization According to Size and Frequency and Suggested Repair

Damage	Size (in)	Location	Frequency	Description
Minimal	Up to ¼	Interior	Less than 5 per 0.9 m² of area	No action needed
		Next to edge	More than 5 per 0.9 m2 of area	Need repair (Epoxy injection)
Minor	1.25≤ and ≤6.00	Any	Less than 5 per 10 in × 10 in of area	Remove defect and replace by a patch of FRP wrappings
Large	>6.00	Any	Any	Indicative of serious condition; complete replacement might be needed

TABLE 7.4
Damage Frequency, Size, and Nondimensional Damaged Area

Damage Diameter (in)	Damage Area (in²)	Number of Occurrences in the Reference Area	Reference Area (ft²)	Nondimensional Damaged Area (%)
0.25	0.049	5	10	0.017
1.00	0.785	3	33	0.050
3.00	7.069	2	67	0.147
6.00	28.274	1	100	0.196

We use the categorizations of Table 7.3 as the basis for quantifying and relating FRP-wrapping damage properties to decision making techniques. We can define a damage measure, Dm_{FRP} as

$$Dm_{FRP} = f\left(S_{FRP}, L_{FRP}, F_{FRP}\right) \tag{7.1}$$

Thus, we are relating the damage measure Dm_{FRP} to the damage size, location, and frequency, S_{FRP}, L_{FRP}, and F_{FRP}, respectively. We can relate S_{FRP}, L_{FRP}, and F_{FRP} to field observations as in Tables 7.5 through 7.7.

We propose, arbitrarily, the form of the function of Equation 7.1 to be

$$Dm_{FRP} = \frac{\left(S_{FRP} + L_{FRP} + F_{FRP}\right)}{3} \tag{7.2}$$

We, thus, have a dimensionless measure of the damage in the range of 0.1 to 1.0.

We have completely linked SHM processes (including visual inspection) to decision making processes via Equation 7.2 and Tables 7.5 through 7.8. Furthermore, we propose a four-limit states of the damage as shown in Table 7.8. The steps are as follows:

1. Using SHM, NDT, or visual inspection techniques, evaluate damage size, location, and frequency of damage
2. From Tables 7.5 through 7.7 evaluate the properties S_{FRP}, L_{FRP}, and F_{FRP}
3. Evaluate Dm_{FRP} from Equation 7.2
4. Estimate damage limit states from Table 7.8
5. Continue the decision making process and choose the optimum course of action

Note that the above procedure can be improved as follows:

TABLE 7.5
Observed Damage Sizes

Observed Damage Sizes (in)	S_{FRP}
≤0.25	0.1
0.25≤ and ≤1.25	0.33
1.25 ≤ and ≤6.00	0.67
>6.00	1.0

TABLE 7.6
Observed Damage Locations

Observed Damage Location	L_{FRP}
Middle: within reasonable distance from the edges of the FRP construct	0.1
Mostly middle, few near edge of FRP wrappings	0.33
Some middle, some near edge of FRP wrappings	0.67
Mostly near or at edge of FRP wrappings	1.0

TABLE 7.7
Observed Damage Frequency

Observed Damage Frequency (one per 10,000)	F_{FRP}
1.7	0.1
5.0	0.33
14.7	0.67
19.6	1.0

TABLE 7.8
Damage Level and Limit States

Value of Dm_{FRP}	Limit State
0.1 (minimal)	No action needed
>0.1 and <0.40 (Limited)	Limited action/repair
>0.40 and <0.75 (Heavy)	Moderate action/repair
≥0.75 (Replacement)	Extensive repair, consider complete replacement

- The limit states can be chosen so that they are continuous, rather than four discrete steps.
- We assumed that the properties S_{FRP}, L_{FRP}, and F_{FRP} are deterministic in Equation 7.2. Such an assumption can be improved and the uncertainties of the damage properties included. Such uncertainties need to be considered when computing the LCA of FRP-wrapping projects. We explore this subject in Sections 7.9 and 7.10.
- The form of Equation 7.2 can be improved through more exhaustive observations and research.

However, for now, we have shown a quantitative method for accurately using SHM/NDT techniques in decision making processes that relate to damage of FRP wrappings.

7.3   NDT METHODS FOR FRP WRAPPING

7.3.1   GENERAL

This section reviews several nondestructive evaluation (NDE) techniques that have been applied to FRP wrappings in this chapter. Most of the section is based on a review by Washer and Alampalli (2005). These authors noted that a common application of FRP materials for civil structures, both steel and concrete, is for retrofitting existing structures to improve strength, repair deficient materials, and increase durability of structure. One of the most widespread applications is to retrofit existing columns to improve earthquake resistance of a structure or to shield it from adverse environment. In these applications, a composite laminate is bonded to the surface of the concrete or steel to provide additional strength and durability. This bond to the existing structure is a critical factor in the performance of the retrofits, and so it has received much attention in the development of NDE techniques. However, over a longer term, the chemical process of deterioration and stress rupture may present unexpected challenges.

Selection of appropriate NDE techniques relies in part on a full understanding of the damage to be detected and is a particular challenge for composites in civil structures due to their limited service experience. In this application, damaged concrete (if present) is removed and repaired by placing a concrete repair material to provide a suitable surface for FRP application. The FRP is then bonded to the surface of the concrete with an adhesive layer of epoxy or similar bonding material. During this repair and installation process, there is a potential for defects in the new construction on account of poor workmanship or procedures. The defects can include

- Delaminations between composite layers
- Debonded areas between the FRP overlay and the concrete surface
- Debond between repair material and original concrete
- Voids and porosity inside the FRP laminates (Figure 7.11)
- Resin-rich/resin-starved areas (Figure 7.12)

Once installed on a structure, the common deterioration mechanisms that may affect a composite in the bridge applications include the following:

- Moisture damage (Figure 7.13)
- Ultraviolet (UV) damage
- Delaminations between composite layers
- Debonding at the interface between the structure and the composite layer
- Unraveling/fibrillation of FRP material

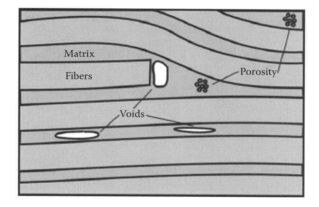

FIGURE 7.11 Voids and porosity inside composite laminates. (From Karbhari, V. et al., Methods for detecting defects in composite rehabilitated concrete structures, Submitted to Oregon Department of Transportation, and Federal Highway Administration, Report No. SPR 336, 2005. Courtesy of Dr. Karbhari.)

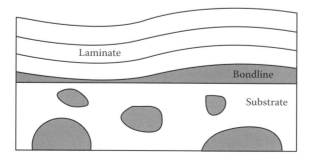

FIGURE 7.12 Moisture rich/moisture-starved regions. (From Karbhari, V. et al., Methods for detecting defects in composite rehabilitated concrete structures, Submitted to Oregon Department of Transportation, and Federal Highway Administration, Report No. SPR 336, 2005. Courtesy of Dr. Karbhari.)

- Concrete delamination beneath repair patch due to ongoing corrosion (Figure 7.14) or other concrete damage mechanisms (Figure 7.15)
- External impact of FRP laminates (Figure 7.16)

A detailed list of FRP-wrapping defects was compiled by Karbhari et al. (2005). The list is reproduced in Section 7.11.

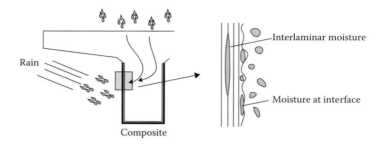

FIGURE 7.13 Moisture damage. (From Karbhari, V. et al., Methods for detecting defects in composite rehabilitated concrete structures, Submitted to Oregon Department of Transportation, and Federal Highway Administration, Report No. SPR 336, 2005. Courtesy of Dr. Karbhari.)

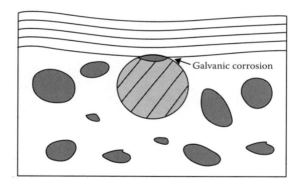

FIGURE 7.14 Corrosion damage. (From Karbhari, V. et al., Methods for detecting defects in composite rehabilitated concrete structures, Submitted to Oregon Department of Transportation, and Federal Highway Administration, Report No. SPR 336, 2005. Courtesy of Dr. Karbhari.)

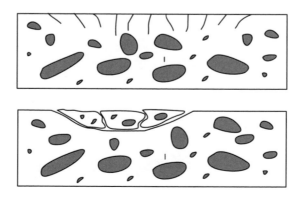

FIGURE 7.15 Concrete microcracking and spalling. (From Karbhari, V. et al., Methods for detecting defects in composite rehabilitated concrete structures, Submitted to Oregon Department of Transportation, and Federal Highway Administration, Report No. SPR 336, 2005. Courtesy of Dr. Karbhari.)

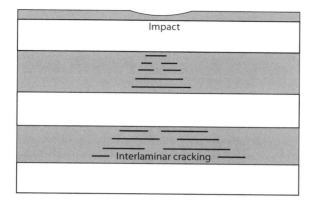

FIGURE 7.16 Impact effects on surface and subsurface lamina. (From Karbhari, V. et al., Methods for detecting defects in composite rehabilitated concrete structures, Submitted to Oregon Department of Transportation, and Federal Highway Administration, Report No. SPR 336, 2005. Courtesy of Dr. Karbhari.)

7.3.2 Main NDT Methods for FRP Wrapping

This section will summarize the main NDT techniques used in FRP wrapping. These include primarily visual inspection, sounding, infrared thermography (IRT), microwave methods, and acoustic methods.

7.3.2.1 Visual Inspection

The most common method of inspection for civil structures, especially bridges, is visual inspection. It offers fundamental data on the performance and condition of the structure and the material, provided adequate procedures are used by experienced/trained personnel. This is also true of structures retrofitted with composite materials.

Microcracking in the outer resin layers, widespread expansion of the matrix materials, or localized blistering may be evident due to water absorption. Loss of bonding between the composite materials and the concrete substrate may occur due to the differences in structural stiffness and response to thermal variation between the concrete and the FRP material and/or moisture damage to the bonded layer. This may generally occur at the corners or edges due to the sudden change in stiffness at the edge of the FRP laminate. As a result, this is a primary area for visual inspection to document for any incipient degradation. Discoloration of the materials can indicate a much more serious problem, such as UV damage or corrosion of the rebars in the concrete. The appearance of moisture-induced deterioration of glass FRP materials can be characterized by white lines appearing on the surface due to cracking between the glass fibers and the matrix (Nishizaki and Meiarashi 2002).

7.3.2.2 Sounding

A common and simple method of detecting delaminations and debonded areas is by mechanical sounding (commonly referred to as coin-tap test). This involves use of a metal or plastic object to strike the surface of the composite material and listening for a hollow tone. The low mass of a coin results in a high-pitched tone that can reveal delaminations between layers of composite and possibly between the composite and the bonded substrate. For deeper features, a larger mass should be used such that the depth of the material is excited by the tapping. For composite retrofits on civil structures, a rock hammer or other suitable impact devices may be used, though care should be taken to avoid damaging the composite material. A ¼- to ½-in steel rod, approximately 6 in long, can be used effectively for civil retrofit applications (Washer 2003). The advantage of using this type of device is that it is readily available, in that it can be formed from a piece of rebar and can provide both high-mass and low-mass impact, depending on orientation of the rod when the impact is made.

7.3.2.3 Thermography

The application of IRT for quality assurance/quality control (QA/QC) inspections of civil structures has been increasing rapidly in recent years thanks to the ease and speed of the inspection process. A record of the inspection provided by the thermographic images also supplies data for future comparisons and document results. The most frequent application is for detecting and quantifying delaminations and debonds immediately after construction. The basic principle behind this method is that the flow of heat through the material being inspected is disrupted by the presence of a defect in the material and manifests in an observable change in surface temperature. While this seems a relatively simple approach, the accuracy is influenced by several factors: dirty, stained, or wet surfaces, coatings, or other materials on the surface, and environmental conditions during the test (ambient temperature, shade, sunshine, wind, etc.). All these have significant effects on the thermal image and may mask subsurface defects. So, inspectors should be well aware of the effects of these conditions and be able to compensate for them.

Both passive and active methods have been used for detecting defects in FRP repairs where FRP is bonded to the surface of a concrete structure (Johnson et al. 1999; Hawkins et al. 2001; Starnes et al. 2003; Washer 2003). For passive thermography, the required temperature gradient is supplied by the ambient environment due to diurnal temperature variations and the difference in the thermal conductivity between concrete and composite materials. A large debonded area of FRP wrap surrounding a concrete column is shown in Figure 7.17, which indicates the importance of QC during the construction of FRP wraps. It also illustrates the usefulness of thermal imaging as a rapid scanning method, which requires minimal interaction with the structure being inspected.

Active thermography utilizes the application of external heat to the structure using heat sources such as heating blankets, heat guns (Mtenga et al. 2001), quartz heaters, and quartz halogen bulbs (Hawkins et al. 2001).

Brown and Hamilton (2004) used both pulse IRT and step-heating IRT in the lab to study the use of IRT for detecting bond between FRP materials and concrete surface.

Pulse IRT involves application of a short burst of high-intensity heat onto the surface to be inspected. Commonly a flash type of source is used and provides a pulse duration of about 15 ms. This method is commonly used in materials with high thermal conductivities containing defects near surface. The required heating and observation times are short, resulting in quick inspection of large areas. But, the method requires high data acquisition rates, making it expensive. At the same time, for the same reasons, deeper defects may not be detected.

Step-heating IRT involves lower energy but longer duration heat sources such as halogen lamps and infrared (IR) heating bulbs. The specimen is monitored during and after heating. The method is

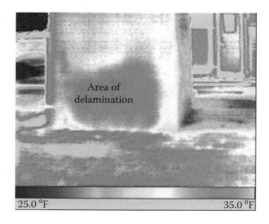

FIGURE 7.17 Thermographic image of a column wrapped with FRP laminate. (Courtesy of New York State Department of Transportation.)

often used in materials with lower thermal diffusivities and deeper defects. But it is often difficult to apply uniform heat to the surface.

Lab experiments showed that FRP thickness and material types significantly affect IRT results. For single-layer carbon fiber reinforced polymers (CFRP), pulse IRT provided sufficient heat to detect air voids at FRP/concrete interface with less than 10 s data acquisition time. For thicker systems, this was not useful. Step heating allowed for inspection of thicker composites, but required heating time and time to maximum signal increases significantly.

Using an array of 250-W IR heat lamps, CFRP-retrofitted exterior girders of a bridge were inspected. Due to working conditions and the time available, the lamps were swept over the surface, and data was recorded as the surface cooled. Thus only qualitative results were obtained. The results indicated the areas of debond, which were verified by the coin-tap method.

7.3.2.4 Microwave Methods

Microwave technologies have also been applied to detect subsurface defects in composite repairs (Hughes et al. 2001). One approach is to use an open-ended wave guide to transmit an electromagnetic (EM) wave that interacts with a material positioned in front of the wave guide (Akuthota et al. 2004). Reflected waves from the material are detected by the probe and analyzed to determine various parameters related to the electronic properties of the materials. Subsurface defects such as delaminations can be detected through changes in these electronic properties.

For a layered material such as a composite, the boundaries between layers reflect and transmit microwaves in proportion to the contrast of the dielectric properties of the materials that form the boundary (Ganchev et al. 1995). As such, measurement of reflected wave properties can be used to infer information about materials at the boundaries. When an FRP composite has a delamination or is debonded from the concrete substrate, a boundary is created between the composite materials and air that provides a reflection due to the contrast in dielectric properties (Feng, De Flaviis, and Kim 2002).

Microwave imaging technology is one of the upcoming methods under investigation for NDE of concrete structures and FRP-retrofitted structures. This method can detect internal damage and debonds even before they become visible to the naked eye. Feng and Kim (2004) showed that focused systems are better, and they developed passive (using dielectric lenses) and active (using antenna arras) systems. They describe an active system (see Figure 7.18). The EM properties (dielectric

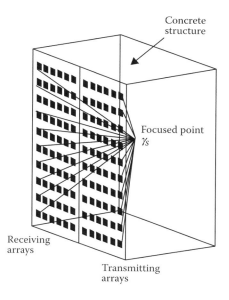

FIGURE 7.18 Use of antenna arrays to focus waves on a subsurface point. (Reprinted from ASNT Publication.)

constant, conductivity, and loss tangent) are essential for modeling and designing microwave-based NDE, and so the authors developed a database of EM properties of concrete and FRP materials with and without water presence.

The active system used transmitting and receiving antenna arrays. Both arrays are focused, using software rather than lenses, so that quick sweep can be achieved, thus making the test fast. Thus, a three-dimensional image can be generated with the structure illuminated by microwaves. The authors also developed a unique bifocusing system to focus both receiving and transmitting systems. Multifrequency technique was proposed instead of single-frequency microwaves to enhance the quality of the image. Both hardware and software systems were developed for such an imaging system and were verified using both numerical and experimental methods.

7.3.2.5 Acoustic Methods

Application of acoustic wave–based NDE methods is widespread in many engineering disciplines. For civil structures, a common application is ultrasonic testing for QC/QA during the fabrication of metal structures. The method generally consists of launching a high-frequency acoustic wave from a transducer and detecting reflections due to internal discontinuities such as cracks, voids, and inclusions. These reflections can be analyzed to estimate size, location, and shape of the discontinuities. This is the most fundamental application of ultrasonic testing, but there are many methods of monitoring the behavior of waves propagating in materials and have been applied for NDE of FRP materials (Dokun, Jacobs, and Haj-Ali 2000; Luangviai, Punurai, and Jacobs 2002; Seifried, Laurence, and Qu 2002, Bastianini et al.). Reflection-based ultrasonic techniques can be used to evaluate the quality of bond between the FRP overlay and the concrete surface. The method consists of launching a high-frequency ultrasonic wave into the surface of the FRP laminate. Reflections occur at the interface of the composite layer and the substrate due to the acoustic impedance mismatch between the two materials.

An acoustoultrasonic approach to evaluate bond quality for composite overlays has also been demonstrated (Godinez-Azcuaga et al. 2004). An FRP composite overlay and a concrete substrate were modeled as a layered media. Energy vectors of a propagating acoustic wave within the layers were predicted by a theoretical model that considers the frequency dependence of reflection coefficients. In this case, the reflection coefficients express the capacity of an acoustic energy to propagate within the layers, not simply between interfaces of material. As such, the propagating wavelength becomes a factor, that is, there is frequency dependence. The authors selected a frequency regime for an ultrasonic pulse such that a debonded area of the FRP overlay (wrap) would produce a reduced amplitude signal as a wave propagated from a transmitter to a receiver.

Godinez-Azcuaga et al. (2004) developed a handheld, acoustic wave–based, field-portable, battery-operated inspection system for FRP-wrapped concrete structures. They showed that the methodology was based on theory and tested it in the lab before embarking on field studies. The only issue is that the method may be hard to use if the surface area of the retrofitted section is too big as the sensor has to be in contact with the surface it is testing. This may be good for QA of the retrofits before accepting the repair work.

The theoretical model simulated propagation of acoustic waves in a concrete structure reinforced with a +/– graphite/epoxy wrap with and without debonds between the concrete and the composite wrap. The results gave an idea of frequencies where the difference between reflection coefficients of the bonded and debonded FRP/concrete sample is large enough to allow for detection of the debond itself (Figure 7.19).

The difference between the reflection coefficients (contrast index, CI) in Figure 7.20 gives an indication of how much amplitude of the reflected acoustic signal will change from one area of the structure to an area where debonds are present. The larger the absolute value of CI, the better the contrast that debond will be present on a C-scan image against the undamaged area background.

A feasibility study, performed using a differential rolling sensor (Figure 7.21) with a 10-cycle 250-kHz square wave ton burst with constant amplitude signal as excitation, showed that the method can successfully detect debonds, delaminations, and cracks.

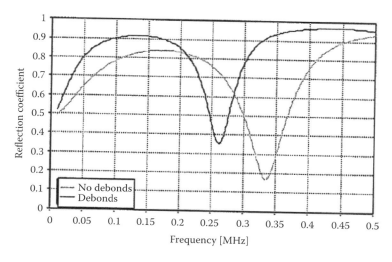

FIGURE 7.19 Frequency shift due to debonding of FRP system. (Reprinted from ASNT Publication.)

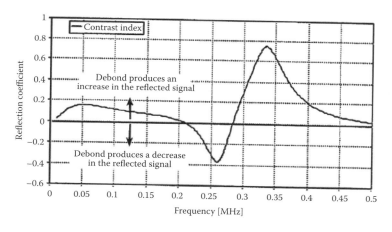

FIGURE 7.20 Contrast index (CI) as a function of frequency. (Reprinted from ASNT Publication.)

A prototype system was developed and tested on a sample with both carbon/epoxy and glass/epoxy sections. The results were successful in detecting debonds simulated by inserting Teflon wafers between the concrete and the FRP, as well as delaminations in FRP. The prototype was then used in a field study to evaluate the FRP placed at critical points on a scaled model fire station subjected to simulated seismic loads. The results from C-scan images showed its capacity to detect defects (see Figures 7.22 and 7.23).

7.3.2.6 Ground Penetrating Radar and Infrared Thermography

The feasibility of ground-penetrating radar (GPR) and IRT was tried by Jackson, Islam, and Alampalli (2000) to assess the performance of some FRP-wrapped reinforced concrete columns on a bridge structure in Owego, NY. It was concluded that both methods can be powerful tools for detecting and assessing various types of deterioration in FRP-wrapped concrete columns.

The GPR technique can be a useful tool for tracking progressive deterioration of the concrete within the FRP-wrapped columns, particularly delaminations, provided a planned monitoring scheme is followed. It can easily detect concrete defects or deterioration virtually at any depth by selection of appropriate frequency antennas (higher frequencies for shallower depths and vice

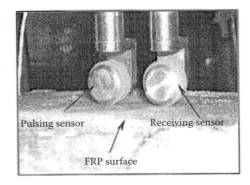

FIGURE 7.21   Rolling sensor during inspection. (Reprinted from ASNT Publication.)

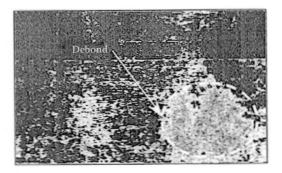

FIGURE 7.22   C-scan image showing debond between concrete and FRP wrapping. (Reprinted from ASNT Publication.)

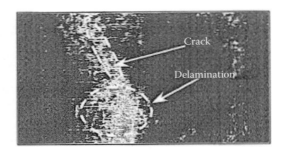

FIGURE 7.23   C-scan image showing delamination between concrete and FRP wrapping, and concrete cracking. (Reprinted from ASNT Publication.)

versa). Figure 7.24 shows the use of the handheld GPR antenna. Figures 7.25 through 7.27 show samples of the results obtained during the experiment.

The IR technique is very effective in detecting disbondment, blisters, and shallow defects (delaminations) in such components. Entrapped moisture between the wrap and the concrete can also be detected. However, defects located deep within the concrete may not be reliably detected. Figure 7.28 shows the visible light photograph of a damaged area, while Figure 7.29 shows the IR image of the same damaged area. Similarly, Figure 7.30 shows a visible light area that contains blisters, disbondment, and delamination, while Figure 7.31 shows the IR view of the same area. The intensity of color differentiates the type of damage (delamination would show in green to red tones and disbondment and blistering in blue tones). Table 7.9 shows quantified IR data from the experiment.

FIGURE 7.24   GRP survey with handheld antenna. (Courtesy of CRC Press.)

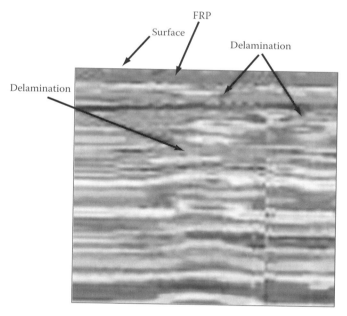

FIGURE 7.25   GPR image obtained with 2-G-Hz antenna. (Courtesy of CRC Press.)

7.3.3   OTHER NDT METHODS

In addition to the main NDT methods described above, there are several other NDT methods that are applicable to monitoring FRP wrappings with varying degrees of efficiency. Karabhari et al. (2005) presented an exhaustive survey of pertinent NDT methods in this field. Table 7.10 shows the observations of that study. Other than NDT methods, global behavior of bridges that was retrofitted with FRP wrapping is usually load tested (see, for example, Hag-Elsafi et al. 2000).

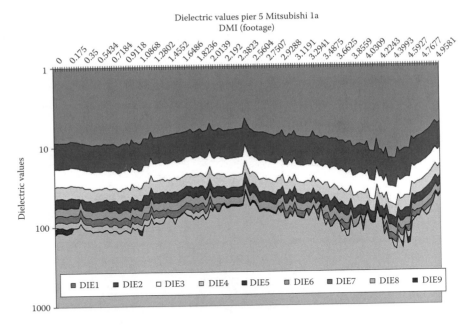

FIGURE 7.26 Plot of dielectric layers as a function of distance along a 12-in strip of an FRP-wrapped column. (Courtesy of CRC Press.)

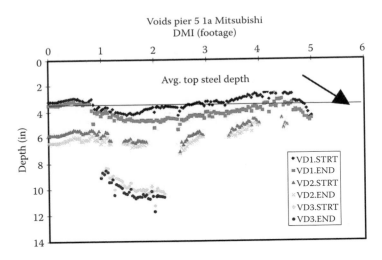

FIGURE 7.27 Location of delamination/voids within an FRP-wrapped column. (Courtesy of CRC Press.)

In another comparative study, the capacity for detecting delaminations between FRP-wrapping retrofits and a damaged AASHTO girder were investigated by Moore et al. (2005). They used three methods for detecting delaminations. The acoustic sounding (AS), the IRT (IR), and pull-off testing methods were used and compared. These methods are described in Section 7.8. Moore et al. (2005) compared the advantages and disadvantages of each of the methods in their capacity to detect delaminations. The summary of their research is shown in Table 7.11.

Although fiber optic sensing (FOS) is not an NDT method, the sensors can be used in a variety of ways to detect damage to FRP wrappings on a short-term or long-term basis. For example,

FIGURE 7.28 Photograph of an area—circled in white—on one of the FRP-wrapped columns that had disbondment and voids. (Courtesy of CRC Press.)

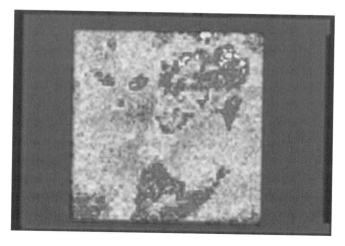

FIGURE 7.29 IR image of same area—dark areas indicate voids. (Courtesy of CRC Press.)

Bonfiglioli et al. (2004) and Zhou (2004) measured global system response of an FRP-wrapped system, using strain measurements that use Bragg grating FOS. Gheorghiu et al. (2004) used Fabry-Perot FOS to monitor fatigue effects in an FRP-wrapped system. For a detailed discussion of the merits of FOS, see Ettouney and Alampalli (2012, Chapter 5).

7.4 APPLICATIONS

7.4.1 STRENGTHENING

FRP material has been used successfully for rehabilitating and maintaining infrastructures. Projects for strengthening and stiffening have shown to be very successful and efficient over the past several years. The projects were applied to different structural components, such as bridge capbeams, beams, box beams, columns, arch bridges, and road sign structures. In addition, FRP applications were used to retrofit whole deteriorated structures such as culverts. They were also used in an innovative fashion to reduce wind pressure on long-span bridges. Details of some of those applications are mentioned below.

FIGURE 7.30   Area with blisters, disbondment, and delaminations on one of the FRP-wrapped columns. (Courtesy of CRC Press.)

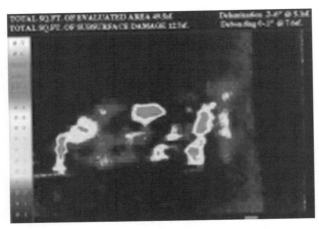

FIGURE 7.31   IR photograph of the area marked by a square in Figure 7.30. (Courtesy of CRC Press.)

TABLE 7.9
Quantified IR Data

Item Quantified	Area (sq ft)
Total area evaluated	4.5
Delamination (0.2- to 0.6-in depth)	5.1
Disbonding/blistering (0- to 0.1-in depth)	7.6
Total subsurface damage	12.7

Source: Courtesy of CRC Press.

7.4.2   MAINTENANCE

FRP products have been used also to contain deterioration of structures. For example, concrete delamination damage has been retrofitted by FRP laminates. Such projects were mostly

TABLE 7.10
Rankings of NDT Methods as Applied to Monitoring of FRP Wrapping

Ranking	NDT Method	Comments
Primary methods	Thermography and shearography	Both methods are proposed for near- and full-field monitoring; however, one may yield higher applicability over the other, which must be confirmed through future experimental investigation
	Ultrasonics	Ultrasonics must be considered applicable for near-field inspection exclusively. Its extensive background and range of instrumentation make it specially favorable
Conditional methods	Radiographic testing (RT) and modal analysis	Applicability of these methods is largely dependent on site conditions and details of inspection. For highly detailed inspection in controlled environments (i.e., access to adequate power supply, no concern for safety) RT remains a favorable technique. In contrast, for changes in structural response through material degradation, modal analysis provides the most conclusive results
Initial qualitative methods	Visual testing and acoustic impact testing	These methods are proposed as supplementary mostly due to their general ease of use and extremely high flexibility. Moreover, they can be applied in a simplistic form and do not necessitate extensive background knowledge of theory and/or application of the technique. The reader should be reminded, however, that neither method is of such transparency or level of sophistication that it can be considered solely for comprehensive *in situ* inspection
Excluded methods	Penetrant testing; Eddy current testing; acoustic emission; strain measurement techniques (fiber optics); ground-penetrating radar; rapid load testing	Low detectability (methods 1, 2, 4, 5, and 6); presumed inadequate for defect localization (methods 3 and 6); presumed inadequate for defect sizing (methods 3, 4, 5, and 6)

Source: Karbhari, V. et al., Methods for detecting defects in composite rehabilitated concrete structures, Submitted to Oregon Department of Transportation, and Federal Highway Administration, Report No. SPR 336, 2005. Courtesy of Dr. Karbhari.

successful because of

- Cost-effectiveness
- Quick installation of retrofits (Figures 7.32 and 7.33).

Of course, using FRP material for structures requires the following:

- Durability: The retrofitted construct needs to be at least as durable as other competing retrofit solutions.
- Number of layers: The number of FRP layers needs to be studied. There is a balance point between cost- and solution-effectiveness.
- Concrete removal strategies: When removing damaged delaminated concrete, an appropriate, safe, and cost-effective strategy is needed.
- Workmanship: Since many of these rehabilitation projects are labor intensive, adequate QA/QC rules need to be enforced.
- Performance of a variety of materials: There are different FRP material types, from different manufacturers, which can be used for these types of projects. When choosing

TABLE 7.11
Attributes of Different FRP-Wrapping Testing Methods

Topic	Comment
Precision	Both IR and AS appear equivalent for larger defects, while AS appears more precise for smaller defects
Accuracy	Both methods (AS and IR) are capable of locating delaminations. However, pull-off testing should be performed to confirm the location of the delaminated areas
Equipment cost	There is little or no cost for the AS equipment, while the rental for the IR equipment is about \$5,000* per month. This cost may include the manufacturer's reporting software
Inspection speed	IR was approximately 1.3 times faster than AS. The speed increase is likely to be more relevant on larger testing areas
AS advantages	Little set-up time, low-cost equipment cost, limited training required, easy to delineate between delaminations and the surrounding area, can be performed with limited access
AS disadvantages	Difficult on noisy construction sites, so off-hour work may be required, no permanent record without field notes, tedious and time consuming when the grid spacing is small, transcription to field sheets or drawings can result in relatively large errors
IR advantages	Relatively fast and easy to perform especially for large areas, provides permanent records
IR disadvantages	High initial cost, affected by surface irregularities, affected by nonuniform heating, apparently affected by number of FRP layers, operator skill is extremely important, requires camera to be approximately 1.8 to 3 m (6–10 ft) from the specimen for imaging, must maintain software to effectively present data, transcription to field sheets or drawings can result in relatively large errors
Pull-off testing	Effective means to verify delamination locations, attachment of steel anchor to test area difficult on vertical and overhead surfaces as well as surfaces that have been coated

*Cost was reported in 2005.
Source: Moore, M. et al., Case studies in the evaluation of FRP bond to concrete, *Proceedings of the 2005 ASNT Fall Conference*, Columbus, OH, 2005. Reprinted from ASNT Publication.

a particular type of FRP material, its adequacy for the project and its cost need to be considered.
- Inspection and monitoring: Frequent inspection and monitoring are needed to ensure adequate long-term performance.

7.4.3 SEISMIC APPLICATIONS

In addition to column-wrapping seismic retrofit applications, there are several other ways FRP material can be used to enhance seismic behavior of bridges. Of these, we present two applications: an innovative way to increase stiffness and damping of systems during seismic events and the application of the wrapping techniques in structural components other than columns.

7.4.3.1 University at Buffalo In-Fill Panels

FRP-based in-fill panels were used to reduce seismic damage to bridge piers and framed buildings. The concept was devised by Aref and Jung (2003) at the University at Buffalo (UB). A typical in-fill panel contains three main parts: an FRP core, an outside FRP shell, and a viscoelastic honeycomb-style layer that separates the two (see Figure 7.34). The assembly of the three components

FIGURE 7.32 Laying FRP layers. (Courtesy of New York State Department of Transportation.)

FIGURE 7.33 Finishing FRP wraps around bridge column. (Courtesy of New York State Department of Transportation.)

constitutes the UB in-fill panels. The panels provide the following performance enhancements during earthquakes:

- Added in-plane stiffness to the base structure; bridge piers, for example
- Added energy dissipating, damping performance

The additions of stiffness and damping to the overall system were shown to reduce the seismic drift of the structural frames; thus reducing seismic damage (see Figure 7.35). The additional advantages of this system are low added mass (easy to transport and low inertial seismic forces), ease of field construction (better QA/QC and lesser labor costs), and scalability (can use multiple panels for large size/massive structures). Figure 7.36 shows a prototype of the UB in-fill seismic panel.

7.4.3.2 Woodlawn Viaduct, NY

Arch bridges offer longer spans, with pleasing esthetics. Because of this, they have been the favorites of bridge builders over the years. One of the reasons for their structurally efficient performance

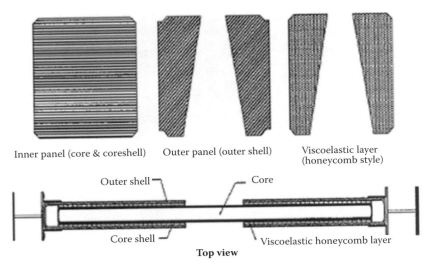

Inner panel (core & coreshell)   Outer panel (outer shell)   Viscoelastic layer
                                                                (honeycomb style)

Top view

FIGURE 7.34   Components of UB in-fill seismic panel. (Courtesy of CRC Press.)

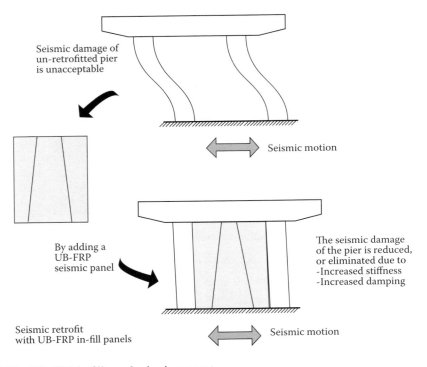

FIGURE 7.35   UB-FRP in-fill panel seismic concept.

is that because of the built-in curvature of the arched geometry, the axial resultant of the total equivalent cross-sectional stress is as pronounced as the flexural resultant of the stresses (the bending moment). This is in contrast to the straight structural systems, such as conventional girders, where axial resultant is minimal or nonexistent. Since axial resultants in a cross section are much more efficient than flexural resultants (both in terms of lower stresses and lower strains), it is logical that systems that can utilize axial resultants more are efficient than mainly flexure systems.

Since most axial forces in arch bridges are compression, the use of wrapping laminates for strengthening, stiffening, or rehabilitating is logical since it provides added confinement to the parent reinforced concrete arch. As discussed earlier, additional confinement would result in increased strength, stiffness, and ductility to the mostly compression structural element.

Several arch bridges were retrofitted using FRP laminates. For example, the capacity of the Woodlawn viaduct bridge, located in Westchester County, NY, which is 75 years old, was not adequate for the increased traffic demands. In addition, the bridge was not designed to meet modern seismic codes. To strengthen the main arches in the bridge, FRP laminates were wrapped around the arches to provide additional confinement, which increased the strength and ductility of the arches to the desired design level. An additional advantage of the FRP-wrapping solution is that it kept intact the original historical appearance of the arch bridge (Figure 7.37).

FIGURE 7.36 Finished UB in-fill panel. (Courtesy of Dr. Amjad Aref.)

FIGURE 7.37 Woodlawn viaduct after strengthening with FRP wraps. (Courtesy of New York State Department of Transportation.)

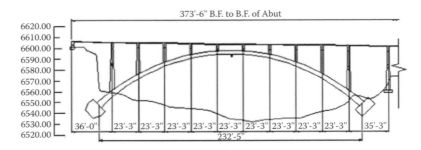

FIGURE 7.38 Elevation view of historic Castlewood Canyon Bridge, CO. (Courtesy of Colorado Department of Transportation: Fafach et al., 2005.)

FIGURE 7.39 Damage in arch before rehabilitation. (Courtesy of Colorado Department of Transportation: Fafach et al., 2005.)

Another arch bridge was retrofitted by FRP laminates in Castlewood Canyon State Park in Colorado. The profile of the historic bridge is shown in Figure 7.38. The bridge was in a bad state of deterioration (Figure 7.39). It was decided to strengthen and retrofit the bridge using FRP laminates. A full description of the project was offered by Fafach et al. (2005). The project steps, as the authors described it, were as follows:

1. Loose materials and debris were removed from the surfaces and from around the reinforcing.
2. Exposed reinforcing bars were sand-blasted clean from rust (Figure 7.40).
3. Penetrating corrosion inhibitor was applied to the surface of concrete arches and struts to address hidden damage (Figure 7.41).
4. Leadline CFRP rods manufactured by Mitsubishi were anchored into the footings with epoxy around the bases of the arches to strengthen the arch-foundation connection.
5. The surfaces of the arches were finished to the original surfaces using hand- and machine-applied mortar. Cracks were sealed using epoxy injection.
6. New pedestals were constructed for new spandrel columns.
7. Fiberwrap, an externally bonded CFRP, was applied to arch ribs to confine concrete and to reinforce and strengthen the arch (Figure 7.42).
8. The arches were then painted to appear like concrete (Figure 7.43).

FIGURE 7.40 Exposed steel during construction. (Courtesy of Colorado Department of Transportation: Fafach et al., 2005.)

FIGURE 7.41 Arches during construction. (Courtesy of Colorado Department of Transportation: Fafach et al., 2005.)

7.4.4 AERODYNAMIC FLOW

When the Bronx-Whitestone Bridge in New York City was built in 1939, the deck was supported by flexible steel girders, which in turn were supported by the suspension cables of the bridge (3770-ft total anchor-to-anchor length with a 2300-ft main span). The tragic failure of the similarly designed Tacoma Narrows Bridge (see Ettouney and Alampalli 2012, Chapter 3) led to a decision to stiffen the bridge with a set of steel-stiffening trusses to ensure better performance during high winds. In 2002, it was decided to remove the stiffening trusses to ensure better overall bridge performance. To accommodate the demands of cross-winds on the bridge in New York City after the truss removal, composite wind fairings were added to the bridge sides. Figure 7.44 shows the newly painted fairings that are ready to be transported to the bridge site. Figures 7.45 through 7.47 show the process of installing the wind fairings on the side of the bridge. The composite wind fairings, designed to streamline wind flow around the bridge cross section (Figure 7.48) weigh more than 890 kips; this

FIGURE 7.42 Applying FRP-wrapping laminates. (Courtesy of Colorado Department of Transportation: Fafach et al., 2005.)

FIGURE 7.43 Bridge after rehabilitation. (Courtesy of Colorado Department of Transportation: Fafach et al., 2005.)

is perhaps the largest use of structural composite in the world, as of the completion of its construction in 2004. In addition to improving the wind pressure demands on the bridge, the FRP construct offers the advantage of light weight over other potential wind fairing materials.

7.4.5 Overhead Sign Structures

Many state transportation agencies have used aluminum trichord overhead sign structures on highways for more than 40 years due to their light weight and anticorrosive properties (see Figures 7.49 and 7.50). But due to fatigue issues that were not well understood 40 years ago, coupled with increasing structural spans and loads, in recent years, these structures have been experiencing cracks in the welded joints of truss diagonals (see Figures 7.51 through 7.54). Failure of these structures can lead to unsafe conditions, including fatalities, and can also cause significant disruptions to traffic. Another issue was that when inspections discover weld problems, it is not easy to do field welding of aluminum when compared to steel. The welding procedure is not easy and is often very expensive. As of the writing of this chapter, an overhead sign structure replacement can cost from $50,000 for small structures to more than $300,000 for a large structure carrying variable message signs depending on location. Recently, it was estimated that about 10% of these trusses had some kind of damage. Due to these issues, some states have stopped using aluminum structures for these applications, but thousands of these structures still remain over the highways and need cost-effective repair

FIGURE 7.44 Newly painted panels before shipping to Bronx-Whitestone bridge site. (Courtesy Weidlinger Associates, New York City, NY.)

FIGURE 7.45 Installation of wind fairings at the side of the Bronx-Whitestone Bridge. (Courtesy Weidlinger Associates, New York City, NY.)

procedures when cracking at joints are noticed until they can be replaced. Figures 7.55 through 7.57 show the steps of FRP retrofitting of damaged joints.

About 10 years ago, the NYSDOT worked with private industry to develop a relatively inexpensive repair method using FRP warp for structures with cracked joints as a temporary fix until they

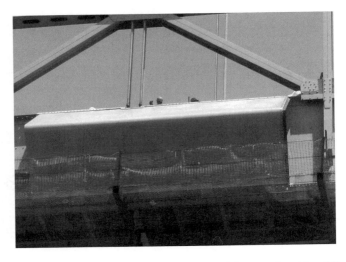

FIGURE 7.46 Lateral view of the installation of the wind fairings at the side of the Bronx-Whitestone Bridge. (Courtesy Weidlinger Associates, New York City, NY.)

FIGURE 7.47 Longitudinal view of the installation of the wind fairings at the side of the Bronx-Whitestone Bridge. (Courtesy Weidlinger Associates, New York City, NY.)

can be replaced in a couple of years. This procedure offers several advantages: (1) It is relatively inexpensive. It was estimated that the repair costs about $3,000 per cracked joint; (2) It requires less time, that is, it takes on an average about 3 hours per joint; (3) It requires only three workers to effectively conduct the repairs; (4) Cracked joint can be restored to the same strength as the original

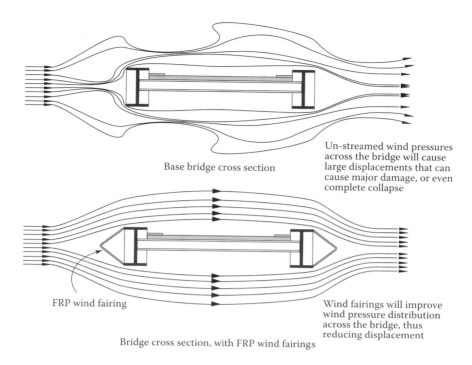

Base bridge cross section

Un-streamed wind pressures across the bridge will cause large displacements that can cause major damage, or even complete collapse

FRP wind fairing

Wind fairings will improve wind pressure distribution across the bridge, thus reducing displacement

Bridge cross section, with FRP wind fairings

FIGURE 7.48 Concept of wind fairings.

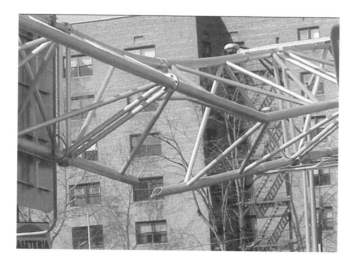

FIGURE 7.49 Damaged road sign structure. (Courtesy of New York State Department of Transportation.)

aluminum weld; (5) Repair can be conducted in the field; and (6) Although the repair is temporary, it gives enough time to schedule the permanent repair or replacement of the structure, thus avoiding emergency replacement at higher costs (see Mooney 2003 for more details).

But on the basis of in-service experience and collaborative research with the University of Utah researchers, NYSDOT now considers these repairs permanent while inspecting them periodically. One issue engineers should be careful with in this solution is that, in case of permanent

FIGURE 7.50   Damaged road sign structure. (Courtesy of New York State Department of Transportation.)

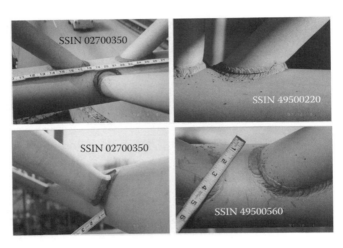

FIGURE 7.51   Details of damaged connections of a road sign structure. (Courtesy of New York State Department of Transportation.)

repairs, good inspection guidelines and periodic inspection of the structures by an experienced engineer are very important, as FRP wrap will cover the cracked joints making them invisible to the naked eye.

7.4.6   Culvert Relining

A large-diameter culvert relining using a composite liner was done by the Erie County Department of Public Works, Division of Highways, in western New York (see Erie County 2002). Bridge inspection revealed severe rust in an existing corrugated 14 by 5 ft, 9-in metal arch culvert built in 1946 under a five-lane collector street (see Figure 7.58). The loss of structural integrity and strength due to the deterioration required that the liner take significant load. Existence of buried utilities in the proximity of the culvert required full excavation for culvert replacement, which was very difficult and expensive since every day 38,000 vehicles used the road carried by the culvert. So, GFRP liners were considered on account of their ease of construction and cost-effectiveness. The GFRP liners can

FIGURE 7.52 Inspection of a damaged road sign structure. (Courtesy of New York State Department of Transportation.)

FIGURE 7.53 Deteriorated base of a road sign structure. (Courtesy of New York State Department of Transportation.)

also carry significant load and improve hydraulic capacity. Thirteen liner sections—each section was made of two parts (an upper and lower half) with tongue-in-groove joints—were manufactured using the hand lay-up method (see Figure 7.59). Once the liners were installed in the host pipe and held in place (Figure 7.60), grout was pumped into the annular gap (Figure 7.61). The construction was done in 2002 with an estimated cost of about $300,000. For more details, see Black (2005).

7.4.7 SHM and Different FRP Applications

This section presents some novel applications of FRP material in the field of civil infrastructures. These applications are subjected to different hazards during their service life. Methods of monitoring the performance of some of these hazards are shown in Table 7.12.

FIGURE 7.54   Details of a cracked welds in a road sign structure. (Courtesy of New York State Department of Transportation.)

FIGURE 7.55   Preparations for FRP retrofits. (Courtesy of New York State Department of Transportation.)

7.5   EAST CHURCH STREET BRIDGE

This FRP applications (Hag-Elsafi et al. 2002) deals with repair of a concrete beam, with flexural and shear cracking due to the addition of a concrete wearing surface and a median barrier, leading to an increase in dead load moment and shear. The bridge was load tested before and after the repairs to investigate the effectiveness of the strengthening system. Effect of the retrofit on structural behavior under ultimate load conditions was studied using a nonlinear finite element program.

The Church Street Bridge, built in 1954, is a two-lane, two-way, 67-m-long, multi-steel-girder, four-span structure. It carries State Route 352 (AADT 9000) in the City of Elmira, NY, over State Route 17 (AADT 20000). It was originally built with a 29-cm-thick composite-concrete deck and no median barrier. During routine inspection of the bridge in 1997, excessive cracks were observed at the capbeams of piers 2 and 3 (see Figures 7.62 and 7.63). The cracking was attributed to an increase in dead load.

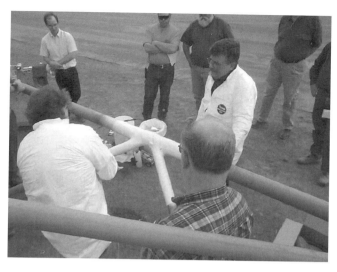

FIGURE 7.56   Applying FRP retrofit. (Courtesy of New York State Department of Transportation.)

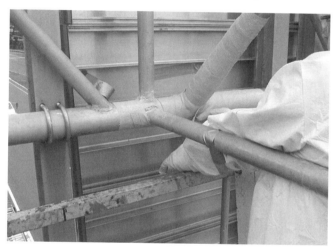

FIGURE 7.57   Finished FRP retrofit. (Courtesy of New York State Department of Transportation.)

Preliminary analysis confirmed the capbeams' deficiency in both moment and shear capacities under current service loads. Two sets of FRP materials were considered: bonded FRP laminates for pier 2 capbeam and bonded FRP plates for pier 3 capbeam. Figure 7.64 illustrates the pier 3 repairs. The FRP repair was designed (Figures 7.64 and 7.65) using carbon glass hybrid materials to increase the capbeam shear and moment capacities by 10% and 20%, respectively. Figure 7.66 shows the process of manually installing the FRP laminates.

Field testing investigated the effectiveness of the strengthening system (Figure 7.67) in reducing current live-load stresses in the beam. Four trucks, each closely resembling AASHTO M18 truck configuration with an average weight of about 196 kN, were used in the load tests. The maximum moments and shear forces developed in the capbeam under maximum loading conditions were close to those produced by an MS18 AASHTO truck. A nonlinear finite element analysis supplemented the load tests. Results indicated that the strengthening system had little effect in reducing service load stresses in the steel rebars in the positive moment region, that is, the FRP plates system moderately reduced live-load stresses in this application. However, the analysis indicated that a 45%

FIGURE 7.58   Longitudinal view of the culvert. (Courtesy of Erie County DPW of New York State.)

FIGURE 7.59   FRP culvert liners before installation. (Courtesy of Erie County DPW of New York State.)

FIGURE 7.60   Installation of FRP culvert liners. (Courtesy of Erie County DPW of New York State.)

FIGURE 7.61 FRP culvert liners in place. (Courtesy of Erie County DPW of New York State.)

TABLE 7.12
Some Potential Structural Health Issues of Different FRP Applications and THEIR SHM/ NDT Solutions

Application	Hazard/Damage	SHM/NDT EFFORT
Rehabilitation of reinforced concrete bridges	See Sections 7.2 and 7.11. For more in-depth treatment, see Karbhari et al. (2005)	See Section 7.3. For more in-depth treatment, see Karbhari et al. (2005)
Wind fairings	Fatigue	FRP has been used to streamline wind flow around suspension bridges (Figures 7.44 through 7.48). Monitoring fatigue performance can be established through (1) recording strain cycles and estimating remaining life (Chapter 5), and (2) using NDT methods to look for defects (thermography, acoustic emission, ultrasound, etc.)
	Environmental degradations	Applicability of different SHM techniques to FRP environmental degradation hazards varies greatly. Karbhari et al. (2005) established detailed tables that discuss these applicability relationships
Traffic sign FRP retrofits	Erection QA/QC	Because of the difficulties in monitoring/inspecting this type of retrofits after installation, creating an erection QA/QC system is recommended. Different NDT methods can be used in such a system in an efficient and cost-effective manner
	Efficiency of concave behavior	Using FRP wrapping to retrofit concave (Figure 7.51) geometries is a challenging effort. Recall that optimal behavior of FRP wrappings is when the material undergoes tension. When the retrofitted surface is concave, such tensile behavior can be achieved only if the surface is doubly curved and the other curved direction is convex. The state of stresses in this double-curved geometry is complex, and delamination can easily result if the state of three-dimensional stresses in the FRP material is not optimal. Monitoring such a complex state of stresses (or strains) can be achieved using conventional NDT methods such as thermography or ultrasound
	Long-term durability	Applicability of different SHM techniques to FRP long-term durability varies greatly. Karbhari et al. (2005) established detailed tables that discuss these applicability relationships
FRP culverts	Efficiency of grout as load- distributing mechanism	Response of the culverts can illustrate the load demands on the FRP culverts. Such responses can be monitored via strain or displacement monitoring

increase in ultimate flexural capacity can be expected. No significant contribution of the shear plates in reducing service load shear in the concrete immediately outside the shear plates was observed. However, for the area covered by the shear plates, a reduction in concrete shear stress of about 16% was noted. The repairs showed the cost-effective ($18,000 compared to the $150,000 required for normal replacement) use of FRP materials in retrofitting deficient concrete bridge members, with minimal traffic disruption.

7.6   TROY BRIDGE

A 70-year-old reinforced concrete T-beam bridge in New York was retrofitted with an FRP composite laminates system. The bridge carries State Route 378 over Wynanskill Creek in Rensselaer County. The bridge is a 12.8-m-long, 36.6-m-wide, reinforced concrete structure consisting of 26 T-beams (see Figure 7.68 for typical details of the beams). The retrofit system was installed based on concerns over the integrity of the steel reinforcing and the overall safety of the structure after observing damage to the beams and severe leakage of water contaminated by deicing salts. Design

FIGURE 7.62   Deficient capbeams. (Courtesy of New York State Department of Transportation.)

FIGURE 7.63   Detail of deficient capbeam. (Courtesy of New York State Department of Transportation.)

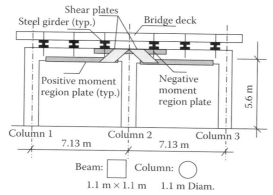

FIGURE 7.64 Schematics of retrofits. (Courtesy of New York State Department of Transportation.)

FIGURE 7.65 FRP retrofits. (Courtesy of New York State Department of Transportation.)

FIGURE 7.66 Installation of laminates. (Courtesy of New York State Department of Transportation.)

FIGURE 7.67   Load test of FRP retrofits. (Courtesy of New York State Department of Transportation.)

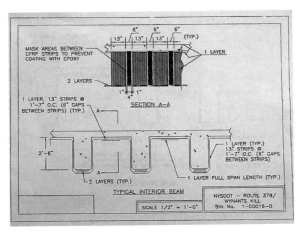

FIGURE 7.68   Typical details of Troy Bridge. (Courtesy of New York State Department of Transportation.)

for shear and flexure was based on a laminates system compensating for an assumed 15% loss of the steel rebar area. The system consisted of Replark 30 unidirectional carbon fibers, having an ultimate strength of 3400 MPa and a guaranteed ultimate strain of 1.5%. Figure 7.69 shows the beams before retrofitting. Figures 7.70 and 7.71 show the retrofitted beams. To evaluate the effectiveness of the retrofit system, 9 out of the 26 beams were instrumented, and the bridge was load tested before and after installation of the FRP laminates system. Baseline analysis and measurements were performed before the retrofitting. Figure 7.72 shows a comparison between baseline measured and computed strains due to truck loads on the bridge.

Two years after the laminate retrofit system in the previous application was installed, the after test was repeated to evaluate in-service performance of the system. The quality of the bond between the laminates and concrete is very crucial in this type of application, and any change in bond quality is expected to impact flexural behavior of the structure. No additional instrumentation was used for this application. The 1999 and 2001 results for the main steel rebars and FRP gauges are compared in Figures 7.73 and 7.74 for selected truck combinations during the test. These results show that the November 2001 strains are generally lower than those recorded in 1999. This can be attributed to an increase in stiffness caused by the newly placed deck and/or an inherent bias in the data acquisition system used in the 2001 test. The relatively proportional reduction in steel rebar and FRP strains in the 2001 results from their 1999 levels suggests that the bond quality has not deteriorated

FIGURE 7.69 Deteriorated beams before retrofitting. (Courtesy of New York State Department of Transportation.)

FIGURE 7.70 Retrofitting FRP laminates. (Courtesy of New York State Department of Transportation.)

during the 2 years between the tests. This was also confirmed by IRT images of the retrofit system (Figure 7.75), which did not show any significant debonding.

7.7 CONGRESS STREET BRIDGE

Fiber-reinforced polymer wrapping is used by bridge owners in several temporary (5–7 years) repairs to control delaminations in reinforced concrete bridge columns. A study by NYSDOT looked into three different surface preparation methods to evaluate their effectiveness, in conjunction with FRP wrapping, to select the optimal method. These methods included (1) removal of unsound concrete to a depth of no less than 25 mm from the rear most point of reinforcement to sound concrete at an estimated cost of about $750/m² (2) removal of unsound concrete to rebar depth at an estimated cost of about $270/m²; and (3) patching depressions and uneven areas with no removal of concrete at

FIGURE 7.71   Retrofitted beams. (Courtesy of New York State Department of Transportation.)

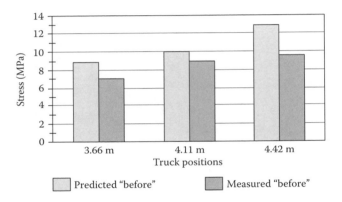

FIGURE 7.72   Analysis versus measurements, before retrofitting. (Courtesy of New York State Department of Transportation.)

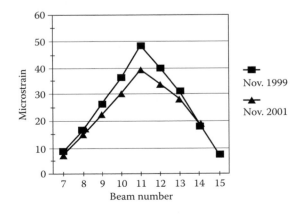

FIGURE 7.73   Effects of retrofits on rebar strains. (Courtesy of New York State Department of Transportation.)

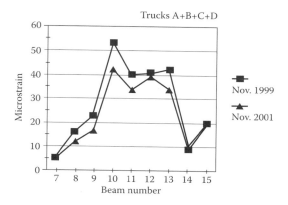

FIGURE 7.74   Effects of retrofits on FRP strains. (Courtesy of New York State Department of Transportation.)

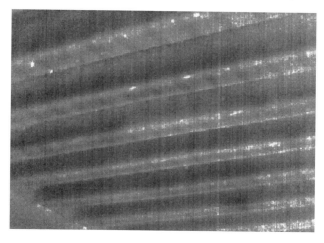

FIGURE 7.75   Thermographic view of retrofitted beams. (Courtesy of New York State Department of Transportation.)

minimal cost. All these costs are for concrete repairs and do not include costs for FRP materials and associated labor, pressure-washing the columns, and sand-blasting the concrete surface.

The test structure is a bridge carrying Route 2 over the Hudson River in Troy, NY (see Figure 7.76) (see Alampalli 2005). This eight-span bridge was built in 1969, is 430 m long and 23 m wide, and carries a concrete deck over steel girders. The columns (three in total) in one of the spans were deteriorated partly by leaking joints in the deck above. These tapered rectangular columns measure 2.28 × 2.28 m at the bottom and 1.89 × 1.28 m at the top. Due to deterioration, these columns were repaired several times before using concrete patchwork in 1991 and 1992. These repairs failed quickly, and repairs were needed again in 1999 for nonstructural purposes. So, FRP materials were used on an experimental basis as a cost-effective way to repair the concrete and to arrest further deterioration.

Three concrete removal strategies (see Figures 7.77 and 7.78) were used and then retrofitted with FRP wrapping. Concrete removal strategy 1 was used in the north column, strategy 2 in the south column, and strategy 3 in the center column. Durability of repairs was evaluated in terms of rate of corrosion and the bond between the concrete and FRP wrap. The corrosion rates of the longitudinal rebar in the column were measured using corrosion probes (see Figures 7.79 and 7.80). The corrosion rates of instrumented rebar are shown in Figures 7.81 through 7.83. Corrosion rates initially went up, then gradually slowed down, and decreased with time. After about 2 years, they converged

FIGURE 7.76   Configuration of columns. (Courtesy of New York State Department of Transportation.)

FIGURE 7.77   Removed delaminated and damaged concrete. (Courtesy of New York State Department of Transportation.)

to about 2 mils/year and stayed constant after that. This data indicates that the FRP wrapping effectively controlled the corrosion rates irrespective of the concrete repair strategy used.

Visual inspections were also performed every time the data was collected. Widespread paint failure was noticed within a year. Several rust marks and lines were noticed, some attributed to leakage of rust water from the inside column surface and others to water leakage from the deck joints (see Figure 7.84). At one location, about 2 years ago, someone removed a portion of the wrap by peeling it off the column surface (see Figure 7.85). Coin tapping was used during bridge inspections every 2 years, and thermography was used on a limited basis to monitor the bond between the concrete surface and the FRP wrapping. Results indicate that, in general, the bond quality did not deteriorate compared to the time of construction in 1999. It can be inferred that for columns similar to these with minor degradation, FRP wrapping with minimal concrete repair work can be considered to control deterioration and delaminations. Thus, FRP wrapping was effective for temporary (5–7 years) repairs in confining the repaired/delaminated concrete columns. Concrete removal strategies

FIGURE 7.78 Application of retrofitting FRP laminates. (Courtesy of New York State Department of Transportation.)

FIGURE 7.79 Installation of corrosion probe. (Courtesy of New York State Department of Transportation.)

did not influence the durability during the 6-year monitoring duration. It should also be noted that this study did not evaluate the long-term durability of FRP wrapping.

7.8 GUIDE TO SHM USAGE IN FRP WRAPPING

This section discusses in detail important performance issues about the use of FRP wrapping to enhance system performance. Figure 7.86 shows the performance (durability and structural issues) issues related to FRP wrapping. SHCE concepts can be of immense use in the following areas.

7.8.1 GLOBAL STRUCTURAL PERFORMANCE

As mentioned above, the use of FRP wrappings can have an effect on the overall (global) structural behavior. It might be desirable to monitor the effects of wrapping on such global performance.

FIGURE 7.80   Placements of corrosion probe. (Courtesy of New York State Department of Transportation.)

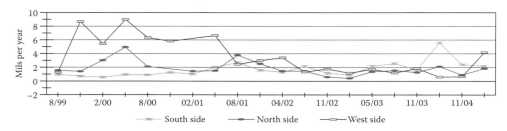

FIGURE 7.81   Corrosion rates in the north column. (Courtesy of New York State Department of Transportation.)

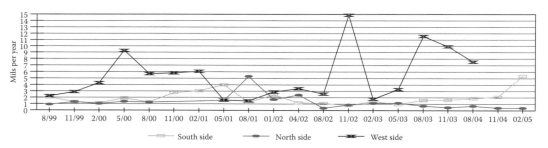

FIGURE 7.82   Corrosion rates in the south column. (Courtesy of New York State Department of Transportation.)

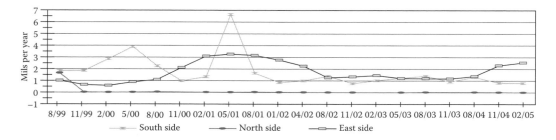

FIGURE 7.83   Corrosion rates in the center column. (Courtesy of New York State Department of Transportation.)

FIGURE 7.84 Rust spots after 2 years of retrofit. (Courtesy of New York State Department of Transportation.)

FIGURE 7.85 Surface peeling after 2 years. (Courtesy of New York State Department of Transportation.)

For example:

- Confinement: If the FRP wrappings are used to provide extra strength to a particular reinforced concrete column through increased confinement, it might be desirable to monitor such confinement on a medium- or long-term basis. Such monitoring would measure the state of triaxial strains in the concrete column. This type of monitoring can be of importance during a severe loading condition, such as a seismic event (see Figure 7.87).
- Increased flexural stiffness and/or increased flexural strength: Adhering FRP layers to the tension side of a flexural component (such as the underside of reinforced concrete bridge girders) is another use of FRP wrapping. It might be desirable to monitor the medium- or long-term stiffness increases through the use of FRP wrapping. This can be accomplished by monitoring the static or dynamic displacements of the girders of interest. Such

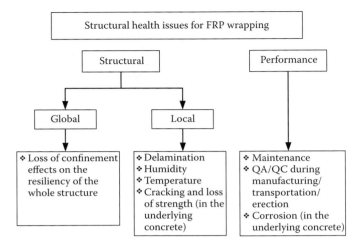

FIGURE 7.86 Some structural health issues for FRP-wrapping constructs.

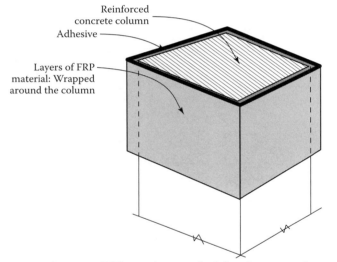

FIGURE 7.87 Components of FRP wrapping for increased confinement.

a medium- or long-term monitoring project can assure the bridge manager of the adequacy of the FRP stiffening/strengthening effects (see Figure 7.88).

7.8.2 LOCAL STRUCTURAL PERFORMANCE

As was mentioned earlier for bridge decks, FRP materials are engineerable materials and are not well understood by the engineering community. Figure 7.89 shows the basic components of local FRP wrapping. Some of the FRP-wrapping local structural issues that can benefit from monitoring techniques as follows:

- **Delaminations:** A typical FRP-wrapping construct contains the FRP layer itself and an adhesive material to connect it to the underlying concrete surface. Thus, proper condition of the adhesive material is essential for the FRP-wrapping project to achieve its objectives.

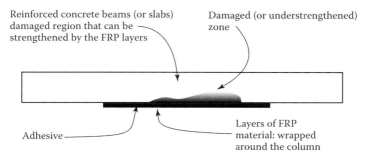

Anatomy of FRP gluing to reinforced concrete surfaces

FIGURE 7.88 Components of FRP wrapping for increased stiffness and strength.

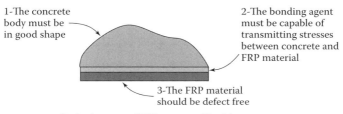

Basic elements of FRP structural health

FIGURE 7.89 Basic elements of structural health of FRP-wrapping local constructs.

Delaminations (loss of adhesion between the FRP material and the underlying concrete) must be monitored and retrofitted as needed. Monitoring delaminations can be achieved by numerous NDT methods. Some of these are described in Section 7.8.

- **Humidity:** One of the uses of FRP wrapping is to inhibit corrosion in the underlying reinforced concrete construction. Because of this, it is important to monitor the humidity inside the wrapping material. Since FRP-wrapping construction would preclude installing humidity sensors after completing the construction, it is essential to install them during or before construction.
- **Temperature:** There can be a correlation between temperature and the corrosion rate in concrete columns that were wrapped by FRP layers. Because of this, it is important to monitor temperature in similar situations. This is particularly important in the long run. Since the FRP-wrapping construction would preclude installing temperature probes after completing the construction, it is essential to install them during construction.
- **Loss of strength of the underlying concrete:** If the concrete surface under the FRP layers is cracked or has lost its strength, the effectiveness of FRP material in meeting its design objectives decreases. Concrete cracking can result from accidental impact, overloading, corrosion of steel reinforcements, and other factors. Monitoring techniques for concrete cracking or loss of strength are similar to those of the delamination conditions above.

7.8.3 DESIGN

Similar to bridge decks, most FRP-wrapping systems available to designers are proprietary systems and are engineered to meet the demands. Some areas where SHCE programs can help in providing better design process (where there is a knowledge gap) include, but are not limited to, the following,

- Concrete-adhesive-FRP layers mechanics
- Ductile versus brittle behavior of the FRP-wrapping construct

- Ultimate deformations under catastrophic loading
- Statistical properties of design parameters

7.8.4 INSTALLATION

Installation of the FRP wrappings is generally a manual process. Therefore, the quality of installation can vary greatly, depending on the skill of the installers. The installation process and the
environment (temperature, etc.) during the process may have a significant influence on the durability
and behavior of the FRP wrapping. So, the installation process should be carefully considered and
made an integral part of any successful SHCE program. Environmental conditions during installation should be defined appropriately, as needed. Temperature, moisture, humidity, and laying the
adhesives and the FRP material themselves should be considered during the design and manufacturing processes and well documented. Any unexpected problems, as well as the impromptu solutions,
might have an effect on future performance and strength; they should be reported on time. All of
this reporting, in a standardized fashion, will assist in developing a standardized erection procedure
for the future. NDT methods should be developed or specified to make sure the FRP-wrapping
material is not damaged during the erection process. This data should be stored, since this baseline
information may be needed to develop future inspection and maintenance procedures.

7.8.5 INSPECTION

Visual inspection, along with limited NDT such as coin tapping or hammering, is routinely used
in FRP-wrapping projects. IRT, microwave, impact-echo, and other advanced NDT methods are
under investigation, but at present not much data is available. Automatic inspection methods, which
rely on a set of sensors and measurements, are under investigation. Measuring different quantities, such as temperature, moisture, strains or displacements, can be an integral part of an SHCE
program. Different algorithms that utilize these types of measurements to identify damage in the
FRP-wrapping construct must be developed and improved. We note that visual inspection/testing
was one of the monitoring methods suggested by Karbhari et al. (2005) as a first line of inspection.
This is due to cost-effectiveness and simplicity of application.

7.8.6 MAINTENANCE

Maintenance issues for FRP-wrapping constructs are fairly similar to those for the FRP bridge
decks described in Chapter 6. Of course, the different ways that FRP decks and FRP wrappings are
constructed and the different ways they perform would result in some differences in maintenance
activities such as

- **Decommissioning:** When FRP bridge decks are to be decommissioned, such a removal
  would not impose overall structural safety concerns. This is because those types of bridge
  decks are not an integral part of the main bridge structural system. On the other hand,
  when an FRP wrapping is to be decommissioned, extreme care should be taken. Recall
  that in many instances FRP wrappings are meant to be an integral part of the structural
  performance components of the system. For example, FRP wrappings are used to retrofit
  degraded concrete or cracked road signs. When the wrappings are removed, during the process of decommissioning, adequate structural safety precautions should be considered.
- **Geometry:** The differences in geometry between FRP wrappings and FRP bridge decks
  create perhaps the major difference in maintenance, at least as far as SHM applications are
  concerned. Given that FRP wrappings are fairly thin, the applicable NDT/SHM methods
  will naturally reflect that. As such, thermography, ultrasonic, and shearography were recommended by Karbhari et al. (2005) to be the most efficient NDT methods for monitoring

FRP wrappings. These methods are also applicable for monitoring the outer layers of FRP bridge decks. This is because all these methods are most efficient for detecting surface or near-surface damage. For monitoring the core of the FRP bridge decks, methods that can monitor deeper into the construct are needed. Such methods include embedded strain sensors, active piezoelectric ultrasound, or acoustic emission sensing.

- **QA/QC:** FRP wrappings might need more strict QA/QC procedures than FRP bridge decks. The main reason is that FRP bridge decks are built in manufacturing plants, where QA/QC procedures are easier to enforce. FRP wrappings, by definition, need to be constructed on site. A good part of the construction process requires manual labor; this would require stricter QA/QC rules.

7.9 DECISION MAKING EXAMPLE: WHEN TO RETROFIT WITH FRP WRAPPING?

FRP wrapping is an emerging technology with the sole aim of repairing or improving the behavior of the structure. Obviously, there are several other techniques and methods that can produce the same results. One of the main challenges of any asset manager it to choose between the different repair/improvement methods available. In many situations, some information is available before the decision is made. For example, SHM data sets from previous projects might be available to the manager, and these can be used to improve the accuracy of the final decision. It is of importance to us in this chapter since it illustrates the quantitative way the SHM data sets can be used to improve accuracy of decisions. The approach is based on a decision-tree analysis and is called preposterior analysis: it requires information *before* the actual decision is made. We describe it with a practical yet simple example.

Consider a situation where a bridge manager is faced with a deteriorating reinforced concrete column situation. Three potential retrofit methods (purely hypothetical) are available:

1. a_1: Use FRP wrapping
2. a_2: Replace steel rebars with larger size
3. a_3: Add more steel rebars

There are obviously two potential outcomes for this project:

1. θ_1: Retrofit successful; that is, it meets long-term performance goals
2. θ_2: Retrofit not successful; that is, it does not meet long-term performance goals

The manager then establishes a simple payoff table as shown in Table 7.13. The payoff table includes the costs of each retrofit options a_i, for each of the potential outcomes θ_j ($i = 1,2,3$ and $j = 1,2$). Table 7.13 also shows the probabilities of successful and failed retrofits of this type, based on previous experiences and observations $p(\theta_j)$. Retrofits, in general, meet their goals, that is, are successful, 70% of the time. They do not meet their goals 30% of the time. Note that these probabilities are for general retrofits of all types; they only address success or failure. They do not specifically address the *type* of retrofit.

A simple prior analysis is performed using the payoff table. The results are shown in Table 7.14. The decision-tree diagram that corresponds to the prior analysis (without SHM support) is shown in Figure 7.90. There are three steps for this decision-tree computation:

1. Step 1: Estimate costs of each method at each event (Table 7.13).
2. Step 2: Moving rightward, multiply the event probabilities by the event costs and add the total weighted costs for each method. Those costs are shown near black circular nodes.

TABLE 7.13
Payoff Table

| | | | Cost ($) | |
| | | | a_2: Replace | |
Outcome	$p(\theta_j)$	a_1: FRP	Rebars	a_3: Add Rebars
θ_1: Retrofit successful	0.7	1000	1200	2500
θ_2: Retrofit not successful	0.3	4000	3500	3000
Totals	1.0			

TABLE 7.14
Simple Prior Analysis

| | Weighted Costs ($) | | |
Outcome	a_1: FRP	a_2: Replace Rebars	a_3: Add Rebars
θ_1: Retrofit successful	700	840	1750
θ_2: Retrofit not successful	1200	1050	900
Totals	1900	1890	2650

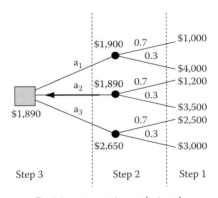

Decision tree—prior analysis only

FIGURE 7.90 Decision tree—prior analysis only.

3. Step 3: Moving rightward, choose the minimum cost and assign to the square node, at the root of the tree. This is the optimum cost of the problem. The path of this optimum cost is the optimum method, a_2: replacing rebars.

Based on the simple prior analysis, it seems that replacing rebars offers a bit less costly solution than using FRP wrappings. Both solutions require much lower costs than adding more steel rebars.

The manager decided that there is a need for additional investigations before taking a decision. The reasons for the additional investigations were as follows:

1. The costs of options a_1 and a_2 were too close to make a definite choice.
2. The probabilities $p(\theta_j)$ were too general; they did not account for the specifics of the available retrofit solutions.

A more in-depth investigation was performed by the manager. The investigations involved database studies of historical SHM results. The question investigated was as follows: in the past, by choosing the obvious retrofit method, in similar deteriorating conditions, was such a decision

1. X_1: a good decision
2. X_2: an intermediate decision, or
3. X_3: a bad decision?

By correlating past situations for the type of decision, X_1, X_2, and X_3 with the actual outcome of those events, θ_1 and θ_2, the conditional probabilities $P(X_i|\theta_j)$ (read: probability that it was felt the decision was X_i, given that the actual outcome was θ_j) can be established, as shown in Table 7.15.

Given the conditional probabilities, $P(X_i|\theta_j)$ and the prior probabilities $p(\theta_j)$, the improved event probabilities, given the SHM observations, $P(\theta_j|X_i)$, can be computed using Bayes theorem as

$$P(\theta_j|X_i) = \frac{p(\theta_j)P(X_i|\theta_j)}{\sum_i p(\theta_j)P(X_i|\theta_j)} \tag{7.3}$$

Note that the term $p(\theta_j)P(X_i|\theta_j)$ is the joint probability of $P(\theta_j$ and $X_i)$. The computations of the joint probabilities are shown in Table 7.16.

The rest of the computations is easier to illustrate graphically using a decision tree. The decision tree incorporating the conditional, joint, and improved probabilities is shown in Figure 7.91. Note that there are five steps in the preposterior decision-tree approach. They are as follows:

1. Step 1: Estimate costs of each method at each event (Table 7.13).
2. Step 2: Moving rightward, multiply the improved event probabilities (Equation 7.1) by the event costs and add the total weighted costs for each method. Those costs are shown near black circular nodes.

TABLE 7.15
Conditional Probabilities for Past Experiences, e.g., SHM Experiments

Outcome	Conditional Probabilities $P(X_i\|\theta_j)$			Total
	X_1	X_2	X_3	
θ_1: Retrofit successful	0.5	0.3	0.2	1.0
θ_2: Retrofit not successful	0.3	0.3	0.4	1.0

TABLE 7.16
Joint Probabilities

Outcome	Joint Probabilities $P(\theta_j$ and $X_i)$			Total
	X_1	X_2	X_3	
θ_1: Retrofit successful	0.35	0.21	0.14	0.7
θ_2: Retrofit not successful	0.09	0.09	0.12	0.3
Total	0.44	0.3	0.26	1

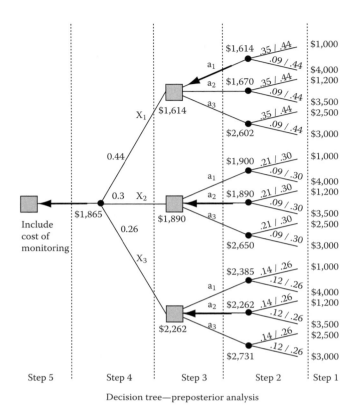

Decision tree—preposterior analysis

FIGURE 7.91   Decision tree for FRP-wrapping example—tree design.

3. Step 3: Moving rightward, choose the minimum cost and assign to the square node. This is the optimum cost of the problem. The path of this optimum cost is the optimum method a_2: replacing rebars.
4. Step 4: Get the weighted costs by multiplying each of the costs assigned to square node by the companion probability (from the totals raw in Table 7.16). The weighted cost is assigned to the black circular node. This is the cost of the project if the SHM investigation results are followed.
5. Step 5: Add the cost of performing the SHM experiment to the cost from step 4. This is the total cost of the project.

From the results shown in Figure 7.91, the cost of the project is $1,865. It is less than the cost with prior only investigation (which is $1,890) by $25. Thus if the cost of the SHM experiment is less than $25, say $20, the total cost of the project is $1,885. Thus, by performing an SHM experiment, and following the trend that the SHM results shows, the manager would save $1,890 − $1,885 = $5. In other words, the value of added information in this case is $5. Conversely, if the cost of the SHM experiment is more than $25, say $40, the total cost of the project is $1,905. Thus, by performing an SHM experiment, and following the trend that the SHM results shows, the manager would spend an extra $1,905 − $1,890 = $15. There is no point in performing an SHM experiment in this case.

We have just reached an important point: there is always a value in investigating past trends before taking decisions. Such an investigation will save costs—either by pointing to the value of performing SHM investigations or by pointing to the fact that SHM investigation will not save costs.

7.10   LCA OF FRP WRAPPING

7.10.1   General

We note that the process described in this section is equally applicable to similar situations where a particular product is installed such as when installing an FRP deck (see Chapter 6).

7.10.2   Life Cycle Cost Analysis (LCCA) of FRP Wrapping

Estimating the life cycle cost of any item involves three components: initial costs, operating costs, and decommissioning costs. FRP wraps are similar. We discuss the three components next.
Initial costs: Initial costs C_{FRP_I} are expressed as

$$C_{FRP\_I} = \sum_{i=1}^{i=N_I} C_{iI} \tag{7.4}$$

In Equation 7.4, N_I is the number of initial cost components. For our immediate purpose, we take $N_I = 4$. We note that N_I can change according to other situations. The four components of initial costs are shown in Table 7.17. The table also shows the role of SHM in accurate estimation of these costs.

   Operating costs: Operating costs C_{FRP_O} are costs that will be incurred during the lifespan (LS) of the FRP wrappings.

$$C_{FRP\_O} = \sum_{i=1}^{i=N_O} C_{io} \tag{7.5}$$

In Equation 7.4 is the number of operating cost components. For our immediate purpose, we take $N_O = 2$. We note that N_O can change according to other situations. The four components of the operating costs are shown in Table 7.18. The table also shows the role of SHM in accurate estimation of these costs.

   Generally, cost of repairs C_{2O} can be estimated using methods similar to those for estimating the costs of other hazards such as seismic hazard

$$C_{2O} = \sum_i \sum_j p_{ij} C_{ijO} \tag{7.6}$$

TABLE 7.17
Sources of Initial Costs For FRP Wrappings

Variable	Comment	SHM Role
C_{1I}	Cost of management. This includes design, analysis, and decision making	STRID tools might be used to ensure optimum designs
C_{2I}	Cost of manufacturing. This includes contracting, processes, QA, QC, and packaging	NDT tools can be used for QA, QC, and inspection
C_{3I}	Cost of transportation	Sensing conditions during different phases of transportation
C_{4I}	Cost of erection. This includes labor, inspection, QA, QC, and certification	NDT tools can be used for QA, QC, and inspection

TABLE 7.18
Sources of Operating Costs for FRP Wrappings

Variable	Comment	SHM Role
C_{10}	Cost of inspection, both scheduled and nonscheduled	NDT tools can be used
C_{20}	Cost of repairs, if needed	STRID tools might be used to ensure optimum designs DMID tools can be used to determine need for repairs, and NDT tools can be used for QA, QC, and inspection

The ith summation is over the different potential damage levels. The jth summation is over the different types of potential damage. The probability of occurrence of jth type of damage at ith level is p_{ij}.

We would like to relate the cost of repairs to the damage measure Dm_{FRP} which was introduced in the earlier sections. We showed that Dm_{FRP} can be estimated directly from SHM/NTD or manual inspection at any given time. For LCA we need to generalize the estimation process of Dm_{FRP} such that

- It spans the whole LS spectrum, not only a single time instant
- It accommodates future uncertainties

We can generalize the expression for C_{2o} to

$$C_{2O} = \int_{LS} p(\text{Dm}_{\text{FRP}}) C_O \big|_{\text{Dm}_{\text{FRP}}} dt \tag{7.7}$$

The integral in Equation 7.7 is over the Lifespan of the product LS. The probability of the occurrence of a damage measure of value Dm_{FRP} at any future time is $p(\text{Dm}_{\text{FRP}})$. The retrofit cost of such an occurrence is $C_O\big|_{\text{Dm}_{\text{FRP}}}$. We can expand 7.7 to

$$p(\text{Dm}_{\text{FRP}}) = \frac{1}{3}\left(\int p(S_{\text{FRP}}) ds_{\text{FRP}} + \int p(L_{\text{FRP}}) d\ell_{\text{FRP}} + \int p(F_{\text{FRP}}) df_{\text{FRP}}\right) \tag{7.8}$$

In Equation 7.8, we considered S_{FRP}, L_{FRP}, and F_{FRP} to be random variables which are time dependent; as such we finally released the deterministic assumption of those damage measures as described in an earlier section. The operational life cycle cost of the FRP wrapping can finally be estimated as

$$C_{2O} = \int_{LS} \frac{1}{3}\left(\int p(S_{\text{FRP}}) ds_{\text{FRP}} + \int p(L_{\text{FRP}}) d\ell_{\text{FRP}} + \int p(F_{\text{FRP}}) df_{\text{FRP}}\right) C_O\big|_{\text{Dm}_{\text{FRP}}} dt \tag{7.9}$$

The probability density functions (PDF) of S_{FRP}, L_{FRP}, and F_{FRP} as a function of time are needed for fully evaluating Equation 7.9. Such PDF can be estimated by one of the following methods:

- Data gathering from SHM experimental projects, such as the ones described in Sections 7.3 through 7.7
- Engineering judgment based on past experiences

Decommissioning costs: Decommissioning of FRP wrappings occurs when they reach the end of their useful life; at that time they would be removed. The removal can be a part of a replacement or a part of major retrofits that involve the parent structure. The decommissioning costs are defined as $C_{\text{FRP\_D}}$.

The life cycle cost of FRP wrappings can now be defined as

$$\mathrm{LCC_{FRP}} = C_I + C_O + C_D \tag{7.10}$$

7.10.3 Life Cycle Benefit Analysis (LCBA) of FRP Wrapping

When a decision maker considers an FRP-wrapping project, its benefits must be considered. The benefits can be immediate or during the LS of the product. Some of the benefits from an FRP-wrapping installation are

- Increasing resiliency (strength, stiffness, etc.)
- Corrosion mitigation

Quantifying those benefits in monetary measures is more difficult than quantifying costs. Table 7.19 includes some guidelines that might be of help in quantifying the life cycle benefits of an FRP wrapping LCB_{FRP}

7.10.4 Lifespan Analysis (LSA) of FRP Wrapping

As usual, estimating the LS of a construct is a controlling factor in both estimations of LCCA and LCBA. For this purpose, we turn to the fragility concept to estimate the LS of FRP wrappings. We assume first that the PDF of $\mathrm{Dm_{FRP}}$, $p(\mathrm{Dm_{FRP}})$ is known as a function of time, according to Equation 7.8. Note that the time dependence of $p(\mathrm{Dm_{FRP}} \geq X_{\mathrm{MAX}})$, where X_{MAX} is the damage level at which the FRP construct is to be replaced, is implied in the variability of the different histograms over the years of observation. We immediately recognize that the latter curve is the fragility of the FRP construct for replacement damage state. The choice of the LS of the construct becomes easy given the fragility curve: it is the time at $p(\mathrm{Dm_{FRP}} \geq X_{\mathrm{MAX}}) \geq p_{\mathrm{MAX}}$. The choice of p_{MAX}, which is the accepted nonexceedance probability, depends on the decision maker and the sensitivity of the project. Probabilities of nonexceedance are usually more than 84%. In important projects, they can be as high as 97%.

As an example, consider a situation where a collective SHM project, which included observation, data collection, damage estimates, and statistical analysis, has produced histograms of probabilities of $\mathrm{Dm_{FRP}}$ as a function of elapsed years. The histograms are shown in Figure 7.92. From the histograms, fragilities can be computed (see Chapter 9) as shown in Figure 7.93. From Figures 7.92 and 7.93, a nonexceedance probability of 84% (16% probability of exceedance), a damage state of replacement will be reached after about 15 years. If the project calls for a nonexceedance probability of 97% (3% exceedance probability), a damage state of replacement will be reached after about 10 years only. Such information is of importance to decision makers, and it clearly shows the importance of SHM in gathering data and observations.

TABLE 7.19
Life cycle Benefit of FRP Wrappings

Benefit		Life Cycle Computation
Increase resiliency strength	Accommodate higher traffic demands	Estimate cost savings from allowing higher traffic
	Accommodate higher demands of natural hazards (earthquake, floods, etc.)	Estimate costs of failure due to the hazard
• Repair damage due to deterioration or abnormal events • Corrosion mitigation		Estimate additional useful life gained from such repair, then estimate cost savings from extension of the LS

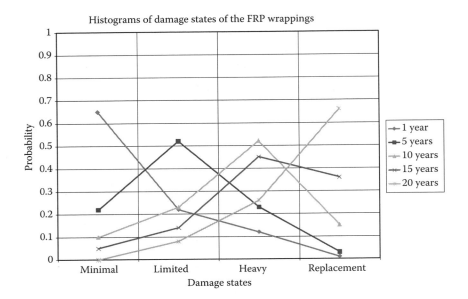

FIGURE 7.92   Histograms of damaged states of FRP wrappings.

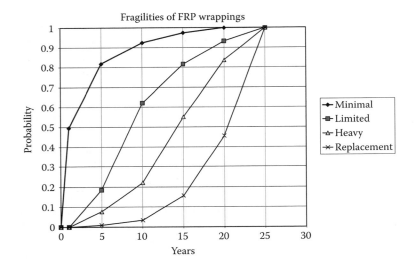

FIGURE 7.93   Fragilities of FRP wrappings.

We end this discussion by observing that the fragility approach can be used in a similar manner for LSA of any construct in this chapter.

7.10.5   SUMMARY

This section illustrated methods to quantify LCCA, LCBA, and LSA of FRP wrappings. Note that the methods can be improved in several ways. For example, note that we did not include discount rates or inflation rates in our analyses. Also, sources of other uncertainties, such as the uncertainty in costs of retrofits or uncertainty of the discount/inflation rates themselves, were not included. Techniques of accommodating those uncertainties are discussed in Sections 7.4 and 7.9.

We can now integrate all the three components of LCA by considering the rate of return of installing FRP wrapping as

$$R_{\mathrm{FRP}} = \frac{\mathrm{LCB}_{\mathrm{FRP}} - \mathrm{LCC}_{\mathrm{FRP}}}{\mathrm{LS}} \qquad (7.11)$$

If $R_{\mathrm{FRP}} > 0$ then the installation is recommended, if $R_{\mathrm{FRP}} < 0$ then the installation is not recommended. Equation 7.11 can also be used to compare different FRP-based strategies. The optimum strategy is the one with the highest R_{FRP}.

7.11   SOURCES OF DAMAGE IN FRP LAMINATES

Karbhari et al. (2005) provided valuable summary of potential defect types in FRP laminates, the potential causes of these defects, and the potential effects of these defects. Defects resulting in raw materials, from situ preparations, during installations, and during in-service are shown in Tables 7.20 through 7.23.

TABLE 7.20
Raw and Material Defects

	Defect Type	Cause/Description	Potential Effect
Resin	Overaged resin	Expired shelf-life	Low strength and modulus. Potential for incomplete cure and nonuniform impregnation
	Resin inclusions	Dirt and/or chemicals	Change in chemical consistency, voids Potential effect and cure
	Resin moisture	Inadequate storage/ environmental exposure	Change in chemical consistency, voids due to evaporation, resin degradation
Fabric	Incorrect fiber/fabric type	Fiber/resin mismatch, human error	Change in strength and modulus, low fiber/ matrix bond (sizing)
	Kinked or wavy fibers	Handling/manufacturing flaw	Fiber breakage, loss in composite properties
	Broken fiber tows	Handling/manufacturing flaw	Stress concentrations
	Fabric contaminations	Environmental exposure/storage	Initiator for debonding and crack propagation
	Fabric wrinkles	Handling/manufacturing flaw	Lower modulus, higher strain at failure, resin-rich encapsulated areas
	Sheared fabric	Handling/manufacturing flaw	Off-axis alignment lower strength and modulus, resin-rich regions
	Damage to free edges	Handling/manufacturing flaw	Loss of integrity, stress concentrations
	Pull-out of fiber tows	Handling manufacturing flaw	Resin richness, localized low strength and stiffness
	Fiber saps	Handling/manufacturing flaw	Resin richness, low crack arresting capability
	Fabric moisture	Inadequate storage/ environmental exposure	Reduced fiber matrix bond effect on composite performance and durability

Source: Karbhari, V. et al., Methods for detecting defects in composite rehabilitated concrete structures, Submitted to Oregon Department of Transportation, and Federal Highway Administration, Report No. SPR 336, 2005. Courtesy of Dr. Karbhari.

TABLE 7.21
Site and Preparation Defects

	Defect Type	Cause/Description	Potential Effect
Resin System	Moisture absorption	Inadequate storage/environmental exposure	Change in chemical consistency, voids due to evaporation. Potential for incomplete cure and decreased performance levels
	Incorrect stoichiometry	Type or proportions of resin and hardener/catalyst	Inadequate matrix strength/modulus, incomplete and/or nonuniform cure
	Incorrect mixing	Low degree mixing, drawing of air	Partial cure, porosity, nonuniform rheology
Substrate	Inadequate primer coating	Over- or under-saturation of substrate	Low stress transfer capability, potential for poor bond
	Lamination on top of "marked-out" regions	Placement on duct tape, crayon layer, etc.	Weak or no bond to substrate
	Degraded substrate	Microcracks, spalled concrete	Lower or no composite action
	Inclusions at imperfections	Dirt, moisture or chemicals in concrete cavities	Low hand of primer/putty to substrate
	Galvanic corrosion	Intimate contact of carbon fibers with steel reinforcement	Deterioration of matrix steel

Source: Karbhari, V. et al., Methods for detecting defects in composite rehabilitated concrete structures, Submitted to Oregon Department of Transportation, and Federal Highway Administration, Report No. SPR 336, 2005. Courtesy of Dr. Karbhari.

TABLE 7.22
Installation Defects

		Defect Type	Cause/Description	Potential Effect
Wet lay-up	Composite-Concrete Interphase	Concrete cavities	In current practice, concrete cavities are not filled with putty	Stress concentrations
		Sagging of infiltrated fabric	Critical in overhead regions	No composite action, potential moisture entrapment at concrete/composite interface
		Resin-rich/poor concrete/composite interface	Nonuniform, primer coating and over- or under-saturation during lay-up	Low stress transfer efficiency
		Porosity and voids	Porous primer, entrapment of air pockets during lay-up	Low stress transfer, stress concentrations, debond and crack initiations sites
		Highly uneven concrete surface	High degree of sand blasting	Voids or air pockets
	Inside composite	Porosity and voids	Air entrainment in resin, entrapment of air pockets during lay-up	Low stress transfer, stress concentrations, decreased performance attributes
		Delamination	Moisture, inclusions	Low or no stress transfer
		Debonding	Fiber contamination, tubular voids	Low stress transfer (localized), sites for wicking of moisture

TABLE 7.22 (continued)
Installation Defects

	Defect Type	Cause/Description	Potential Effect
	Incorrect stacking sequence	Misplaced fabric, human error	Alteration of strength and stiffness
	Resin-richness/poorness	Nonuniform infiltration	Low crack arresting capability, decreased stress transfer capabilities, locally weak zones
	Indentations	Handling damage	Damaged fibers, stress concentrations
	Missing layers	Human error	Entirely different (and decreased) laminate properties
	Damaged edges	Fiber pull-out during infiltration	Stress concentrations, she for crack initiation
Prefabricated material	Voids at concrete/adhesive interface	Adhesive applied to highly porous concrete substrate	Stress concentrations, moisture accumulation, local zones of weakness
	Disbonding at adhesive/composite interface	Smooth surface of prefabricated strip	Low stress transfer/inadequate bond strength

Source: Karbhari, V. et al., Methods for detecting defects in composite rehabilitated concrete structures, Submitted to Oregon Department of Transportation, and Federal Highway Administration, Report No. SPR 336, 2005. Courtesy of Dr. Karbhari.

TABLE 7.23
In-Service Defects

	Defect Type	Cause/Description	Potential Effect
Concrete/composite interface	Penetration of moisture and chemicals	Exposure to aggressive environments	Degradation of adherent layer, plasticization, reduced stiffness, potential for premature failure through peel and or lamination
	Heat damage	Exposure to sun or fire damage	Softening/degradation of matrix, peel and/or separation from substrate
Composite	Penetration of moisture and chemicals	Exposure to aggressive environments	Plasticization, reduced stiffness, degradation of composite
	Heat damage	Exposure to sun or fire damage	Softening/degradation of matrix
	Matrix cracking	Interlaminar crack formation	Initiator for delamination and/or splitting
	Surface scratches	Traffic, hail, etc.	Fiber breakage and initiator for premature local failure
	Impact damage	Traffic, hail, etc.	Delamination

Source: Karbhari, V. et al., Methods for detecting defects in composite rehabilitated concrete structures, Submitted to Oregon Department of Transportation, and Federal Highway Administration, Report No. SPR 336, 2005. Courtesy of Dr. Karbhari.

REFERENCES

Akuthota, B., Hughes, D., Zoughi, R., Myers, J., and Nanni, A. (2004). "Near-field microwave detection of disbond in carbon fiber reinforced polymer composites used for strengthening cement-based structures and disbond repair verification." *Journal of Materials in Civil Engineering*, Vol 16, No 6, 540–546.

Alampalli, S. (2005). "effectiveness of frp materials with alternative concrete removal strategies for reinforced concrete bridge column wrapping." *International Journal of Materials and Product Technology*, Interscience Publishers, 23(3/4), 338–347.

Aref, A. and Jung, W-Y. (2003). "Energy-dissipating polymer matrix composite-infill wall system for seismic retrofitting," *Journal of Structural Engineering, ASCE*, 129(4), 440–448.

Astrom, B. (1997). *Manufacturing of Polymer Composites*, Chapman & Hall, London, UK.

Bastianini, F., Ceriolo, L., Di Tommaso, A., and Zaffaroni, G. (2004). "Mechanical and nondestructive testing to verify the effectiveness of composite strengthening on historical Cast Iron Bridge in Venice, Italy." *Journal of Materials in Civil Engineering*, 16(5), 407–414.

Black, S. (2005). "Repair of Infrastructures with Composites," Composites Technology, http://www.compositesworld.com/articles/ct.aspx. Accessed March 31, 2011.

Bonfiglioli, B., Pascale, G., Strauss, A., and Bergmeister, K. (2004). "Strain Measuring on Fiber Composites by Bragg Grating Sensors," *Proceedings of 2nd International Workshop on Structural Health Monitoring of Innovative Civil Engineering Structures*, ISIS Canada Corporation, Manitoba, Canada.

Brown, J. and Hamilton, H. (2004). "Infrared Thermography Inspection Procedures for the Non-Destructive Evaluation of FRP Composites Bonded to Reinforced Concrete," *Proceedings, NDE Conference on Civil Engineering*, ASNT, Buffalo, NY.

Chahrour, A. and Soudki, K. (2004). "Analysis of Corroding Steel Expansion that Leads to Crack Initiation and Propagation in CFRP Wrapped Concrete Cylinders," *Proceedings of 2nd International Workshop on Structural Health Monitoring of Innovative Civil Engineering Structures*, ISIS Canada Corporation, Manitoba, Canada.

Dokun, O. D., Jacobs, L. J., and Haj-Ali, R. M. (2000). "Ultrasonic monitoring of material degradation in FRP composites." *Journal of Engineering Mechanics*, 126(7), 704–710.

Erie County. accessed May 2011. "Trenchless Culvert Rehabilitation," http://www.liquiforce.com/images/assets/Projects/USA/DickRd-CulvertRehab-ErieCounty-NewYork.pdf?phpMyAdmin=574b844a15441e088619089ebe7b8a03.

Ettouney, M. and Alampalli, S. (2012). *Infrastructure Health in Civil Engineering: Theory and Components*, CRC Press, Boca Raton, FL.

Fafach, D., Shing, B., Chang, S., and Xi, Y. (2005). "Evaluation Of The FRP-Retrofitted Arches In The Castlewood Canyon Bridge," Colorado Department of Transportation Research Report No. CDOT-DTD-R-2005–01, Denver, CO.

Feng, M. Q., De Flaviis, F., and Kim, Y. J. (2002). "Use of microwaves for damage detection in fiber reinforced polymer-wrapped concrete structures." *Journal of Engineering Mechanics*, 126(2), 172–183.

Feng, M. Q. and Kim, Y. J. (2004). "NDE of Concrete Structures using Microwaves," *Proceedings, NDE Conference on Civil Engineering*, ASNT, Buffalo, NY.

FHWA. (2008). "Composite Bridge Column Retrofitting in California," http://www.fhwa.dot.gov/BRIDGE/frp/calcomp.cfm. Accessed March 31, 2011.

Ganchev, S. I., Qaddoumi, N., Ranu, E., and Zoughi, R. (1995). "Microwave detection optimization of disbond in layered dielectrics with varying thickness." *IEEE Transactions on Instrumentation and Measurement*, 34(4), 326–328.

Gheorghiu, C., Labossiere, P., and Proulx, J. (2004). "Fatigue and Post-Fatigue Reliability of Fabry-Perot FOS Installed on CFRP-Strengthened RC-Beams," *Proceedings of 2nd International Workshop on Structural Health Monitoring of Innovative Civil Engineering Structures*, ISIS Canada Corporation, Manitoba, Canada.

Godinez-Azcuaga, V. F., Gostautas, R. S., Finlayson, R., Miller, R., and Trovillion, J. (2004). "Nondestructive Evaluation of FRP Wrapped Concrete Columns and Bridges," *Proceedings, NDE Conference on Civil Engineering*, ASNT, Buffalo, NY.

Hag-Elsafi, O., Alampalli, S., and Kunin, J. (2004). "In-service evaluation of a reinforced concrete T-beam bridge FRP strengthening system." *Journal of Composite Structures*, Elsevier Science, 64(2), 179–188.

Hag-Elsafi, O., Kunin, J., and Alampalli, S. (2000). "Evaluating Effectiveness of FRP Composites for Bridge Rehabilitation through Load Testing," Structural Materials Technology: an NDT Conference, ASNT, Atlantic City, NJ.

Hag-Elsafi, O., Lund, R., and Alampalli, S. (2002). "Strengthening of a bridge pier capbeam using bonded FRP composite plates." *Journal of Composite Structures*, Elsevier Science, 57(1–4), 393-403.

Hawkins, G. F., Johnson, E. C., and Nokes, J. P. (2001). "In-Field Monitoring of the Durability of Composite Materials." Aerospace report No. ATR-2001 (7595)-1, prepared for California Department of Transportation, Sacramento, CA.

Hughes, D., Kazemi, M., Marler, K., Myers, J., Zoughi, R., and Nanni, A. "Microwave detection of delaminations between fiber reinforced polymer (FRP) composite and hardened cement paste." Review of Quantitative Nondestructive Evaluation, Brunswick, ME, 512–519.

Jackson, D., Islam, M., and Alampalli, S. (2000). "Feasibility of Evaluating the Performance of Fiber Reinforced Plastics (FRP) Wrapped Reinforced Concrete Columns Using Ground Penetrating Radar (GPR) and Infrared (IR) Thermography Techniques," Structural Materials Technology: an NDT Conference, ASNT, Atlantic City, NJ.

Johnson, E. C., Nokes, J. P., and Hawkins, G. F. (1999). "NDE of composite seismic retrofits to bridges." *Nondestructive Characterization of Materials*, 367–372.

Karbhari, V., Kaiser, H., Navada, R., Ghosh, K., and Lee, L. (2005). "Methods for Detecting Defects in Composite Rehabilitated Concrete Structures," Submitted to Oregon Department of Transportation, and Federal Highway Administration, Report No. SPR 336.

Luangviai, K., Punurai, W., and Jacobs, L. (2002). "Guided Lamb wave propagation in composite plate/concrete component." *Journal of Engineering Mechanics*, 128(12), 1337–1341.

Mooney, P. (2003). "A fix for aluminum overheads," *Public Roads*, 67(3).

Moore, M., Green, T., Kahn, L., and Zureick, A. (2005). "Case Studies in the Evaluation of FRP Bond to Concrete," *Proceedings of the 2005 ASNT Fall Conference*, Columbus, OH.

Mtenga, P. V., Parzych, J. G., and Limerick, R. (2001). "Quality assurance of FRP retrofit using infrared thermography." *Structures, 2001.*

Nishizaki, I. and Meiarashi, S. (2002). "Long-term deterioration of GFRP in water and moist environment." *Journal of Composites for Construction*, 6(1), 21–27.

Priestley, M. J. N. and Seible, F. (1991). "Seismic Assessment and Retrofit of Bridges." Struct. Sys. Res. Proj., Rep. No. SSRP-91/103, University of California at San Diego, CA.

Sanyal, D. (2007). "Laser Induced Ultrasonic Characterization of FRP Composites," *Proceedings, International Conference on Advanced Materials and Composites*, ICAMC.

Seifried, R. J., Laurence J., and Qu, J. (2002). "Propagation of guided waves in adhesive bonded components." *NDT & e International*, 35(5), 317–328.

Starnes, M. A., Carino, N. J., and Kausel E. A. (2003). "Preliminary thermography studies for quality control of concrete structures strengthened with fiber-reinforced polymer composites." *Journal of Materials in Civil Engineering*, 15(3), 266–273.

Taheri, F., Bell, S., and Halpin, B. (2002) "Static Response of FRP Strengthened Sawn Timber and Glulam Beams," *Proceedings of 1st International Workshop on Structural Health Monitoring of Innovative Civil Engineering Structures*, ISIS Canada Corporation, Manitoba, Canada.

Washer, G. and Alampalli, S. (2005) "NDE Technologies for Condition Assessment of FRP Retrofits," ASNT Fall Conference and Quality testing Show, Columbus, OH.

Washer, G. A. (2003). "Nondestructive Evaluation for Highway Bridges in the United States." International Symposium on Nondestructive Testing in Civil Engineering, Berlin, Germany.

Zhou, Z., Wang, B., and Ou, J. (2004). "Local Damage Detection of RC Structures with Distributive FRP-OFBG Sensors," *Proceedings of 2nd International Workshop on Structural Health Monitoring of Innovative Civil Engineering Structures*, ISIS Canada Corporation, Manitoba, Canada.

8 Load Testing

8.1 INTRODUCTION

As we have seen in this document so far, there are numerous ways to preserve structural health through monitoring and decision making techniques. Perhaps the most popular way is load testing. There are many ways to load test a bridge; there are also many reasons for it. Generally, the testing involves placing a load on the bridge; as the bridge responds, the experiment monitors the responses by a set of sensors and instrumentation. Relating those measurements to the bridge system, using the techniques of structural identification and damage assessment, is the next step. Finally, the appropriate decisions are made by the bridge official. Figure 8.1 shows the steps of load-testing procedures.

8.1.1 THIS CHAPTER

The main goal of this chapter is to present different aspects of load testing from the structural health in civil engineering viewpoint. Specifically, the objectives of this chapter are

1. Present load testing as a structural health tool by showing that structural health tools and steps are used in load testing
2. Present some new concepts in load testing, again using structural health procedures.
3. Formalize load-testing steps (by using structural health components of Chapter 2 by Ettouney and Alampalli 2012)
4. Point to some knowledge gaps in the load-testing field

We first discuss the many categorizations of load tests. We also introduce two new concepts applicable to load tests: dynamic influence lines (DIL) and monitoring (or testing) techniques for bridges prone to brittle failure. Sensors and instrumentation issues of load tests are discussed next and several practical guidelines offered. The guidelines can help practitioners plan their tests. Issues regarding STRID and DMID during and after load tests are offered next. Bridge ratings and their interrelationships with load testing are discussed in depth. Decision making examples of load testing and life cycle analysis (LCA) are also discussed, and we show that there is a direct relationship between load testing and the life cycle of a bridge. Finally, we offer two detailed case studies of load tests, which highlight many practical issues that confront the professional when embarking on a load-testing project. Figure 8.2 illustrates the main subjects of this chapter.

8.2 GENERAL CONSIDERATIONS FOR LOAD TESTING

8.2.1 GOALS OF LOAD TESTS

Load testing of bridges aims at determining how a bridge responds to known live loads in actual conditions. Knowing the actual live loads, which are usually standard truck loads, the measured responses at some critical locations can be used in a variety of ways to achieve certain objectives. The objectives include analytical model validations and/or determining bridge rating (see Figures 8.3 through 8.5). One of the advantages of bridge testing is that it can help to establish bridge rating for

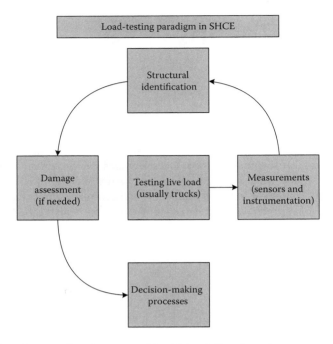

FIGURE 8.1   Load-testing paradigm in structural health in civil engineering.

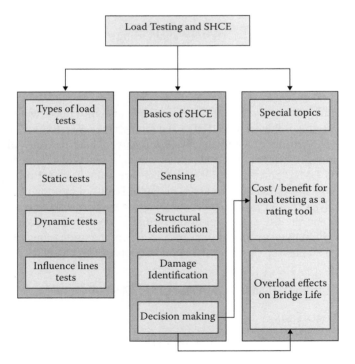

FIGURE 8.2   Overview of load-testing chapter.

complex bridges when analytical rating might be difficult. Accurate ratings that might result from testing can produce higher rating by avoiding the excessive conservatism built into analytical ratings. In some situations, ratings based on testing might produce lower ratings than analytical ratings, thus

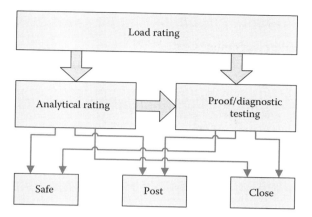

FIGURE 8.3 Load-rating process.

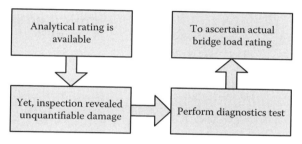

FIGURE 8.4 Diagnostics load test.

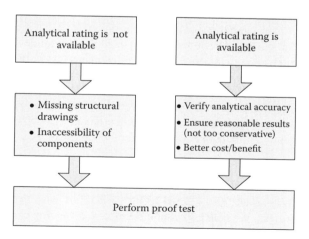

FIGURE 8.5 Proof load testing.

leading to safer decisions by the bridge official. Lagace (2004) showed the potential cost savings from load testing. He described the use of load testing in one of the New York State Department of Transportation (NYSDOT) regions, with four bridge examples, to remove or avoid load postings. The cost of load testing of about $23,000 versus $100,000 yearly diversion costs by trucking industry if posting was required. Table 8.1 shows the estimated replacement costs. Table 8.2 shows the changes in load rating as results from the load tests. Thus the bridges did not have to be replaced as a direct result of the load testing. The savings are self-evident. Prader et al. (2006) provided an overview of

TABLE 8.1
Detour Lengths and Replacement Costs

BIN	Carried	Crossed	Detour Length (miles)	Estimated Replacement Cost ($)
1004850	Route 8	Sacandaga Lake Outlet	54	2,000,000
1008000	Route 10	West branch Sacandaga River	68	550,000
1002490	Route 5	Zimmerman Creek	20	300,000
1039060	Route 171	Moyer Creek	6	400,000

Source: Reprinted from ASNT Publication.

TABLE 8.2
Load-Rating Results

BIN	Carried	Crossed	Previous Load Rating (HS Tons)	Load Test Load Rating (HS Tons)
1004850	Route 8	Sacandaga Lake Outlet	43	58
1008000	Route 10	West branch Sacandaga River	49	64
1002490	Route 5	Zimmerman Creek	37	73
1039060	Route 171	Moyer Creek	51	81

Source: Reprinted from ASNT Publication.

the adequacies and limitations of load testing as a means of load rating. McCaffrey (2006) observed that bridge load ratings are typically determined by analytical methods based on information taken from bridge plans, supplemented by information gathered from field inspections or field testing. Bridge load ratings are used in a variety of program areas that include, but are not limited to, overload permit review, bridge load posting determination, and capital program development. These results are also used in conjunction with other bridge inventory and inspection information to determine the Federal Bridge Sufficiency Rating, which, in turn, is a factor used to determine the eligibility of a project for the Highway Bridge Replacement and Rehabilitation Program.

In the rest of this section we discuss different objectives of load tests. At the end of the section we present a practical case study showing how maximum efficiency can be obtained by setting well-defined goals at the outset of a project in addition to good planning and execution.

8.2.1.1 Load Rating/Condition Assessment

Load testing can be used to evaluate bridge load rating. See NYSDOT (2005), NYSDOT RR 163, and 153, TRB (1998), and Barker (2001). AASHTO MCEB (2001) also describes in detail the test steps for evaluation of bridges. We note here that inaccurate load testing can result in an inaccurate load rating, so adequate preparation before the test and accurate test analyses are needed to ensure correct bridge rating.

Inventory versus Operating Ratings: Inventory rating is the live load that can safely utilize the bridge for an indefinite period of time where as operating rating is the maximum permissible live load that can be placed on the bridge. It should be noted that allowing unlimited load crossings at the operating rating level can reduce bridge life.

8.2.1.2 Other Operational Goals

There are several other situations where load testing can help operational bridge decisions. For example, if a bridge's maximum load capacity is below the maximum legal load, the bridge may be

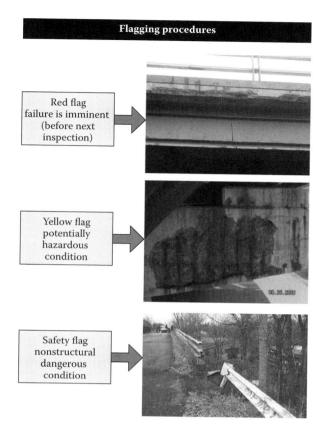

FIGURE 8.6 Flagging procedures. (Courtesy of New York State Department of Transportation.)

restricted to some loads (posting), closed, or owner may be alerted of the critical finding (known as "flags" in NYSDOT terminology) for further evaluation (see Figure 8.6).

8.2.1.3 Model Verification (STRID)

Load testing can be used to verify analytical models or design assumptions for bridges. Some popular design assumptions are (1) lack of composite action, (2) load distribution over multiple girders, and/or (3) actual boundary conditions of girders, frames, or columns. To achieve this goal, structural identification techniques (Chapter 6 by Ettouney and Alampalli 2012) need to be used. Dynamic load testing might be needed for modal identification of the bridge.

8.2.1.4 Maintenance, Retrofit, and Rehabilitation

As load testing might reveal damage or deteriorating conditions, accurate decisions can be made on the need for specific maintenance, retrofit, and rehabilitation (MR&R) activities.

8.2.1.5 Innovations

As new construction materials, new structural systems, or new design paradigms are introduced, load testing becomes an invaluable method to assess the immediate and long-term performance of those innovations. See Chapters 6 and 7 for testing of new fiber-reinforced polymer (FRP) components and bridge decks.

8.2.1.6   Overloads—DMID

Increasing truck load demands need to be investigated both analytically and experimentally. The behavior of the bridge under higher loading demands should be monitored first by load testing; it also needs to be verified analytically. When an infrequent overload condition is required, proof testing might be performed. Such infrequent demands by overloads from a bridge system can also occur if the bridge is to be used as a detour route.

8.2.1.7   Quality Assurance/Quality Control

Bridge load testing can also be beneficial for quality assurance (QA)/quality control (QC) operations. NYSDOT (2005) has a useful definition of QA as "The use of sampling to verify or measure the level of the entire bridge inspection and load-rating program." As such, load testing a few bridges within the bridge network can be used as a reasonable QA program.

8.2.1.8   Improving Inventory Database

Any load test will produce information about bridge behavior due to prescribed loading. The results and observations gathered during the test can then be added to the bridge inventory database. The database can be used to produce trends and statistics for future decision making. There are several examples in this chapter of the use of statistics of databases (see, for example, Chapter 8 by Ettouney and Alampalli 2012).

8.2.1.9   Importance of Identifying Objectives, Good Planning, and Execution of Test

Limited goal experiments with well-defined goals can be very successful. Good planning and execution will improve the rate of return of the experiments. Structural health and monitoring (SHM) does not need to be too grand or far reaching, as exemplified by the load testing of the Coeymans Creek Bridge offered by Hag-Elsafi and Kunin (2002). Coeymans Creek Bridge in New York was monitored during the crossing of superload permit trucks. The bridge is an integral-abutment structure consisting of prestressed concrete box beam members and a composite concrete deck. The superload permits were approved based on conservative analyses, recommending crossing of the bridge in two patterns: steering a crabbed trailer across the bridge and driving a trailer in a diagonal crossing fashion. Crabbing was recommended for trailers with gross weights approaching 1775 kN and diagonal crossings for lighter weights. The low rating of the structure and the unusually heavy loads prompted a need for investigating actual stress levels in the bridge beams. The bridge was instrumented and strain data collected during the crossing of superload trucks.

The bridge was built in 1985 as a single-span, integral-abutment structure, consisting of eleven 32.6-m-long, adjacent prestressed concrete box beams and a 15.2-cm-thick composite concrete deck. The three center (interior) beams are 0.91 m wide and the remaining eight (exterior) are 1.22 m wide. The interior and exterior beams, respectively, were prestressed using 34 and 42, 1862-MPa 13-mm strands. The operating and inventory ratings of the bridge are 86.1 and 33.5 metric tons, respectively.

Analysis of midspan strains for a crabbed and a diagonally driven superload trucks (Figure 8.7), indicated that, for this bridge, there were no clear advantages for one crossing pattern over the other, because the trailers carrying the loads assumed similar orientation with respect to the bridge centerline during the crossings. Figure 8.8 shows concrete stresses at the midspan of Beam 8 (one of the most stressed beams), based on monitoring results (actual), assuming simply supported conditions, and adjusted simply supported conditions (adjusted by multiplying the simply supported stresses by a factor reflecting the effect of fixity). From this (Figure 8.8), it is clear that the moments in the bridge beams induced by the superload trucks remained well below the beam's cracking moment. That is, bridge integrity was not compromised.

FIGURE 8.7   Superload on the bridge. (Courtesy of New York State Department of Transportation.)

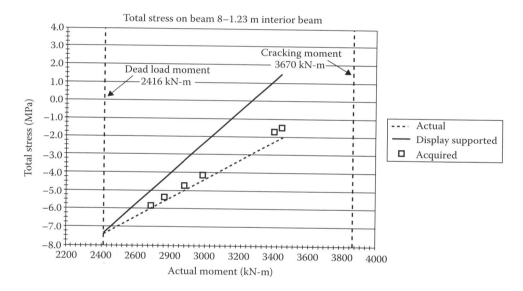

FIGURE 8.8   Actual and adjusted stresses. (Courtesy of New York State Department of Transportation.)

8.2.2   Typical Load Test

8.2.2.1   Pretest Design

Pretest design and planning include the following issues:

- Detailed definitions of test goals and main objectives, secondary objectives, and potential nondesigned side-effects (see the serendipity principle in Section 4.7 in Chapter 4 by Ettouney and Alampalli 2012)
- Number, type, and location of sensors
- Adequate hardware, instrumentations, and software
- Type and quality of information: manual, digital, photographs
- Labor, material, and equipment

- Costs and benefits of the test
- Other plans for decision making to utilize all test information
- Readjust plans if needed
- Assembling and disassembling the test equipment
- Detours, if needed
- Coordination with disciplines (local authorities, police, etc.)
- Contingency plans

8.2.2.2 Performing the Test

The test process integrates all test components such as sensors, wiring (if any), instrumentation, loading processes as well as any QA/QC plans. Truck-loading processes include loading combinations and spatial locations. Also, the loadings can be static or dynamic. The plan should also include measures for redundancy and repeatability of measurements. Alampalli and Hag-Elsafi (1994) described a typical static load test procedure which we reoffer, with some modification, as follows

Prepare test instrumentation: A multiplexer is used to scan data from a sensor group or from all sensors during a test, usually with one signal conditioner allocated for each sensor group. Although this practice is cost-effective, it may not be possible to set different excitation voltages, gains, and filters for individual channels. Also, when mixed types of sensors are used, it may not always be possible to set optimum input voltages for individual sensors. For example, even when the same type of strain gauges are used, when mounted on different materials, optimum input voltages depend on the characteristics of individual material thermal conductivity. Before the test, data collection sequence has to be specified, and parameters such as input voltage and gain for an individual or a group of channels have to be determined. Once all the circuitry is checked, zero readings are recorded. This ensures a QA of recorded measurements.

Test procedure: The structure is loaded in steps (Figures 8.9 and 8.10) according to the test plans and data collected at the end of each step in digital form, using an analog-to-digital (A/D) converter. Data is normally stored in a file. Then, the second load increment is applied, and the collected data is appended to the previous file. This process is repeated for each load increment. After the final predetermined load is reached, the structure is unloaded by removing the loads, generally in reverse order, while recording the response. Data is monitored continuously to observe structural behavior and to note any permanent deformations under the applied loads. Such tests are repeated to eliminate random

FIGURE 8.9   Truck loading process. (Courtesy of New York State Department of Transportation.)

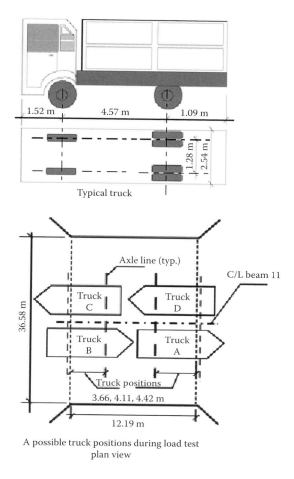

FIGURE 8.10 Typical load during a bridge load test. (Courtesy of New York State Department of Transportation.)

noise effect by averaging and checking for nonlinearities and permanency. Whenever possible during the test, zero readings (data recorded with no load on the structure) are recorded to correct for drifts and experimental bias. Reported strains usually range from 1 to 400 microstrains) and could be much lower than 1 microstrains when monitoring is conducted during the construction stages. The testers need to try to resolve any apparent inconsistent data in the field before winding down.

8.2.2.3 After-Test Steps

Alampalli and Hag-Elsafi (1994) observed that once static load testing is completed, the collected data is normally sorted by channels using a computer program, and the generated data file is imported to a spreadsheet. Load effects are calculated after subtracting earlier recorded zero values from the data acquired at the end of each load step. Collected data is manipulated in the spreadsheet environment to account for gain and excitation voltages for each channel. Engineering parameters (such as maximum strains, stresses, and inclinations) of interest to the test are calculated by applying appropriate formulae.

As of the writing of this chapter, most of the available software do not provide online data display and analysis capabilities. If the results at any intermediate stage between load increments are required during the test, the test has to be stopped and the data imported to the spreadsheet environment to obtain the desired values. Also, for relatively long test times, factors such as drift and environmental conditions may influence test data. Most of the software do not have provisions to account for these factors.

8.2.2.4 Test Manpower and Duration

Manpower during the test varies greatly depending on the goals and complexity of the test. Different experiences are needed during the test. They include experienced sensors and instrumentation professionals, drivers loading trucks, and overall managers of the test. Sometimes representatives of bridge owners and overseers are present during the test. Proof tests require experienced engineers and test planners to be present so as to make real-time decisions on loading increments.

Duration of load tests varies from a few hours to a few days, depending on the complexity of the bridge and the test goals.

8.3 CATEGORIES OF LOAD TESTS

8.3.1 GENERAL

Because of their versatility, there are many ways to categorize bridge load tests. For example, several types of loads can be applied to the bridge during the test. Also, load tests can be categorized by the magnitude of those loads and the test goals themselves. Two broad categories are diagnostics and proof load tests. Another category is the type of results that can be produced by a load test. A versatile type of result that is not well utilized is static influence lines. We explore all those load test categories as shown in Figure 8.11 in this section. We also introduce the concept of DILs and show that it is closely related to the well-known frequency response function (FRF). Finally, we propose a new load test category: brittle failure monitoring. We show that such a phenomenon should be of concern to the bridge community. We also suggest further research and testing to make such a test a regular structural health-monitoring activity.

8.3.2 LOADS DURING TESTS

8.3.2.1 Loading Types

Applying loads to the bridge during a load test can be achieved in many ways. In all situations, the following guidelines apply:
- Magnitude and location of the loads must be accurately measured
- Dynamic effects of the loads must be minimized during static tests

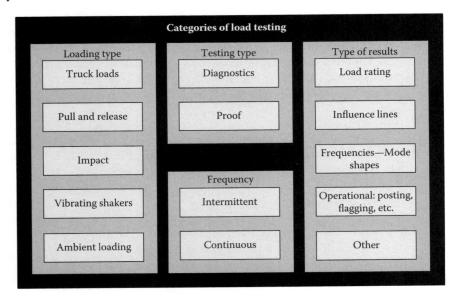

FIGURE 8.11 Categories of load testing.

- Temperature and other environmental effects such as humidity must be monitored and recorded.
- Loads should not cause permanent damage to the bridge.

Some of the important loading types are discussed next.

8.3.2.1.1 Truck Loads

Using truck loads in a load test is the most common type of loading. This is due to the ease of deployment and the versatility of applying the loads. The trucks can also be placed at almost any location along the bridge, thus simulating live-load conditions in an accurate manner. Factors that should be carefully monitored during truck loading are speed, location, and exact weight of trucks.

8.3.2.1.2 Pull and Release Loading

Pulling the bridge down by a set of cables and hydraulic jacks has been used in load testing. The method is useful only for smaller bridges; it is not clear if it can be classified as a nondestructive testing (NDT) method (see NCHRP 1998).

8.3.2.1.3 Impact

Biggs (1964) discussed the theoretical basis of the impact factor used in modern bridge designs. He showed that the factor results from the dynamic interactions between the bridge and the moving vehicle. Lih et al. (2004) studied the effects of different parameters on the impact coefficient of bridges. Among their findings were that (1) the impact coefficient is not linearly related to the vehicle frequency, (2) increased vehicle damping should result in a reduced impact coefficient, and (3) the impact coefficient increases as the speed of the vehicle increases. The latter result is consistent with the earlier Biggs (1964) findings. They also found that the road surface has direct and nonlinear effect on the impact coefficient: it is obvious that road surfaces should be kept as even as possible to reduce such impact effects.

In dynamic (modal load testing, the load can be impacted to the bridge by using either a simple impact hammer or a hydraulic hammer. Because of the weight of hammers required, the method is practicable only for short-span bridges. Green (1995) recommends that a 30-m-span might be an adequate limit for testing bridges with drop hammers. In general, background noise problems need to be resolved, since the duration of the input loading (impact) is much shorter than the recorded output (see Salawu and Williams 1995).

8.3.2.1.4 Vibrating Shakers

Hydraulic shakers can be used to impart loads on bridges. The shakers are versatile in the type of wave forms they can generate: harmonic, random, or a prespecified range of spectral input forces. This method of loading is compatible with many STRID methods. The force amplitudes vary in the range 5 to 90 kN (Green 1995). The use of shakers is labor intensive and costly.

8.3.2.1.5 Ambient Loading

Measuring only output structural responses under normal traffic conditions, without any needed effort for measuring input forces, has several advantages: (1) suitable for continuous monitoring, (2) suitable for all bridge sizes, (3) traffic friendly, and (4) no cost of measuring input loads. There are several STRID methods compatible with this form of load testing (see Chapter 6 by Ettouney and Alampalli 2012).

8.3.2.2 Comparison of Loading Methods in Load Tests

Table 8.3 compares different loading methods in load tests. It is based in part on the works by Karbhari et al. (2005).

TABLE 8.3
Comparison of Loading Methods in Bridge Load Tests

Method	Advantages	Disadvantages
Truck loads	Widely used, suitable for bridge rating, inexpensive, versatile in application, can be static or dynamic	Limited dynamic applications, require care in truck speed and locations
Pull and release	Suited for smaller bridges	Can damage the bridge, difficult to produce three-dimensional modes, not suited for large bridges
Impact	Simple to implement, compatible with many STRID methods	Not suited for large bridges, not suited for continuous monitoring, ambient noise can affect accuracy
Vibrating shakers	Compatible with many STRID methods	Can be expensive, labor intensive
Ambient loading	Compatible with continuous monitoring, long-span bridges, cost-effective	Analytical methods can be difficult to implement, some modes of the structure cannot be generated by ambient vibrations alone

8.3.3 Diagnostics Load Test

8.3.3.1 General

Diagnostics load tests are a versatile class of load testing of bridges. The loads, generally trucks, are applied to the bridge in a controlled fashion. The response of the bridge is monitored through a set of sensors. There are several reasons for conducting diagnostics load testing:

- More realistic rating factor: NCHRP (1998) recommended that diagnostics test be conducted to improve rating factors lower than HS20. In addition, bridges that have high redundancies, such as multigirder steel or concrete bridges, are good candidates for diagnostics load tests. NCHRP (1998) suggested caution in trying to improve analytical rating factors for two-girder bridges or low-redundancy bridges.
- Improve knowledge of important parameters such as load distribution or impact factors.
- Improve assumptions about material properties, boundary conditions, effectiveness of repair, unintended composite actions, and/or deterioration effects.
- Improve understanding of nonstructural component effects such as parapets or noncomposite bridge decks.
- Validate design and/or analysis models and assumptions.

Bridge rating methods based on diagnostics load test are discussed in Section 8.5.

8.3.4 Proof Load Test

8.3.4.1 General

Proof load tests are performed in one of the following conditions:

- When analytical (calculated) ratings are low and more realism is needed
- If dead load to live-load ratio is unusually large
- It is difficult to model the bridge analytically; for example, older bridges, insufficient information, large uncertainties, lost documentation, and so on

Proof load tests involve estimating a target load to be used. The target load is achieved by increasing the test load incrementally on the bridge and monitoring the bridge performance. The test continues until one of two conditions is reached:

1. The estimated target load is reached
2. Signs of distress are observed

In what follows, we discuss several aspects of proof load tests. For detailed discussions, the reader is referred to Fu (1995) or NCHRP (1998).

8.3.4.2 Target Loads

One of the most important decisions in proof tests is defining target proof load L_{TARGET}. We explore two different expressions for L_{TARGET} as follows:

Reliability-based L_{TARGET}: Fu (1995) suggested a target proof load as

$$L_{TARGET} = \frac{1}{\phi}\left(\alpha_L L_n \left(1 + I_n\right) + \alpha_D D_a\right) \tag{8.1}$$

The parameters of Equation 8.1 are

ϕ	=	Resistance reduction factor
α_L	=	Target live-load factor
L_n	=	Nominal design live load
I_n	=	Nominal impact factor
α_D	=	Additional dead load factor (usually taken as 1.25)
D_a	=	Additional dead load effect that might be added to the structure after proof load test.

Factor α_L depends on the analytical rating factor, R; it was computed by Fu (1995), using a large sample of bridge inventory. Its value is shown in Table 8.4.

Experience-based L_{TARGET}: NCHRP (1998) offered a target proof load as

$$L_{TARGET} = \left(\alpha_L L_n \left(1 + I_n\right)\right) \tag{8.2}$$

The target live load factor is defined as

$$\alpha_L = \alpha_{L0} \prod_{i=1}^{i=6} \alpha_i \tag{8.3}$$

TABLE 8.4
Values of Target Live-Load Factor α_L

Condition	$R \geq 0.7$	$R < 0.7$
Low-volume roadways, with reasonable enforcement and apparent control of overloads	1.35	1.45
High-volume roadways, with reasonable enforcement and apparent control of overloads	1.45	1.55
Low-volume roadways, with significant sources of overloads without effective enforcement	1.80	1.90
High-volume roadways, with significant sources of overloads without effective enforcement	1.90	2.00

Source: Fu, G. Highway bridge rating by nondestructive proof-load testing for consistent safety. New York State Department of Transportation, NYSDOT, Research Report No. 163, Albany, NY, 1995. Courtesy of New York State Department of Transportation.

The factor αL_0 is

$$\alpha_{L0} = 1.4 \tag{8.4}$$

The factors in the sum in Equation 8.3 are computed from Table 8.5 and used only whenever valid.

8.3.4.3 Acceptance Criteria

There are several qualitative and quantitative guidelines for the termination point of proof load tests; for example, Fu (1995) suggested one of four conditions for the termination:

1. Target proof load is reached
2. 10% or more nonlinearity is observed
3. Significant movements or settlements are observed
4. Appearance of signs of distress (crack widening or significant crack developments)

8.3.4.4 Bridge Rating

Assuming that the final proof load during the test is L_{PROOF}, the bridge rating, according to Fu (1995) is

$$R = \frac{\phi L_{\text{PROOF}} - \alpha_D D_a}{\alpha_L L_n \left(1 + I_n\right)} \tag{8.5}$$

NCHRP (1998), on the other hand, offered an operating level rate factor as

$$R_{\text{OPERATING}} = \frac{L_{\text{PROOF}}}{\alpha_L L_n \left(1 + I_n\right)} \tag{8.6}$$

An inventory level rating is

$$R_{\text{INVENTORY}} = 0.73 \left(\frac{L_{\text{PROOF}}}{\alpha_L L_n \left(1 + I_n\right)} \right) \tag{8.7}$$

TABLE 8.5
Values of Target Live-Load Factor α_i

i	Condition	α_i
1	One lane governs the bridge response	1.15
2	There are fracture critical details or no redundant load paths	1.10
3	Routine inspections are conducted in accordance with AASHTO C/E manual	1.10
4	The structure is rateable and the calculated rate factor exceeds 1.0	0.95
5	The test is stopped because of signs of distress before reaching L_{TARGET}	0.88
6	Additional factors, including traffic intensity and bridge condition (see Moses and Verma 1987)	

Source: NCHRP manual for bridge rating through load testing, National Cooperative Highway Research Program, NCHRP, Research Results Digest, No. 234, Washington, DC, 1998.

8.3.5 Influence Lines

The concept of influence lines has been in use for bridge analysis and design for many years. This is due to its simplicity and usefulness. The simplicity comes from the fact that a single loading condition can reveal the behavior of the whole structural system; hence its usefulness. The concept of static influence lines has been applied recently to bridge load-testing projects. This section discusses the concept of static influence lines and their applications to load testing. We then introduce the concept of DIL: a generalization of their static counterparts. We then discuss the potential of applying DIL to load testing. We finally observe the similarities and differences between DIL and Frequency Response Function (FRF).

8.3.5.1 Influence Line Load Test (Static)

Static influence line test is performed by moving a load, generally trucks, along the bridge while measuring the desired influence line metric. The load movement is usually slow so that inertia does not affect the measurements. The concept of influence lines is based on the reciprocity principle in the theory of structures and is illustrated in Figure 8.12. Influence lines can be useful in showing maximum value of the influence line metric and the location of the load that would produce such maximum value. They can be computed analytically. Also, they can be measured during load tests. For example, Alampalli and Fu (1992) used the concept of influence lines to study stress distributions in isotropically reinforced concrete slabs on steel bridges. By moving the test truck slowly along the slab, and continuously measuring the strains in the desired location, it is possible to establish the stress influence line as shown in Figure 8.13. The influence line can then be used in design assessment as needed. Note that there are two peaks in the influence line of Figure 8.13 due to the two axles of the test truck.

Traditionally, influence lines are produced using only a single moving concentrated load. So, the loads with two or more axles would produce nontraditional influence lines. Hirachan and Chajes (2005) introduced a method that can extract single load influence lines from multiaxle influences lines.

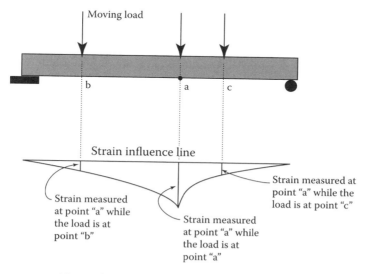

Measured static influence line during load test
(no inertial effects)

FIGURE 8.12 Concept of static influence lines.

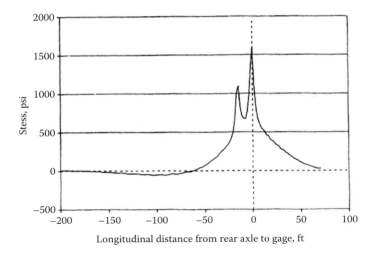

FIGURE 8.13 Stress influence line from a two-axle truck loading. (Courtesy of New York State Department of Transportation.)

8.3.5.2 Dynamic Influence Lines

Ettouney (1980) generalized the concept of static influence lines into DILs where inertia effects are included. For a steady-state motion at a driving frequency of Ω, the reciprocity principle still holds. It can be shown that

$$u_{ij} = u_{ji} \tag{8.8}$$

Where u_{ij} is the harmonic displacement at degree of freedom i due to a unit load at degree of freedom j and u_{ji} is the harmonic displacements at degree of freedom j due to a unit load at degree of freedom i. Based on the above, the DILs can be generated by a slowly moving unit harmonic load with a driving frequency of Ω. The resulting influence lines will have complex values $U = u_r + \sqrt{-1}\, u_i$, with u_r and u_i as the real and imaginary components of U. The phase angle is $\phi = \tan^{-1}(u_i/u_r)$ representing the phase difference between U and the applied unit load (Figure 8.14).

Dynamic influence lines U is obviously continuous function in space and frequency such that

$$U = U(x,\Omega) \tag{8.9}$$

Note that x is the direction of the moving unit harmonic load. If the influence line is measured at N_x discrete points and N_Ω discrete frequencies, it can be represented in the matrix form $[U]$. The matrix has N_x rows and N_Ω columns. The jth column of this matrix $\{U\}_j$ represent the dynamic influence at the jth frequency, $j = 1,2...N_\Omega$.

8.3.5.3 FRF and DIL

Upon further inspection, we observe that the DIL is a special case of the system's FRF (see He and Fu 2001). Recall that for a given multidegrees of freedom system, the ith and jth component in an N order FRF, $\alpha(\Omega)_{ij}$ is defined as the displacement in the ith direction due to a unit force in the jth direction, with a driving frequency of Ω. This is the exact definition of DIL. The main difference is that the points of interest of DIL are along a straight line, usually the moving force, or traffic in case of bridges.

This interrelationship between DIL and FRF is important since there is a large body of knowledge about FRF, their properties, and how to measure them in practice. For example, He and Fu (2001)

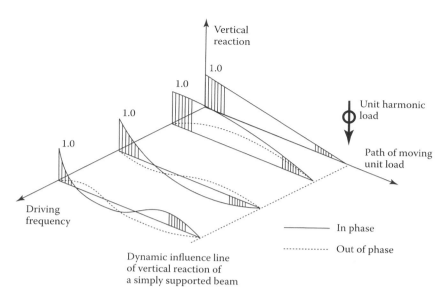

FIGURE 8.14 Dynamic influence lines.

described at length the methods of measuring FRF for linear systems. FRF can be measured from sinusoidal, random, pseudorandom, and impact excitations. There are needed precautions and data checking when measuring FRF; both were described at length by He and Fu (2001). The use of these techniques and procedures in generating DIL for bridges can be of value for bridge STRID as well as bridge design.

8.3.5.4 Measuring Influence Lines during Load Tests

Hunt and Helmicki (2002) describe a load testing conducted to check if a superload can cross the bridge as the routine analysis showed possible overload conditions in connection with this load/structure combination. Field test with known loads projected to superload showed that the structure can withstand the load safely. This was verified by monitoring the structure during the superload passage.

The bridge was a two-lane, six-span, 650-ft steel stringer bridge with reinforced concrete deck, built in 1963, which crosses Muskingum River. The structure was instrumented with strain transducers, and a series of controlled load tests were conducted using two dump trucks of known weight. The results showed that the distribution factor was 0.56 as compared to 0.75, according to AASHTO specifications. *Static influence lines* were generated from load tests and then used to simulate the overload response with good results. Simulations were within 10% of actual stresses for bottom flange of the critical regions, which exhibited good predictive capabilities and consistent linear behavior.

Finite-element method was then used to predict response in other locations after calibrating with field data. Based on these simulations and field data, appropriate recommendations were provided. Tests were repeated the day before superloads, during, and after the loads passed, to verify the capacity ratings. The tests did show that the bridge capacity was negatively affected by these overloads.

8.3.6 Brittle Failure Test (Monitoring)

After the disaster of the I-35 bridge failure (see Chapter 3 by Ettouney and Alampalli 2012), it is important to monitor bridges or any other type of structure susceptible to sudden or brittle failure. We define brittle failure as the type of failure that occurs suddenly with little or no warning signs. There is a similar concept that has been well known in the bridge community for a long time, which is called fracture critical condition (see FHWA, BIRM 2002). Fracture can occur in members or bridges that satisfy two criteria:

- Axial stresses or axial forces are tensile within the member
- Nonredundant system

Criteria for fracture criticality are shown in Figure 8.15.
  Redundancy is categorized by FHWA, BIRM (2002) as one of three types:

- Load path redundancy: This type of redundancy is satisfied when the bridge system has three or more girders (see Figure 8.16).
- Structural redundancy: This type of redundancy is generally satisfied when the number and type of supports of single girder is of more than certain value. For example, continuous girders (Figure 8.17) can be considered as having structural redundancy.
- Internal redundancy: Depending on the geometry of the bridge frame or truss, it can exhibit internal redundancy.

It was argued that fracture critical bridges or members that satisfy the above two criteria are susceptible mostly to fatigue; hence the tensile stress equipment. Because of the danger posed by fatigue damage and the suddenness of the potential failure, the bridge community has always been careful about fatigue monitoring (Chapter 5) and fracture critical bridges and members in general.
  We would like to point out that there are other classes of structures (including bridge structures) that are also susceptible to brittle failure. We remind the reader of the damaging effects of compressive stresses and compression members. Remember that the overall stability of members or whole systems is dependent on the inequality

$$\sigma_{cr} \geq \sigma_a \qquad (8.10)$$

The critical buckling stress (capacity) is σ_{cr}, and the applied compressive stress (demand) is σ_a. As long as the above inequality is satisfied, the member or the whole system is stable. If the above

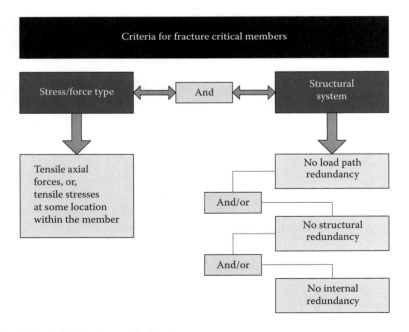

FIGURE 8.15 Criteria for fracture critical systems.

FIGURE 8.16 Load path redundant bridge. (With permssion from National Highway Institute.)

FIGURE 8.17 Structurally redundant bridge continuous girders. (With permission from National Highway Institute.)

inequality is breached, then the member or the whole system becomes unstable. In such a situation, the member or the whole system will fail in a brittle fashion.

Unfortunately, detecting brittle failure due to compression is much more difficult than detecting brittle failure due to tension. This is because both sides of Equation 8.10 need to be monitored: the capacity and demand need to be compared almost constantly when there is susceptibility of compression brittle failure.

The capacity of a compression member, a member with mostly compressive stresses (such as gusset plates), or a whole system, such as bridge bent or columns, can be reduced over the years. Some examples of reduced capacity is loss of area due to corrosion or delaminations, construction or design errors, or loss of footing supports due to scour. The demands can increase also. For example, changed traffic patterns, construction loads, temperature effects, or uneven soil settlement can cause demand increase.

This leads us to suggest the importance of devising monitoring tests that can detect compression-based brittle failure. As of the writing of this chapter, there is no ongoing effort to address such a serious knowledge gap.

8.4 SENSORS, INSTRUMENTATIONS, HARDWARE, AND SOFTWARE

8.4.1 GENERAL

Ettouney and Alampalli (2012, Chapter 5) discussed sensors and instrumentations in general. In this section we further discuss the subject as it applies specifically to bridge load testing. We offer many practical guidelines and observations that can be of help to professionals.

8.4.2 TYPE OF MEASUREMENTS

There are many types of measurements (TOM) during load testing; this is due to the large possible objectives of load testing. For example, most load tests aim at measuring strains at different bridge components. In addition, measuring displacements is also part of many load tests. During displacement measurement, one must accommodate the potential of linear, recoverable, and nonlinear displacements. Also, relative displacements and rotation of components can be of interest. Other TOM can be ambient temperature, forces, rigid body motion, slippage, crack openings, and humidity.

8.4.3 SENSORS

Static bridge load testing uses strain gauges, inclinometers, load cells, linear variable, displacement transducers (LVDTs), and demountable gauges (see Figure 8.18). For dynamic tests, accelerometers might be used. These sensors were discussed in Chapter 5 by Ettouney and Alampalli (2012).

Selection of sensors and test planning require examination of the bridge and the objectives of the load test to identify the important response parameters to be measured. In this section we reiterate the sensor selection guidelines described by Alampalli and Hag-Elsafi (1994).

Purpose of the sensor: Although many sensors work on similar principles, they are not equally effective in measuring certain parameters. Thus, purpose of sensor is an important factor in sensor selection.

Test type: Test type is an important factor since it is not always possible to use the same sensor in each test. For example, many of the commercially available inclinometers require a longer response time to obtain stable data; therefore, sampling speed or load cycle time is important and should be given special consideration. Sensor selection depends on loading type and expected structural response in each test. In many cases, different installation procedures and adhesive types may be required, depending on whether static or dynamic load effects are measured, so that the gauges may survive structural response duration.

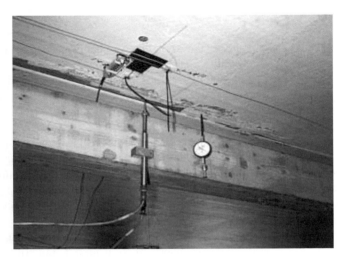

FIGURE 8.18 Typical sensor arrangement during load test (strain gauges, LVDT, and dial gauges). (Courtesy of New York State Department of Transportation.)

Test environment: Test environment plays a major role in choosing a sensor, especially when one considers the fact that many of the available sensors are designed for laboratory applications and may not sustain inclement weather conditions without proper protection. However, there are a few rugged sensors specially manufactured for use in variable or hostile environments which are more suitable in bridge testing. If these sensors are unattainable, proper weatherproofing (against temperature, dust, water, wind, vibration, etc.) should be specified to meet the necessary requirements for sensor operation. Even though they are relatively expensive, rugged gauges are a must for long-term monitoring. When weatherproofing is required, the entire measurement system, including connecting cables, should be properly protected.

Data range and resolution: Response range during the entire test period has to be estimated before the sensor selection is made. Accuracy of some sensors varies when they are used to measure data at different ranges. Some sensors behave linearly only for a small part of their operable range and may behave differently for positive and negative responses. Hence, the sensor's operating range and required accuracy should be chosen with the expected response range in mind.

Response resolution: Resolution is often coupled with data range. Normally, the minimum possible range has to be used to obtain maximum resolution. Most available instrumentation converts analog response to digital response using A/D converters in a fixed number of steps. For example, 20 microstrain data range with 12-bit A/D gives a resolution of 0.078 microstrain, and when the same A/D is used with 200 microstrain range gives a resolution of 0.78 microstrain. If expected structural response is less than 10 microstrain, a sensor with a range close to 10 microstrain should be chosen. At the same time, when the data range is around 200 microstrain, and if a resolution of 0.078 microstrain is required, an A/D with more number of steps has to be used.

Loading conditions: Some sensors do not operate effectively under certain types of loading. Each sensor takes a brief while (reaction time) to stabilize, depending on its principle of operation. For example, many LVDTs and inclinometers require longer reaction time compared to strain gauges. If structural response changes faster than the sensor reaction time, accurate data cannot be collected even when high sampling speeds are used. If transient or dynamic response is to be measured, sensors with short reaction times should be chosen. Also, when long-term monitoring is planned, possible instrument drift should be determined.

Technical skills available: Sensor selection also depends on available technical skills. While a large number of sensors can be installed with relative ease, a few installations need more experience and technical skills to obtain reliable and consistent data. For example, some strain gauges are manufactured and sold with cables which eliminate the need for field soldering and connections. However, they are more expensive than their counterparts without cables—a trade-off for the time saved and the lesser need for gauge installation expertise.

Software and hardware compatibility: It is very important to select sensors that are software and hardware compatible. The excitation voltage, output voltage, wiring scheme, overload handling conditions, and resetting time of the hardware and software should be considered when selecting a sensor. It is possible to obtain sensors of very high accuracy, but if the software and hardware are less reliable, it may be more economical to choose sensors with the same order of reliability. For example, hardware is manufactured with either a fixed output range or a capacity to choose different ranges. If both systems have the same 12-bit A/D boards, the latter system will give better accuracy and resolution; therefore, the sensor data range may not be very important in this case.

Structure type: The type of test structure is also critical in choosing a proper sensor type in some cases. Different adhesives and bonding procedures have different effects on material types. Also, the prevailing standards and specifications should be considered. In steel bridges, drilling holes for gauge installations in tension zones of bridge girders may not be permitted. Strain gauges are made with materials to match thermal conductivity of structural materials to increase accuracy and reliability of measured data.

Miscellaneous: These guidelines do not cover the entire criteria for sensor selection. Several factors such as cable types, distance between test structure and instrumentation, cable lengths, surrounding environment, and electrical installations are also important in deciding the right sensors for the test. For example, if the cable lengths between instrumentation and sensors are very long, then sensors with built-in amplifiers or wireless sensors should be considered to reduce noise/signal ratio.

These specific guidelines for load-testing environment, coupled with the general guidelines of Chapter 5 by Ettouney and Alampalli (2012), can ensure accurate, reliable and cost-effective sensor selection. We conclude this section by reemphasizing the role of past experiences and good engineering judgment in sensor selection.

8.4.4 INSTRUMENTATION

With emerging technology, many new instrumentation, data acquisition systems, and processing software are becoming available. Many of these systems are built for general testing and offer a wide range of choices such as customized analysis procedures, controlled sampling speeds, and improved accuracy. The selection of appropriate features is often governed by available technical skills, frequency of testing, test type and conditions, and availability of funds. Alampalli and Hag-Elsafi (1994) described the general technical guidelines and recommendations on the features to be considered when purchasing a test system. We reiterate their recommendations next.

Commercially available data acquisition and analysis systems can generally be divided into two main categories (Ashour et al. 1993): (1) Several independent instrumentation devices grouped together as a test and measurement system with display and computation capabilities usually controlled by a personal computer. However, there are integrated systems as well in this category which are capable of interfacing with microcomputers and accessing their drives for additional data processing; and (2) Systems consisting of several add-on cards installed inside a computer and controlled by resident software packages, primarily to collect data from sensors. This data can then be analyzed by a host computer. Normally, these systems have two modules of cards, referred to as daughter and mother/master boards. The mother/master board resides inside the computer. The daughter board includes an A/D converter, usually housed in an external box and connected to the mother board through cables. This is done to protect the A/D converter from the computer electric noise and to increase signal/noise ratio by keeping it close to the sensors. These systems are relatively less expensive than the former. However, they are not easily expandable (expandability is limited by the number of boards that can be supported by the computer).

The maximum sampling speed for each channel is also very important when conducting dynamic tests. Required sampling speed can be calculated by multiplying desired maximum sampling speed/ channel by the maximum number of channels. These channels should include any possible future expansion. If additional channels are added later on, by adding cards or by supplementing the system with one or more units, problems may arise if higher sampling speeds are desired and were not initially considered. Updating the system with add-on cards for high sampling rates is sometimes quite expensive.

Necessary sources of excitation depend on the type of sensors being used. Also, it is preferable to have both constant voltage and constant current for each channel and have the hardware capable of detecting simple wiring connection errors. It is necessary, in some applications, to have a few D/A conversion channels and interrupts as they may be needed for setting external triggers needed in bridge testing. The system should also be suitable for providing software selectable gains, filters, excitation voltages, and high-speed, high-volume data sampling.

A/D converter selection should be based on required accuracy in test measurements, which also depends on data range and data resolution described earlier in the paper. Currently, A/D converters with 16-bit resolution (65 536 steps) and systems that use the entire A/D conversion range in selected data ranges are available in the market.

Test hardware should accommodate mixed types of sensors such as strain gauges, LVDTs, and inclinometers although each one may have its own requirements. For example, strain gauges require electric bridge completion circuits (depending on strain gauge resistance and number of active arms).

Finally, systems should be able to communicate with host computers and other instrumentation, using standard interfaces such as IEEE, RS-232, USB, and serial ports. This gives the flexibility to upgrade host computers in the future to meet changes in the complexity of postprocessing needs. It is also desirable to have high RAM "memory" for stand-alone instrumentation, especially during high-speed and high-volume data sampling.

8.4.5 SOFTWARE

Software in load testing plays a major part in collecting, displaying, analyzing, and storing data. As such, it should be selected with extreme care. Software selection is also very important while choosing a data acquisition system. For efficient interfacing and operation, the software should be user-friendly, menu-driven with online help, and be compatible with the computer operating system in order to take full advantage of available storage and memory. Alampalli and Hag-Elsafi (1994) offered the following guidelines for selecting software to be used in bridge load testing:

- Provide real-time display and scan at various scanning speeds
- Have the ability to support different test setups
- Be able to adjust gains, filters, excitation voltages, and so on for each channel
- Be able to save test setups and definitions together with collected data
- Be able to save, recall, and replay the data and test setup
- Provide pseudochannels to manipulate data from different channels using user-
- Defined formulae for on-site real-time display
- Provide intelligent trigger modes and sources
- Have the capability to store data in local computer disk drives in real-time during
- High-speed and high-volume data acquisition
- Have autobalancing features
- Be able to provide alarm conditions to detect abnormalities in collected data
- Have a built-in formulae library to solve rosette info, and so on online at the test site
- Have provisions to account for gauge drift, temperature, and so on
- Have internal and external calibration functions
- Provide overload protection
- Record pre- and posttrigger data
- Be able to export data in spreadsheet format
- Save raw as well as processed data with formulae
- Check test setup for inconsistencies
- Save all or only requested channels data
- Have good, high-resolution graphics
- Show data with time or with any other reference channel
- Have autoscaling features for display
- Have an efficient filing system
- Possess on-site printing and plotting capabilities
- Allow for customization
- Have provision to write comments during each phase of the test
- Generate test reports at the end of the test

8.4.6 APPLICATIONS

8.4.6.1 General Applications

Shenton et al. (2000) offered an SHM technique developed for detecting strains from live load. Such a system can help in fatigue evaluation, more precise bridge load rating, and an early warning for any excessive bridge overload. A general discussion of the use of NDT in assessing bridge capacity/ rating was presented by Chajes et al. (2000). They opined that load testing is required during design because of the following:

1. Parameters (such as load and resistance factors [LRFR]) given in the AASHTO LRFD Specifications used in design are by necessity conservative.
2. Many secondary sources of stiffness and strength are either neglected or are too difficult to compute.
3. Some sources of stiffness are too bridge-specific to be included without field testing.
4. Material properties are typically based on design specifications rather than tests conducted on the *in situ* materials.
5. The assumptions commonly used in bridge load rating, which may not be completely accurate in terms of the behavior of the as-built bridge are

   - The nature of composite action
   - Estimates of load distribution
   - Behavior of simply supported spans (no fixity at supports) and fixed supports (full fixity at supports)
   - Section properties including the effect of section loss due to corrosion
   - Impact factor
   - Effect of parapets, railings, and so on
   - Material properties
   - Effect of fill
   - Effect of skew
   - Continuity effects
   - Participation of secondary members
   - Load carried by the deck
   - Effect of arching action

The authors concluded that for existing bridge analysis to determine a safe and accurate load-carrying capacity for a bridge, the bridge itself should be used/tested.

There are several techniques to measure bridge behavior during load tests. For example, Zeng et al. (2002) used fiber optic sensors to measure bridge performance during a load test. In the rest of this section we offer in more detail different sensing techniques during load tests.

8.4.6.2 Impact-Echo

The impulse response (also known as impact-echo) method is a surface reflection technique that relies on the interpretation of observed compressive wave reflections at the top of a drilled shaft. Before, during, or after load tests, it can be used to detect damage, if any (see Finno and Chao 2000). Transient vibrations are introduced by impacting the top of a shaft with an impulse hammer, while measuring the force and particle velocity on the impacted surface. Results of these tests provide a measure of the quality of concrete in the shaft and information about the shaft performance. Fast Fourier transforms of force and the velocity signals are used to get mobility as a function of frequency.

Quantitative information concerning the shaft length and impedance can be obtained from a mobility versus frequency plot. Knowing shaft geometry and using the theory of elasticity, one can estimate the theoretical mobility for the shaft as

$$N = \frac{1}{\rho_c v_c A} \tag{8.11}$$

The density and Poisson's ratio of concrete are ρ_c and ν_c, respectively. The cross-sectional area is A. If the measured N is greater than the theoretical value, it is likely that there is a defect in the shaft due to a smaller-than-expected cross-section or poor concrete quality (i.e., low ρ_c or ν_c). Typically, the method can be used with confidence when the length-to-diameter ratio of a shaft is less than 20 or 30, depending on the shear wave velocities of the surrounding soils and rocks.

Impulse response tests were conducted on 3- and 8-ft-diameter drilled shafts used for support of bridge piers at the Central Artery/Tunnel Project in Boston. Shaft length-to-diameter ratio varied from 12 to 34. Results showed that mobilities are within the range of those expected for shafts constructed with the design diameters specified for each shaft. However, the portions of the shaft embedded in the rock are not sensed significantly by the induced stress waves.

8.4.6.3 Accelerometers

Strain gauges are a norm for load ratings to estimate distribution factors. Chowdhury (2000) suggests using accelerometers also to validate strain gauge test data independently and also to avoid using longer gauge length strain gauges in the presence of cracks to prevent false data. It is also relatively easy to install accelerometers, and they can be used with relatively small loads.

The strain gauges and accelerometers were used on two bridges for load testing: a continuous-span multigirder steel bridge and a single-span reinforced concrete T-beam bridge with four different vehicle configurations. Bridges were instrumented with rapid-strain transducers and accelerometers, and the tests were conducted by driving the test vehicle across the bridge at different speeds.

Time-domain accelerations and strains were collected. Using strain data, the level of composite action was determined. For observed composite action of the slab-girder system, the DF for the ith girder among the load-sharing n girders can be computed using

$$\mathrm{DF}_i\left(t\right) = \frac{\ddot{y}_i\left(t\right)}{\sum_{j=1}^{n}(I_j / I_i)\,\ddot{y}_j\left(t\right)} \tag{8.12}$$

In Equation 8.12, $\ddot{y}_i(t)$ or $\ddot{y}_j(t)$ is the acceleration of the ith or jth girder, and I is the moment of inertia for the participating girders. This equation neglects the effects of the ratios of natural frequencies for respective girders. Equation 8.12 is based on the ratio of moments for each beam to the total moments as presented in

$$\mathrm{DF}_i = \frac{M_i}{M_{\mathrm{TOTAL}}} \tag{8.13}$$

In which M_i is the bending moment shared by the ith girder.

Finally, we compute the dynamic amplification of acceleration as

$$\mathrm{DLFA} = \frac{\ddot{y}_{\mathrm{FAST}}\left(t\right)}{\ddot{y}_{\mathrm{SLOW}}\left(t\right)} \tag{8.14}$$

where DLFA is the dynamic amplification of acceleration and is different from the dynamic amplification factor for moment. It can be shown that for a simple-span bridge, the dynamic amplification for moment is about 82% of that for deflection.

Results indicate that the accelerometer can conveniently extract the DF for bridge load ratings. Accelerometer data provide additional information to measure the vehicle-bridge dynamic interaction through DLFA.

8.4.6.4 Strain Gauges

Yost and Assis (2000) used strain gauges during a bridge load test. The South Norwalk Railroad Bridge is located in downtown Norwalk, CT, on the heavily traveled northeast corridor mainline. Designed in 1895, the bridge is a single-span steel structure consisting of three parallel through trusses, each separated by two tracks. Because of its age, importance, and heavy service load, the bridge was load tested to verify structural performance and integrity. Load testing was performed by collecting strains using demountable strain gauges under in-service live loads.

The results were used to make some general conclusions:

1. A wide load distribution range was noted among multiple eyebar tension members. Also, several eyebars are carrying no load. This was attributed to fabrication error or damage at the eyebar-pin interface.
2. A degree of translational resistance exists in the roller bearing supports. Stresses were recorded at locations that should be zero-force members.
3. The floor system appears to contribute longitudinal stiffness to the truss assembly. The mechanism for this load transfer appears to be the cross bracing.

8.4.6.5 Ultrasonics

Fuchs, Washer, and Chase (2000) experimented with a full-scale bridge has been erected and was tested in the TFHRC structures laboratory. The structure consists of three 90-ft-long curved steel bridge girders joined together by several cross-frames. The cross-frames were specifically designed for this project and are of a nontypical configuration. Steel pipes with an outer diameter of about 4.5 in are bolted together to make up the cross-frames. Each end of the various sections of pipe is welded to a flange that contains bolt holes for connection to the bridge girder. Installation of the cross-frames between two girders is difficult due to alignment problems. As a result, the cross-frame members have to be forced into place. This installation process causes unwanted stresses to be introduced into the cross-frame members.

As part of the bridge testing, a method to determine the amount of locked-in stresses in these cross-frames was desired. Ultrasonic stress measurement techniques and Barkhausen noise stress measurement techniques were used to measure the cross-frame stresses. Both methods have the potential to allow for quick, convenient measurements at various points during the curved girder bridge testing.

Ultrasonic measurements were on two cross-frame members (Figure 8.19). On one cross-frame, measurements were made at seven locations, and on the second cross-frame measurements were made at five locations. At each location a total of four measurements were made around the circumference

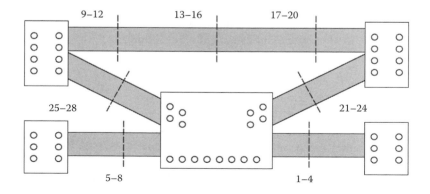

FIGURE 8.19   Cross-frame measurement locations. (Courtesy of CRC Press.)

of the pipe. The ultrasonic stress measurements were converted to strain and compared with the conventional strain gauge values (Figure 8.20).

The ultrasonic birefringence data showed good correlation with conventional strain gauges. Since a baseline measurement was possible with the cross-frames not installed in the bridge, the ultrasonic measurements were relative instead of absolute. On the basis of the off-line calibration data, it would appear that the resolution of the ultrasonic measurements would be of the order of ±5 ksi. This resolution seemed consistent with the actual test data since the strain gauge data (used as a comparison to the ultrasonic data) was probably only accurate to 60 microstrains.

Most useful measurements with ultrasonic stress measurement techniques rely on making differential measurements as opposed to an absolute measurement. This has potential implications for SHM, as one can measure relative stresses between different members under the same live loads to make sure the data is valid in a global sense.

Ultrasonic stress measurement technique

A polarized shear wave (SH) electromagnetic acoustic transducers (EMAT) was used for the ultrasonic measurements (Figure 8.21). The EMAT is built in a cylinder-shaped enclosure that has an outer diameter of 3.5 in and a measurement aperture of 0.5 in. The nominal frequency of operation for the transducer is 2MHz. To make the birefringence measurements, the change in phase (or time of flight) of the SH wave is found as the EMAT is rotated 360°. For this reason, the EMAT used for testing was equipped with motorized rotation. A sinusoidal-type pattern is observed as the EMAT is rotated, and the peak-to-peak value of this waveform can be related to the stress in the specimen.

Stress is related to the peak-to-peak value of the waveform (birefringence value) with the following equation.

$$B = B_0 + k \cdot \sigma \tag{8.15}$$

where B is the birefringence, B_0 is the unstressed birefringence, k is the stress acoustic constant, and σ is the stress. To measure stress, a valid value for B_0 and k must be obtained. For this reason an off-line calibration was performed on a section of pipe from the cross-frame members.

8.4.6.6 Vibration Sensing

The objective of a study by Wacker et al. (2006) was to refine forced-vibration dynamic testing to evaluate overall stiffness of timber bridge superstructures. The first bending mode of several

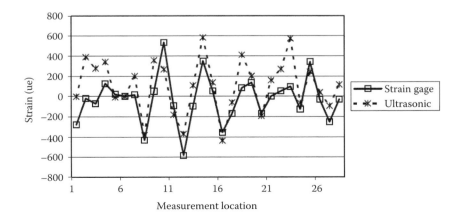

FIGURE 8.20   Cross-frame stresses due to insertion into bridge. (Courtesy of CRC Press.)

FIGURE 8.21   EMAT on calibration pipe. (Courtesy of CRC Press.)

simple-span sawn timber stringer and plank deck bridges located in St. Louis County, Minnesota, was measured using a motor attached to the deck planking near midspan. Static live-load tests were also conducted. The first bending frequency was also measured by placing additional 2000 lb sandbags on the deck to simulate live loads. The results showed that the measured stiffness from static load tests correlates with the measured frequency, the longer bridge length lowers the frequency (Figure 8.22), and additional weight increases the frequency (see Figure 8.23).

Beam theory seems to apply more than the plate theory. It may suggest that timber decking is not really contributing much to the stiffness.

8.4.6.7   Laser Measurements

The Florida Department of Transportation has been proactive in using diagnostic and proof load testing of bridge structures for improving load ratings, avoiding postings, and evaluating capacity for permit vehicle routing. As part of the load tests (Roufa 2006), normally strains and displacements are measured. The deflection measurements are often needed for load testing normal reinforced concrete bridges, as microcracking within the limits of the strain gauges might produce erroneous data yielding inaccurate structural behavior. In these cases, displacement transducers are used to measure displacements. These pose several problems, including lane shifts and/or closure of bridge to traffic and wiring the structure (which may require additional work for structures over water and traffic) posing safety hazards for the public as well as the instrumenting personnel. The sensors are also affected/damaged by high winds, debris, and temperature effects, and they can thus yield inaccurate results and be expensive and time-consuming. So, FLDOT used rotating laser systems to measure deflections during a couple of load tests to measure deflections.

These systems need no wiring and are sufficiently accurate for most bridge tests. They may not be suitable for long-span bridges where large displacements (over 3 in) are expected. Field evaluation by FLDOT showed that these laser systems present a viable option for many load-testing applications where moving load test data is not required. They have shown to provide more reliable and accurate data than displacement transducers, particularly for structures over flowing waters (see Figure 8.24).

In another set of tests, Fuchs and Jalinoos (2006) describe the laser system the FHWA has developed in its TFHRC laboratory and show its application in the curved girder bridge experimental laboratory. The laser system essentially is described by the authors as an extremely accurate,

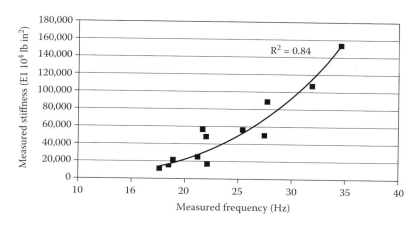

FIGURE 8.22 Relationship between measured stiffness and measured frequency. (Reprinted from ASNT Publication.)

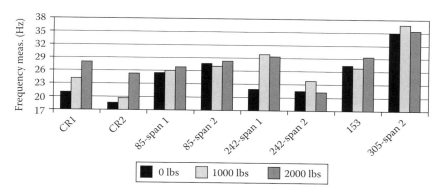

FIGURE 8.23 Effect of adding live load during forced-vibration testing. (Reprinted from ASNT Publication.)

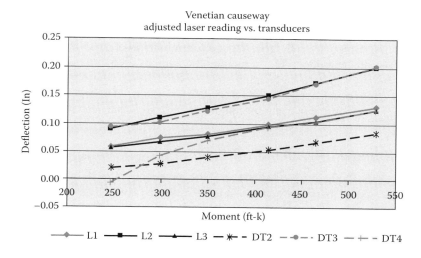

FIGURE 8.24 Adjusted laser data versus transducer data. (Reprinted from ASNT Publication.)

Measuring web panels

Laser scanner

FIGURE 8.25   Laser system positioned on the laboratory floor for web panel measurements. (Reprinted from ASNT Publication.)

large-volume, three-dimensional coordinate mapping device. It can scan across the surface of an object and achieve sub-mm accuracy measurements. Measurement accuracy is less than 0.2 mm out of a total range of 24 m., with a field of view of nearly 360 degrees.

This was applied to obtain vertical deflections of a curved girder bridge during the erection to during different load steps (see Figure 8.25). The laser system was used to get data from more than 2500 points, which is not possible with conventional systems. The system was also useful as the movements of curved girders are complex and involve vertical displacements and rotations, which are difficult to measure accurately with conventional sensor systems.

8.5   STRID IN LOAD TESTING

8.5.1   General

When a load test is performed on any type of bridge, a STRID process is typically used. Structural identification in load testing is used to compare the load test measurements with analytical/numerical analysis of the bridge. This is usually done for validation of analytical methodologies and assumptions of the subject bridge. Also, structural modeling is used as a basis of analytical bridge rating. The effort can vary from performing simple calculations to development of detailed finite element models. This section explores several STRID issues related to bridge load testing. We also introduce STRID methods that account for important uncertainties during proof tests.

8.5.2   Load Rating and Bridge Load Testing

One of the major uses of diagnostic load testing is the accurate evaluation of bridge rating. NCHRP (1998) provided a method that relates theoretical and test-based rating. The method starts by observing that the theoretical load-rating equation is

$$RF_C = \frac{\text{capacity} - \text{factored dead load effect}}{\text{factored live load effect} + \text{impact}} \tag{8.16}$$

The details of Equation 8.16 are also discussed in detail in Section 8.7.2. We only note here that the components of Equation 8.16 are computed on the basis of analytical modeling of the bridge and the design loads, as pertinent. When these design loads are used in load testing, the resulting measurements can be used to improve the theoretical rating RF_C such that

$$RF_T = K \cdot RF_C$$

(8.17)

The new rating RF_T is based on the load-testing results, and factor K is a dimensionless factor computed on the basis of test results and conditions. NCHRP (1998) suggested that

$$K = K_a K_b + 1$$

(8.18)

Factor K_a relates the measured strains, ε_C, and computed strains, ε_T, such that

$$K_a = \frac{\varepsilon_C}{\varepsilon_T} - 1$$

(8.19)

Thus, K_a can be computed for any components that need to be rated.

Factor K_b is a qualitative factor that considers test and bridge conditions such that

$$K_b = K_{b1} K_{b2} K_{b3}$$

(8.20)

The values of K_{b1}, K_{b2}, and Kb_3 are computed from Tables 8.6 through 8.8. Note that T and W in Table 8.6 represent the test vehicle weight and the gross rating weight, respectively. The answer to the question in Table 8.6 can be established either analytically or by proof load testing.

NCHRP (1998) included also a method that related proof load test bridge rating to analytical bridge rating.

8.5.3   QUANTIFYING BRIDGE BEHAVIOR

There are several factors that cause differences between analytical and test results. NCHRP (1998) discussed qualitatively some of those factors, as shown in Table 8.9.

TABLE 8.6
Values of Kb_1

Can Member Behavior be Extrapolated to 1.33W?	Magnitude of Test Load			
	$T/W < 0.4$	$0.4 \le$ $T/W \le 0.7$	$T/W > 0.7$	Kb_1
Yes	Yes			0.0
Yes		Yes		0.8
Yes			Yes	1.0
No	Yes			0.0
No		Yes		0.0
No			Yes	0.5

Source: NCHRP manual for bridge rating through load testing, National Cooperative Highway Research Program, NCHRP, Research Results Digest, No. 234, Washington, DC, 1998. With permission from NCHRP.

TABLE 8.7
Values of Kb_2

Inspection		Kb_2
Type	Frequency	
Routine	Between 1 and 2 years	0.8
Routine	Less than 1 year	0.9
In-depth	Between 1 and 2 years	0.9
In-depth	Less than 1 year	1.0

Source: NCHRP manual for bridge rating through load testing, National Cooperative Highway Research Program, NCHRP, Research Results Digest, No. 234, Washington, DC, 1998. With permission from NCHRP.

TABLE 8.8
Values of K_{b3}

Fatigue Control?	Redundancy		Kb_3
Yes	No		0.7
Yes		Yes	0.8
No	No		0.9
No		Yes	1.0

Source: NCHRP manual for bridge rating through load testing, National Cooperative Highway Research Program, NCHRP, Research Results Digest, No. 234, Washington, DC, 1998. With permission from NCHRP.

TABLE 8.9
Parameters That Result in Differences between Analytical and Testing Results

	Type of Bridge			
Effects	Beam and Slab	Concrete Slab	Truss	Box Girder
Unintended composite action	P, I/T	N/A	S, I/T	P, I/T
Nonstructural components (parapets, curbs, railings, etc.)	P, A	P, A	N/A	P, A
Discrepancies in material properties (actual and assumed)	S, I/T	S, I/T	S, I/T	S, I/T
Effects of bracing/secondary members	S	N/A	S	S
Unintended continuity	S, I/T	S, I/T	S, I/T	S, I/T
Support conditions (bearing restraints)	S, I/T	S, I/T	S, I/T	S, I/T
Load distribution effects	P, A	P, A	P, A	P, A
Analysis assumptions/methods	P, A	P, A	P, A	P, A
Skew effects	S, A	P	N/A	S, A

Source: NCHRP manual for bridge rating through load testing, National Cooperative Highway Research Program, NCHRP, Research Results Digest, No. 234, Washington, DC, 1998. With permission from NCHRP.
Notes: A, include in conventional analysis; I/T, inspection and/or testing is needed to verify; N/A, not applicable; P, primary factor; S, secondary factor

Barker (2001) quantified factors that contribute to the differences between analytical and experimental steel bridge rating. In what follows, we restate his important work. The basis of the development is a slightly different rating equation based on allowable stress rating (ASR)

$$R_{A\_I} = \frac{\left(0.55 F_y - \sigma_D\right)}{I_A (M_{WL} / S_A) \text{DF}_A} \text{RVW} \tag{8.21}$$

R_{A_I} = Inventory ASR
F_y = Yield stress
σ_D = Stress from dead load
RVW = Rating Vehicle Weight
I_A = Analytical impact factor
M_{WL} = Bending moment resulting from RVW truck
S_A = Analytical section modulus
DF_A = Analytical distribution factor

The author then identified seven contributing dimensionless factors as shown in Table 8.10. The table also shows the ratios of the contributions of each of these factors in the rating formula.

The ratio of experimental inventory rating R_{E_I} to the analytical inventory rating R_{A_I} is shown to be

$$\frac{R_{E\_I}}{R_{A\_I}} = \left(\alpha_I\right)\left(\alpha_S\right)\left(\alpha_k\right)\left(\alpha_{DF}\right)\left(\alpha_{\text{BEARING}}\right)\left(\alpha_M\right)\left(\alpha_{\text{COMPOSITE}}\right) \tag{8.22}$$

Using load rating tests and their equivalent analytical rating evaluations, Barker (2001) quantified the ratios at three locations along the bridge (as shown in Table 8.11). Careful examinations of those ratios produced the acceptable experimental inventory ratings shown in Table 8.12. The corresponding analytical ratings and the ratio between the two ratings are also shown in Table 8.12. It is of interest to note that the product totals of Table 8.11 are more than the ratios of analytical ratings to accepted experimental ratings of Table 8.12. The reason for this is that the authors of the method eliminated some of the nonrealistic artifacts of the experiments, using the analytical factors of Table 8.11. This shows that (1) experimental load rating can improve analytical ratings, and (2) load testing rating results need to be examined carefully; any nonreasonable test behavior needs to be eliminated from the resulting experimental rating. This is needed to ensure both realistic and safe bridge rating.

TABLE 8.10
Important Factors That Cause Differences Between Analytical and Experimental Bridge Ratings

Factor	Description
α_I	Actual impact factor
α_S	Actual section dimension
α_k	Unaccounted system stiffness (curbs, railings, etc.)
α_{DF}	Actual lateral live-load distribution
α_{BEARING}	Bearing restraint effects
α_M	Actual longitudinal live-load distribution
$\alpha_{\text{COMPOSITE}}$	Unintended composite action

Source: Barker, M.C., *ASCE, J. Bridge Eng.*, 6, July/August, 2001. With permission from ASCE.

TABLE 8.11
Experimental Factors

Factor	Noncomposite Positive Moment		Noncomposite Negative Moment		Composite Positive Moment	
	Exterior Girder	Interior Girder	Exterior Girder	Interior Girder	Exterior Girder	Interior Girder
α_I	1.016	1.016	1.000	1.000	−.984	0.984
α_S	1.033	1.033	1.022	1.025	1.064	1.087
α_k	1.120	1.280	1.224	1.098	1.045	1.204
α_{DF}	1.186	1.238	1.125	1.436	1.264	1.216
$\alpha_{BEARING}$	1.054	1.009	1.082	1.022	1.038	1.012
α_M	1.059	1.043	1.139	1.248	1.007	0.928
$\alpha_{COMPOSITE}$	1.313	1.301	1.392	1.286	1.042	1.008
Product Total	2.040	2.280	2.420	2.650	1.510	1.480

Source: Barker, M.C., *ASCE, J. Bridge Eng.*, 6, July/August, 2001. With permission from ASCE.

TABLE 8.12
Comparison between Experimental And Analytical Ratings

Condition	Noncomposite Positive Moment		Noncomposite Negative Moment		Composite Positive Moment	
	Exterior Girder	Interior Girder	Exterior Girder	Exterior Girder	Interior Girder	Exterior Girder
Analytical inventory rating, tons	14.7	19.2	23.9	38.4	9.7	17.8
Acceptable experimental inventory rating, tons	21.6	33.4	38.4	77.6	13.5	25.9
Ratio	1.47	1.73	1.61	2.02	1.39	1.45

Source: Barker, M.C., *ASCE, J. Bridge Eng.*, 6, July/August, 2001. With permission from ASCE.

The approach proposed by Barker (2001) can be then generalized to objectively relate analytical and experimental ratings for other types of bridges.

8.5.4 PROOF TESTING

8.5.4.1 General

Proof testing presents a particular challenge to professionals. The load on the bridge is increased gradually until it reaches a predetermined level or some early signs of distress are observed. In either situation, there remain some obvious questions: suppose there was a damage not expected while predetermining the load level? How about the damage that occur during the test that cannot be observed visually? A perhaps more important question is the one about brittle failure modes. What if there is an unknown brittle failure mode that might occur at a particular loading level? As was discussed in Chapter 3 by Ettouney and Alampalli (2012), brittle failures occur suddenly without adequate warning; they can have catastrophic results and should be prevented.

Two general methods are used to help professionals in managing proof tests; they are the pretest analysis and during-test observations. We discuss the two methods next. For each method we discuss the prevailing practice first. We then explore some potentially useful additional methods.

8.5.4.2 Pretest Analysis

Pretest analyses are usually performed before proof testing to estimate how the bridge will behave as the test loading increases. The analyses vary in their complexity as shown in Table 8.13.

Perhaps the most limiting factor in pretest analysis is the presence of uncertainty at all of the analysis steps. Some obvious uncertainties are geometry (actual dimensions and configurations), connectivity (how different components are interconnected), and material properties (especially for degraded or old constructions). The use of deterministic analysis that uses average estimates of input parameters will lead only to an estimate of an average of the bridge response to test loadings. When applying load factors to accommodate for uncertainties in general, such a load factor approach would have three obvious limitations:

* Load factors are generic by nature; it does not account for the specific uncertainties of the bridge under consideration.
* Load factors do not permit the professional to choose the level of uncertainty in both input parameters and output parameters.
* Load factors are generally obtained using linear-static methods. In some proof-loading pretest analysis, it might be desirable to use dynamic, nonlinear static or nonlinear-dynamic methods. Accuracy of load factors in these situations is not very clear.

We, thus, established the need for an approach that can accommodate uncertainties in pretest analysis of proof-loading tests. Figure 8.26 shows conceptually how the uncertainties propagate as the loading increases in such tests. Propagation of uncertainties is not an easy subject to handle. However, one simple approach is to use a hybrid Markov process/ Monte Carlo technique to compute uncertainties in proof testing in a simple and accurate manner.

The hybrid Markov process/Monte Carlo process is based on computing transition probabilities at different loading steps of the proof load test. We discussed probability transition matrices $[T]_{a \to b}$ in Chapter 8 by Ettouney and Alampalli (2012) as a part of the Markov processes. The components of the matrix p_{ij} constitute the probability that the state of the subject of interest changes from the ith state at a particular instant of time "a" to the jth state at a later instant of time "b." We propose that structural identification techniques can be used to generate transition matrices of proof-testing

---

TABLE 8.13
Considerations of Pretest Analysis

Type of Analysis	Comments
Linear-Nonlinear	For diagnostics tests, it is usually expected that the main components of the bridge system are modeled as linear components. Some nonlinearity might be modeled such as slippage between surfaces or friction-type behavior. For proof tests, it might be advisable to model the behavior in a nonlinear manner such that several limit states are investigated
Inertia effects	Static or dynamic modeling would depend on the nature and the goal(s) of the test
Resolution	It is recommended to have as high a geometric and dynamic resolution of the model as possible. Care is needed to ensure that the model resolution accurately reproduces the desired bridge performance. For example, the resolution needed to evaluate the model behavior of a bridge is different from the resolution of a model that aims at evaluating stress state near a bridge bearing

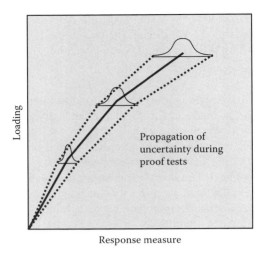

FIGURE 8.26 Propagation of uncertainty during proof tests.

stages for the bridge, or any type of infrastructure, that needs to undergo proof testing. This can be done in the following steps:

1. During the planning of the proof test, compute the target-loading level.
2. Subdivide the target loading into $N_{LOAD} \geq 2$ steps. The inequality indicates that the target loading should not be applied in one step: it should be applied incrementally.
3. Model the bridge analytically using an adequate numerical model such as the finite-element (FE) model. Pay special attention to bridge components that might be of concern during the test. Also, components that can contribute to the performance of the bridge as well as those components that have highly uncertain properties should be included in the model. The analytical method can be linear or nonlinear, static or dynamic.
4. Assign probabilistic properties to the uncertain parameters in the model. This can include means μ_k, variances V_k, and probability distributions PDF_k. The subscript k is a counter for all uncertain parameters. Note that for $V_k = 0$, the kth parameter is deterministic. If $V_k = 0$ for all parameters in the model, the problem reverts to a deterministic problem.
5. The means μ_k can be estimated using one or more of the STRID methods in Chapter 6 by Ettouney and Alampalli (2012). The methods include modal identification, parameter identification, or neural networks. Obviously, to use STRID techniques, some earlier testing (diagnostic) results should be available. If no testing results are available, then personal (engineering) judgments can be used to assign appropriate values for μ_k.
6. Similarly, the variances V_k can be estimated from (1) historical data, and/or (2) personal (engineering) judgment.
7. Record all pertinent states of all pertinent components at the start of the process.
8. Simulate the first loading step using Monte Carlo technique and evaluate the statistical properties (means and variances) of all pertinent output measures at the end of the loading step.
9. Compute probabilities of different states of different components. The transition matrix of this loading step can now be filled using archived information from #7 above.
10. Repeat # 8 and #9 for additional loading steps until the target load level is reached.

All transition matrices are now on hand. The analyst can use those transition matrices to reach different decisions during the proof test of interest. The process is illustrated in Figure 8.27.

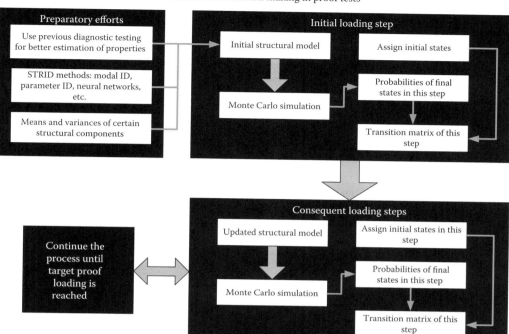

FIGURE 8.27 Role of STRID and decision making in proof tests.

8.5.4.3 During-Test Observations

As the loads are increased during a proof test, the professionals observe the behavior of the bridge closely. This observation is made either visually or by using real-time sensing (displacements, strains, tilt, position, etc.). They stop the testing process if any sign of distress or damage is observed, for immediate analysis. The results of such an analysis can cause the test procedure to be modified or even the whole test to be terminated. Such a subjective approach is reasonable, given the very nature of proof testing. However, we need to ask: is there an objective manner to approach this situation?

Let us survey the problem further, which can be done by examining Figure 8.28. The figure shows a load-response relationship of a proof test. Let us assume that the test proceeded satisfactorily up to load level at point "a." When the load increased to point "b," a severe distress was observed. Obviously, the test would be terminated at such a point. However, a specific damage has already been done, and the cost of the damage might be unacceptable to such an extent that it would erase any potential benefit that might have been gained from the information gathered by the test results up to that point. Now we can restate the previous question: Given a test state at point "a," can we devise a method to call the test off at point "a," without proceeding to point "b?" The advantage of such a method is that we would gain all the benefits of the test information gathered up to point "a," without the potential cost of the damage that can occur if we proceed to point "b."

We can use the previously generated transition matrices to help reach a decision. Let us define point "a" in Figure 8.28 by its state pair (S_i, F_i). The load level at "a" is F_i and the bridge state is S_i. Not that S_i can be a displacement, strain, natural frequency, or any other response measure that is being used for the proof test on hand. The next step is to define the limit state S_j that defines "a." This can be accomplished by the condition

$$S_j \leq S_i < S_{j+1} \tag{8.23}$$

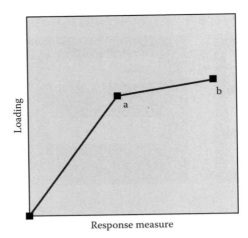

FIGURE 8.28   Decision-making need in proof testing.

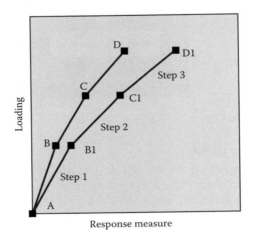

FIGURE 8.29   Use of transition probabilities in proof testing.

From a previously computed transition matrix, $[T]_{a \to b}$ we identify the jth row vector, $\{P\}^T$ with components $p_k|_{a \to b}$. This is the probability that the bridge will be in kth state when the load increases from level "a" to level "b." Note that the order of $\{P\}^T$ is N_{STATE} and $k=1,2,\ldots,N_{STATE}$. Also

$$\sum_{k=1}^{k=N_{STATE}} p_k\Big|_{a \to b} = 1.0 \tag{8.24}$$

If we define the last state, $k=N_{STATE}$ as an undesirable state, the probability that the bridge will be in that state when loaded from "a" to "b" is $p_{k=N_{STATE}}\big|_{a \to b}$.

Armed with the knowledge of this probability, the test official can now decide whether or not to proceed with increasing the loading on the bridge from "a" to "b."

As an example, let us consider a proof test designed for three loading steps as shown in Figure 8.29. Five limit states are used for this test: pristine, very light, light, moderate, and undesirable damage. Structural identification analysis as described earlier produced three transition matrices as shown in Tables 8.14 through 8.16.

TABLE 8.14
Transition Probabilities Matrix: First Loading Step

	Pristine	Very Light	Light	Mod	Undesirable
Pristine	0.95	0.04	0.01	0	0
Very light	0	0.9	0.08	0.02	0
Light	0	0	0.8	0.15	0.05
Mod	0	0	0	0.7	0.3
Undesirable	0	0	0	0	1

TABLE 8.15
Transition Probabilities Matrix: Second Loading Step

	Pristine	Very Light	Light	Mod	Undesirable
Pristine	0.75	0.19	0.05	0.01	0
Very light	0	0.7	0.22	0.08	0
Light	0	0	0.6	0.3	0.1
Mod	0	0	0	0.6	0.4
Undesirable	0	0	0	0	1

TABLE 8.16
Transition Probabilities Matrix: Third Loading Step

	Pristine	Very Light	Light	Mod	Undesirable
Pristine	0.6	0.25	0.1	0.05	0
Very light	0	0.59	0.27	0.12	0.02
Light	0	0	0.5	0.35	0.15
Mod	0	0	0	0.58	0.42
Undesirable	0	0	0	0	1

At the start of the test, it was estimated that the initial state of the bridge is pristine. This means that the first row in Table 8.14 would govern the first step. There is zero probability that the first loading step would result in an undesirable limit state; thus, the first loading step is performed. After the first loading step, going from point "A" to point "B" in Figure 8.29, the measurements at point "B" showed that the bridge is still in pristine state. Again this indicates that the first row in Table 8.15 would also govern the second step. There is zero probability that the second loading step would result in an undesirable limit state; thus, the second loading step is performed: going from "B" to "C" in Figure 8.29. After inspecting state at point "C," it was revealed that the bridge is now in a very light damage state. This indicates that going from "C" to the final test goal at point "D," the second row in the governing transition matrix in Table 8.16 will apply. There is a 2% probability that the final loading step would produce an undesirable bridge state. The decision to continue or not would depend on management policies for that particular situation. A risk assessment analysis might be needed to estimate potential risks.

Let us assume that the initial state of the bridge is lightly damaged at the start of the test. The second row in Table 8.14 would govern the first step. There is still a zero probability that the first loading step would result in an undesirable limit state; thus the first loading step is performed.

After the first loading step, going from point "A" to point "B1" in Figure 8.29, the measurements at point "B1" showed that the bridge is now in a light damage state. This indicates that the second row in Table 8.15 would now govern the second step. There is 10% probability that the second loading step would result in an undesirable limit state, assuming that the second loading step is performed: going from "B1" to "C1" in Figure 8.29. After inspecting state at point "C1," it was revealed that the bridge is now in moderate damage state. This indicates that going from "C1" to the final test goal at point "D1," the third row in the governing transition matrix in Table 8.16 will apply. There is a 15% probability that the final loading step would produce an undesirable bridge state. Probabilities of undesirable outcome have obviously increased due to the changes in the initial bridge state.

8.6 DAMAGE IDENTIFICATION IN LOAD TESTING

8.6.1 Expected Damage during Proof and Diagnostic Testing

As the bridge is loaded during a test, it responds accordingly: this load-response process should be done carefully. The loading should be applied without causing any "damage" to the bridge or any of its components. In diagnostic testing, where the loading is generally within design levels, damage is not expected. In proof-loading tests, where loading levels are generally high, damage might occur. In both situations, the testers should watch for signs of damage and stop the test as soon as any damage sign is observed. Because of the importance of this issue, we need to explore it further.

All bridge components, component assemblies, and the whole bridge system will reach different limit states as the test loading increases. Some of these limit states are recoverable and other limit states are permanent. We need to subdivide the behavior of the bridge and its components during load tests into ductile and brittle behavior. The limit states, the nature of bridge behavior, and monitoring them during tests are discussed in this section.

8.6.2 Types of Potential Damage—Ductile Behavior

8.6.2.1 General

Consider Figure 8.30 that shows a load-deflection relation of a structural system during a loading test. The system has several limit states as follows:

- Linear limit state "a": The test load is F_a. No permanent deflections or damage occur. The unloading is linear elastic.
- Elastic limit state "b": The test load is F_b. No permanent deflections or damage occur. The unloading is elastic.
- Nonlinear limit state "c": Large deflections occur; any increase in loads would generate measurable deflections. The unloading is nonlinear and permanent deformations and damage occur.
- Failure limit state "d."

During either diagnostic or proof tests, it is not desirable to load the bridge beyond point b (or Point "a") since any modest increase of loading would prompt larger deflections that are not recoverable, that is, an undesired permanent damage would occur. All damage states are usually computed a priori. Thus, the loading during the test is always less than F_b. Figures 8.31 and 8.32 show highly redundant timber and steel bridges. If either of these bridges is overloaded during a load test, signs of overloads will probably show as excessive deformations or deflections, due to the highly redundant systems.

Given the uncertain nature of computations and the state of bridge itself, sometimes the actual test load F_{LOAD} exceeds F_b. In such a case, the bridge would respond in a ductile manner: the deformation will increase in a rate higher than the load increase. Permanent (nonrecoverable) deformation

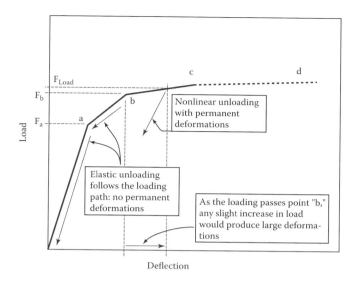

FIGURE 8.30 Ductile behavior during load test.

FIGURE 8.31 Potential ductile behavior during a load test of a timber bridge. With permission from National Highway Institute.

FIGURE 8.32 Potential ductile behavior during a load test of a steel bridge. With permission from National Highway Institute.

or damage would occur. This is an undesirable situation; however, because of the large magnitude of deformation, it can be easily monitored. The test can then be stopped immediately. The lessons are clear:

- Ductile behavior during diagnostics or proof tests is desired to help reduce any potential undesired damage.
- It is important to monitor behavior (strains, displacement, etc.) of components susceptible to permanent deformation or damage. Monitoring only midspan displacements might not be sufficient. Also, visual observations might not be accurate enough to detect varied limit states.

Obviously, analysis, good engineering judgment, and past experiences are all needed to limit damage during load testing.

8.6.3 Types of Potential Damage—Brittle Behavior

Brittle behavior of bridge components or bridge systems is a much more serious situation that should be addressed carefully during load tests. Consider Figure 8.33. It shows a load-deflection relation of a brittle system. Only two limit states are present:

- Linear limit state "a": The test load is F_a. No permanent deflections or damage occur. The unloading is linear elastic.
- Failure limit state "b": The system fails suddenly if the loading exceeds F_a.

This failure occurs when the pretest analysis shows that the system (or component) will fail in a brittle fashion at load level F_a. Thus, the testers allowed for a safety margin larger than 1.0 and thus, allowed the maximum loading to be a fraction of F_a. Let us assume that environmental changes, material degradation, geometric changes, or other factors have reduced the brittle failure limit to be less than F_{LOAD}. In such a situation, the system (or component) will fail in a brittle fashion during the test. Such a failure will occur suddenly, without any visual indications or signs. Yanev (2007) discussed how safety factors that protect structures from instability can be reduced or can vanish.

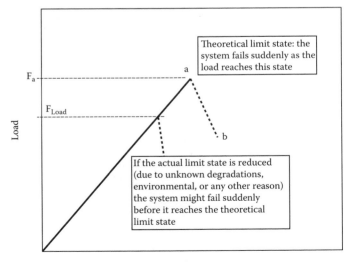

FIGURE 8.33   Brittle behavior during load test.

He offered the example of a bridge column. Figure 8.34 shows a potential unstable foundation due to scour erosion. Figure 8.35 shows a potential brittle shear failure in a floor beam.

In case of potential of brittle failure during load tests, a two-pronged strategy is needed, a pretest and a during-test strategy:

For pretest strategy, we offer the following guidelines to detect reduction in capacity:

- When analyzing the bridge, a stability analysis should be performed.
- As-built properties should be used in the analysis, not the design properties.
- Environmental and state changes should be accommodated in the analysis.
- Uncertainties should be included in the analysis.

These guidelines should be enforced, especially if there are compression components in the system (columns, axially loaded plates, etc.)

During the test, it is advisable to monitor both spatial behavior (strains, displacement, etc.) and temporal behavior (natural frequencies). It was shown (Chapter 8 by Ettouney and Alampalli 2012) that an instability condition can be detected by monitoring natural frequencies of components. To get the full benefit of such monitoring, a real-time data analysis is obviously needed.

FIGURE 8.34 Potential brittle behavior during a load test: foundation problem. (Courtesy of New York State Department of Transportation.)

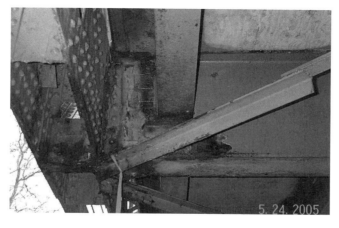

FIGURE 8.35 Potential brittle behavior during a load test steel beam problem. (Courtesy of New York State Department of Transportation.)

8.7 DECISION MAKING IN LOAD TESTING

8.7.1 GENERAL

Load testing is becoming a popular tool for the bridge management community. As we have seen, it can accomplish many important goals that can help bridge managers ensure safety and security at reasonable costs. We observe that several uncertainties are present in load testing. The uncertainties involve the bridge itself, the testing procedure, and the results of the test. Because of these uncertainties, decision-making tools need to be used to ensure that the decisions taken under such uncertainties are optimal ones.

One of the main by-products of load testing is bridge rating. As we have seen, the analytically based bridge rating can be improved by using load-testing results. Traditionally, load rating was based on deterministic equations, load factor rating (NCHRP 1998). In recognition of the uncertainties involved in the rating methods, a probabilistic-based method, LRFR, has been developed (see Mertz 2005). The LRFR is based on statistical evaluations of many bridges, and it uses adequate reliability factors that account for the observed uncertainties during the statistical studies. In this section, we generalize the LRFR approach by introducing a closed-form probabilistic method for computing the rating factor. Thus, the bridge engineer can tailor-make the load factor to fit exactly the bridge under consideration. Another advantage of this generalized approach is that it can use previous information to improve the accuracy of the probabilistic load factor equation. Finally, the use of information gathered from any SHM project can be used to improve the load factor equation: this is a direct interaction between load rating and SHM projects.

We also offer a method to evaluate the value of load testing. The impetus of such a method is an observation by NCHRP (1998) that, in some situations, load testing is not recommended. One of those situations is, of course, when the value of such a testing is negligible or even negative. This section will conclude with a method used by Fu (1995), which evaluates the relationship between proof load testing and cost–benefit. We recommend that the bridge engineer evaluate the value of load testing before executing the project in order to fully achieve the project's full potential.

8.7.2 CASE STUDY: TAYLOR SERIES AND PROBABILISTIC EVALUATION OF BRIDGE RATING

Computation of bridge rating is a process that is well documented (see, for example, *AASHTO* 1994 and 2001. The process defines the structural member rating value as R, which can be computed from

$$R = \frac{C - A_1 \cdot D}{A_2 \cdot L \cdot (1 + I)} \cdot W \qquad (8.25)$$

where

C = Estimated capacity of the structural member
D = Estimated dead load effect on structural member
L = Estimated live-load effect on structural member
W = Weight (tons) of vehicle used to determine live-load effect
I = The impact factor used with the live-load effect
A_1 = Factor for dead load
A_2 = Factor for live load

The factor for dead load, A_1, is a constant that is prescribed by the bridge owner, usually taken as $A_1 = 1.3$ all different rating levels. The factor for live load, A_2, is a constant that is prescribed by the bridge owner, usually taken as $A_2 = 2.17$, 1.3, 1.3 for inventory rating level, for operating rating level, and for posting rating level, respectively. The impact factor, I, accounts for the dynamic effects of the live loads. As such, it is fairly difficult to accommodate. However, it is usually prescribed for rating problems. For the sake of simplicity, we will assume that I is a constant with a value of 0.30.

In practice, the values of C, D, and L are obtained using analytical modeling of the bridge. This implies that C, D, and L are presumed deterministic. No uncertainties are accommodated in applying the rating Equation 8.25. Obviously, such an implied deterministic character of C, D, and L is not accurate. Consider, for example, the following sources of uncertainties:

- Uncertainties in structural modeling assumptions (connections, linear behavior, degree of composite action, contact behavior, etc.)
- Uncertainties in material behavior (local plastic effects, elastic constants, soil-structure interfaces, etc.)
- Uncertainties regarding wear and tear, that is, degradation effects.

Because of all these uncertainties, it is reasonable to assume that C, D, and L are random variables. Such an assumption will result in the fact that the bridge rating, R, is also a random variable. It is of interest then to establish the following confidence (probabilistic) statement:

There is an $x\%$ chance that $R \leq \bar{R}$

where x and $\bar{R}$ are computed using the probabilistic method coupled with conventional analytical techniques. Contrast the probabilistic statement with the conventional (deterministic) statement:

There is an 100% chance that $R = \bar{R}$

where $\bar{R}$ is computed using conventional analytical techniques only. Clearly, the deterministic statement is not as realistic as the probabilistic statement. On the other hand, a probabilistic statement would offer the decision maker a performance-based spectra of decisions that the deterministic approach cannot offer. In addition, the probabilistic approach can utilize the results of an SHM project in a formal/analytical manner, accommodating SHM results. We will first introduce a probabilistic framework for evaluating bridge rating; then we explore how SHM projects can be used in a formal manner to improve the accuracy of such ratings.

8.7.2.1 Probabilistic

The theoretical background of using the Taylor series method to obtain probabilistic properties of a function of random variables is described in Chapter 8 by Ettouney and Alampalli (2012). We consider the rating $R = R(C, D, L)$ as a function of three random variables C, D, and L. As a first step in the solution, we need the partial derivatives of R. They can be expressed as

$$\frac{\partial R}{\partial C} = \frac{W}{B} \tag{8.26}$$

$$\frac{\partial R}{\partial D} = \frac{W}{B}\left(-A_1\right) \tag{8.27}$$

$$\frac{\partial R}{\partial L} = \frac{-W}{B^2}\left(C - A_1 \cdot D\right)\left(A_2\left(1 + I\right)\right) \tag{8.28}$$

$$\frac{\partial^2 R}{\partial C^2} = 0 \tag{8.29}$$

$$\frac{\partial^2 R}{\partial D^2} = 0 \tag{8.30}$$

$$\frac{\partial^2 R}{\partial L^2} = \frac{2W}{B^3}\left(C - A_1 \cdot D\right)\left(A_2\left(1+I\right)\right)^2 \tag{8.31}$$

with

$$B = A_2 \cdot L \cdot \left(1+I\right) \tag{8.32}$$

It is assumed that the probability distributions of the three random variables C, D, and L are known. Thus, it is assumed that the expected values (means) and variances of C, D, and L are known and are defined as $\bar{C}$, $\bar{D}$, and $\bar{L}$ for the expected values (means) and V_C, V_D, and V_L for the variances, respectively. Since the Taylor series is expanded about ε_C, ε_D, and ε_L such that

$$\varepsilon_C = C - \bar{C} \tag{8.33}$$

$$\varepsilon_D = D - \bar{D} \tag{8.34}$$

$$\varepsilon_L = L - \bar{L} \tag{8.35}$$

Note that, by definition, $E(\varepsilon_C) = 0$, $E(\varepsilon_D) = 0$, and $E(\varepsilon_L) = 0$. Also, from 8.33 through 8.35

$$V_C = E\left(C^2\right) = E\left(\varepsilon_C^2\right) \tag{8.36}$$

$$V_D = E\left(D^2\right) = E\left(\varepsilon_D^2\right) \tag{8.37}$$

$$V_L = E\left(L^2\right) = E\left(\varepsilon_L^2\right) \tag{8.38}$$

It can be shown that $E(\varepsilon_C) = 0$.

Applying the Taylor series method of Chapter 8 by Ettouney and Alampalli (2012) to this problem, and limiting the order of the series to $O\left(\varepsilon_i^2\right)$, the expected value (mean) of the rating can be expressed as

$$E(R) = \bar{R} + \frac{1}{2}\left(V_C \cdot \overline{\frac{\partial^2 R}{\partial C^2}} + V_D \cdot \overline{\frac{\partial^2 R}{\partial D^2}} + V_L \cdot \overline{\frac{\partial^2 R}{\partial L^2}}\right) \tag{8.39}$$

And the expected mean square is

$$E\left(R^2\right) = \bar{R}^2 + \left(\begin{array}{c} V_C \cdot \left(\bar{R} \cdot \overline{\frac{\partial^2 R}{\partial C^2}} + \left(\overline{\frac{\partial R}{\partial C}}\right)^2\right) + V_D \cdot \left(\bar{R} \cdot \overline{\frac{\partial^2 R}{\partial D^2}} + \left(\overline{\frac{\partial R}{\partial D}}\right)^2\right) \\ + V_L \cdot \left(\bar{R} \cdot \overline{\frac{\partial^2 R}{\partial L^2}} + \left(\overline{\frac{\partial R}{\partial L}}\right)^2\right) \end{array}\right) \tag{8.40}$$

The over bar in Equations 8.39 and 8.40 indicates that the function is evaluated at the mean of the random variables. The mean, variance, and the standard deviation of R can now be evaluated as

$$\mu_R = E(R) \tag{8.41}$$

$$V_R = E(R^2) - E^2(R) \tag{8.42}$$

$$\sigma_R = \sqrt{V_R} \tag{8.43}$$

$$\text{COV} = \frac{\sigma_R}{\mu_R} \tag{8.44}$$

To illustrate the applications of the uncertainty of bridge rating, consider a simple case of a bridge deck rating. The gross weight W is assumed to be 36 tons. Using analytical techniques, the analyst computed C, D, and L as shown in Table 8.17. The rest of the parameters of Equation 8.25 are also shown in Table 8.17. Also, Table 8.17 shows the resulting deterministic deck rating, R.

Recognizing that the computed values of C, D, and L are based on many uncertain factors, the analyst decides to perform probabilistic analysis of the deck rating. To start with, the analyst decided to use the computed values of C, D, and L in Table 8.17 as the means (expected values). The analyst also made a reasonable estimate for the coefficient of variation (COV) of C, D, and L. The standard deviation and the variances of each parameter were then computed using Equations 8.43 and 8.44, as shown in Table 8.18.

TABLE 8.17
Bridge Deck Example Parameters

Parameter	Operational	Inventory
C (Kips. ft.)	12.81	12.81
D (Kips. ft.)	0.77	0.77
L (Kips. ft.)	5.24	5.24
I	0.3	0.3
A_1	1.3	1.3
A_2	1.3	2.17
R - Tons	62.4	37.4

TABLE 8.18
Statistical Properties of Input Rating Parameters

Parameter	Mean (Kips. ft)	COV	Standard Deviation (Kips. ft)	Variance (Kips. ft)^2
C	12.8	0.15	1.92	3.6922
D	0.77	0.15	0.12	0.0133
L	4.03	0.15	0.6	0.3656

Utilizing Equations 8.26 through 8.40, the expected value and the expected mean square for the inventory and operational deck rating can be computed. Utilizing Equations 8.41 through 8.44, the rest of the statistical properties of the inventory and operational deck rating can be computed. The statistical results are shown in Table 8.19.

Some interesting observations can be made on the resulting statistics. First, note that the expected value (mean) of R is slightly higher than the deterministic value of R (in Table 8.17). This is due to the effects of the variance terms in Equation 8.39. Additionally, note that the COV in Table 8.19 is higher than all the COV of the input random variables in Table 8.18. We note that the COV for both operational and inventory ratings are identical. This is due to the linear dependence of R on A_2. It is of interest to note that the COV have increased for 0.15 for all input random variables to 0.21 for the deck ratings. Thus, we reach an important conclusion: even if the uncertainties in the input parameters of the bridge rating (Equation 8.25), such uncertainties do increase when computing the ratings. Such an increased uncertainty needs to be addressed seriously by the analysts, especially when the ratings are near posting rating levels.

To produce a probabilistic statement that can be used for decision making, the analyst would need more information than the statistical parameters of Table 8.19. In fact, the probability distribution function (PDF) of R is needed to make the probabilistic statement. There is no simple way, short of a full Monte Carlo simulation analysis, to obtain the needed PDF. Instead, the analyst might choose to make a reasonable assumption regarding the PDF. One possible assumption is that the PDF is simply a normally distributed function. Such an assumption seems reasonable and simple. Until further research proves otherwise, there seems to be no reason not to use it in our current example.

For normal distribution PDF with the statistical properties of Table 8.19, the nonexceedance probabilities of a given deck rating can be computed. For example, if the closing operational and inventory ratings for the deck of interest are 40 and 30 tons, respectively, then the probabilities that the operational and inventory deck ratings will not be less than those closing ratings will be 96% and 84%, respectively. Figure 8.36 illustrates the concept of computing those probabilities (a normal distribution table is needed for such computations). A different way to make the probabilistic statement is: there is a 4% and 16% chance that the operating and inventory deck ratings will be less than the closing ratings, respectively. If, on the other hand, the closing operational and inventory ratings for the deck are 45 and 35 tons, respectively, then the probabilities that the operational and inventory deck ratings will not be less than the closing ratings will be 91% and 65%, respectively. This means that there is a 9% and 35% chance that the operating and inventory deck rating will be less than the closing ratings, respectively. Armed with these probabilities, the decision maker can make an informed decision as to the need for declaring this deck as unsafe. One possible method for reaching an informed decision is the use of the payoff table approach, discussed in other sections in this manuscript. It will also be discussed next in conjunction with the use of SHM for better estimates of bridge rating.

TABLE 8.19
Statistical Parameter of Deck Rating

Statistical Parameter	Operational	Inventory
Expected value (Kips. ft)	63.81229	38.22856
Expected mean square (Kips. ft)^2	4261.417	1529.4
Mean (Kips. ft)	63.81229	38.22856
Variance (Kips. ft)^2	189.4088	67.97783
Standard deviation (Kips. ft)	13.76259	8.244867
COV	0.215673	0.215673

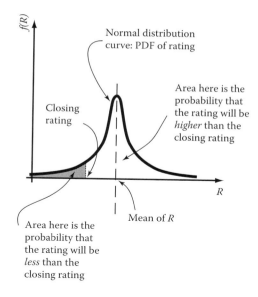

FIGURE 8.36 Exceedance probabilities and normally distributed PDF.

8.7.2.2 SHM and Bridge Rating

Due to the importance of bridge rating in improving public safety, some important questions arise, given the results of the previous section:

1. Is there a way to improve on the computed probabilities of the last section?
2. If so, what is the value (benefit/cost) of such an improvement?

We will address both questions in this section.

We will try to improve the computed probabilities of the previous section by using Bayes' theorem for conditional probabilities. The reader will recall that the computed probabilities can be improved upon by conducting experiments and observing the results. The experiment in the current context would simply mean an SHM project. In reality, SHM projects have been used in conjunction with bridge rating in several forms. For example, SHM testing was used to directly estimate the bridge rating (see Lichtenstein 1993). Also, SHM testing was used to improve the rating equation parameters (see Barker 2001). Note that each of those utilization modes of SHM is deterministic in nature. The methods we describe next differ from those methods in three ways:

1. It combines both SHM and analytical methods: it does not aim at change, nor does it ignore either method
2. It formalizes the decision making process by adding a cost–benefit mechanism
3. It is probabilistic; it thus accommodates uncertainties involved in the rating process

The first step in improving the previous probabilities by using observations from SHM testing is to define limit states of the rating R. We already have indirectly identified a limit state of R. The reader will recall that we used the posting rating as a decision making metric. We can simply keep the posting rating as a single limit state in the bridge-rating problem. However, since the safety and cost implications of bridge rating are high, we propose to use two limit states (three states of R) as follows:

1. Closing-rate limit state, R_P: this limit state is the conventional limit state. Any rating that falls below it, $R \leq R_P$, would trigger specific and well-known actions from the bridge official. We define this state as the *closing* state θ_1.
2. A posting-rate limit state R_A: this limit state is a bit higher than the closing limit state. If the rating lies below this limit state, but higher than posting, $R_P \leq R \leq R_A$, then the bridge official would post this bridge (component). We define this state as the *posting* state θ_2. If the rating is higher than this limit state, $R_A \leq R$, then no action is needed; the bridge (component) is safe. This is the *safe* state, θ_3. Figure 8.37 illustrates the two limit states and the corresponding three states.

Establishing a reasonable value for R_A is a qualitative effort. The only requirement is that $R_A > R_P$. Continuing the previous example, we offer some suggested values of R_A and the corresponding values of R_P in Table 8.20. From the two limit states, we identify three states of R.

From Tables 8.19 and 8.20, and recalling the assumption of normally distributed PDF of R, we compute Table 8.21.

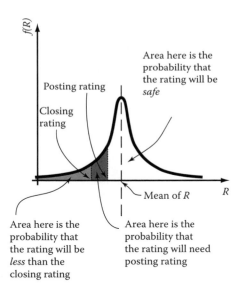

FIGURE 8.37   Bridge-rating limit states.

TABLE 8.20
Rating limit States for Example Problems

	Case 1		Case 2	
Rating loading	R_P (Tons)	R_A (Tons)	R_P (Tons)	R_A (Tons)
Operating	40	45	45	50
Inventory	30	35	35	40

TABLE 8.21
Probabilities of Rating States—Purely Analytical
Computations

	Case 1		Case 2	
State	**Operating**	**Inventory**	**Operating**	**Inventory**
$P(\theta_1)$	0.04	0.16	0.09	0.35
$P(\theta_2)$	0.05	0.19	0.07	0.23
$P(\theta_3)$	0.91	0.65	0.84	0.42
Total	1.0	1.0	1.0	1.0

In Table 8.21, we define $P(\theta_i)$ as the probability that R is in one of the three bridge-rating states, θ_i. Note that changing states from two to three did not have any effect on the probabilities of the closing state, as expected.

We return now to the matter of improving the state probabilities of Table 8.21 by using the results of an SHM project. In particular, three steps are needed: (1) establish an SHM approach to evaluate bridge (or component) rating, (2) use the results to establish a conditional probability statement about bridge rating, and (3) use these conditional probabilities in conjunction with the state probabilities of Table 8.21 to improve the accuracy of the decision making process.

Use of SHM techniques to compute bridge or component ratings is discussed extensively throughout this chapter. Two in-depth methods are also presented. So, we will not address this issue here. We only note that the results of any SHM project would be a set of experimentally based ratings for the bridge, or any of its components, R_i, where $i = 1,2, ..., M$. The number of repeated experiments within a specific SHM-rating project is M. Note that the SHM experimental sample results can have three possible types of results: x_1, x_2, and x_3. They correspond to safe, posted, and closed. Those sample types correspond to the three states θ_1, θ_2, and θ_3. The analyst needs first to decide, given R_i, as to what prevailing state x_J the sample (experimental) results is indicating. After that, the conditional probabilities $P(x_J|\theta_i)$ need to be computed. The term $P(x_J|\theta_i)$ defines the probability that R is in sample-rating state x_J given the actual rating state levels θ_i.

Finding the prevailing sample state, x_J, can be done by one of two methods. A qualitative method would be for the analyst to use personal judgment. A bit more quantitative approach would be to evaluate the sample mean

$$\bar{R}_S = \frac{1}{M} \sum_M R_i \tag{8.45}$$

Using rating limit states, similar to those in Table 8.20, the sample prevailing state, x_J, can be evaluated.

The conditional probabilities $P(x_J|\theta_i)$ can be established using one of two methods. The first method is to establish a PDF from the sample R_i. Using the PDF, the different probabilities can be computed (see Benjamin and Cornell 1970) for methods of computing PDF from available samples. We will use a simpler approach for this example. The sample standard deviation is computed as

$$\sigma_S = \frac{\sum_M \left(R_i - \bar{R}_S \right)^2}{M} \tag{8.46}$$

We then conveniently assume that the PDF of the experimental sample is a uniform distribution function. With the sample mean and standard deviation computed, the probabilities $P(x_J|\theta_i)$ can easily be computed using uniform distribution tables. Finally, using Bayes' theorem, the conditional probabilities for the rating being in a natural state θ_i, given the results of the SHM project, can be expressed as

$$P\left(\theta_i \middle| x_J\right) = \frac{P\left(\theta_i\right) P\left(x_J \middle| \theta_i\right)}{\sum_j P\left(\theta_j\right) P\left(x_J \middle| \theta_j\right)} \qquad (8.47)$$

Let us continue the earlier example to show how bridge testing can improve the purely analytical rating probabilities of Table 8.21. Consider that a bridge-testing project was performed for the bridge on hand. We assume that four sample points were taken in the course of the test ($M = 4$). The ratings were then computed on the basis of the results of the tests; we will refer to those ratings as *sample* ratings. Table 8.22 shows the sample ratings for the inventory and operational levels. Table 8.22 also shows the sample mean and standard deviation as computed from Equations 8.46 and 8.47.

The prevailing sample state, x_J, is found (see Table 8.23) by comparing the sample mean from Table 8.22 with the rating limit states of Table 8.20.

From the rating limit states of Table 8.20 and the statistical properties of Table 8.22, also taking advantage of assuming a uniformly distributed PDF for the samples, the conditional probabilities of different rating states can be computed as shown in Table 8.24.

We have now all the components needed to compute improved (posterior) probabilities for rating states of different situations of our example. The posterior probabilities are simply an application of Equation 8.47. They are shown in Tables 8.25 through 8.28.

The results in Tables 8.25 through 8.28 show clearly the importance of performing SHM experiments to improve estimations of probabilities of rating states. For example, note case 1: operating conditions now show clear-cut safe ratings. The probabilities of closing and posting ratings have dropped from 4% and 5% to 0.2% and 1%, respectively. The probability of safe rating has increased from 91% to 99%. More importantly, for case 1, inventory loading, the probability of posting rating has increased from 19% to 42%. The probabilities of closing and safe ratings have dropped from 16% and 65% to 2.6% and 54%, respectively. It is clear that case 1 calls for posting decision. Without the SHM experiment, the decision might have been to close the bridge, since the probability of closing (inventory weight) was 16%. Even worse, at 65% probability of safe rating for inventory weight, the analyst might have decided to keep the bridge open, which is clearly an unsafe decision.

TABLE 8.22
Sample Ratings from SHM Experiment

Point #	Operating (Tons)	Inventory (Tons)
1	50	32
2	44	45
3	56	43
4	47	37
Mean (average)	49.25	39.25
Standard deviation	5.12	5.90

TABLE 8.23
Prevailing Sample State for Different Situations

Rating Load	Operating	Inventory
Case 1	Safe, x_3	Safe, x_3
Case 2	Post, x_2	Post, x_2

TABLE 8.24

Conditional Probabilities of Prevailing Sample States—Based on Load-Rating Tests

Condition	Case 1		Case 2	
	Operating, $J = 3$	Inventory, $J = 3$	Operating, $J = 2$	Inventory, $J = 2$
$P(x_J \mid \theta_1)$	0.032	0.05	0.203	0.24
$P(x_J \mid \theta_2)$	0.171	0.69	0.352	0.307
$P(x_J \mid \theta_3)$	0.797	0.26	0.445	0.453
Total	1.0	1.0	1.0	1.0

TABLE 8.25

Computations of Posterior (Improved) Probabilities (Case 1, Operational)

Rating State θ_i	Prior Probability (Pure Analytical) $P(\theta_i)$	Conditional Probability $P(x_J \mid \theta_i)$	Joint Probability $P(\theta_i)P(x_J \mid \theta_i)$	Posterior Probability (Improve through SHM Testing) $P(\theta_i \mid x_J)$
Closing, θ_1	0.04	0.032	0.00128	0.001741
Posting, θ_2	0.05	0.171	0.00855	0.011631
Safe, θ_3	0.91	0.797	0.72527	0.986628
Total	1		0.7351	1

TABLE 8.26

Computations of Posterior (Improved) Probabilities (Case 1, Inventory)

Rating State θ_i	Prior Probability (Pure Analytical) $P(\theta_i)$	Conditional Probability $P(x_J \mid \theta_i)$	Joint Probability $P(\theta_i)P(x_J \mid \theta_i)$	Posterior Probability (Improve through SHM Testing) $P(\theta_i \mid x_J)$
Closing, θ_1	0.16	0.05	0.008	0.025966
Posting, θ_2	0.19	0.69	0.1311	0.425511
Safe, θ_3	0.65	0.26	0.169	0.548523
Total	1		0.3081	1

TABLE 8.27

Computations of Posterior (Improved) Probabilities (Case 2, Operational)

Rating State θ_i	Prior Probability (Pure Analytical) $P(\theta_i)$	Conditional Probability $P(x_J \mid \theta_i)$	Joint Probability $P(\theta_i)P(x_J \mid \theta_i)$	Posterior Probability (Improve through SHM Testing) $P(\theta_i \mid x_J)$
Closing, θ_1	0.09	0.203	0.01827	0.043843
Posting, θ_2	0.07	0.352	0.02464	0.05913
Safe, θ_3	0.84	0.445	0.3738	0.897027
Total	1		0.41671	1

Similarly, case 2 reaffirms the importance of improving rating probabilities. The probabilities of operating loading show an increase of safe-rating probability from 84% to 90%. Inventory weight probabilities showed an increase of safe-rating probability from 42% to 55%. However, the combined closing- and posting-rate probabilities are sizable, with the closing-rate probability slightly higher at 24%. The analyst would be wise to declare the bridge unsafe. The decision to post it or to

TABLE 8.28
Computations of Posterior (Improved) Probabilities (Case 2, Inventory)

Rating State θ_i	Prior Probability (Pure Analytical) $P(\theta_i)$	Conditional Probability $P(x_J \mid \theta_i)$	Joint Probability $P(\theta_i)P(x_J \mid \theta_i)$	Posterior Probability (Improve through SHM Testing) $P(\theta_i \mid x_J)$
Closing, θ_1	0.35	0.24	0.084	0.2439
Posting, θ_2	0.2332	0.307	0.071592	0.207874
Safe, θ_3	0.4168	0.453	0.18881	0.548226
Total	1		0.344403	1

close it would need further investigation. Perhaps additional testing or refined analysis would lead the analyst to a more clear decision in this case.

A final note on the example on hand: we recall from Table 8.17 that the conventional deterministic and analytically based rating method yielded operational and inventory ratings of 62.4 and 37.4 tons, respectively. If only these ratings were used in the decision making process, then (after consulting Table 8.20) the decision would have been that the bridge is safe for both case 1 and 2, for both inventory and operational weights. Such decisions are clearly neither safe nor economical. We have just shown the importance of accommodating uncertainties in evaluating bridge rating. We also showed the need for improving the all-analytical rating processes by performing an adequate SHM project.

8.7.2.3   Closing Remarks

A final note on the assumed independence of random variables of Equation 8.25 is needed. We assumed earlier that the impact factor I is constant. This assumption can be generalized to

$$I = I(L) \tag{8.48}$$

To accommodate the more accurate assumption that I is a function of L. Obviously, that would change the above probabilistic developments, since it would change the nature of Equation 8.25. The resulting equations, while more complex, are still simple to apply using the Taylor series approach. The readers are encouraged to explore the effects of the function $I = I(L)$ on their own.

We also make another important remark concerning the importance of SHM in the bridge-rating process. In the above rating examples, we showed how the decisions can be influenced by using probabilistic analysis and experimental testing (SHM) as complements to the conventional analysis-only procedures. However, we have not offered a quantitative procedure to estimate the value of the experiment to the rating process. Such a procedure is described next. Also, a procedure to quantify the value of testing and information is described in Chapter 2 by Ettouney and Alampalli (2012).

8.7.3   Value of Load Testing

Earlier, we defined the total load test costs to be C_{LT}. We need to establish the total experimental load rating, CDI_1 as

$$CDI_1 = C_{LT} + C_{Rating} \tag{8.49}$$

with C_{Rating} as the cost of computing the rating from the processed experimental data. Similarly, we identify the total analytical load rating, CDI_2 as

$$CDI_2 = C_{Analysis} + C_{Rating} \tag{8.50}$$

The analysis costs, $C_{Analysis}$ includes all analytical efforts that compute bridge responses, for example, stresses, that are needed for load-rating evaluations. For ease of discussion, let us relate experimental and analytical costs as

$$CDI_1 = \alpha_{EA} CDI_2 \tag{8.51}$$

Where α_{EA} is the experimental to analytical costs of load-rating the bridge. We propose that the decision of load-rating the bridge, based on analytical or experimental means, should depend on the value of α_{EA} such that

$$\text{if } \alpha_{EA} > 1.0 \tag{8.52}$$

then perform analytical load rating, and

$$\text{if } \alpha_{EA} < 1.0 \tag{8.53}$$

then perform experimental load rating. If $\alpha_{EA} = 1.0$, then the decision maker can go either way. As of the writing of this manuscript, it is typical that $\alpha_{EA} > 1.0$ for most situations. Thus, if the decision is based on costs alone, the load ratings are always based on analytical rather than experimental methods. Is this an appropriate conclusion for such an important subject matter? Let us consider this issue a bit further.

The key to exploring this issue further is to address the validity of the total costs, CDI_2 and CDI_1 as defined earlier. We note that there are the *direct* costs of the load rating. To have a more meaningful comparison, we need to consider also the *results* (value of load rating) and *consequence* (safe, post, or close) costs of the load-rating results. In Sections 8.3 and 8.4, we showed that the load-rating method (analytical vs. experimental) affects the load-rating results and consequences. In general, there are three consequences of any load-rating effort: safe, post, or close. We submit that there is a cost associated with each of these consequences. The costs of safe rating, postrating, and close rating can form vector crs_i, with $i = 1,2,3$ as explained in Table 8.29. The close rating is the cost of closing the bridge to traffic. This includes cost of repairs, or replacement, traffic rerouting, as well as any other social/economic costs. Posting costs include costs for traffic rerouting and repair. Safe rating costs are more complex to ascertain; at first glance, there seems to be zero consequence costs for a safe rating such that

$$crs_1 = 0 \tag{8.54}$$

Accounting for results and consequence costs of load rating is not an easy task. This is because the decision maker does not know a priori the results of the load rating, making impossible the exact comparison between analytical and experimental load-rating costs that include the *results* and *consequences*. The decision under uncertainty methods of Chapter 8 by Ettouney and Alampalli (2012) can be used to resolve this problem. We first assume that records show that the probability of the

TABLE 8.29
States of the ith Counter

i Counter	State
1	Safe rating
2	Post rating
3	Close rating

TABLE 8.30
States of the *j*th Counter

j Counter	State
1	Experimental load rating
2	Analytical load rating

results of load rating is p_{ij} with $i = 1,2,3$ and $j = 1,2$. The definitions of i and j are shown in Tables 8.29 and 8.30, respectively.

The load rating *results* and *consequences* cost vector CRS_k, $k = 1,2$, are computed as

$$CRS_k = \sum_{i=1}^{i=3} crs_i p_{ik} \tag{8.55}$$

with

CRS_1 = Cost of experimental load rating results and consequences
CRS_2 = Cost of analytical load rating results and consequences

We now add all direct results and consequence costs as follows

$$CLR_k = CRS_k + CDI_k \tag{8.56}$$

We can generalize Equations 8.52 and 8.53 to the more realistic inequalities

$$\text{If} \left(CRS_2 + CDI_2 \right) < \left(CRS_1 + CDI_1 \right) \tag{8.57}$$

then perform analytical testing, and

$$\text{If} \left(CRS_1 + CDI_1 \right) < \left(CRS_2 + CDI_2 \right) \tag{8.58}$$

then perform experimental testing

8.7.4 Cost–Benefit of Proof Load Testing

Fu (1995) observed that proof load tests offer special benefit-to-cost advantages. Among the costs of proof tests are

- Equipment costs: These include loading trucks, concrete blocks (for added weights), transport trucks, sensors, instrumentations, and traffic control trucks
- Labor costs: These include load test crews and traffic control crews

Among the benefits are

- Saving of analytical rating costs
- Reduced mileage resulting in restricted truck travel
- Reduced accidents

It was observed that the annual benefit-cost ratio of proof testing is about 2.1.

8.8 COST, BENEFIT, AND LCA OF BRIDGE LOAD TESTS

8.8.1 GENERAL

In this section we offer quantitative models for evaluating costs and benefits of load tests. We then evaluate the effects of load tests on the LCA of bridges.

8.8.2 COST–BENEFIT MODEL FOR LOAD TESTS

8.8.2.1 Costs

There are numerous sources for load-testing costs. Generally, single-span bridges offer the simplest load test projects. This is due to the simplicity of placing sensors, ease of controlling traffic, and the relative short time required to load the bridge with trucks. As the geometry of bridges becomes more complex, the operation becomes more complex, and the costs increase accordingly. It should be noted that evaluating costs of load testing, as opposed to different evaluations of costs presented in Chapters 1 through 7, is more specific. These costs deal with more specific situations, that is, the particular testing on hand; thus the confidence in estimating load-testing costs should be high. We discuss some sources of load-testing costs next.

Traffic: Since the loading in a load test is the standard trucks that would be required to travel along the bridge while monitoring the bridge response, the normal bridge traffic needs to be controlled during the loading operation. The costs of traffic control depend on the size of bridge and time required for the test. We define the cost of controlling traffic as C_T.

Sensors and other instruments: Costs of sensors and instrumentation are another important source of costs. The main sensors during the load tests are strain sensors. However, other types of sensors might be used depending on the test objectives. We define the cost of sensors and instrumentations as C_S.

Data analysis: Engineering services, such as analyzing data, analyzing bridge responses and comparing them with the test results or any other required analysis or design tasks, can add to the costs. We define the cost of data and engineering analysis as C_D.

Labor costs: Labor costs include placing sensors and other instrumentations, any needed temporary construction, and the time of all those involved in the test. These costs also include cost of management. We define the cost of labor as C_L.

Loading trucks: The cost of trucks, their drivers, and operations are also included in the overall load test costs. We define the cost of loading trucks as C_W.

Turnkey load tests: We note that several turnkey load tests are available for small bridges. Turnkey operations offer simplicity and efficiency to the bridge owners. As such they are cost-effective. Many commercial outfits offer such turnkey solutions. As the bridge geometry and/or the load test objectives become more specialized or complex, the turnkey solutions will obviously need to be redesigned to fit the needs of the load test.

The total cost of a bridge load test can now be offered as

$$C_{\text{LOAD\_TEST}} = C_T + C_S + C_D + C_L + C_W + C_{\text{MISC}} \tag{8.59}$$

Note that C_{MISC} includes any other cost source not described specifically in the above discussion. For the purposes of this chapter, we start a hypothetical and simple example where the total cost of a load test is estimated at $2,500.00. We will use this cost in a cost–benefit analysis later in this section.

8.8.2.2 Benefits

As usual, quantifying benefits is not an easy task. To avoid this problem, we try a different approach for quantifying load testing. We subdivide each potential benefit into one of two categories: a quantifiable benefit and a qualitative benefit that is not quantifiable or very difficult to quantify. We then devise a logical method to produce a total benefit for the load rating. We note that the method is general enough to be applied to any other situation.

8.8.2.2.1 Quantifiable Benefits

Operational decision making aid: Operational decisions such as flagging, posting, rating, and permits can all be aided by bridge testing. Both proof and diagnostic testing can be used to aid any of the operational decisions. The process is already being performed by all bridge owners in a qualitative fashion (e.g., based on visual inspection or solely on analytical methods): how can we quantify the load-testing benefit?

First, we observe that all the above operational processes contain distinct limit states, with each limit state being i, with $i=1,2,\ldots,N_{Oj}$. The number of the limit states for the jth process is N_{Oj}. Let us assume that the qualitative decision for the jth process is

$$i = I_{0j} \tag{8.60}$$

Let us assume that the actual limit state, k, as predicted by the load test, is I_{kj}. There can be two situations:

$I_{kj} = I_{0j}$: The load test confirms the conventional DM approach. In such a case, the benefit of the test is qualitative (see discussion below).

$I_{kj} \neq I_{0j}$: The load test shows that the conventional DM approach is either too conservative or unsafe (depending on the process). In either situation, it is easy to estimate the benefit of the load test B_{Oij} as the cost of deviation between state 0 and state k, C_{O_k0j}, such that

$$B_{Oij} = C_{O\_k0j} \tag{8.61}$$

Unfortunately, estimating the load test benefit in this manner can take place only after the test is performed. We can estimate the benefit before the test is performed by accounting for previous experience such that

$$B_{Oij} = \sum_{k=1}^{k=N_{Oj}} \left(p_{O\_kj} C_{O\_k0j} \right) \tag{8.62}$$

The probability that the actual load test is p_{kj}. Note that, by definition

$$\sum_{k=1}^{k=N_{Oj}} p_{kj} = 1.0 \tag{8.63}$$

The element of vector p_{kj} can be estimated either objectively from previous similar test results or subjectively from personal experience of the bridge manager.

Maintenance decision making aid: Load testing can also be helpful in decisions regarding MR&R. Two specific categories of benefits can be observed: (1) the load testing result can prioritize the MR&R decision, and (2) the load testing can establish the level of MR&R effort. The quantification process of the benefit for both categories (prioritization, B_{MP_ij}, and level of effort, B_{ML_ij}) can follow a similar development as above such that

$$B_{MP\_ij} = \sum_{k=1}^{k=N_{MP\_j}} \left(p_{kj} C_{MP\_k0j} \right) \tag{8.64}$$

$$B_{ML\_ij} = \sum_{k=1}^{k=N_{ML\_j}} \left(p_{kj} C_{ML\_k0j} \right) \tag{8.65}$$

As before

C_{MP_k0j} = Cost of deviation between state 0 and state k for prioritizing MR&R effort
C_{ML_k0j} = Cost of deviation between state 0 and state k for performing MR&R effort

The number of possible prioritization and level of effort limit states are N_{MP_j} and N_{ML_j}, respectively.

Allowing higher performance: Proof testing to explore higher bridge capacities can be helpful for allowing even higher performance, such as truck overloads. Benefits of exploring higher performance can be quantified in a fashion similar to the above such that

$$B_{HP\_ij} = \sum_{k=1}^{k=N_{HP\_j}} \left(p_{kj} C_{HP\_k0j} \right) \tag{8.66}$$

As before

C_{HP_k0j} = Cost of deviation between state 0 and state k for prioritizing MR&R effort

The number of possible high performance limit states is N_{HP_j}.

Total quantitative benefits: The total quantitative benefits can now be evaluated as

$$\text{BQN} = B_{Oij} + B_{MP\_ij} + B_{ML\_ij} + B_{HP\_ij} + B_{OTHER} \tag{8.67}$$

Any additional benefits that have not been accounted for in an explicit manner can be included in B_{OTHER}.

Example: Consider an example of a load test where the number of operational limit state was estimated as $N_{Oj} = 3$, which is shown in Table 8.31. The table also shows the probabilities of each state (from past experiences).

The conventional procedures indicate that the bridge should be flagged, $k = 2$. The bridge official made some cost analysis. The cost of doing nothing is null, the cost of posting, C_{O_20j}, is \$3,000.00, while the cost of closing the bridge, C_{O_30j}, is \$10,000.00. Applying Equation 8.62 would produce a benefit, B_{Oij}, of \$3,500.00.

Similar studies indicate quantitative benefits as shown in Table 8.32. The total quantitative benefits of this particular load test is \$5,000.00.

8.8.2.2.2 Qualitative (Nonquantifiable) Benefits

Estimating benefits qualitatively can be fairly simple. Table 8.33 shows different sources of load test qualitative benefits. For each of these benefits, the decision maker would assign a relative benefit, B_{Qj}, on a scale 0–100. The previous example is continued, with its results as shown in Table 8.34. The example shows that the bridge officials placed medium value on improving their inventory

TABLE 8.31
Limit Operational States and Their Probabilities

k	Description	p_{kj} (%)
1	Do nothing	30
2	Post	50
3	Close	20

TABLE 8.32
Monetary Values of
Quantitative Benefits

Benefit	Value ($)
B_{Oij}	3,500.00
B_{MP_k0j}	500.00
B_{ML_k0j}	500.00
B_{HP_k0j}	0.00
B_{OTHER}	500.00
BQN	5,000.00

TABLE 8.33
Qualitative Benefits of Load Tests

Counter, j	Type of Qualitative Benefit
1	Improve inventory database
2	QA & QC: check over conditions of just-finished MR&R projects, new construction materials such as FRP (see Chapters 6 and 7), high-performance concrete or steel systems
3	Confirming decisions regarding detours
4	Validation of designs/analysis
5	Other?

TABLE 8.34
Example of Qualitative Load Test Benefits

Counter, j	Score (0–100)
1	50
2	30
3	90
4	20
5	10 (overall value of additional information)
Total	200

database. Also, very high value was placed on exploring the potential of using that load test to support detouring decisions. The particular test under consideration is not meant to validate immediately an analysis method, nor is it meant for an immediate QA/QC validation. These are reflected in the relative scores in Table 8.34.

The total qualitative benefits can be expressed as

$$\text{BQL} = \sum_j B_{Qj} \tag{8.68}$$

From Table 8.34, the total qualitative benefits in our example are 200 units.

8.8.2.2.3 *Total Benefits*

The final step in estimating the total benefits of load tests it to combine BQN and BQL. The problem we face is the mismatch in units. While BQN is monetary, we have BQL in a nondimensional form. One possible way of logical combination of the two mismatched units is to transform one unit into another and then combine the two. For example, the bridge official can assign a logical weight, W_{BQN}, for all the quantitative benefits by comparing all benefits together. A monetary value for the qualitative benefits can then be computed as

$$BQL_{Monetary} = BQL\frac{BQN}{W_{BQN}} \tag{8.69}$$

Now the qualitative benefits are expressed in monetary units. The total benefits of load test are

$$B_{LOAD\_TEST} = BQN + BQL_{MONETARY} \tag{8.70}$$

To continue the above example, let us assume that the official estimated that

$$W_{BQN} = 500 \tag{8.71}$$

Such an estimation would reflect that the official felt that the quantitative benefits for this test far outweigh the qualitative benefits (ratio of 500–200). Applying Equation 8.69 would result in an estimated monetary value of qualitative benefits of $2,000.00. Thus, the total benefits of the load test are estimated at $7,000.00.

8.8.2.3 Closing Remarks

We can now compare the costs and benefits of load test and reach appropriate conclusions. For example, by comparing the costs and benefits of the above example, we find that the benefits ($7,000.00) outweigh the costs ($2,500.00). The appropriate decision is self-evident. (Note that the load test cost was estimated to be pretty low in this case for illustration purposes.)

8.8.3 LCA AND LOAD TESTS

We have established a quantitative basis for load-testing cost and benefit models. We turn our attention now to the overall LCA of a bridge and the effects that a load test, or more than a load test, might have on it.

An incorporation of the costs and benefits of load tests, as described above, might seem to be a straightforward matter; for example,

$$LCCA_{LOAD\_TEST} = \sum_{i=1}^{i=N_{LOAD\_TEST}} \left(C_{LOAD\_TEST}\right)_i \tag{8.72}$$

And

$$LCBA_{LOAD\_TEST} = \sum_{i=1}^{i=N_{LOAD\_TEST}} \left(B_{LOAD\_TEST}\right)_i \tag{8.73}$$

with

$LCCA_{LOAD\_TEST}$ = Total life cycle cost of all load tests over the whole lifespan of the bridge (ignoring discount rates, for the sake of simplicity)
$LCBA_{LOAD\_TEST}$ = Total life cycle benefits of all load tests over the whole lifespan of the bridge (ignoring discount rates, for the sake of simplicity)
N_{LOAD_TEST} = Total number of all load tests over the whole lifespan of the bridge

Upon further reflection, we should ask ourselves a question: *would load tests have an indirect effect on the lifespan of a bridge?* Such an effect cannot be measured in the direct cost and benefit models presented earlier. However, it seems logical to assume that there should be a definite effect of load tests on the lifespan of a bridge. It is logical that the information gathered about a bridge would give the decision maker a better understanding of its state. Such an understanding should improve the potential decisions about the bridge; this in turn should improve the deterioration rates of the bridge, thus improving its lifespan.

Unfortunately, as of the writing of this manuscript, there is no information about the correlation between load tests and the lifespan of the bridge. Such correlation, when proved, should increase the value of load testing and increase the frequency of its use.

8.8.4   Extending Lifespan as a Result of Load Test

Another approach to estimate the effects of load tests on the life cycle of a bridge is by using the deterioration curve approach. The deterioration curves relate a measure of bridge performance such as condition rating and the elapsed time. Agrawal et al. (2009) investigated different methods of generating deterioration curves. Figure 8.38 shows a typical deterioration curve. At time $t = 0$, the time at which the bridge starts its functional life, it is expected that the condition rating, y_0, of the bridge is maximum. The rating, y, decreases, until it reaches a theoretical zero-condition rating. The theoretical lifespan of the bridge is thus

$$T_0 = t\big|_{y=0} \tag{8.74}$$

The general shape of the deterioration curve can be convex, concave, or nearly linear. Parameters that affect such a shape were investigated by many authors (see Chapter 9). For our immediate purposes, let us assume that load tests were performed at a point (t_1, y_1) on the deterioration curve. As a result of this test, it was decided to perform retrofits that would increase the condition rating from y_1

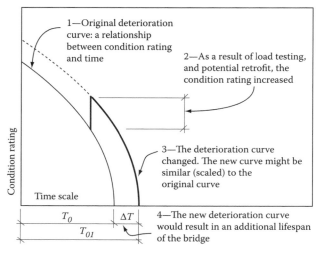

FIGURE 8.38   Extending lifespan as a result of load tests.

to y_{11}. Therefore, the deterioration path would change. Thus, the theoretical lifespan would increase to T_{01}. The net increase in theoretical lifespan is

$$\Delta T_0 = T_{01} - T_0 \tag{8.75}$$

This increase in theoretical lifespan can be considered a potential benefit of the load test.

We give an example of the increase lifespan analysis. First we make the following assumptions:

- The original deterioration curve, $y = f(t)$, is known.
- The retrofitted deterioration curve, $y = f(t)$, is known.

For the current example, we approximate the original deterioration curve by a second-order curve in the form

$$y = a + bt + ct^2 \tag{8.76}$$

The constants a, b, and c are obtained from the three known points on the curve $(0, y_0)$, $(t_0, 0)$, and (t_1, y_1). After the load test and retrofit, we assume that the new deterioration curve will be *similar* to the original curve. Thus, it can be described as

$$y = \beta a + bt + \frac{c}{\beta}t^2 \tag{8.77}$$

with β as the scaling factor. The scaling factor can be obtained from

$$\Delta y = y_{11}\left(\beta, t = t_1\right) - y_1\left(t = t_1\right) \tag{8.78}$$

Finally, the increase in lifespan is

$$\Delta T_0 = \left(\beta - 1\right)t_0 \tag{8.79}$$

which is the benefit of the load testing.

A numerical example with properties is given in Table 8.35. Note that the deterioration test is performed after 20 years. After performing the computations, it was found that the lifespan is increased by 4.79 years as a result of condition rating increase of 1.0. Figure 8.39 shows the original and the retrofitted deterioration curves. When the load test is performed a bit later, at 32 years, the

---

TABLE 8.35
Values for Life Increase Case 1

Maximum original rating, y_0	7
Original bridge lifespan, T_0 (years)	40
Intermediate rating, y_1	5
Intermediate time, t_1 (years)	20
Improved rating, Δy	1
β	1.119913
ΔT_0 (years)	4.79

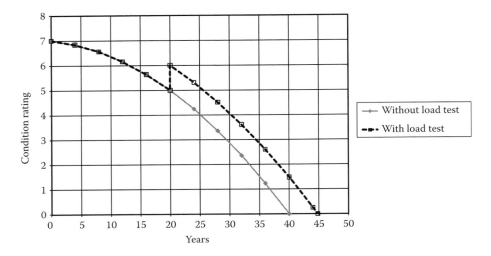

FIGURE 8.39   Effect of retrofits on deterioration, Case #1.

TABLE 8.36
Values for Life Increase Case 2

Maximum original rating, y_0	7
Original bridge lifespan, T_0 (years)	40
Intermediate rating, y_1	2.36
Intermediate time, t_1 (years)	32
Improved rating, Δy	1
β	1.095181
ΔT_0 (years)	3.80

lifespan benefit is decreased from 4.79 to 3.80, even though the increase in condition rating is the same. The diminishing benefit is not surprising due to the nonlinearity of the deterioration curves. Table 8.36 and Figure 8.40 show the controlling values of the case and the original and the retrofitted deterioration curves, respectively.

8.9   MONITORING AND LOAD TESTING OF COURT STREET BRIDGE

8.9.1   INTRODUCTION

This section explores in detail the load-testing project of a bridge in upstate New York as reported by Hag-Elsafi et al (2006). The Court Street Bridge replaced an old bridge at the same location and carries Route 96 (Court Street) over Route 17 and the Susquehanna River into Owego (Figure 8.41). It is 338 m long with six spans (52, 65, 65, 65, 65, 52, and 39 m) and 14.45 m wide, including a 12.40-m center-to-center spacing between two supporting trusses and two 1.02-m cantilever overhangs. It carries three lanes of traffic, a northbound, a southbound, and a turning lane with an estimated ADT of 6000. It is a continuous steel structure, consisting of stringers, floor beams, two trusses, and a lightweight concrete deck. The concrete deck was built composite with the stringers as well as the top chords of the trusses that is not very common and complicating structural behavior by introducing secondary moments in the truss members. Moments in the truss members are also influenced by the behavior of the bolted connections—acting as pinned, semirigid, or rigid. Several

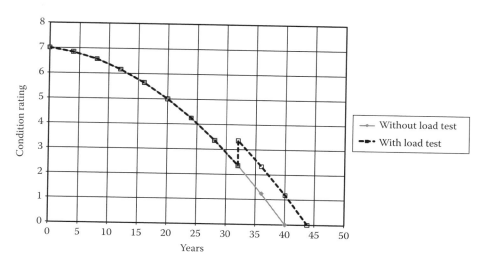

FIGURE 8.40 Effect of retrofits on deterioration, Case #2.

FIGURE 8.41 Court Street Bridge. (Courtesy of New York State Department of Transportation.)

design assumptions were made. The project was initiated to investigate axial forces and moments in the main truss members due to service deck dead load and service live load, to compare the values used in the design, and to better understand the structural behavior under dead and live loads.

The bridge structure was analyzed using the NYSDOT Bridge Load Rating System (BLRS) computer software for truss analysis and rating, and STAAD III, a general structural analysis FE software package. The bridge was designed on the basis of the NYSDOT Standard Specifications for Highway Bridges, with all provisions in effect as of April 2001 (NYSDOT 2001), the AASHTO Standard Specifications for Highway Bridges, 16th edition 1995 (AASHTO 1996), including 1997 and 1998 Interim Specifications, and the AASHTO Guide Specifications for Strength Design of Truss Bridges (Load Factor Design) 1985 (AASHTO 1985), including 1986 Interims. The bridge superstructure steel conforms to ASTM A709M Grade 345W (non-HPS) or ASTM A709M Grade 485W (HPS). An elastic modulus of 2×105 MPa was specified for the design of both the steel-type members (NYSDOT 2001). For design purposes, compressive strength and elastic modulus of concrete for the substructure and deck slab at 28 days were specified at 21 and 1.64×104 MPa, respectively.

8.9.2 INSTRUMENTATION AND LOAD TEST PLANS

The bridge instrumentation plan was designed to provide data for investigation of axial forces and secondary moments in the downstream truss members. The gauges were mounted in pairs near members' ends to collect strain and temperature data during the first three deck pours and for a postconstruction load test. Vibrating wire gauges (Model 4000 Vibrating Wire Strain Gauges, manufactured by Geokon of Lebanon, New Hampshire) were selected for this project as they have long-term durability and do not require correction for drift. The gauges were read using a Geokon Model GK-403 Readout Box that reads one gauge at a time, giving the gauge's strain in $\mu\varepsilon$ and temperature in °C. The vibrating wire gauges have a gauge correction factor of 0.945.

Five of the downstream truss members were instrumented with vibrating wire gauges to record strains in the members during the first three deck pours and to collect additional data during a load test conducted after the bridge construction was completed. The instrumented members were located near the Pier 1 side of Span 2 on the downstream truss (see Figure 8.42). All gauges were arc-welded to the members, except for the top chord (U16-U17) where they were mounted using a quick-set epoxy resin. In the transverse directions, all the gauges were mounted 13 mm from a member's nearest edge, except for Gauge 11 which was mounted 35 mm below the edge next to the concrete deck (see Figure 8.42). All gauges remained operational throughout the project, except for the two mounted on the top chord (Gauges 11 and 12), which debonded before the live-load tests were performed.

8.9.3 MONITORING RESULTS AND DEAD–LOAD ANALYSIS

Strain data was collected for deck pours 1, 2, and 3 (see Figure 8.43) during October and November 2002. There was a pause in construction activities during the winter season before the rest of the deck was cast.

Time histories of the stresses, after compensating for the temperature, were calculated for each paired gauges on a member during the three pours (see Figure 8.44 for typical plot) and show the general quantitative changes in a gauge's stress as the deck is poured in the direction (beginning to end) depicted by the arrows in Figure 8.43. The horizontal axes in these plots are indicative of selected areas of the deck being poured and the corresponding pour numbers. The plots also show the bending experienced by the members during the three deck pours, which is indicated by the separation between the two time-history lines for any paired gauges mounted on a member.

Axial forces and secondary moments for the FE analysis and monitoring results were calculated. The graphical presentation establishes a relationship between the test (actual) and FE analysis results for axial forces and moments; linear relationships (see Figures 8.45 and 8.46) illustrate the consistency of the test data.

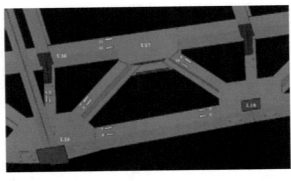

FIGURE 8.42 Instrumented downstream truss members. (Courtesy of New York State Department of Transportation.)

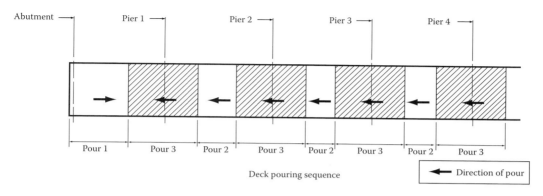

FIGURE 8.43   Deck pour sequence. (Courtesy of New York State Department of Transportation.)

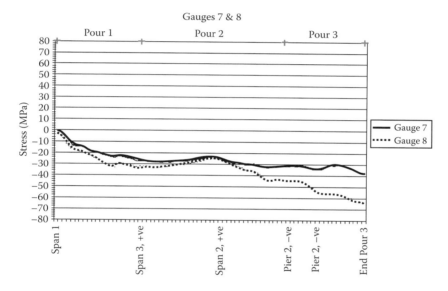

FIGURE 8.44   Deck pour stresses for a typical paired gauges. (Courtesy of New York State Department of Transportation.)

Based on the results from a two-dimensional FE analysis created for the bridge design, the concrete deck load effects were determined to be about 65% of the total service dead–load effects. The FE analysis dead–load secondary moments were obtained by proportioning those shown on the bridge plans by the 65% factor and adjusting for the lightweight concrete deck. Incremental FE analysis results of the structure considering the actual deck pouring sequence confirmed the validity of the above approach to estimate concrete deck axial forces and moments. The unpredictable out-of-plane bending results showed the importance of three-dimensional FE analyses in investigating the structural behavior of a bridge of this type.

8.9.4   Dead–Load Analysis

Axial forces and secondary moments in the instrumented members under total service dead load, using the service deck load analysis results, were determined. This was achieved by applying the 65% factor to the FE and monitoring results in Table 8.37 to generate a similar set of data for predicting total service dead load axial forces and moments (Table 8.38). It is important to note that this approach for predicting total dead–load effects is based on the assumptions that the structural characteristics of the

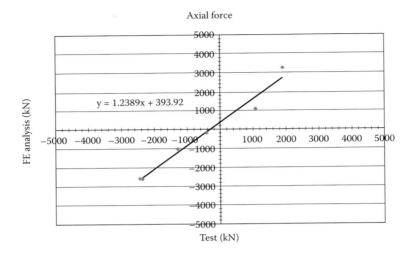

FIGURE 8.45 Axial force (FA): deck service load (FE analysis) versus load test. (Courtesy of New York State Department of Transportation.)

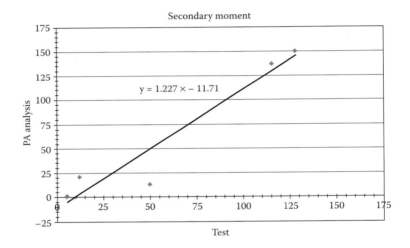

FIGURE 8.46 Bending moment (FA): deck service load (FE analysis) versus load test. (Courtesy of New York State Department of Transportation.)

TABLE 8.37
Deck Load Axial Forces and Secondary Moments

Gauge Number	Member Mounted on	Axial Force (kN)		Secondary Moment (kN-m)	
		FE	Test	FE	Test
1, 2	U16-L16	−175	−428	1	6
3, 4	L16-U17	−1068	−1291	21	12
5, 6	L16-L18	−2581	−2444	137	115
7, 8	L16-L18	−2581	−2394	137	115
9, 10	U17-L18	1074	1072	12	50
11, 12	U16-U17	3258	1903	150	128

TABLE 8.38
Total Dead Load Axial Forces and Secondary Moments

Gauge Number	Member Mounted on	Axial Force (kN)		Secondary Moment (kN-m)	
		FE	Test	FE	Test
1, 2	U16-L16	−267	−658	1	9
3, 4	L16-U17	−1672	−1987	33	19
5, 6	L16-L18	−3962	−3759	–	–
7, 8	L16-L18	−3962	−3683	212	177
9, 10	U17-L18	1642	1650	19	76
11, 12	U16-U17	5022	2927	231	196

FIGURE 8.47 Three- and four-axle trucks used in the load tests. (Courtesy of New York State Department of Transportation.)

bridge were not affected by the concrete-curing process and the subsequent loss of moisture and the possibility of the poured concrete acting compositely with the steel and temperature changes between concrete pours were uniform and the deck load was applied incrementally, in a manner resembling actual pours sequence, in the FE solution for forces and moments due to deck pours.

8.9.5 LIVE–LOAD ANALYSIS

Axial live-load forces in the bridge members were determined mainly on the basis of the BLRS program, using a truss model and an adjusted AASHTO HS-20 load to reflect an HS-25 line load. A two-dimensional STAAD III frame model, assuming end fixity of the truss members, was loaded with combination lane loadings to produce maximum axial forces in chord members at midspan and at the piers. The maximum stresses resulting from the BLRS and STAAD III analysis (two-dimensional model) were compared, and the higher of the two stresses was used in the final design.

The load tests included loading the bridge with four trucks of known weights and configurations to maximize forces and moments at the gauge locations near Pier 1. All the trucks had three axles, except for Truck I, which had four axles (Figure 8.47). Four load test cases were used for the testing, and strains from the gauges were recorded during the tests (Hag-Elsafi et al. 2006). A typical influence line is shown in Figure 8.48.

A three-dimensional STAAD III model, loaded in a manner replicating the actual truck loads on the bridge during the load test, was used to obtain FE results for axial forces and moments. Using these results together with the member and material properties, member stresses at gauge locations

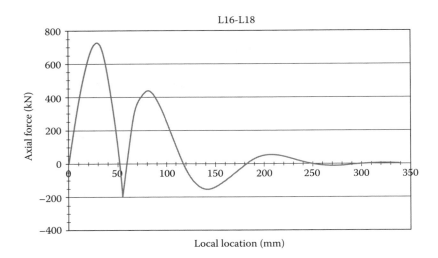

FIGURE 8.48   Influence line for member L16-L18. (Courtesy of New York State Department of Transportation.)

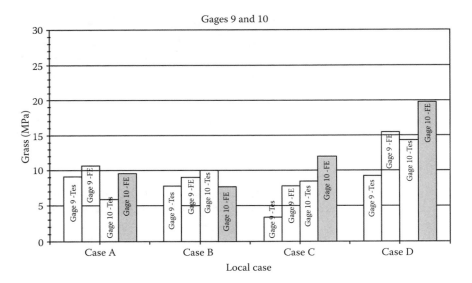

FIGURE 8.49   Comparison of test and FE stresses for member U17-L18. (Courtesy of New York State Department of Transportation.)

were determined. Comparison of test and FE results for a typical instrumented member are shown in Figure 8.49 (Hag-Elsafi et al. 2006). The relationships between the test and FE results were linear, indicating the consistency of the load test data.

To determine actual axial forces and secondary moments in the instrumented members under service live load, the load test effects (moment and shear) were compared/proportioned to those due to the service live load used in the members' design. Owing to the differences between the two results sets noted (see Figure 8.49), it is important to evaluate the design interaction equations for combined stresses using actual axial forces and moments. This evaluation was performed, and the comparisons indicated that actual axial forces and moments satisfied the interaction equations. This confirms the adequacy of the structural design. For the vertical member (U16-L16), the interaction equation was marginally exceeded and was attributed to an overestimation of the members' axial forces during the deck pour monitoring (Hag-Elsafi et al. 2006).

8.9.6 RESULTS

The results (Hag-Elsafi et al. 2006) indicated that the members' actual service dead load axial forces and moments were overestimated by about 20%, and service live-load axial forces were overestimated by about 30% in the design. They also showed that the service dead load moments were within 25% of those used in the design, and service live-load moments were underestimated by about 55%. Adequacy of the structural design was confirmed by checking the AASHTO interaction equations for members under combined stresses using actual axial forces and moments in the bridge members. The study also recommended that three-dimensional FE models be used for analysis of structures similar to Court Street Bridge, due to their ability to predict out-of-plane bending, which may result under some loading situations.

8.10 LOAD TESTING FOR BRIDGE RATING: ROUTE 22 OVER SWAMP RIVER

The bridge carrying Route 22 over Swamp River in Dover in Duchess County, NY, is a simple-span multigirder steel structure built in 1938 (Figure 8.50). The bridge has seven beams built as noncomposite with an 8.5-in. concrete deck over it and also short fascia beam overhangs. In 1981, to meet the increased traffic demands, the original deck was replaced with a new concrete deck, and the bridge was widened by increasing the length of the fascia overhangs to 3 ft 4 in. The overhang depth was also increased from 8.5 to 13 in. to meet the structural demand.

This rehabilitation of the structure added considerable dead load and thus reduced the live-load capacity. Fascia beam load-rating capacity was mainly reduced and controlled the system load-rating capacity after the rehabilitation. Based on the analysis, and taking the assumptions made during the design, and the current structural condition, the bridge was restricted to legal loads only, that is, no vehicles above legal loads were permitted.

The bridge was on a busy route having heavy truck traffic and posed hardships to trucking traffic. A diagnostic load testing was conducted to evaluate the true load capacity of the bridge to take advantage of the reserve strength (Hag-Elsafi and Kunin 2006). The reserve strength was expected from unintended composite action between the deck and the girders, the bearing fixity, if any, and actual distribution factors that could be less than the distribution factors used in the design.

The load test involved instrumenting the bridge with strain gauges and measuring the strains under known loading (see Figures 8.51 and 8.52). On the basis of the expected contributions described above, the bridge was instrumented appropriately to measure midspan bottom flange stresses for all seven beams to estimate distribution factors, level of fixity at the ends of two beams, midspan neutral axis locations for four of the beams, and end neutral axis locations for two beams. All bottom gauges were mounted on top of the bottom flanges, and all top gauges were mounted on the bottom of the top flanges. A general-purpose strain gauge measurement system was used to collect data.

FIGURE 8.50 Bridge floor plan and section. (Courtesy of New York State Department of Transportation.)

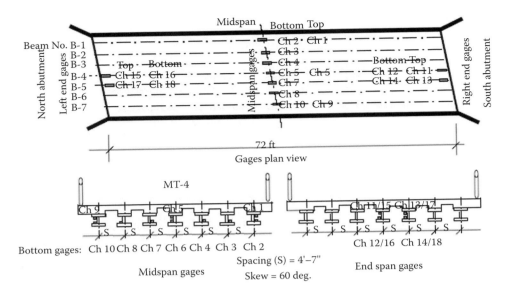

FIGURE 8.51   Instrumentation plan. (Courtesy of New York State Department of Transportation.)

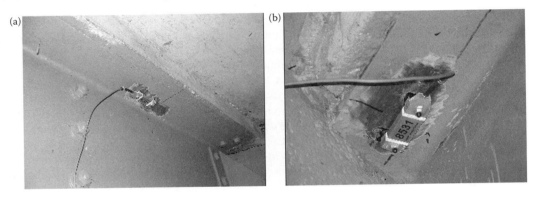

FIGURE 8.52   Strain gauges on the girders (a) setting on the beam and (b) closer view. (Courtesy of New York State Department of Transportation.)

Four trucks, each resembling an AASHTO H-20 truck, of known weights were used in the load tests; with a typical truck weighing 21 tons (see Figure 8.53). Load tests were optimized to determine distribution of live load to the beams, neutral axis location for the fascia beams, fixity of the beams at the abutments, and dynamic allowance factor. Four trucks were used for the static tests with a predetermined sequence of loading. For the dynamic test, one truck was used driven at 15, 30, 45, and 55 mph during high-speed data collection. Shears and moments due to the various truck combinations from static tests were obtained, assuming both simply supported and fixed end conditions. At the beginning of the test, one of the trucks was also driven at 5 mph while collecting data to make sure that the test system was working as expected.

Transverse load distribution and moments on the structure from test data indicated that fascia beams were two of the three most loaded beams carrying about 43% and 33% of the total moment on the structure when neighboring lanes to the beams were loaded. The actual midspan moments were only 50% of those calculated for a simply supported structure, indicating higher flexural capacity of the bridge (see Figure 8.54). A high level of fixity (more than 65%) was observed at both abutments during the load testing and remained present under the large loads applied during the testing (see Figures 8.55 and

FIGURE 8.53   Typical load test. (Courtesy of New York State Department of Transportation.)

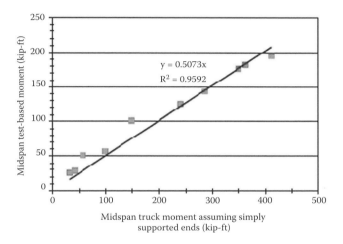

FIGURE 8.54   Test versus calculated midspan moments for beam 4. (Courtesy of New York State Department of Transportation.)

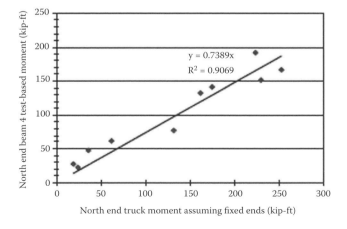

FIGURE 8.55   Test versus calculated north end moments for beam 4. (Courtesy of New York State Department of Transportation.)

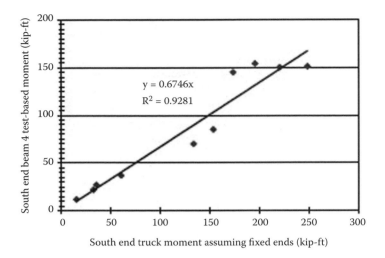

FIGURE 8.56 Test versus calculated south end moments for beam 4. (Courtesy of New York State Department of Transportation.)

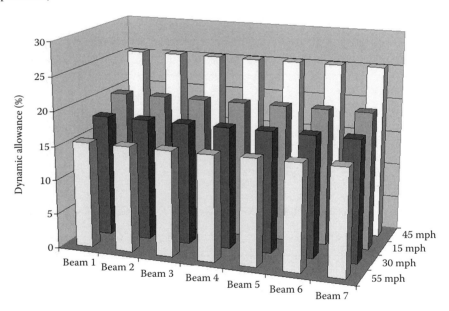

FIGURE 8.57 Dynamic allowances observed from load tests. (Courtesy of New York State Department of Transportation.)

8.56). Presence of composite action was observed consistently even under the very high loads used during the testing, and thus it was determined that the theoretical section modulus can be safely increased by 15% to account for the presence of composite action. Dynamic tests showed that the dynamic allowance factors were very close to those specified by the AASHTO specifications (see Figure 8.57).

Thus, the test confirmed that the beams drew their strength from a considerable fixity at both abutment ends, unintended composite action between the bridge girders and the concrete deck, and a lower level of transverse load distribution than that specified by the AASHTO specifications. The analytical load rating was adjusted on the basis of these test data and indicated that the rating factors can be doubled, that is, the bridge posting can be eliminated.

REFERENCES

AASHTO. (1985). *Guide Specifications for Strength Design of Truss Bridges (Load Factor Design)*, 1st edition including 1986 Interims, American Association of State Highway and Transportation Officials, Washington, DC.

AASHTO. (1994). *Manual for Conditions Evaluation of Bridges*, American Association of State Highway and Transportation Officials, Washington, DC.

AASHTO. (1996a). *Standard Specifications for Highway Bridges*, 16th edition including 1997 and 1998 Interim Specifications, American Association of State Highway and Transportation Officials, Washington, D.C.

AASHTO. (1996a). *Standard Specifications for Highway Bridges*, 16th edition, American Association of State Highway and Transportation Officials, Washington, DC.

AASHTO. (2001), *Manual for Condition Evaluation of Bridges*, American Association of State and Highway Transportation Officials, Washington, DC.

Agrawal, A., Kawaguchi, A., and Qian, G. (2009). "Bridge Element Deterioration Rates: Phase I Report," New York State Department of Transportation, Project # C-01-51, Albany, NY.

Alampalli, S. and Elsafi, O. (1994). "Sensors and Instrumentation for Bridge Testing," 40th *International Instrumentation Symposium, Instrumentation Society of America*, Baltimore, Maryland.

Alampalli, S. and Fu, G. (1992). "Influence Line Tests of Isotropically Reinforced Bridge Deck Slabs," NYSDOT report, Albany, NY.

Ashour, H., Al-Harami, A., and Rublah, A. (1993). "Building a general purpose microcomputer based mechanical measurement system," *Experimental Techniques, 17*(3), 34–37.

Barker, M. C. (2001). "Quantifying field-test behavior for rating steel girder bridges," *ASCE, Journal of Bridge Engineering*, 6(4), July/August, 254–261.

Biggs, J. M. (1964). *Introduction to Structural Dynamics*, McGraw-Hill, New York, NY.

Chajes, M. J., Shenton, H. W., and O'Shea, D. (2000). "Assessing Bridge Capacity Using NDT Methods," Structural Materials Technology: an NDT Conference, ASNT, Atlantic City, NJ.

Chowdhury, M. R. (2000). "A comparison of Bridge Load Capacity Using Accelerometer and Strain Gauge Data," Structural Materials Technology: an NDT Conference, ASNT, Atlantic City, NJ.

Ettouney, M. and Alampalli, S. (2012). *Infrastructure Health in Civil Engineering: Theory and Components*, CRC Press, Boca Raton, FL.

Ettouney, M. M. (1980). "Dynamic Influence Lines with Application to the Design of Turbine Generator Pedestals," Prepared for the Standards Division, Burns and Roe, Inc., Woodbury, NY.

FHWA (2002). *Bridge Inspector's Reference Manual (BIRM)*, Vols. I and II, FHWA-NHI-03-001, Federal Highway Administration, U.S. Department of Transportation, Washington, DC.

Finno, R. J. and Chao, H. C. (2000). "Nondestructive Evaluation of Selected Drilled Shafts at the Central Artery/Tunnel Project," Structural Materials Technology: an NDT Conference, ASNT, Atlantic City, NJ.

Fu, G. (1995). "Highway Bridge Rating by Nondestructive Proof-Load Testing For Consistent Safety," New York State Department of Transportation, NYS-DOT, Research Report No. 163, Albany, NY.

Fuchs, P. A. and Jalinoos, F. (2006). "Laser System Measurements on the FHWA Curved Girder Bridge," NDE Conference on Civil Engineering, ASNT, St. Louis, MO.

Fuchs, P. A., Washer, G. A., and Chase, B. (2000). "Ultrasonic Stress Measurements on Curved Girder Bridge Cross Frame Members," Structural Materials Technology: an NDT Conference, ASNT, Atlantic City, NJ.

Green, M. F. (1995). "Modal Test Methods For Bridges: A Review." *Proceedings of the 13th International Modal Analysis Conference (IMAC)*, Volume 1, Nashville, Tennessee.

Hag-Elsafi, O. and Kunin, J. (2002). "Integrity Evaluation of Coeymans Creek Bridge," *Proceedings, NDE Conference on Civil Engineering*, ASNT, Cincinnati, OH.

Hag-Elsafi, O. and Kunin, J. (2006). "Load Testing for Bridge Rating: Route 22 Over Swamp River," Special Report 144, Transportation R&D Bureau, NYSDOT, Albany, New York, NY.

Hag-Elsafi, O., Kunin, J., and Alampalli, S. (2006) "Court Street Bridge Monitoring and Load Testing," Special Report 143, Transportation Research and Development Bureau, New York State Department of Transportation, Albany, NY.

Hirachan, J. and Chajes, M. (2005). "Experimental influence lines for bridge evaluation," *Journal of Bridge Structures*, 1(4), 405–412.

Hunt, V. J. and Helmicki, A. J. (2002). "Field Testing and Condition Evaluation of a Steel-Stringer Bridge for Superload," *Proceedings, NDE Conference on Civil Engineering*, ASNT, Cincinnati, OH.

Karbhari, V., Kaiser, H., Navada, R., Ghosh, K., and Lee, L. (2005). "Methods for Detecting Defects in Composite Rehabilitated Concrete Structures," Submitted to Oregon Department of Transportation, and Federal Highway Administration, Report No. FHWA-OR-RD-05-09.

Lagace, S. (2004). "Practical Use of Load Testing to Eliminate Permit Vehicle Restrictions," *Proceedings, NDE Conference on Civil Engineering*, ASNT, Buffalo, NY.

Lichtenstein, A. G. (1993). "Bridge rating through nondestructive testing." Final Draft, NCHRP Proj. 12-28(13) A, Transp. Res. Board, National Research Council, Washington, DC.

Lih, Z., Zhou, M., and Liu, J. (2004). "The analysis on the Impact Coefficient of Bridge with Response Spectrum," *Proceedings of the 23rd Southern African Transport Conference (SATC 2004)*, Pretoria, South Africa.

McCaffrey, B. (2006) "Bridge Load Rating Practices in New York State," *NDE Conference on Civil Engineering*, ASNT, St. Louis, MO.

Mertz, D. (2005), "Load Rating By Load and Resistance Factor Evaluation Method," NCHRP Project 20-70/ Task 122, Washington, DC.

Moses, F. and Verma, D. (1987). "Load Capacity Evaluation of Existing Bridges," NCHRP Report No. 301, Washington, DC.

NCHRP. (1998). "Manual for Bridge Rating through Load Testing," National Cooperative Highway Research Program, NCHRP, Research Results Digest, No. 234, Washington, D.C.

NYSDOT. (2001), Albany, NY. "New York State Department of Transportation Standard Specifications for Albany, Highway Bridges" and provisions in effect as of April 2001, New York State Department of Transportation, Albany, NY.

NYSDOT. (2001). Project Design file, unpublished, Structures Division, New York State Department of Transportation, Albany, NY.

NYSDOT. (2005). "Load Rating/Posting Guidelines for State Owned Highway Bridges," New York State Department of Transportation Engineering Instruction Report No 05-000, Albany, NY.

NYSDOT Research Report 153 "Proof Testing of Highway Bridges." Available from the Transportation Research and Development Bureau, New York State DOT, State Campus, Albany, NY. 12232.

NYSDOT Research Report 163 "Highway Bridge Rating by Nondestructive Proof-Load Testing for Consistent Safety." Available from the Transportation Research and Development Bureau, New York State DOT, State Campus, Albany, NY. 12232.

Prader, J., Grimmelsman, K, Jalinoos, F., Ghashemi, H., Burrows, S., Taylor, J., Liss, F., Moon, F., Aktan, E. (2006). "Load Testing and Rating of Undocumented Reinforced Concrete Bridges," NDE Conference on Civil Engineering, ASNT, St. Louis, MO.

Roufa, G. J. (2006). "Use of Rotating Laser Systems in Load Testing of Bridges," *NDE Conference on Civil Engineering*, ASNT, St. Louis, MO.

Salawu, O. and Williams, C. (1995). "Review of full-scale dynamic testing of bridge structures." *Engineering Structures,* 17(2), 113–121.

Shenton, H.W., Chajes, M. J., and Holloway, E. S. (2000). "A System for Monitoring Live Load Strain in Bridges," Structural Materials Technology: an NDT Conference, ASNT, Atlantic City, NJ.

Wacker, J. P., Wang, X., Brashaw, B., and Ross, R. (2006). "Estimating Bridge Stiffness Using a Forced Vibration Technique for Timber Bridge Health Monitoring," NDE Conference on Civil Engineering, ASNT, St. Louis, MO.

Yanev, B. (2007). *Bridge Management*, John Wiley & Sons, Hoboken, NJ.

Yost, J. R. and Assis, G. (2000). "Load Testing of the Washington & Main Street Railroad Bridge," Structural Materials Technology: an NDT Conference, ASNT, Atlantic City, NJ.

Zeng, X., Yu, Q., Ferrier, G., Bao, X., Steffen, R., and Bowman, M. (2002). "Strain Measurement of the Load Test on the Rollinsford Bridge Using the Distributed Brillouin Sensor," *Proceedings of 1st International Workshop on Structural Health Monitoring of Innovative Civil Engineering Structures*, ISIS Canada Corporation, Manitoba, Canada.

9 Bridge Management and Infrastructure Health

9.1 INTRODUCTION

9.1.1 OVERVIEW

Infrastructure management philosophy has changed considerably in the recent years. In the past, the emphasis was mainly on safety, and simplified design methods were accepted. Such methods were based on the safety factor. They were also based heavily on component design practices.

However, recent management efforts, even while concentrating on safety, emphasize reliability. Reliability-based designs consider inherent uncertainties and aim at uniform reliability designs (which imply uniform failure probability). There is more emphasis on global and component behavior of systems. Also, there is an emerging trend toward considering multihazard designs. Short-term and long-term considerations also play a role in management as is evident from life cycle analysis (LCA) and performance-based engineering. Recent infrastructure management trends emphasize environment, energy, and economy. Social concerns of security and mobility are now routinely regarded as integral components of infrastructure management.

Besides, infrastructure management includes various elements such as resources (personnel and fiscal) and data management. Technology also plays a central role. Finally, an important parameter in infrastructure management is differing types of infrastructure owners. Infrastructure ownership can be private, local, State, or federal. For example, Figure 9.1 shows the varied ownerships of bridges in New York. The differing management rules and cultures of these owners would only add to the complexity of network management of infrastructures. The interrelationship between all those components is controlled by an intricate set of decision making rules. The guiding objectives in infrastructure management can be summarized as ensuring of safety and optimal functionality at reasonable costs (see Figure 9.2). Another way of attaining these objectives is by stretching the useful (functionally adequate) and healthy service life of the infrastructure as long as possible at reasonable costs.

An integrated structural health monitoring (SHM) and bridge management methodology has been developed by Medina et al. (2005). It integrates various nondestructive evaluation (NDE) models, SHM techniques, and bridge life management models into a comprehensive method and software. It provides cost–benefit assessments of the appropriate NDE tools and methods. Among the SHM techniques that the method considers are the cost–benefit SHM experiments, including reliability of sensors and instrumentation. Inspection issues are included in the method, as also the optimum interval between inspections and the value of using different NDE and SHM techniques during inspection. All components of the method are optimized using single- and multiobjective function techniques. Uncertainties are also included in the method, using Monte Carlo simulation techniques. The integrated method and its front-end software (called Virtual NDE or simply VNDE) is a powerful tool that combines the potential of NDE and SHM to help bridge officials make cost-effective and safe management decisions.

9.1.2 THIS CHAPTER

As discussed above, useful and healthy life at reasonable cost can be considered the objective of modern infrastructure management. This chapter explores different management issues that can

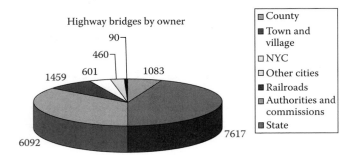

FIGURE 9.1 Composition of bridge owners. (From NYSDOT, SFY 2006–07: Annual Report of Bridge Management and Inspection Programs, NYSDOT, 2007. With permission.)

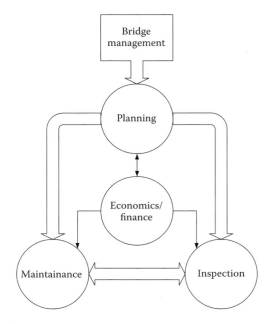

FIGURE 9.2 Basic components of bridge management.

affect infrastructure health. First, different infrastructure management strategies as they relate to structural health are discussed. The following section is devoted to the all-important subject of deterioration of infrastructures. Techniques for quantifying deterioration and their relationship with structural health are discussed. The next three sections explore the three major bridge management activities: inspection, maintenance, and repair/rehabilitation. Several objective techniques as well as decision making methods that can be of help to infrastructure managers are offered. This chapter concludes by discussing management tools and their relationship to structural health techniques. Figure 9.3 shows the flow of this chapter.

9.2 BRIDGE MANAGEMENT STRATEGIES AND SHM

9.2.1 INTRODUCTION

This section explores different aspects of bridge management strategies and how they relate to infrastructure (bridge) health. We offer first the general concepts of infrastructure management.

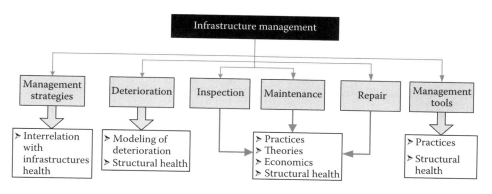

FIGURE 9.3   Arrangement of Chapter 9.

We then discuss briefly different elements of infrastructure management. Some examples of management strategies are then discussed. We conclude by presenting in detail an emerging concept in infrastructure management: multihazard considerations. We show that multihazard considerations have applications in almost all aspects of infrastructure management. Moreover, by following a multihazard approach, the manager will ensure safer operations at reduced long-term costs.

9.2.2   General Concepts

9.2.2.1   What Is Management?

A reasonable definition of management is that it is a process that follows some sort of decision-making rules combined with some business rules. Specifically, in bridge (or any other type of civil infrastructure) management, the manager aims at

- Developing right goals (safety, security, cost)
- Effective use of resources (optimization)
- Balance short-term versus long-term and project versus network level goals
- Sound engineering solutions (design analysis that conforms to acceptable design codes and guidelines)
- Risk considerations
- Economic and cost-effective solutions (life cycle considerations)

In short, the manager wishes to have a smooth and continued operation of the infrastructure subject to

- Safety and security of users
- At reasonable costs

That is an optimization process that seeks to minimize costs, subject to safety requirement constraints.

9.2.2.2   Infrastructure Management

Infrastructure management, in addition to above requirements, needs to accommodate further demands such as

Change in customer expectations: Customers (the public in case of bridges) change their expectations and demands according to the changes in social, economic, and political environments. The manager needs to satisfy the expectations/demands as much as possible.

Change in technology: Technology changes at an increasingly rapid pace. Use of newer technologies needs to be considered and their benefits balanced against safety and cost factors. Several examples in this chapter offer methodologies that highlight such considerations.

Globalization: Globalization has a direct and indirect influence on the manager. Economic models, material and labor costs, technological sources, and availability of labor are some of these factors.

Limited resources: Limited resource (including finance and labor) is perhaps the most important factor that can affect infrastructure management. This is simply an optimization process, and using limited resources efficiently is the key to the success of management strategy.

Uncertainties: Another major impediment to infrastructure management is the predominant uncertainties along most of the processes. In many situations, the manager confronts such uncertainties with a subjective approach. We provide in this chapter several examples that might aid in facing uncertain management conditions objectively.

Because of all these difficulties, constraints, and at times conflicting demands, infrastructure managers require a great deal of flexibility while facing difficult situations.

9.2.2.3 Executive Management

At a higher level, the infrastructure manager needs to have a larger perspective; some important concepts here are

- Define clearly the mission and vision. Make periodic changes if needed.
- Know the customer base well. Try to engage them whenever possible. One of the most important, and least used, risk management tool is *risk communication* (see Ettouney and Alampalli 2012)
- Understand customer expectations and try to meet them. If meeting these expectations is not possible, communicate with the customers and try to reach an understanding. In some situations the expectations can be complementary, in some other situations conflicting. Try to address the situations as they arise.
- Prioritize meeting customer expectations and goals in appropriate order. Many objective techniques for such prioritizations are offered throughout this chapter.
- Keep in view the overall global perspective all the time.

9.2.2.4 Assets

The infrastructure manager needs to understand well the attributes of the managed assets; some of these are

- Asset functions
- Interactions between assets, if any
- Asset value. This may differentiate between utility value and actual value
- Estimated lifespan and expected useful life

Besides, details of the managed inventory need to be well understood. This includes (1) number of assets in the inventory, and (2) condition of the assets.

9.2.2.5 Planning

After addressing the overall picture of assets, the manager need to understand fully the detailed attributes of the assets; this includes

- Current condition of the assets
- Potential future condition of the assets
- Current and future resources available

- Balancing use of resources to meet future goals
- Studying alternatives to see if the available resources are adequate to meet the goals

9.2.3 ELEMENTS OF INFRASTRUCTURE MANAGEMENT

Infrastructure management has many components. Several authors have studied these in the past. For example, the use of internet-based monitoring systems was described by Miyamoto and Motoshita (2004). They used the internet to monitor cable-stayed bridges. Takeda et al. (2004) introduced several analytical variables that link actual monitored and inspected data. They offered an example of automated maintenance of bridges using SHM techniques. Ulku, Attanayake, and Aktan (2006) studied the suitability and power of data-mining techniques to predict conditions. Data from field investigations of 20 PC-I beam bridges were used. The data was collected to investigate the beam end conditions, which are susceptible to corrosion due to leaking joints. The authors used the "classification" procedure for data mining, which is defined as learning a function that maps/classifies a data item into one of several predefined classes. Decision trees with various meta-learners were used due to the ease of interpreting the results. Taljsten, Hejll, and Olofsson (2002) studied the changing reliability of bridges, as utilized in estimating bridge safety, by measuring in-service performance. They offered several practical examples of using SHM in bridge management. Frangopol and Yang (2004) introduced a method that links safety and health issues.

We try to formalize several elements of infrastructure management and discuss each of them next. Figure 9.4 shows those management components with different structural health in civil engineering (SHCE) components that might be of benefit to them. Note that many of these elements are discussed in more detail throughout this book.

9.2.3.1 Analysis and Design

In a particular situation, such as rehabilitation, new construction, or simple maintenance, the manager needs to evaluate and/or design the potential course of action.

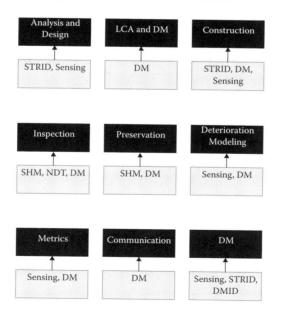

FIGURE 9.4 Elements of infrastructure management and their relationship to SHCE.

9.2.3.2 Life Cycle Analysis and Decision Making

Cost and benefit implications need to be considered at almost every step of the management process. Among the components of costs implications are

- Useful life
- Materials
- Hazards to consider
- Current and future needs
- Constructability and time of construction
- Repairability
- Inspectability

9.2.3.3 Construction

The construction phase, either on a limited or large scale, needs to be considered carefully. The best and most effective methods must be investigated. The interrelationship of construction methods and future performance needs to be studied. Finally, interruption of services during construction and quality assurance (QA)/quality control (QC) of construction must be well understood and implemented.

9.2.3.4 Inspection

Inspection (manual or automatic) is a basic tool of bridge management. It aims at estimating the condition of the infrastructure at the time of inspection. The manager's role is to decide on the type and quality of data to be collected during inspection to carry out the planning, design, and preservation efforts. Also, the frequency of inspection, the qualifications of the inspectors, documentation and archiving of the data are all decisions that the manager needs to take. Accuracy of inspection is emerging as an important subject. One of the main goals of SHM is to improve the accuracy and optimize the frequency of bridge inspections.

9.2.3.5 Preservation of Infrastructure (Maintenance, Rehabilitation, and Retrofit)

The levels and details of maintenance efforts need to be considered. The types of actions that need to be taken at a given time and the value of each need to be balanced to ensure cost-effective maintenance efforts.

9.2.3.6 Deterioration Rate as a Bridge Management Tool

Accurate objective deterioration rates can be of great help to bridge managers. By understanding deterioration rates and the factors that can control them, managers can take objective decisions.

9.2.3.7 Metrics

Infrastructure performance metrics need to be well understood and established. Moreover, metrics need to be established at every stage of the management/activities pyramid. If there are links between activities, such links need to be reflected in the setting up of the performance metrics. Also, when setting up performance metrics, such metrics need to be easily quantified. Finally, well-detailed action plans need to be in place if performance metrics are not met.

9.2.3.8 Communication

As suggested earlier, *risk communication* is an essential component in infrastructure management. The manager needs to (1) communicate internally with the asset owners, (2) communicate with the customers, (3) measure customer satisfaction/dissatisfaction accurately, and (4) apply the lessons learned to future situations.

9.2.3.9 Decision Making

All management elements have one thing in common: they all require a decision. Most of those activities begin with a decision to make and so should end with arriving at a decision. The decision making process is acknowledged in this chapter as one of the main bases of infrastructure health. The methods can be simple or involved. Methods of decision making are presented throughout the chapter.

9.2.4   Different Management Strategies

Several bridge management systems (BMS) have been reported. For example, the Finnish BMS, Söderqvist and Veijola and Veijola (1998), includes a network BMS that leads to project-level BMS. Both systems interact with a bridge register module (the bridge register is similar to the database module). The Finnish BSM system relies on probabilistic prioritization/optimization techniques for the bridge network; it then uses deterministic methods for annual repair and/or reconstruction of individual bridges (project level).

Yanev (1998), in his discussion of BMSs, identified three key areas for bridge management from a practical viewpoint. The areas are (1) collection, archiving, and management of qualitative and quantitative data sets that describe bridge conditions, (2) deterioration of structures (bridges), and (3) preservation and the underlying economics.

9.2.4.1   Example: NYSDOT

We use the NYSDOT practice as an example of a BMS component. Our main goal is to establish how the needed monitoring techniques and decision making tools are used or can be used to support BMS. The New York State BMS components are shown in Table 9.1. Note that in the last column in Table 9.1 we evaluated the potential level of health-monitoring for each of these BMS key components.

9.2.4.2   Convenient Definitions

For our purpose we are interested in bridge management from the point of view of health monitoring and decision making as a result of health monitoring. As such, we opt for the rather limited definition of bridge management as follows: bridge management, from a SHM viewpoint, comprises economic optimization of inspection, deterioration, maintenance, and repair of bridges. Such a

TABLE 9.1
NYSDOT Bridge Management System Structure

Module		Key Component	Level of Health Monitoring Role?
Decision support system	Network level	Condition assessment	High
		Strategy selection	Some
		Cost estimation	Some
		Prioritization	Some
		Optimization	Some
		Forecasting	Medium
	Project level	Individual bridge needs	High
		Individual bridge life cycle strategy	High
		Individual bridge work strategy selection	Medium
Database		Construction and maintenance.	High
		Inventory and inspection	High
		Safety assurance	High
		Project planning and programming	Medium
Engineering support system		Load rating	High
		Drafting	Low
		Bridge design	Low

definition is limited in scope, but it will help us focus on the interaction between SHM activities and bridge management.

9.2.5   Multihazards, Bridge Management, and SHM

9.2.5.1   Overview

We stated earlier that safety and cost-effectiveness are two major infrastructure management goals. Safety includes issues such as (1) ensuring adequate capacity for service loads, and (2) accommodating different hazards demands. Note that some hazards have a low probability of occurrence during the service life of the bridge, whereas others, such as overloads, have a higher probability of occurrence. All these issues have multihazard implications. Cost-effectiveness, which is the second major objective, includes costs of inspection, repair/retrofit, and cost of operation during the service life of the bridge. All those cost issues also have multihazard implications.

We can thus conclude that bridge (or any other type of infrastructure) management needs to include multihazard considerations to ensure safety at reasonable cost. The theory of multihazards by Ettouney and Alampalli (2012) can be restated for bridge systems as follows. As hazards affect the bridge, they interact together through its different aspects. By identifying the different ways in which the hazards affect the bridge, optimizing those issues can result in achieving management goals of (1) improved safety, and (2) efficient expenditure. Figure 9.5 shows the overall multihazard strategy for bridge management. In what follows we explore how some management components are affected by multihazard considerations.

9.2.5.2   Inspection

Visual/conventional bridge inspection is either scheduled (looks over wear-and-tear signs, corrosion, cracks, etc.) or hazard specific, which aims at estimating vulnerability ratings for specific hazards such as earthquakes and scour. The hazard-specific inspection is usually performed before an event. Another type of inspection is special (unscheduled) inspection, which is usually performed after an event such as flooding. Since we theorized that hazards interact through the system, an obvious question arises: Can we change the inspection procedures to accommodate multihazard considerations? The answer is yes! One potential way of providing a multihazard inspection process is by recognizing multihazard behavior in bridge-rating manuals and inspection guides. Figure 9.6 shows schematically such a procedure. We discuss this concept next in more detail.

9.2.5.3   Bridge Rating

There are currently several bridge-rating manuals and guides for different hazards such as seismic bridge condition ratings, load and resistance factor rating (LRFR) guidelines, scour vulnerabilities, and so on. Again, since there are some interactions between the hazards in the manner they affect the bridge, it is reasonable to assume that the guides/manuals will include some measure of interaction between the hazard effects. Moreover, if such an interaction is accounted for in various guides/

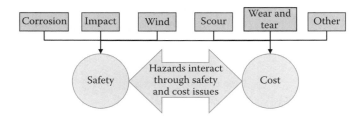

FIGURE 9.5   General multihazard approach in bridge management.

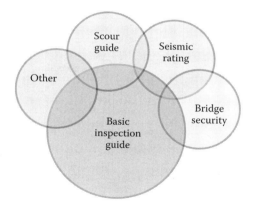

FIGURE 9.6 Interaction between hazards through inspection and rating guides.

manuals, there would be a net increase in efficiency in the cost of ownership without any loss of safety. Note that the cost of modifying guides/manuals is a one-time cost. The efficiency benefits of these modifications are recurring; thus, the net multihazard consideration costs should be expected to be beneficial in the long run.

9.2.5.4 Repair/Retrofit (R/R)

We ask similar questions for repair/retrofit efforts: Can we change repair/retrofit procedures to accommodate multihazard considerations? To answer this question accurately, we need to observe how repair/retrofit efforts are conducted

The general steps of retrofit/repair efforts are

1. Identify need/cause
2. Identify level
3. Secure budget
4. Perform R/R

Each of the above steps includes potential of multihazard considerations as follows:

9.2.5.4.1 Need and Cause

While trying to identify the need for retrofit, or the cause of damage, a one-hazard-at-a-time approach will cost extra for analysis (structural and damage identification). Also, there are benefits for accurate identification of source of defect by using a multihazard approach (see Figure 9.7 for an example of such benefits). Figure 9.8 shows the process of identifying sources of defects in a multihazard environment.

9.2.5.4.2 Level of Retrofit

While trying to identify the level of a retrofit or the cause of damage, a one-hazard-at-a-time approach will cost extra for analysis (structural and damage identification). There are benefits in accommodating interaction between hazards during retrofit for one of them (see Figure 9.9 for an example of such benefits).

9.2.5.4.3 Budget

Three main components of the budget of any repair/retrofit are (1) initial cost, (2) discount rate, and (3) service life of the bridge and the particular repair/retrofit under consideration. All three components are dependent on multihazard considerations, as shown in Figure 9.10.

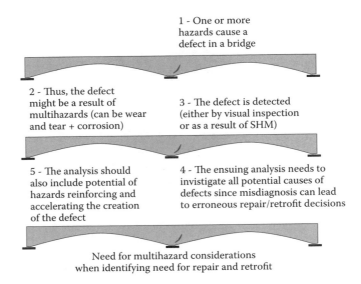

FIGURE 9.7   Multihazards and repair decisions.

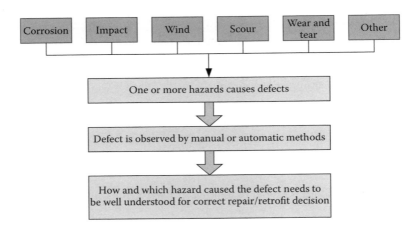

FIGURE 9.8   Process of identifying sources of defects in a multihazard environment.

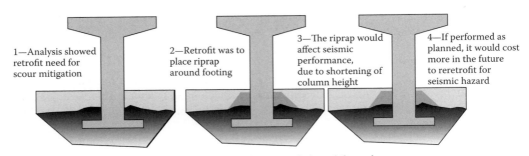

FIGURE 9.9   Interaction of hazards through retrofit efforts.

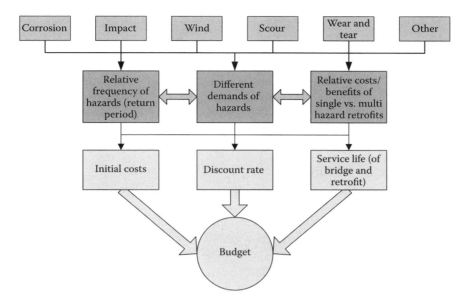

FIGURE 9.10 Decision making processes and multihazard considerations.

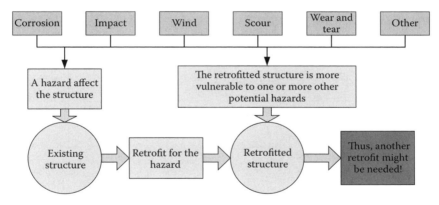

FIGURE 9.11 Accommodating multihazards during retrofit process.

9.2.5.4.4 Concluding Remarks

We have just argued that there is a cost-effectiveness of multihazard considerations during the repair/retrofit steps. A potential multihazard repair/retrofit is shown in Figure 9.11.

9.3 DETERIORATION

9.3.1 GENERAL

Bridge deterioration is one of the major components of bridge management. Because of this, issues that are related to bridge deterioration have received considerable interest from researchers, practicing engineers, and bridge managers. To start with, we need to define deterioration of bridges. We propose to define deterioration as the reduction of the *state* of the bridge (or any type of infrastructure) from a given *baseline*. Usually, such a *baseline* is defined as the initial desired state of the bridge, for example, at conclusion of construction. Definition of the *state* is more difficult. Many define the *state* as the *capacity* to resist specific *demands*. Such a definition can indicate that the desired *state* varies as *capacity* and/or *demand* varies. Following such logic indicates that deterioration is related to current bridge capacity and demand. Thus, determining deterioration can be an involved and complex process. We continue this line of discussion in Sections 9.3.5 and 9.3.6.

Given the complexity of the considerations of bridge deterioration, we start this section by discussing some sources of bridge deterioration (as defined above) in Section 9.3.1.1. We then summarize current models that determine deterioration in Section 9.3.1.2. Sections 9.3.2 through 9.3.4 will present in detail some popular deterioration models. Section 9.3.5 will discuss bridge deterioration in more general manner. Finally, Section 9.3.6 will present additional possible ways to determine deterioration in an objective manner.

9.3.1.1   Sources of Deterioration

Sources of bridge deterioration are numerous. In fact many of those sources have been discussed throughout this book and the companion volume (Ettouney and Alampalli 2012). Some deterioration sources are categorized in Table 9.2. Obviously, effects of these sources are cumulative and many of them have cascading effects. As such, quantitative estimates of such effects are difficult. However, many models are used to estimate objectively the resulting deterioration: these models are described next.

9.3.1.2   Deterioration Models

Bridge managers need to estimate future states of bridges for appropriate maintenance decisions. In order to do that objective analysis, bridge deterioration as a function of time is needed. We assume that the bridge (system) is composed of one or more subcomponents. Obviously, the deteriorated state of the bridge (or any other type of infrastructure) is a function of the deteriorated states of its subcomponents such that

$$D(t) = f(s_1, s_2, \cdots, s_n) \tag{9.1}$$

TABLE 9.2
Sources of Deterioration of Bridges

Type	Time scale	Source
Man-made source	Long	• Construction defect (for example, inadequate grouting of posttensioned tendons or inadequate water proofing, inadequate expansion joints) • Inadequate designs • Overloads
	Sudden/short	• Construction defects • Overload • Vehicle/barge collision • Fire
Natural source	Long	• Corrosion • Fatigue • Thermal effects • Freeze-thaw • Chemical processes such as alkali-silica interaction or carbonation • Inadequate drainage • Paint deterioration • Creep/shrinkage • Settlement of supports/foundations
	Sudden/short	• Earthquake • Wind • Flood/scour/debris

where

$D(t)$ = A measure of the deteriorated state of the desired component, or system, at time t.

s_i = Deteriorated state of the ith subcomponent of the system, with $i = 1,2, \cdots, n$.

n = Number of components of the system, or component, under consideration.

Note that $n = 1$ if there is only one component in the system under consideration. The function $f()$ is a suitable function.

Further, the deteriorated state of the ith component can be described as

$$s_i = g_i (x_1, x_2, \cdots, x_{mi}) \tag{9.2}$$

where

x_j = The jth source of the deterioration of ith component, with $j = 1,2, \cdots, m_i$.

The sources for deterioration of different components can be found from studying Table 9.2. Obviously, the function $g_i()$ is a difficult function to quantify. Because of this, evaluating $D(t)$ using formal relations between the deteriorating state and the sources of deterioration is beyond the current state of the art.

A simple method to estimate the deteriorating state of bridges is possible by using the concept of bridge rating, R, as a deterioration measure, such that

$$D(t) \simeq R(t) \tag{9.3}$$

Thus,

$$R(t) = f_1 (R_1, R_2, \cdots, R_n) \tag{9.4}$$

where R_i is the rating of the ith component of the system. Rating bridge components is done using visual inspection in a qualitative manner. The function $f_1()$ is a simple function that accounts for all major components in the system under consideration.

For the bridge manager, it is important to know the current rating, R (t = Now), as well as the future rating, R (t > Now). In order to achieve this extrapolation to the future, the deterioration rate (assuming that $D(t) \simeq R(t)$) needs to be estimated. Several models for estimating such rating is currently used by bridge managers. For example, Yanev (1996, 1997, and 1998) explored regression modeling for evaluating deterioration rates. These methods can be described as deterministic in nature. We describe in detail two of the stochastic-based deterioration rate models next.

9.3.2 MARKOVIAN MODELS

9.3.2.1 Overview

The Markovian model of predicting deterioration is simple. First, it uses condition-rating states as the measure of bridge deterioration. Depending on the application, the number of the condition ratings for a component N can differ. Benjamin and Cornell (1970) used five states for a pavement condition. In the Arizona Pavement Management System, a total of 45 condition states were used in the Markovian modeling of pavement deterioration. A transition probability matrix $[P]_k$ is then defined for the component under consideration. Matrix $[P]_k$ is a square matrix of size N. The components of matrix p_{ijk} (ith row and jth column) are defined as the probability that the condition of the component will change from the ith condition state to the jth condition state at the end of the kth time period. The final piece in the deterioration Markov modeling is to define the state probability vector $\{S\}_k$; the size of $\{S\}_k$ is also N. The components of the vector s_{ik} (ith row) are defined as the probability that the condition rating is $R_k = i$ at the end of the kth time period. More details of the generic Markov process method are given in Chapter 8 by Ettouney and Alampalli (2012).

Another practical use of Markovian Model in estimating deterioration is a simple yet effective computer code developed for the City and County of Denver. The computer code has a built-in database of information about bridges managed by the City and County of Denver Department of Public Works. The database includes such information as the year the structure was built, average daily traffic (ADT), available historical condition ratings of the deck, superstructure, and substructure. The computer code utilizes the built-in Markovian Model method to extrapolate the deterioration of different components in the future. The deterioration of the components is represented by condition ratings of the component. Two types of deterioration models can be utilized by the method; they are the age of the component and the ADT.

We also note that Markov processes are also used in Pontis (2005a and 2005b) and BRIDGIT (1998).

9.3.2.2 Step-by-Step Evaluation

Following Agrawal et al. (2008a and 2008b), the steps for using Markov processes in detailing bridge deterioration rate can be summarized as:

Collect data: The first step is to collect all pertinent rating data over appropriate time spans. After the data is collected, it should be filtered and then grouped in a logical fashion. Filtering of data must be performed to avoid any misleading results. Factors that must be included in data filtering are shown in Table 9.3.

Accurate grouping/classification of data is needed, since different factors that affect deterioration must be accounted for in the grouping. Inaccurate groupings can lead to misleading results. An appropriate grouping used by the authors is shown in Table 9.4.

Group data according to components: The data is subdivided further into separate sets according to the rated components. The need for this step is self-evident.

Form rating vectors: After filtering, classification, and grouping, the rating for each bridge component is then arranged in a separate data set. Such data set can be, for example, in the vectors $\{R\}$: the rating of the components of interest. The kth component of $\{R\}$; $R_{k\_OBSERVED}$ is the component rating at the kth time period. The size of $\{R\}$ is N which is number of time periods. This necessitates that $1 \leq k \leq N$.

We note that $R_{k\_OBSERVED}$ is usually an integer number whose range depends on the bridge owner. For example, in New York, the range of $R_{k\_OBSERVED}$ is $1 \leq R_{k\_OBSERVED} \leq 7$. The qualitative description of the rating is shown in Table 9.5.

The range of R_k does not impact the process to evaluate deterioration rates.

Form Markov relationships: We note that the rating range N can be considered as the number of states in the Markov process. Thus,

$$\{S\}_{k+1} = \{S\}_k^T [P]_k \tag{9.5}$$

A basic assumption for Equation 9.5 is that $[P]_k$ is constant, so we drop the subscript k in the following developments. The T superscript indicates transpose.

For a seven-rating states process, Agrawal showed that the transition matrix should have the form

$$[P] = \begin{bmatrix} p(7) & 1-p(7) & 0 & 0 & 0 & 0 & 0 \\ 0 & p(6) & 1-p(6) & 0 & 0 & 0 & 0 \\ 0 & 0 & p(5) & 1-p(5) & 0 & 0 & 0 \\ 0 & 0 & 0 & p(4) & 1-p(4) & 0 & 0 \\ 0 & 0 & 0 & 0 & p(3) & 1-p(3) & 0 \\ 0 & 0 & 0 & 0 & 0 & p(2) & 1-p(2) \\ 0 & 0 & 0 & 0 & 0 & 0 & 1 \end{bmatrix} \tag{9.6}$$

TABLE 9.3
Filtering Parameters for Rating Data Set

Parameter	Comment
Duplications	Duplication of data can occur for of several reasons, e.g., methods of archiving the data might lead to some duplication. Obviously, duplicate data should not be used
Structures other than bridges	Culverts behave differently from bridges; so, they should not be included in a database that archives bridge deterioration behavior
Miscoded data	Data with incomplete records should not be included in the database, since they can bias the database analysis results
Unusual rating	Data that show unusual rating would bias the analysis and should be dropped from the database. This is applicable, but not limited, to bridges with unusual rating drop

Source: Agrawal, A. et al., Case study on factors affecting deterioration of bridge elements in New York. NDE/NDT for Highways and Bridges: Structural Materials Technology (SMT), ASNT, Oakland, CA, 2008b.

TABLE 9.4
Logical Grouping of Rating Data

Grouping	Comments
Climate	Bridges are classified according to three climate metrics: temperature, precipitation, and Palmer Drought Severity Index. Additionally, the geographical location of the bridge is accounted for in the grouping
Truck traffic	Bridges were grouped according to annual average daily truck traffic (AADTT). The range of each group was 1000 AADTT. This resulted in five groupings
Material	Bridges were grouped according to main span construction material. The groupings included reinforced concrete, steel, post- and prestress concrete, timber, and masonry. The authors of the document also included a single material category that included aluminum and wrought iron or cast iron. Since periodically the reinforcing status, or the lack of it, of an old concrete bridge is not known, these concrete bridges with unknown status are included as a separate material group in the database
Structural type	Eighteen groups were identified according to bridge structure types. These included (1) slab, (2) stringer/multibeam or girder, (3) girder or floor plan system, (4) tee beam, (5) box beam or box girder, (6) frames, (7) orthotropic, (8) truss, (9) arch, (10) suspension, (11) stayed girder, (12) movable, (13) tunnel, (14) culvert, (15) mixed type, (16) segmental box girder, (17) channel, and (18) other
Management region	This grouping type indicates the ownership of the bridge
Features	This grouping has two subgroups. The first subgroup indicates the type of traffic *carried* by the bridge; the second subgroup includes the type of crossing *under* the bridge. The *carried* grouping has three primary classifications: state, local, and all other. The *under* grouping has five primary classifications: state, local, navigable waterway, nonnavigable waterway, and all others
Snow accumulation	Similar to climate grouping technique, the bridges are also grouped according to yearly snow accumulation
Salt usage	Similar to climate grouping technique, the bridges are also grouped according to yearly salt usage

Source: Agrawal, A. et al., Case study on factors affecting deterioration of bridge elements in New York. NDE/NDT for Highways and Bridges: Structural Materials Technology (SMT), ASNT, Oakland, CA, 2008b.

The probability that the rating will change from i to i, that is, remain constant, is $p(i)$. The probability that the rating will change from i to $i + 1$ is $1 - p(i)$. Note that the sum of the rows in Equation 9.6 is 1.0, as it should be. Similar transition matrices can be evaluated for other number of rating states.

From Equation 9.5

$$\{S\}_{k+1} = \{S\}_1^T [P]^k \tag{9.7}$$

TABLE 9.5
New York State Rating System

Rating	Definition
1	Totally deteriorated or in failed condition
2	Used to shade between ratings of 1 and 3
3	Serious deterioration or not functioning as originally designed
4	Used to shade between ratings of 3 and 5
5	Minor deterioration, but functioning as originally designed
6	Used to shade between ratings of 5 and 7
7	New condition. No deterioration

Vector $\{S\}_1$ is the initial probability vector. Matrix $[P]^k$ is the transition matrix raised to the power of k. If the initial rating, $k = 1$, of the component is i, then

$$S_{i,1} = 1.0 \tag{9.8}$$

Where $S_{t,1}$ is the ith component of $\{S\}$. The rest of the initial probability vector is null.
    We can estimate the rating at time step k as

$$R_k = \{S\}_k^T \{\overline{R}\} \tag{9.9}$$

with

$$\{\overline{R}\} = \begin{Bmatrix} 7 \\ 6 \\ 5 \\ 4 \\ 3 \\ 2 \\ 1 \end{Bmatrix} \tag{9.10}$$

We recognize Equation 9.9 as a form of averaging (first moment) of the rating as a random variable.
    The rating at any time k can be evaluated by

$$R_k = \{S\}_1^T [P]^{k-1} \{\overline{R}\} \tag{9.11}$$

We need now to evaluate the components of $[P]$.
    Compute transition matrix: There are six unknown probabilities in $[P]$; $p(i)$ with $i = 2,3,\dots,7$. Agrawal et al. (2008b) suggested that those probabilities can be computed by minimizing the function

$$C = \sum_{k=1}^{N} |R_{k\_\text{OBSERVED}} - R_k| \tag{9.12}$$

The minimization process is subjected to the conditions

$$0.0 \le p(i) \le 1.0 \tag{9.13}$$

A nonlinear programming technique can be used in the minimization process.

The transition matrices for each of the bridge components can now be used to evaluate ratings of different bridge components.

9.3.2.3 Merits of Markov Process

There are several disadvantages in using the Markov process for estimating deterioration and deterioration rates. Among the disadvantages:

1. The constant transition probability matrix is not realistic.
2. The assumption that the transition from one state to the next is independent of past histories is not accurate in many situations.
3. It is not clear how the method for retrofitted or rehabilitated bridges can be used.
4. Since it is component-based, the potential interdependencies of deterioration between different components are not accounted for.

There is one compelling reason for the popularity of the Markov process in determining deterioration rates of bridge components: simplicity. The method is fairly simple as seen above. In addition, there is some evidence of its validity (see Madanat and Ibrahim 1995).

9.3.3 PROBABILITY DISTRIBUTION FUNCTION APPROXIMATION METHODS

9.3.3.1 Overview

Another technique in quantifying and studying deterioration rates of bridges or bridge components is by using probability distribution function (PDF) models. In this technique, available historical records of bridge or bridge component ratings are arranged in logical histograms. The histograms are then approximated using appropriate PDF. The resulting PDFs can then be used in decision-making processes that involve the changes (deterioration) in bridge or component ratings. In this section we present the process of building the histograms from a given ratings dataset $\{R\}$. Next we study the approximation of the histograms by an analytic PDF (or the corresponding cumulative distribution functions, CDF). Finally, we examine some of the merits of this popular method in bridge management.

9.3.3.2 Step-by-Step Evaluation

Establish histograms: In general, historical records of ratings are accumulated as time passes. We assume that these records for a given component or a bridge are contained in the dataset $\{R\}$. Now, subdivide the data set $\{R\}$ into a two-dimensional subset with the sizes of N_{RATING} and N_{AGE}, respectively. The number of ratings in the system under consideration is N_{RATING}. The age range of interest is subdivided into N_{AGE} ranges. For example, if the age range of interest is 150 years, and $N_{\text{AGE}} = 10$, then each age subrange, Δk, is $\Delta k = 15$ years. The subset vectors, $\{R\}_{ij}$, with $1 \le i \le N_{\text{AGE}}$ and $1 \le j \le N_{\text{RATING}}$ is chosen from $\{R\}$ such that the date of the rating k is

$$\left(\Delta k\right)\left(i\right) \le k < \left(\Delta k\right)\left(i+1\right) \tag{9.14}$$

Also, the components of each $\{R\}_{ij}$ are constants such that

$$\{R\}_{ij} = \begin{Bmatrix} j \\ j \\ \vdots \\ j \\ j \end{Bmatrix} \tag{9.15}$$

The size of each $\{R\}ij$ is then counted as N_{ij}. A matrix $[HIST]$ is constructed with components of N_{ij}. The size of $[HIST]$ is $N_{RATING} \times N_{AGE}$.

The matrix $[HIST]$ is a two-dimensional histogram that contains occurrences of component ratings at a given component age.

Evaluate parameters of PDF: The histograms are now approximated by adequate PDF. Several possible analytical distributions can be used for this purpose. Specifically, lognormal and Weibull distributions are argued to be the most suited (see Mishlani and Madanat 2002 and DeLisle, Sullo, and Grivas 2002). Agrawal et al. (2008b) used the two-parameter Weibull PDF to model bridge deterioration rates in New York. They started by introducing a time random variable, T_i, with $i = 1,2,\dots$ N_{RATING} (in Agrawal et al. 2008b, study, $N_{RATING} = 7$). Random variable T_i expresses the time duration at which the bridge component will be in the ith rating. They expressed the probability that the $T_i > t$, with t as the time variable, as an analytical Weibull cumulative distribution function (CDF)

$$F_i(t) = e^{-(t/\eta_i)^{\beta_i}} \tag{9.16}$$

The two-parameter Weibull function includes η_i and β_i. Factor $\eta_i > 0$ is a scaling factor. Factor $\beta_i > 0$ is a shape factor. The shape factor controls the rate of failure. If $\beta_i < 1$ then the failure rate is decreasing. If $\beta_i = 1$ then the failure rate is constant. Finally, if $\beta_i > 1$ the failure rate is increasing. Agrawal et al. (2008b) explained the situation of increasing failure rate as the increased possibility that a particular component rating will degrade as the time such a component is staying in that rating increases.

For each component, using the histogram of ratings and time, $[HIST]$, a simple fitting routine can establish the factors η_i and β_i. This will completely identify the Weibull CDF for the component of interest. The mean of such a function can finally be expressed as

$$E(T_i) = \eta_i \; \Gamma\left(1 + \frac{1}{\beta_i}\right) \tag{9.17}$$

Equation 9.17 expresses the average time that the component is expected to stay in an ith rating. With this expression, it is easy to generate deterioration of components as a function to time (see Figure 9.12). Decision makers can also use such an expression in forecasting bridge and component behavior, and making optimal decisions accordingly.

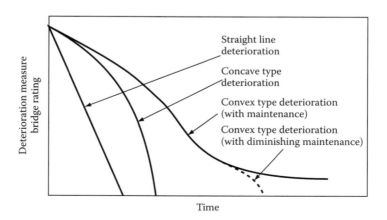

FIGURE 9.12   Types of deterioration in a bridge network.

9.3.3.3 Merits of PDF Methods

Using Weibull or any other suitable PDF/CDF in establishing bridge deterioration rates is a simple and efficient decision making technique. On the other hand, we note that in general the histograms of the historical records might be largely at variance with the simple analytical functions of $F_i(t)$ and $E(T_i)$. Specifically, we are bringing the issue of errors of estimating uncertainties of the histogram by the analytical PDF/CDF models. For accurate results this issue needs to be investigated further.

9.3.4 Reliability Models

The use of condition rating as a measure of deterioration has been criticized for its qualitative nature (But, by using SHM techniques the condition rating can be more quantitative, as is discussed elsewhere in this document). Instead of using deterioration to describe the adequacy of the bridge state, the bridge capability to carry load, that is, safety, is used. The time-changing reliability index of the bridge β is then used as a measure of the adequacy of the bridge to perform its function as it ages. Frangopol and Yang (2004) suggested five states for the bridge, shown in Table 9.6, based on its reliability index β

The first step in utilizing the reliability index approach is to define a target reliability β_{target}. The target reliability is the reliability figure at which a major bridge rehabilitation effort should commence. We note that the choice of β_{target} is mostly qualitative; it should be based on the experience of decision makers and bridge managers.

The typical reliability index versus age is shown in Figure 9.13. It assumes the following steps during the lifespan of a bridge:

1. For new bridges $t = 0$, the reliability index is β_0.
2. The reliability index remains constant at $\beta = \beta_0$, until an initial damage occurs, at time $t = T_1$.
3. After the initial damage occurs, the reliability index is reduced at a constant rate of α.
4. If the bridge is to have no maintenance, the reliability index will keep on dropping until it reaches the minimum reliability target, β_{target} at time T_g: the rehabilitation time for the bridge if there is no maintenance.
5. If, after time T_{PI} a maintenance/repair effort is performed on the bridge, the reliability index will increase by γ. The rate of reduction of the reliability index will decrease to θ for a period of T_{PD}, after which the reduction rate will revert to α.
6. If a periodic maintenance/repair is to continue, at a period of T_P then step #5 above will be repeated.
7. Eventually, the reliability index will reach β_{target} at time $T_{RP} > T_g$: the increased time for bridge rehabilitation, due to periodic maintenance and repair, is $(T_{RP} - T_g)$.

TABLE 9.6
Reliability Index and States of Bridge

Reliability Index - β	State	Description
$\beta \geq 9.0$	State 5	Excellent
$9.0 > \beta \geq 8.0$	State 4	Very good
$8.0 > \beta \geq 6.0$	State 3	Good
$6.0 > \beta \geq 4.6$	State 2	Fair
$4.6 > \beta$	State 1	Unacceptable

Source: With permission from ASCE.

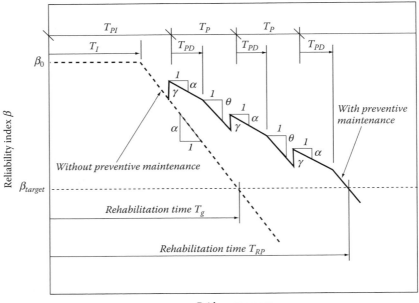

Bridge age, years

FIGURE 9.13 Lifetime reliability with and without preventive maintenance. (From Kong, J.S. and Frangopol, D.M., *J. Struct. Eng.*, *ASCE*, 129, 2003. With permission from ASCE.)

A total of ten parameters are needed to build a reliability index chart similar to Figure 9.13. All those parameters are random variables (Frangopol and Yang 2004).

Thus, the evaluation β as a function of bridge age can be done using the probability distributions developed by the authors of the study, the above-mentioned steps, and a Monte Carlo simulation approach. The β diagram can be used by the decision maker to achieve several goals, such as

1. Prioritizing maintenance activities (level and period)
2. Optimizing life cycle cost of the bridge or network of bridges
3. Comparing and understanding the behavior of bridges in different environments

For our immediate purpose, the obvious question as usual is: How would SHM help in developing β? We answer this question next.

9.3.4.1   SHM and Reliability Index β

The main disadvantage of the reliability index approach is the evaluation of accurate magnitudes of the parameters that control the reliability index. This is where SHM techniques can be of great help. Let us consider a simple cast-in-place slab bridge (Figure 9.14). The bridge is single span. Thus, two failure modes (flexure and shear) can be applied easily to the slab (even though the slab is a plate, we will assume, for the sake of simplicity, that it behaves as a beam). The loss of load-carrying capacity of this type of bridge can be attributed to many factors such as

- Corrosion of reinforcing steel
- Overload
- Normal wear
- Cracking
- Delamination spalling

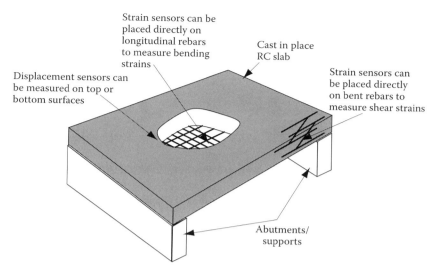

Strain sensors can be
placed directly on
longitudinal rebars
to measure bending
strains

Cast in place
RC slab

Strain sensors can
be placed directly
on bent rebars to
measure shear strains

Displacement sensors can
be measured on top or
bottom surfaces

Abutments/
supports

FIGURE 9.14 SHM scheme for simple cast-in-place bridge.

Invariably, all those hazards can be directly related to displacements and/or strains in the slab. Thus, we can assume that the displacement and/or strains within the slab are an accurate measure of the load-carrying capacity of the bridge, that is, β. Formally

$$\beta = f(u) \tag{9.18}$$

or

$$\beta = g(\varepsilon) \tag{9.19}$$

where u and ε are the displacement and strains at appropriate locations. For example, displacement u can be the vertical displacement at the center of the slab. Strains ε can be shear strains near the support or axial rebar strains at either the middle of the bridge or near the supports, as shown in Figure 9.12.

Equation 9.18 can be simplified to

$$\beta = a u \tag{9.20}$$

Note that factor a is a constant that relates reliability with displacement.

9.3.4.2 Multicomponent Reliability Considerations

Note that when considering multicomponent reliability, the need for reliability of each component must be evaluated first; then the overall bridge reliability can be computed. In this situation, SHM must be used at a component level.

9.3.5 Load Capacity, Deterioration, and Age

Chase and Gáspár (2000) showed that there is a functional relationship between load-carrying capacity and bridge deterioration. Figures 9.15 through 9.17 explain this functional relationship.

9.3.6 Deterioration and SHM

Time-deterioration measure, $D(t)$, can be used in management decisions by relating the measure into maintenance costs either for a single bridge, or for a cumulative bridge network as shown in Figure

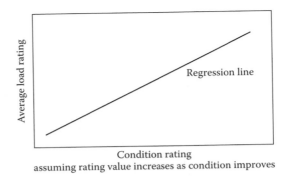

FIGURE 9.15   Condition rating (of superstructure) versus average load rating.

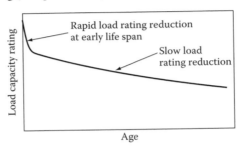

FIGURE 9.16   Load capacity rating versus age.

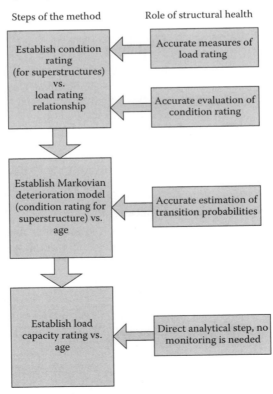

FIGURE 9.17   Role of structural health in accuracy of method.

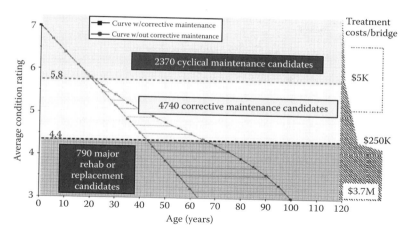

FIGURE 9.18 Deterioration and corrective maintenance. (From NYSDOT, SFY 2006–07: Annual Report of Bridge Management and Inspection Programs, NYSDOT, 2007. With permission.)

9.18. Such use indicates the need for accurate estimation of $D(t)$. This brings a question: how accurate are the current deterioration rate models? We showed in Table 9.2, several sources for bridge deterioration. We note that most popular deterioration rate predictive models do not account for the formal relationships between deterioration sources and the state of observed deterioration. This makes most of current models to be fairly qualitative. We now ask the obvious question: is there a way to improve the state of deterioration modeling such that the model can account for the cause (deterioration sources) and effect (resulting deterioration) relationship in an objective manner? At first glance, given the complex nature of such relationship, it seems to be a daunting analytical task. However, we believe that with appropriate and careful planning of SHM projects, several cause–effects can be quantified. Such quantifications can improve the efficiency and accuracy of bridge management. We discussed some of the potential methods of utilizing SHM for identifying such cause–effect relationships in Chapters 1 through 5 of this volume. Obviously, further studies are needed before SHM can realize its potential in truly quantifying deterioration and deterioration rates in an objective manner.

9.4 INSPECTION

9.4.1 STATE OF BRIDGE INSPECTION

9.4.1.1 Overview

We start this section by introducing the definition of a bridge. A bridge is a structure erected over a depression or an obstruction, such as water or highway, for carrying traffic and having an opening (a span) of 20 ft or more (see Figures 9.19 and 9.20). On the basis of this, we identify the inspection process as a process of visualizing and reporting the condition of the bridge. This section will address different issues of bridge inspection as they relate to SHM/SHCE. We first present a brief history of National Bridge Inspections (NBIS). We then ask the obvious question: "Why Inspect Bridges?" Different inspection issues and the future of inspections are discussed. Finally, we present an Economic Theory of Inspection (EIT). The EIT would provide for quantification of inspection value and help in optimizing the cost–benefits of the inspection processes. In addition, the EIT can help in providing a link between SHM/SHCE and manual inspection.

9.4.1.2 History

Following the Silver Bridge collapse in 1967, NBIS were first established in 1971 to improve bridge safety by an Act of Congress in 1970. They were modified several times, as recent as 2004, to reflect the lessons learned from bridge failures and changes in practices. The main aim of these standards was to improve bridge safety, create a national inventory of all highway bridges, and make some bridge management decisions at national level. The current program generally serves the nation well for its intended purpose.

Perhaps the first step in instituting formal bridge inspection programs was the 1916 Act: Federal Aid to Highways. Inspections of highway structures were part of maintenance work by States and others. Later a more detailed program under Public Roads Administration during 1930–40s was introduced. As an outcome of the Ohio River bridge collapse in 1967 (see Chapter 3 by Ettouney and Alampalli 2012), President Johnson initiated a Taskforce to determine procedures available to preclude future disasters and implement changes, if needed. In March 1968, FHWA issued a memo that required (1) review and inventory all existing structures to be completed by January 1970, (2) all structures to be reviewed once in 5 years, and (3) a 2-year inspection interval for important structures. One of the implications of the memo was the immediate need for qualified personnel to meet all those requirements. The AASHTO (1964) guide for maintenance personnel did help in specifying the inspection requirements. Those steps helped in creating a complete inventory of bridges, and in identifying and fixing serious deficiencies.

The NBIS offered for the first time uniform guidelines and criteria for bridge inspection. This included (1) the need for licensed engineers in each organization for bridge inspection, (2) a 2-year

FIGURE 9.19   Nonbridge structure. (Courtesy of New York State Department of Transportation.)

FIGURE 9.20   Highway bridge. (Courtesy of New York State Department of Transportation.)

inspection cycle (the first cycle started in July 1973), (3) detailed reporting format, appraisal ratings (present vs. current desirable), and sufficiency rating, (4) inspection types: inventory, routine, damage, in-depth, and interim, and (5) load rating and appropriate measurements. The NBIS was revised in 1988 to include the following: (1) states can vary frequency of routine bridge inspections when certain conditions are met, (2) establishment of fracture and scour critical bridges requiring a maximum inspection interval of 2 years, (3) special requirements for fracture critical member inspections and appropriate NBI designations, and (4) underwater bridge inspection requirements. Another revision in 1993 to NBIS limited the maximum inspection interval to 4 years.

Another major development was the passing of the 1978 Surface Transportation Act, which helped in (1) the establishment of dedicated funding for bridge improvements Program, (2) improvement of significantly important and unsafe bridges, (3) repair and retrofit based on structural deficiencies, physical deterioration, and functional obsolescence, (4) extension of inspection program to nonfederal aid systems, and (5) classification of bridges for prioritization. Another revision in 1993 to NBIS limited the maximum inspection interval to 4 years.

The 2004 the NBIS revisions, which became effective in January 2005, required (1) state DOT to be responsible for ensuring that inspections are done for all highway bridges on public roads except for bridges owned by federal agencies, (2) more ways to qualify to be a team leader, (3) 2-year interval defined as 24 months, (4) maximum inspection interval cannot exceed 48 months, (5) maximum interval for underwater inspection is 72 months, (6) follow-up on critical findings was mandated, (7) Special requirements for complex bridges, (8) specifications for QA/QC, and (9) requirement for refresher training.

9.4.1.3   Current Bridge Inspection Programs

Several issues were noted with the current NBIS in making better bridge management decisions. These standards were created for routine bridges, and they do not adequately address special and complex bridges. They also rely mostly on visual inspections that may not give adequate details of bridges with concealed elements. At the same time, no rational basis exists for current inspection intervals. This data may also not be enough for individual bridge owners, and so most owners collect data beyond what is required by NBIS to make appropriate planning, preservation, rehabilitation, and replacement decisions (bridge management decisions).

Different statutory requirements for bridge inspection include NBIS (2004), FHWA Bridge Inspector's Reference Manual, BIRM (2003), and FHWA (1995). The main features in current inspections programs are described in Table 9.7.

9.4.1.4   Why Inspect Bridges?

There are several reasons for inspecting bridges. These can be put into four categories: safety, database upkeep, federal and local requirements, and bridge management issues. Table 9.8 shows the details of each of these categories. In addition, Table 9.8 indicates whether SHM/SHCE can be of help in achieving inspection goals. In short, bridge inspection aims at increasing the understanding of structural behavior and failure mechanisms.

9.4.1.5   Inspection Issues

Several issues affect inspection procedures. We discuss some of these next with their implications for SHM/SHCE.

9.4.1.5.1   FHWA Intended Use

The current program serves the FHWA intended purposes but bridge owners need more information for effective management of their bridges. Different FHWA requirements from the inspection process can be aided by adopting some form of automation provided by SHM techniques, as shown in Table 9.9.

TABLE 9.7
Features of Current Inspection Programs

Issue	Comments
Coverage	All publicly owned highway bridges
Frequency of inspection	At least once in 24 months. Diving inspections at least once in 60 months
Inspector qualifications	Well-defined qualifications for team leaders. Refresher courses and training required
Ratings	Evaluate the entire structure to as-built condition
	Rate a few elements, indicative of entire structure, not for localized deterioration. This includes superstructure, deck, substructure, channel and channel protection, culverts, and capacity
	Several States go beyond FHWA requirements and conduct element-level inspections. This varies significantly from State to State
Outcome	Safety assurance
	Appraisal ratings: indicative of level of service compared to new one built to current standards
	Sufficiency rating for funding eligibility (100 indicates completely sufficient bridge, and 0 indicates entirely insufficient or deficient)
	Used for funding needs by planners

TABLE 9.8
Goals for Bridge Inspection

General Goal	Details	Can SHM Help?
Assuring safety	At inspection time	Yes, all components of SHCE can contribute to all the safety goals
	Until next inspection	
	Critical findings (structural and safety related)	
	Capacity	
Database upkeep	Inspection reporting keeps bridge records up to date	Decision making techniques can help organize and analyze data
Federal and local requirements	Inspection satisfies different federal and local requirements	
Bridge management goals	Preventive maintenance	Yes, all components of SHCE can contribute to all the safety goals
	Corrective maintenance	
	Replacement and rehabilitation assistance	
	Funding eligibility determination	
	Permitting operations	
	Postevent assessment	

TABLE 9.9
FHWA Intended Goals for Inspection

Intended Use	SHM/SHCE Role
Assuring safety	All components of SHCE
Inventory and statistics	Decision making processes
Planning at national level	Decision making processes

9.4.1.5.2 Stakeholder

The current inspection practices are supposed to meet the needs and requirements of several stakeholders: federal, local, and so on. As such, it can be difficult to adjust them and make them respond quickly to changes in environment. One the other hand, if the process is more automated, such changes can be adopted faster.

9.4.1.5.3 Routine versus Complex Bridges

Inspection processes are generally designed for routine, conventional bridges (Figures 9.21 and 9.22). Generally, additional inspection efforts might be needed for situations shown in Table 9.10. The table also shows that SHM/SHCE techniques can be used to complement those situations.

9.4.1.5.4 Concealed Components

One of the most obvious issues with manual inspection is its reliance on visual processes. Because of this, the state of any hidden elements and/or hidden damage cannot be revealed during inspection. This limitation can be overcome by a careful use of NDT or SHM methods, when practical.

FIGURE 9.21 A conventional bridge. (Courtesy of New York State Department of Transportation.)

FIGURE 9.22 A conventional bridge. (Courtesy of New York State Department of Transportation.)

TABLE 9.10
Nonconventional Situations

Nonconventional Situations	SHM/SHCE Use
Special bridges	Special bridges, such as lift bridges or suspension bridges, might require additional inspection. Nonconventional load paths and additional mechanical equipment in lift bridges require additional inspection routines. The complex geometry and material behavior of high strength cables in suspension bridges necessitates special inspection efforts
Complex bridges	Complex bridge structural systems which have nonconventional load paths need special inspection routines (see Figure 9.23)
New materials and designs	Using new materials in bridge construction will always require special inspection practices. Fiber-reinforced polymer represents a new material that has been used in bridge construction (see Chapters 6 and 7)

FIGURE 9.23   Complex bridge structural systems need special inspection procedures.

9.4.1.5.5 Arbitrary Inspection Interval

We discussed earlier the different mandated inspection intervals. We note that there was no rational basis for specifying the intervals. We offer later an ETI that discusses this issue and presents a quantitative approach to optimize the inspection interval. The theory will also help in quantifying the value of SHM/SHCE utilization in the inspection process.

9.4.1.5.6 Qualitative Nature of Appraisal Ratings

An inherent nature of the inspection process is that it is invariably behind the curve of the state of knowledge. An obvious example is the qualitative nature of rating definitions that are not consistent with the quantitative nature of bridge conditions available elsewhere. Advances in the state of the art, which are usually reached in research centers, in State DOT offices, or by consulting engineering firms usually take time before they are instituted into inspection practices. One way of reducing this knowledge time lag is by streamlining management practices to ensure that pertinent information and procedures are available to inspectors within a reasonable time.

9.4.1.5.7   Limited Data

Manual inspection inherently produces limited dataset. Since efficient bridge management relies on availability of a complete set of data, these limitations can result in a less-than-optimal management practice. Some examples are

1. Ratings are generally global; they do not extend to component level, except in special situations. When they do, the inspection process becomes slow and costly.
2. Inspection results are qualitative by nature. They do not produce quantitative deterioration rate estimations for components.
3. It is very difficult to estimate the extent of damage in many situations. This is usually needed for financial estimations and decision making processes.
4. Directs links between bridge maintenance practices and inspection data are not available.

In contrast, SHM/SHCE tools can provide almost limitless dataset that can overcome many of these limitations. But, this comes with a cost and not possible for network level management.

9.4.1.5.8   Single Hazard versus Multihazards

Special inspection processes are usually performed after the occurrence of hazards, for example, earthquakes or floods. Although the processes are well thought out and complete, they are inefficient. A perhaps more efficient approach is to acknowledge the inherent multihazard inspection processes. Such processes are not available as of the writing of this chapter.

9.4.1.5.9   Load (Demand) Data not Available

Manual inspection observes the state of the system; thus, it concerns itself mainly with the capacity side of structural reliability. Inspecting demands, for example, truck loads or wind pressures, cannot be performed by manual inspection (Figure 9.24). Specific SHM/SHCE techniques are needed for this.

9.4.1.5.10   Durability Definitions

Agrawal et al. (2008b) mentioned some of the sources of bridge deterioration as corrosion, concrete degradation, creep, shrinkage, cracking, and fatigue. It is possible to visually observe some of the effects. However, it is impossible to get an accurate, objective estimate of the effects only by visual inspection.

9.4.1.5.11   Reactive versus Proactive

Inspection processes by definition are reactive. However, it can be used as part of a larger process that includes SHM/SHCE, where trends and observations, coupled with decision making tools,

FIGURE 9.24   Demands on this bridge cannot be estimated by manual inspection. (Courtesy of New York State Department of Transportation.)

can be used to estimate future problems proactively and mitigate them. An example is the bridge deterioration rate work by Agrawal et al. (2008b). By using decision making techniques (Markov processes), they were able to utilize bridge-rating records to produce deterioration rates. Such rates can then be used in bridge management.

9.4.1.5.12   *Types of Inspection*

Inspection processes can be subdivided into two categories, scheduled and special, as follows:

9.4.1.5.12.1   Scheduled Inspection   These inspections are aimed at determining the normal deterioration effects on the bridge. They are usually done at least once in 24 months as mandated federally. In some situations the frequency of scheduled inspection can be longer or shorter. The optimal frequency of scheduled inspection is addressed by ETI later.

9.4.1.5.12.2   Special Inspection   Special inspections are those that are performed after abnormal events, such as floods or earthquakes. Sometimes special inspections are performed after major changes or findings in federal or local guidelines. SHM, NDT, and decision making tools can be of major assistance in this type of inspection.

9.4.2   Future Trends in Inspection

9.4.2.1   Overview

As the bridge infrastructure deteriorates, due to a lag in preservation efforts and constrained resources, bridge inspection will become increasingly important in order to manage the available resources in a cost-effective fashion (Alampalli and Jalinoos 2009). Bridge inspection has evolved considerably in the last 30 years from as-needed maintenance inspections to periodic inspections with qualified personnel at fixed intervals with requirements for QC, QA, follow-up to critical findings, and so on. NDE/NDT and SHM are used often to supplement visual inspections by most owners on an as-needed basis. This trend is expected to continue.

But, the current bridge inspection process is still reactive in nature as it gives the current condition of the structure without much emphasis on the reasons for reaching that condition. Several issues are noted in the previous sections. After the 2006 bridge failure in Minnesota, a joint ad hoc group consisting of members of the American Society of Civil Engineers/Structures Engineering Institute and the American Association of State Highway and Transportation Officials (ASCE/SEIAASHTO) discussed the inspection and rating methods and practices that are used to ensure the safety of highway bridges across the United States. The group concluded that, in general, the current NBIS and programs developed to address those standards have adequate policies and procedures in place to ensure public safety (ASCE 2009). The group also concluded that the current system can be improved, and it identified the gaps in bridge safety and the requirements for improving it and ensuring uniformity, consistency, and reliability of inspections nationwide. As developed by the ad hoc group, this white paper describes gaps, needs, and issues associated with the current practices and policies for assessing bridges. These were divided into ten general categories from which the following concepts were highlighted:

- A more rational, risk-based approach to determining the appropriate inspection intervals is needed, as opposed to a set 24-month cycle for all bridges. This approach would consider factors such as the design, details, materials, age and loading of specific bridges to determine the inspection intervals.
- New and more assertive types of QC/QA, such as performance testing of inspectors, could be used to encourage consistency of inspection practices.
- The consistency and effectiveness of inspection nationally could be improved if inspector qualifications were matched to the bridge type, condition, and complexity in a more uniform manner.

- A bridge inspection manual for nationwide use should be developed with expanded use of photographs, illustrations, and detailed drawings indicating specific deterioration conditions and methods of reporting deterioration.
- There is a need to have close collaboration between those responsible for maintenance and repair of bridges and those responsible for inspections.
- The load-rating process should be reliable, uniform, and consistent across the States.
- The development and maintenance of a centralized system for documenting critical deterioration in bridges, as experienced by bridge owners, is needed to support exchange of information and provide a resource for bridge owners.
- There is a need to develop standardized procedures for special inspections involving NDT, for example pin testing, to provide more guidance to bridge owners.
- Terms such as "structurally deficient," "functionally obsolete," and "fracture critical" require accurate definitions in the public arena such that public perception of bridge safety is consistent with facts.
- A mechanism should be developed to ensure that the critical conditions identified during bridge inspections are addressed in time.

Several research and technology transfer projects are already being developed to address these issues, and they should further improve inspection methodologies and effectiveness, and they may set the trend for the near future. But the long-term trend is to develop network-based inspection and monitoring technologies, as a first step, using noncontact methods such as remote sensing, followed by individual bridge-level inspections. Individual inspections will be based on the data collected during the first step while accounting for risks posed by individual structures to public safety and the network that the bridge is part of.

SHCE can play a major role in helping to achieve the above goals:

1. Evaluation of various methodologies, concepts, and technologies for network-level and project-level inspections
2. Identifying and recording data needed to evaluate and improve performance. These data could include environmental data, operational data, load data, material data, maintenance and rehabilitation data, so as to correlate the current condition identified by the bridge inspections with bridge history
3. Evaluating how data can be used more effectively to improve the inspection as well as entire bridge management process by identifying elements requiring improvement well ahead, so that maximum benefit can be achieved with associated inspection costs
4. Accounting for structure type and complexity to decide inspection interval, personnel qualifications, extent of inspections, data to be collected, and the NDE/NDT method required to supplement routine inspections
5. Developing proactive inspection and assessment using the multihazards approach, designing for inspectability, and leveraging current sensor and computing technologies

Overall, SHCE can play a major role such that, in the long run, for many structures visual inspection could be secondary to remote inspection based benefit–cost analysis.

9.4.2.2   Manual Inspection versus Automatic Monitoring

We are interested in answering the question: For a given subject to be inspected in a bridge setting, what are the merits of manual inspection versus automatic monitoring? Before we try to answer this question, let us clarify the basis of the analysis:

- A given subject, as used in the question, is the subject to be inspected or monitored. For example, fatigue cracks, corrosion extent, or soil erosion.
- We assume that the subject under consideration can be inspected *and* monitored. For example, surface fatigue cracks can be observed visually, and they can also be detected

automatically by, say, an NDT electromagnetic flux experiment. Similarly, the state of surface corrosion of an exposed reinforced concrete column can be observed visually and monitored automatically by half-cell corrosion measurements.
* We define the value in the above question by the amount of information that the stakeholder can get from the experiment and how important such information is to the subsequent decisions to be made.

On the basis of these two clarifications, we submit that the value of manual inspection versus automatic monitoring should be a function of cost of both efforts, as shown in Figure 9.25. The figure shows that the value of manual inspection increases rapidly at low cost until it reaches a certain limiting level, beyond which the manual inspection cannot offer any more value. This limiting level is due to the limits of human senses and the highly qualitative nature of manual inspection. For example, beyond detecting fatigue cracks that are large enough to be observed visually, and perhaps the size of those cracks, manual inspection cannot detect the depth of the cracks or whether there are subsurface cracks. This information is certainly valuable if it can be had at low costs; however, there are diminishing value at higher costs.

Automatic inspection methods would require higher expenditure before they can produce any value. They have also a limiting value for automatic monitoring. Such limiting value, however, is higher than that of manual inspection. This is expected, since automatic monitoring can produce very detailed damage and behavioral information that manual inspection cannot.

Studying Figure 9.25 a bit further reveals that there is a cross-over point at which manual inspection and automatic monitoring would produce similar value at the same cost. It is of great interest to stakeholders to identify such a cross-over point for different inspection subjects, as defined in the question above. Identification of such cross-over points can aid decision makers in choosing the most efficient ways to inspect/monitor their bridge network. Below are couple of studies that discuss use of virtual reality and impact echo method for inspection.

9.4.2.3 Inspection through Virtual Reality

Virtual reality is proposed for improving visual inspection, where depth perception is provided to the inspector through "immersive environment." This may be helpful as two-dimensional images cannot provide good inspection of fracture-related details. Some preliminary investigations of the concept was described by Baker, Chen, and Leontopoulos (2000).

Head-mounted display (HMD) or stereo shutter glass is suggested to generate stereo vision of otherwise two-dimensional images (Figure 9.26). Here, perceived depth is created through the

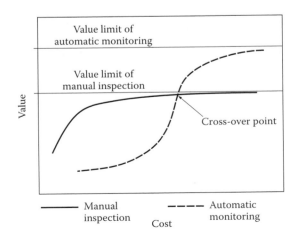

FIGURE 9.25 Manual inspection versus automatic monitoring.

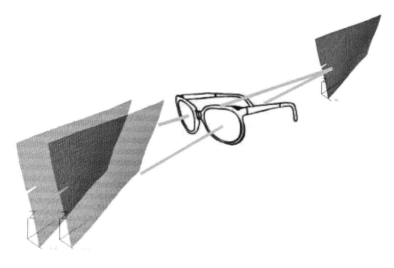

FIGURE 9.26 Concept of stereosposis. (Courtesy of CRC Press.)

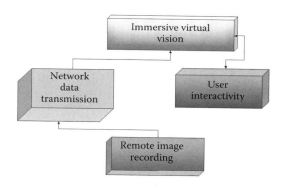

FIGURE 9.27 Schematic concept of a remote visual inspection system. (Courtesy of CRC Press.)

convergence of lateral disparity in the two separate viewing points. Thus, the relation between the lateral disparity and depth is critical.

In case of bridge inspection, a continuous video recording is suggested with a permanent on-site recording setup. The collected image is then sent to a computer vision lab, and stereo images of bridge components are generated for viewing through HMD or shutter glass. Through remote viewing it is suggested that the inspector does not have to be in dangerous situations. In their work, finite-element method–based modal analysis was used to show the concept.

This concept, illustrated in Figure 9.27, seems more applicable to SHM rather than bridge inspection. As bridge inspection is done at fixed intervals, having a permanent setup does not seem cost-effective. At the same time, the entire bridge video cannot be generated from a single location due to geometry, and so on, and thus complete inspection is not feasible. But, for SHM where focus is on a small area/component, with the advance in computer technology, this system can be used for detection of anomalies when combined with good processing tools.

9.4.2.4 Inspection through Impact-Echo

A numerical study was conducted by Shoukouhi et al. (2006) to study the limitations of impact-echo (IE) methodology in detecting delaminations in bridge decks (see Figures 9.28 and 9.29). Factors

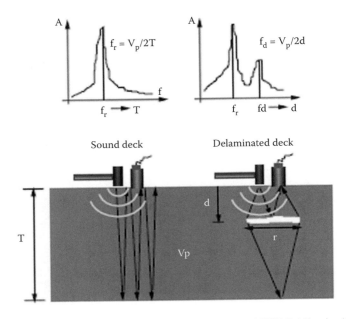

FIGURE 9.28 Deck evaluation using impact echo. (Reprinted from ASNT Publication.)

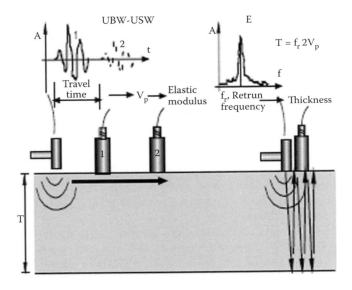

FIGURE 9.29 UBW-USW test setup. (Reprinted from ASNT Publication.)

studied were delaminations depth and size and location of the sensor relative to delaminations location in a plan view. The results indicated the following:

1. Detectability depends on all three components, that is, delaminations depth, delaminations size, and position of the IE sensor.
2. Displacement spectra are advantageous when low frequency content is of interest for detecting deep delaminations and flexural frequency peaks.
3. Acceleration spectra are advantageous and critical for detecting shallow delaminations.

4. Shallow delaminations can be detected and their depth estimated as long as the IE receiver is located above the delaminated area.
5. Deeper delaminations are detected only when the delamination diameter is greater than the depth of the delaminations from the slab top.
6. If larger impact durations are used with acceleration spectra, shallow and smaller delaminations can also be detected.
7. If test setup is not located directly above the delaminations area, the test can give erroneous results.

Experimental and analytical results agree very well.

9.4.2.5 NDT Role in Bridge Inspection

One of the FHWA studies (Jalinoos 2008) on bridge inspection variability showed that significant variability was observed in routine inspection tasks from one inspector to another and from one State to another for the same bridge. This is illustrated in Figure 9.30. The study concluded that, on an average, between four and five different condition-rating values were observed for each primary component. Therefore, there is growing consensus that visual inspection needs to be complemented by more objective measures if accurate estimates of both component and overall system reliability are to be expected throughout the nation.

Table 9.11 presents a brief synopsis of typical bridge elements and some of the standard inspection techniques that complement visual inspection as well as some of the NDE tools available to obtain more information.

Several commonly available nondestructive testing systems, as well as their advantages and disadvantages, are described briefly below.

1. Ultrasonic testing—Uses high-frequency sound energy to assess flaws (surface and subsurface) and dimensional measurements; typically used on metals with untreated or cleaned surfaces.

2. Eddy current—Uses electromagnetic induction to assess surface flaws, material thickness, and coating thickness; typically used on metals with painted or untreated surfaces.

3. Ground-penetrating radar—Uses electromagnetic waves to assess subsurface flaws and to image embedded reinforcement or tendons; typically used in concrete, masonry, and timber structures.

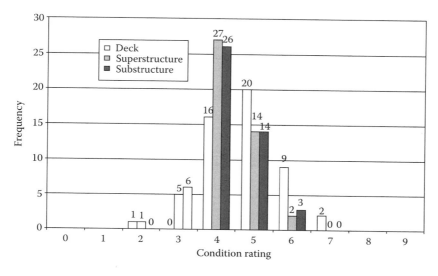

FIGURE 9.30 Bridge condition-rating statistics. (Reprinted from ASNT Publication.)

TABLE 9.11
Brief Synopsis of Typical Bridge Elements and Inspection Practices

Bridge Element	Main Cameras	Standard Practice	Example NDE Tools
Concrete deck	Delamination/rebar corrosion	Chain drag/hammer	Ground-penetrating radar Impact echo Infrared thermography
Pins/hangers/eyebars	Fatigue cracks	Dye penetrant/magnetic particle	Ultrasonic
Steel girders/trusses	Fatigue cracks	Dye penetrant/magnetic particle	Eddy current Ultrasonic Infrared thermography Radiography acoustic emissions
Concrete prestressed girders	Tendon corrosion	Hammer	Magnetic flux Leakage strain gauges
Concrete posttensioned girders	Corrosion, grout holes	Hammer	Impact/ultrasonic echo Ground-penetrating radar
Bearing	Movement, lack of movement	-	Tilt meters Remote sensor bearings
Concrete columns	Rebar corrosion	Hammer	Ground-penetrating radar Ultrasonic pulse velocity
Foundation	Integrity and scour	Probing	Sonar cross-hole conic logging Time domain reflectometry Parallel seismic

Source: Reprinted from ASNT Publication.

4. Impact echo/ultrasonic pulse velocity—Uses impact-generated stress waves to assess sub-surface flaws and material thickness; typically used in concrete and masonry structures.

5. Infrared thermography—Measures the amount of infrared energy emitted by an object to calculate temperature; used in all bridge types to assess deterioration, flaws, and moisture intrusion.

Table 9.12 indicates a basic comparison of some of the representative NDE technologies as used in Bridge Inspectors NDE Showcase.

9.4.2.6 Closing Remarks

We close this section by recapping some possible improvements suggested earlier as (1) accounting for structure type and complexity in defining the inspection interval, personnel qualifications, data collected during inspections, and so on (2) obtaining better consistency and uniformity in ratings collected as part of NBIS inspections through better training, better manuals, QA/QC procedures, introduction of reference bridges, and so on (3) proactive inspections by collecting the data based on the decisions to be made and the risk or various hazards to which the structures are subjected, and (4) leveraging new technologies and practices when needed.

Finally, we emphasize that safety aspects cannot be forgotten while improvements are made for collecting better data for bridge management. Decision making processes should govern any changes in future inspection programs, including introduction of new technologies and health monitoring.

9.4.3 Economic Theory of Inspection

9.4.3.1 Theory

In this section, we suggest various ways in which SHM/SHCE are interrelated to the bridge inspection process. Admittedly, in most of those situations, such interrelationships are qualitative in nature. We desire to explore the interrelationship between traditional (manual) inspection and SHM in a more formal manner. To do that, we need to have (1) as wide a view as possible of the two processes (inspection

TABLE 9.12
Comparison of Representative Technologies Used in Bridge Inspectors NDE Showcase

Method	Advantages	Limitations
Ultrasonic testing 	Ultrasonic testing makes use of mechanical vibrations similar to sound waves but of higher frequency. Used for pin inspection, penetration welds (plate girder flanges, circumferential welds in pipe, etc.) Length and thickness measurements	Surface condition critical. Permanent record has limited value
Eddy current 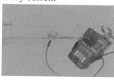	Can detect near-surface defects through paint	Magnetic properties of weld materials can influence results Orientation of probe during scanning can affect results
Ground-penetrating radar 	A technique that uses electromagnetic waves to examine concrete and other nonferrous materials. Used for detecting embedded metals, thickness of materials, mapping of reinforcement location and depth of cover	Environmentally sensitive to the presence of moisture, road salts, electromagnetic noise
Impact echo/ultrasonic pulse velocity 	Gives information on the depth of the defect and concrete quality	Best applied for determining member thickness
Infrared thermography 	A global technique that covers greater areas than other test methods, making it cost-effective. Provides an indication of the percentage of deteriorated area in a surveyed region	Proper environmental conditions are required for testing. Anomalies are difficult to detect, the deeper they are in the concrete

Source: Reprinted from ASNT Publication.

and SHM/SHCE), and (2) a quantitative universal common denominator that can be related to both processes. Upon further reflection, we suggest that LCA of a bridge (or any infrastructure) represents as wide a view as possible. LCA, in addition, has direct applications and use in both inspection process and SHM/SHCE. Furthermore, monetary considerations can be used as a natural common denominator for SHM/SHCE and inspection process. On the basis of this, we offer the following ETI:

The value of manual inspection can be derived directly from its effect on bridge life cycle analysis; as such, maximizing the value of manual inspection involves the optimization of all manual inspection parameters that affect bridge life cycle analysis.

In the above, LCA includes life cycle costs, life cycle benefits, and lifespan (see Chapter 10). By linking directly the value of inspection to LCA of bridges, we aim at offering quantifiable methods of the value of the inspection process, thus making all of its parameters a subject of optimization that are linked directly to a quantifiable objective function: LCA. We next offer a proof of ETI, followed by a practical discussion and some practical corollaries; then we offer some simple examples of the use of ETI in practical situations.

9.4.3.2 Proof of Theory

Consider the cost C_s of a particular structure as a function of time t. There are a few logical assumptions that can be made about the relationship $C_s = C_s(t)$

1. The cost is always increasing, or constant, as a function of time: it is never going to decrease as time passes.
2. The cost function can exhibit sudden jumps. These jumps can be due to sudden expenditure of funding to retrofit, maintain, or inspect the structure.

Figure 9.31 shows an example of the cost function C_s.

The first assumption above can be expressed as

$$\frac{d\,C_s}{d\,t} \geq 0 \tag{9.21}$$

Let us focus on a specific time span of the structure t_I. Further, let us assume that the cost of an uninspected, hence not maintained (or retrofitted), structure can be approximated as a straight line, as in Figure 9.32. Note that this assumption is needed only for the sake of simplicity of the theory proof. The theory can be proved for arbitrary cost functions as long as these cost functions adhere to Equation 9.21.

We need to introduce more parameters: the cost of periodic inspection, ΔM, the elapsed time between inspections, Δt, and the number of inspections N_I. Again, for the sake of simplicity, we assume that the cost of periodic inspection and the elapsed time between inspections are constant during t_I. On the basis of this we can state that

$$t_I = N_I\,\Delta t \tag{9.22}$$

When a structure is inspected, we can reasonably expect that the cost function of the bridge would change from C_s to C_{sI}. The new cost function would reflect the improved conditions of the structure that resulted from the inspection process. We can express this as

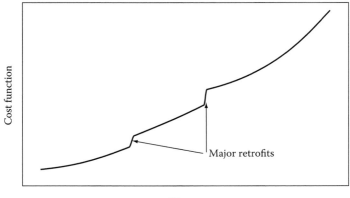

FIGURE 9.31 Typical cost function.

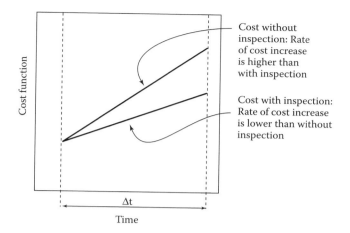

FIGURE 9.32 Details of cost function.

$$\frac{d C_s}{d t} \geq \frac{d C_{sI}}{d t} \geq 0 \tag{9.23}$$

Again let us assume, for the sake of simplicity but without loss of generality of theory, that C_{sI} is a linear function. Functions C_s and C_{sI} during t_I are shown in Figure 9.32. We can express the two linear cost functions as

$$C_s = \alpha_s t \tag{9.24}$$

$$C_{sI} = \alpha_I t \tag{9.25}$$

$$\alpha_I = \left(\beta_I \Delta t^k \right) \tag{9.26}$$

The constants α_s and α_I represent the rate of cost with respect to time for both functions. This indicates $\alpha_s \geq \alpha_I$. We expressed the rate of cost for the inspected structure to be dependent on Δt. For longer Δt, it is expected that the rate of cost will be higher than situations with shorter Δt. The constants β_I and $k \geq 0$ will result in a reasonable behavior of α_I and C_{sI}.

We now have all the tools to prove the theory. Let us define the value (in monetary terms) of N_I inspections during that period as V_I.

From Figure 9.32, the value of inspection V_I can be expressed as

$$V_I = \alpha_s t_I - N_I \left(\alpha_I \Delta t^k \right) \Delta t - N_I \Delta M \tag{9.27}$$

Substituting Equation 9.22 in Equation 9.27

$$V_I = \alpha_s t_I - \frac{t_I}{\Delta t} \left(\alpha_I \Delta t^k \right) \Delta t - \frac{t_I}{\Delta t} \Delta M \tag{9.28}$$

For maximum inspection value we need $dV_I/d\Delta t = 0$, thus

$$-\frac{t_I \Delta M}{\Delta t^2} + k t_I \Delta t^{(k+1)} = 0 \tag{9.29}$$

The equation shows that the value of inspection can indeed be maximized by choosing the inspection period Δt in accordance with a unique relationship of cost functions and inspection costs. The inspection period should not be arbitrary for maximum inspection value. Thus the theory is proved.

Note that the proof just provided considered simplifications of cost functions. Benefits are implied in assumption #1 above. Thus, the proof of the ETI is general.

9.4.3.3 Applications of Theory

9.4.3.3.1 *Relationship Between Bridge Life Cycle Analysis and Manual Inspection*

Perhaps one of the most important implications of ETI is that it relates in a quantitative manner the value of inspection to bridge life cycle analysis (BLCA) through costs of ownership as a function of time and the effects of inspection and all parameters that relate to and affect inspection on BLCA. It is through acknowledging that inspection can be addressed to optimize bridge management goals.

9.4.3.3.2 *What to Inspect?*

Inspection is, by default, a retroactive process. It might lead to proactive measures, such as retrofitting, as a result of some observations during the inspection. However, in general, the inspector will observe the state of the structure at the particular time of the inspection, that is, *after* the fact.

9.4.3.3.2.1 Capacity
Most inspections are made to observe the current *capacity* of the structure. Let us consider the state of corrosion in a partially submerged bridge pier. When an inspector sees that corrosion has actually occurred to a specific degree, such an observation can be correlated directly with the capacity of the pier. The status (cracking, spalling) and the extent of the damaged area can be used to assess the current capacity of the pier. Similarly, when an inspector notices a hairline crack at a particular welded connection in a plate girder bridge, the observation can be related directly to the current *capacity* of the plate girder. Some common inspection objectives related to *capacity* are listed in Table 9.13. Note that Table 9.13 is for illustration purposes only; the actual observations for each situation can be much more extensive than shown.

9.4.3.3.2.2 Demand
Observing *demand*, as part of an inspection effort, is not done as often as observing capacity. This is mainly because of the nature of both the inspection and the demands. Let us consider first the nature of inspection. By definition, inspection occurs at a particular time instant, typically a few hours at the most, when compared with the lifespan of the structure. During that time, the inspector is capable of observing the demands exerted on the bridge. Only during the inspection times can those demands be observed. This is shown in Figures 9.33 and 9.34. Figure 9.33 shows a situation when the demands vary irregularly, and the inspection process occurs also irregularly such that they occur at exactly the right instances to observe this irregularity. Scheduling inspections in

---

TABLE 9.13
Inspection and Bridge Capacity

Objective	Observations	Affected Capacity?
Deck cracking	Crack developments and sizes, spalling. Extent of damage	This would result in an unacceptable rough traffic condition. Ultimately the capacity of the deck itself might be affected
Corrosion	Concrete cover conditions, presence of cracking or spalling, extent of corroded areas	Rusting would result in loss of area. Reduced strength capacity of steel reinforcement can degrade overall capacity of component
Fatigue	Small rust stains in a steel girder	Such stains are indications of small crack formations. Such cracks can result in a brittle failure of components
Scour	Soil erosion behind an abutment	Reduction of load-carrying capacity of the abutment
Earthquakes	Large deformations of joints/connections	Postyielding conditions would reduce available ductile behavior of the affected component

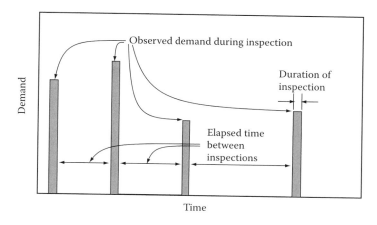

FIGURE 9.33   Observed demands during inspection.

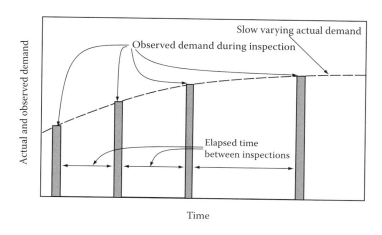

FIGURE 9.34   Actual observed demands during inspection.

this fashion is not realistic. Figure 9.34 shows a situation where the demands vary slowly such that the observations can easily detect such demands.

This brings us to an interesting fact: for slowly varying demands, inspection is effective in observing them; for sudden or rapidly changing demands, inspection is not as effective, as displayed in Figure 9.35. Table 9.14 shows the attributes of some common demands and the effectiveness of inspection in observing them.

9.4.3.3.3   Inspection Frequency

Another direct outcome of ETI is the establishment of the fact that there is an optimum inspection frequency that would maximize the value of manual inspection. As of the writing of this chapter, there are no quantitative studies that address optimum inspection frequency.

9.4.3.3.4   Inspection Costs

We have addressed the benefits (value) of inspection. We have also addressed the costs of not performing inspection. We need to state here that there are direct costs of inspection that need to be accommodated in any quantitative analysis of inspection. These are mainly labor and management costs.

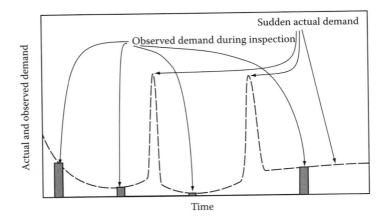

FIGURE 9.35   Observed demands for noneven inspection intervals.

TABLE 9.14
Inspection and Bridge Demands

Type of Demand	Behavior in Time	Effectiveness of Inspection
Traffic	Slow	Effective
Corrosion	Slow	Effective
Fatigue	Slow to sudden	Little to effective
Impact	Sudden	Little
Scour	Medium to sudden	Little to effective
Wind	Sudden	Little
Earthquakes	Sudden	Little

9.4.3.3.5   Inspection Efficiency

We introduce a parameter called inspection efficiency I_e where

$$0 \le I_e \le 1.0$$

The parameter I_e measures the level at which the inspection achieves its stated goals. It is a qualitative measure assigned by those in charge of the inspection process. The magnitude of I_e depends, among other factors, on

1. The experience of the inspectors
2. The allotted time for completing the inspection
3. The difficulty of what is being inspected
4. Other case-dependent factors

It is clear that the value of an inspection relates directly to I_e. It is thus important to keep I_e as close to unity as possible. Some possible means to achieve that are

1. Continued training of inspectors
2. Assignment of reasonable time for inspection
3. Matching of the level of experience of inspectors with level of difficulty of what is being inspected
4. Use of inspection teams that have an adequate mix of experience

TABLE 9.15
Suggested Values of Parameter I_e

Situation			I_e
Level of Experience	Difficulty	Allotted Time	
High	Difficult	Adequate	0.67
Medium	Difficult	Adequate	0.52
Low	Difficult	Adequate	0.35
High	Simple	Adequate	0.95
Medium	Simple	Adequate	0.74
Low	Simple	Adequate	0.49
High	Difficult	Not adequate	0.34
Medium	Difficult	Not adequate	0.26
Low	Difficult	Not adequate	0.18
High	Simple	Not adequate	0.48
Medium	Simple	Not adequate	0.37
Low	Simple	Not adequate	0.19

Table 9.15 shows possible magnitudes of I_e. It should be stressed that this is an arbitrary assignment for illustration only; for actual projects the magnitudes of I_e should be assigned on a case-by-case basis.

Inspection efficiency parameter can be used while estimating the quantitative value of inspection, as well as the direct and indirect costs of inspection.

9.5 MAINTENANCE

9.5.1 INTRODUCTION

Bridge maintenance is a major component of activities required for upkeep of bridge infrastructure, to maintain the required safe operational levels with minimum disruptions, and extend the life of structures cost-effectively. So, maintenance is an integral part of any bridge management effort. It is universally agreed that proper maintenance (DeLisle, Shufon, and Adams 2003; Testa and Yanev 2002) would result in a longer service life for the bridge. However, the exact relationship between specific maintenance efforts and their role in extending durability is not well documented and, in some cases, is extremely difficult to quantify. There are several studies on the effects of maintenance, both general and specific, on service life of bridges. The effects of maintenance activities on the life cycle costs of bridges have been studied recently (see Hawk 2003). This can be summarized, numerically,

$$C_{LC} = f\left(x_1, x_2, \cdots x_i, \cdots x_{N-1}, x_N\right) \qquad (9.30)$$

where C_{LC} is the life cycle cost of the bridge, x_i are the parameters that affect the life cycle costs, and N is the total number of parameters. If the jth parameter in Equation 9.30 is the contribution of maintenance costs to the overall life cycle cost, we can further write

$$x_j = \sum_{k=1}^{k=N_M} B_{jk} \qquad (9.31)$$

where B_{jk} is the contribution of the kth maintenance efforts to the total life cycle costs. The total number of possible maintenance efforts is N_M. A discussion of the details of Equation 9.31 can be found in Hawk (2003).

Recognizing the value of maintenance to bridge management efforts, the following deserve attention:

- Definition or categories of maintenance
- Components requiring maintenance
- Type and scope of maintenance efforts
- Frequency of various maintenance efforts
- Optimization of maintenance efforts and the associated cost, without affecting the required performance levels and durability

With the advent of information technology, innovative sensors, and reliable instrumentation and data acquisition systems for monitoring (for structural, environmental, security, traffic purposes), SHM techniques can be used effectively to address some of these issues.

It should be noted that there is no universal agreement in the bridge community over the definition of maintenance and associated categories. This section discusses some published opinions on the issue. A quantitative cost/benefit approach to maintenance efforts is presented next, followed by the use of SHM, coupled with a cost/benefit approach. Note that Ettouney and Alampalli (2000) advocated that any SHM project not incorporating decision making/cost–benefit ideas in all its tasks cannot be a successful project and should not be pursued. A simple SHM application is given in the paper to show how the techniques coupled with quantitative methodology can be used to optimize maintenance activities.

9.5.2 MAINTENANCE CATEGORIES

We observed earlier that maintenance is a common component in all BMS. Any successful BMS must properly address the subject. It is natural then to seek the help of structural health concepts for maintenance. How can concepts of structural health help the owner in a bridge maintenance program?

Defining maintenance and categorizing various maintenance activities are very important for bridge management. Maintenance can be defined as the work required to keep bridges in proper condition, to preserve/keep in a given existing condition, or to defend against danger or attack. DeLisle, Shufon, and Adams (2004) have subdivided maintenance efforts in a subjective manner to cyclic or corrective classes, as shown in Table 9.16.

A less qualitative and more general maintenance categorization scheme was presented by Testa and Yanev (2002), who recognized that maintenance for a bridge varies with the bridge component and illustrated such interdependence (see Table 9.17). Table 9.17 shows what was defined as the importance factor $0.0 \leq I_{ij} \leq 1.0$, where i and j represent maintenance and bridge component, respectively. For $I_{ij} \rightarrow 0.0$, the ith maintenance activity for the jth component is not relevant. Conversely, when

---

TABLE 9.16
Categorizing Maintenance

Maintenance Class	Examples
Cyclic	Washing, painting, etc.
Corrective	Repair of delaminated/spalled concrete (piers, columns, beams, abutments, etc.)
	Structural steel or concrete repairs
	Bearing replacements
	Bridge deck-wearing surface repairs

Source: DeLisle, R. et al., Development of network-level bridge deterioration curves for use in NYSDOT's asset management process. Presented at the 2004 Transportation Research Board Annual Meeting, Washington, DC, 2004.

TABLE 9.17
Relationship between Maintenance Activities and Bridge Components

j		1	2	3	4	5	6	7	8	9	10	11	12	13
i		Bear.	BWall	Abut.	Wwall	Seats	Prim. Mem.	Sec. Mem.	Curbs	Side walk	Deck	Wear. Surf.	Piers	Joints
1	Debris removal	0.7	0.5	0.2	0.1	0.8	0.5	0.5	0.8	0.8	0.8	0.9	0.1	0.8
2	Sweeping	0.2	0.1	0.1	0	0.5	0.5	0.5	1	0.8	0.9	1	0.1	1
3	Clean Drain	0.9	0.9	0.9	0.8	1	1	1	1	1	1	1	0.5	1
4	Clean abutments/ piers	1	1	1	0.9	1	0.8	0.8	0	0	0.5	0.5	1	0.5
5	Clean gratings	1	0.5	0.7	0.1	1	1	1	0.1	0.1	0.8	1	1	0.9
6	Clean expansion joints	1	0.8	1	0.5	1	1	0.8	0.5	0.5	0.9	0.9	0.9	1
7	Wash deck, etc.	0.5	0.3	0.2	0	0.6	0.4	0.4	1	0	1	1	0.4	1
8	Paint	$1/0^a$	0.5	0	0	1	$1/0^a$	$1/0^a$	0	0	0.4	0	$1/.1^a$	0.5
9	Spot paint	$1/0^a$	0.5	0	0	$1/0^a$	$1/0^§$	$1/0^a$	0	0	0	0	$1/.1^a$	0
10	Patch walks	0	0	0	0	0	0	0	1	1	0.1	0.1	0	0.5
11	Pavement and curb seal	1	1	1	0.5	1	1	1	1	1	1	1	0.5	0.5
12	Electric device maintenance	0	0	0	0	0	0	0	0	0	0	0	0	0
13	Oil mechanical components	1	0.5	0.5	0.2	1	1	1	1	0	0.5	0	1	1
14	Replace wearing surface	0	0.1	0	0	0.1	0.1	0.1	0.5	0.5	1	1	0.1	1
15	Wash underside	1	1	1	0.5	1	1	1	0	0	0.8	0	1	0.9

Note: [a] Indicates steel/concrete.

Source: The column and row headers are based on Testa, R.B. and Yanev, B.S., *Comput. Aided Civ. Infrastruct. Eng.*, 17, 358-367, 2002.

$I_{ij} \rightarrow 1.0$, the ith maintenance activity for the jth component becomes highly relevant. This categorization scheme is attractive because it can be used in open-ended database (additional maintenance and/ or bridge components can be added without affecting the whole categorization scheme). The categories can also be easily linked with SHM systems, which is the main objective of this chapter.

9.5.3 Cost–Benefit (Value) of Maintenance

The focus of this paper is the use of SHM techniques for maintenance, which has not received much attention owing to lack of data and an approach. Quantification of costs and benefits of maintenance activities are needed before evaluating the usefulness of SHM.

Cost of maintenance: The cost of any maintenance can be identified as the direct cost of labor, equipment, and materials. Another source of costs is the indirect cost such as traffic routing or lane closings during maintenance and also user costs. The total maintenance cost for the ith activity and jth component is C_{mij}. For N activities over a time period T and with a discount rate of I per activity period, the total cost of maintenance C_{Mij} (in current dollars) can be given as

$$C_{Mij} = C_{mij} \left(\sum_{k=0}^{k=N} (1+I)^k \right) \tag{9.32}$$

Benefit of maintenance: Quantifying benefits of maintenance is more difficult than estimating costs. One way to define maintenance value is as the cost of *not performing maintenance activities*. For example, in a period of time T_{sm}, if the drains were not cleaned from the bridge deck ($i = 3$ and $j = 10$ in Table 9.17), the accumulating rainwater might cause traffic delays; the cost of such delays would constitute part of the value of maintenance. Other maintenance values include potential of accidents, health hazards (from still water), or corrosion (if accumulated water has salt in it). Clearly, this has to be done case by case, and discussing it at length is beyond the scope of this paper. For now, maintenance value V_{mij} is defined as the estimated value of maintenance (cost of no maintenance) activity i for bridge component j. As such

$$V_{mij} = \sum_{k=1}^{k=L} v_{ijk} \tag{9.33}$$

In Equation 9.33, it is assumed that there are L constituents that add to the value (cost of no maintenance). v_{ijk} is the value of the kth constituent. Over a period T, the total value of maintenance is

$$V_{Mij} = V_{mij} \left(\sum_{k=0}^{k=N_1} (1+I)^k \right) \tag{9.34}$$

Note that $N_1 = T/T_{SM}$.

Cost–benefit (value) of maintenance: Equations 9.33 and 9.34 provide the basis for making objective maintenance decisions. A breakeven point is achieved when,

$$V_{Mij} = C_{Mij} \tag{9.35}$$

Situation $V_{Mij} < C_{Mij}$ is not desirable, since it indicates that the cost exceeds the value. Some possible remedies in such a situation are to reduce N or reduce C_{mij}. Conversely, if $V_{Mij} > C_{Mij}$, the maintenance efficiency is high, and so no changes are needed.

We observe that Equation 9.35 is only for a specific combination of i and j (i.e., specific bridge component and maintenance). Of course, in practice, there is an interaction between various maintenance activities and bridge components, which should be taken into account with the entire bridge network included in the cost–benefit evaluation.

9.5.4 MAINTENANCE INSPECTION AUTOMATIONS

There are several situations where automatic monitoring has been done. For example, Furuta et al. (2004) offered an SHM application that uses digital photos. Such techniques show the potential of SHM application in bridge maintenance.

The thickness of a tunnel is directly related to its health. Because of this, Kreiger and Friebel (2000) presented (IE) and laser-based methods for measuring and checking the thickness of the inner tunnel lining as well as the results of measurements with a high-speed, two-channel, laser-scanning device for inspecting road tunnels that have grown in considerable number in Germany (Figure 9.36). This effort is described below.

Tunnel thickness measurements: Drilled (closed) tunnels have two shells separated by a membrane to make the tunnel waterproof. Flaws, cavities, and variations of the shell thickness in the area of the joints between segments have a severe influence on the correct function of the membrane. Therefore, it was necessary to check the thickness of the inner shell for flaws, cavities, and areas with insufficient thickness shortly after the construction is finished (QA).

Amplitude versus time signal was recorded and then transformed into the frequency domain to get a plot of the amplitude versus frequency. The thickness of the structure or construction part can then be calculated using

$$f = 0.96 \frac{C_p}{2T} \tag{9.36}$$

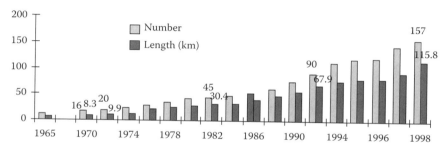

FIGURE 9.36 Number and total length of tunnels on federal roads. (Courtesy of CRC Press.)

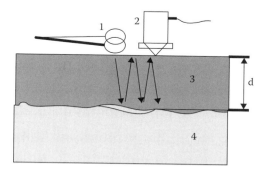

FIGURE 9.37 Schematics of impact-echo method. (Courtesy of CRC Press.)

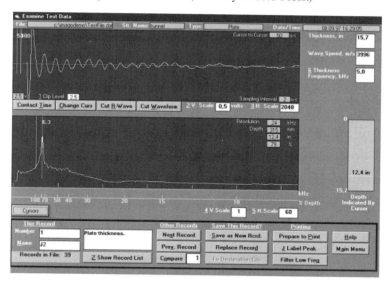

FIGURE 9.38 Results of IE: time and frequency measurements. (Courtesy of CRC Press.)

where f is the frequency, C_p the compression wave speed, and T the tunnel thickness. Figures 9.37 and 9.38 illustrate the IE concept and its results in the time and frequency domains.

Using a grid of 40×40 cm, IE was used to generate thickness plots to decide areas with defects (see Figure 9.39). Defects were fixed, and then IE was used to verify the repair effectiveness.

Laser scanner for inspection: Inspecting large areas in tunnels can be time-consuming, leading to considerable traffic disruptions. So, the use of automatic devices was examined using "tunnel-scanner," a scanning system for digitally recording high-resolution visible and infrared pictures of

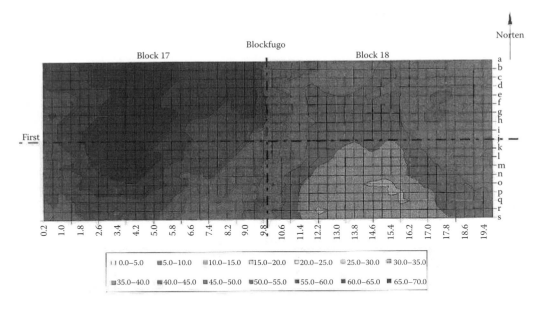

FIGURE 9.39   Thickness measurements using impact-echo. (Courtesy of CRC Press.)

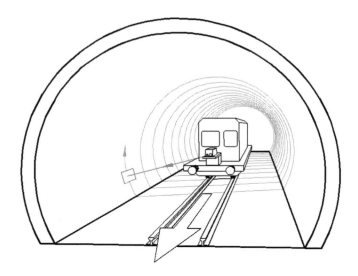

FIGURE 9.40   Schematic of the scanner (SPACETEC). (Courtesy of CRC Press.)

the inner tunnel lining. The schematic of the tunnel scanner and its use are shown in Figures 9.40 and 9.41. The maximum resolution is 10,000 points/360° for the visible picture and 2,500 points/360° for the infrared picture.

By the analysis of the two visible or infrared pictures, cracks, flaws, cavities, and water leakage can be detected. Figure 9.42 shows a comparison of infrared pictures of two measurements from 1991 and 1997 to analyze tunnel deterioration.

Laser shearography: Laser shearography is based on electronic speckle interferometry principles. It used two shearograms, one in the original condition and one after it has been stressed by some means (see Figure 9.43). By subtracting one image from the other, the relative displacement of neighboring points is estimated to detect the cracks. The method has the potential to detect

FIGURE 9.41 Scanner mounted on a vehicle (SPACETEC). (Courtesy of CRC Press.)

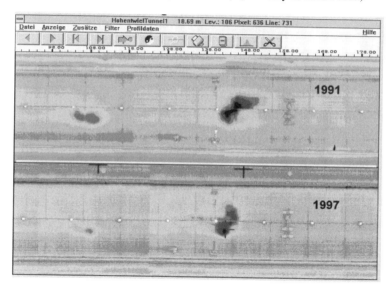

FIGURE 9.42 Water leakage effects: 1991 and 1997 measurements. (Courtesy of CRC Press.)

cracks with a resolution of about 10 microns and is used in aerospace applications successfully. So, a prototype system was developed for civil engineering applications. It consists of both hardware and software and is intended to be portable for field applications. The software was also capable of providing automated crack characterization, using automated image analysis algorithms to extract statistics about cracks. The system was used by Livingston et al. (2006) on a highway bridge in Maryland, and all systems seemed to have performed satisfactorily.

9.5.5 SHM for Maintenance Applications

This section discusses possible use of SHM for maintenance operations by using the maintenance categorization scheme shown in Table 9.17 in a health-monitoring scheme. The first and most obvious use would be to eliminate nonrelevant maintenance activities. We define nonrelevant activity as that where $I_{ij} \leq I_{NR}$, where I_{NR} is an arbitrary cut-off value below which the bridge owner will

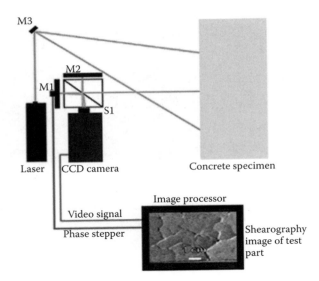

FIGURE 9.43 Schematic diagram of laser shearography applied to concrete. (Reprinted from ASNT Publication.)

TABLE 9.18
Key for Maintenance Structural Health-Monitoring Factor

Mshmij	Comment
3	Activity possible with current technology, cost effective
2	Activity possible with current technology, somewhat cost effective
1	Activity possible with current technology, not cost effective
0	Activity does not lend itself to structural health monitoring
−1	Activity possible, but technology not available yet

assume that health monitoring of the ith maintenance activity for the jth bridge component is not cost-effective or nonrelevant.

To illustrate this, assume $I_{NR} = 0.75$. We now introduce the factor $Mshm_{ij}$, which is a SHM measure for the ith maintenance activity and the jth bridge component. We further assume that $Mshm_{ij}$ is a discrete integer number, as shown in Table 9.18.

We now insert different values of $Mshm_{ij}$, based on our judgment, in Table 9.19. Note that Table 9.19 has the same structure as Table 9.17. For cells with $I_{ij} \leq I_{NR} = 0.75$, no value is entered (i.e., they are left blank).

As noted earlier, objective maintenance decisions can be made by evaluating quantitatively the maintenance cost–benefit values. Note that SHM can affect both costs and benefits due to possible changes in maintenance schedules and scopes. Let us define the period T_M as the new time between maintenance activities due to the use of SHM, whereas the conventional maintenance period is T_{SM}. Taking $\alpha = T_{SM}/T_M$, we can establish the benefits of SHM in such cases as follows:

1. When $\alpha < 1$, there is a need to perform the maintenance task in a shorter period than in a conventional period, thus improving performance.
2. When $\alpha > 1$, there is a need to perform the maintenance task in a longer period than in a conventional period, thus saving costs.

Factor α is thus the effect of using SHM techniques for maintenance. By evaluating α, the SHM effects and benefits on maintenance activities, or the lack thereof, can be quantified.

TABLE 9.19
Recommended Values of $Mshm_{ij}$

j	1	2	3	4	5	6 Prim. Mem.	7 Sec. Mem.	8	9 Side walk	10	11 Wear. Surf.	12	13
i	Bear.	BWall	Abut.	Wwall	Seats			Curbs		Deck		Piers	Joints
1 Debris removal					3			3	3	3	3		3
2 Sweeping								3	3	3	3		0
3 Clean drain	3	3	3	3	3	3	3	3	3	3	3		3
4 Clean abutments/piers	3	3	3	3	3	3	3					3	
5 Clean gratings	3				3	3	3			3	3	3	3
6 Clean expansion joints	3	3	3		3	3	3			3	3	3	3
7 Wash deck, etc.								3	0	3	3	3	3
8 Paint	2				2	2	2	2	2			2	
9 Spot paint[a]	2				2	2	2					2	
10 Patch walks								2	2				
11 Pavement and curb seal	2	2	2		2	2	2	2	2	2	2		
12 Electric device maintenance													
13 Oil mechanical components	1				1	1	1	1				1	1
14 Replace wearing surface										1	1		1
15 Wash underside	0	0	0		0	0	0			0		0	0

Note: [a]Indicates steel only.

Source: The column and row headers are based on Testa, R. B. and Yanev, B. S., Bridge maintenance level assessment, *Comput. Aided Civ. Infrastruct. Eng.*, 17, 2002. With permission.

Figure 9.44 shows the steps needed for using SHM techniques. We observe that care must be given to the choice of appropriate SHM technologies meant to enhance maintenance activity. In the next section, we introduce some practical cases for such use.

9.5.6 CASE STUDIES

We first present the use of the cost/benefit method given above. We will consider both situations where $\alpha < 1$ and $\alpha > 1$. Due to space constraints, we will consider only the maintenance that involves cleaning-type activities; activities $i = 1$ to $i = 7$ ($\alpha > 1$) in Tables 9.16 and 9.18. Similar treatment can be applied to other activities. We then introduce several cases where SHM techniques have been used to enhance various maintenance activities.

9.5.6.1 Cost of Cleaning-Type Maintenance Activities ($i = 1$ to $i = 7$), $\alpha > 1$

We note that all these activities involve debris removal, washing, cleaning, and so on. Consider, for the purpose of illustration, the use of a closed-circuit television (CCTV) placed in a strategic location that requires maintenance. From a remote site, trained personnel can observe the CCTV

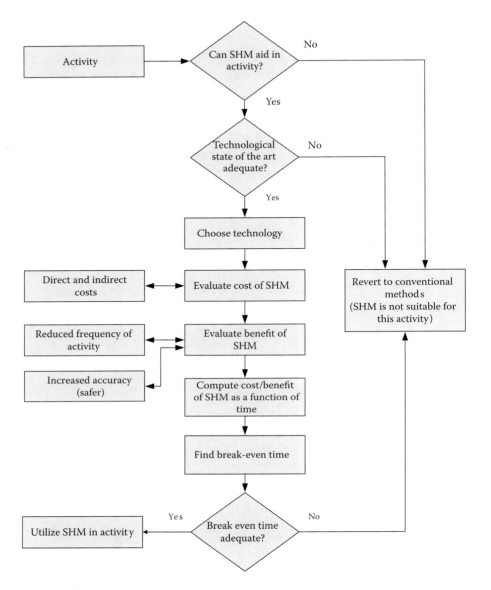

FIGURE 9.44 SHM utilization for maintenance activities.

periodically, perhaps once a day or once a week, as deemed practical or needed by the owner. The decision of when to perform a maintenance task can then be made by appropriate personnel.

The cost savings, C_{Sij} (in current dollars) for ith maintenance and the jth bridge component in a time period T can be estimated as

$$C_{Sij} = C_{mij} \left(\sum_{k=0}^{k=\alpha M} \left(1+I_M\right)^k - \sum_{k=0}^{k=M} \left(1+I_{SM}\right)^k \right) - C_{CCTV} \tag{9.37}$$

where

$$T = M\,T_{SM} = \alpha\,M\,T_M \tag{9.38}$$

C_{mij} = Cost of single ith maintenance activity and the jth bridge component
I_{SM} = Discount rate for period T_{SM}
I_M = Discount rate for period T_M
C_{CCTV} = Cost of installing a CCTV system

$$I_{SM} = \alpha\, I_M \ \text{ or } \ T_{SM} = \alpha\, T_M \tag{9.39}$$

Further, assume that a CCTV system can be used by more than one maintenance activity and for more than one bridge component. Assuming that the SHM system can be shared by K functions, Equation 9.37 can be written as,

$$C_{Sij} = C_{mij}\left(\sum_{k=0}^{k=\alpha M}\left(1+I_M\right)^k - \sum_{k=0}^{k=M}\left(1+I_{SM}\right)^k\right) - \frac{C_{CCTV}}{K} \tag{9.40}$$

Let us consider that a bridge owner is interested in developing such an SHM scheme for a midsize bridge to observe maintenance needs such as debris removal and cleaning of drains, expansion joints, and gratings. Assume that the total cost per activity is C_{mij} = $2,000.00 and a discount rate of 4%, α is estimated to be 1.5. Further, assume that the cost of installing a CCTV is $10,000.00 in current dollars, and it is designed to monitor only one of the above activities ($K = 1$). Figure 9.45 shows a comparison of cash outlays (in current dollars) for a conventional maintenance schedule at one activity per month and an SHM-assisted activity that is determined to be at 1.5 months. The breakeven point is only after 15 months. After 3 years, the SHM-assisted maintenance would save almost $15,500.00.

For a different outlook on the potential cost savings of SHM-assisted approach, assume that for a small increase of the cost of CCTV to $30,000.00, the outreach of the monitoring system will increase to include more maintenance activities. This would mean $K = 4$. Figure 9.46 shows the comparisons. The breakeven point is sooner, at 10 months. The savings after 36 months increased to almost $18,800.00. Similarly, if such a system is already available on the bridge for traffic monitoring or security purposes and can be used to monitor locations requiring maintenance, by controlling

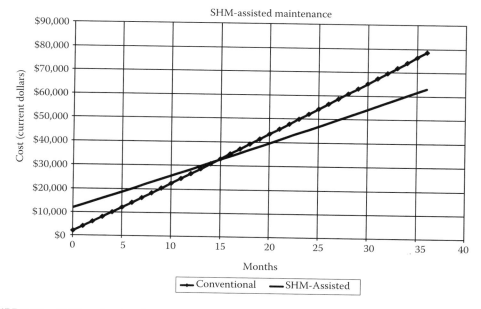

FIGURE 9.45 SHM-assisted maintenance cost—$K = 1$.

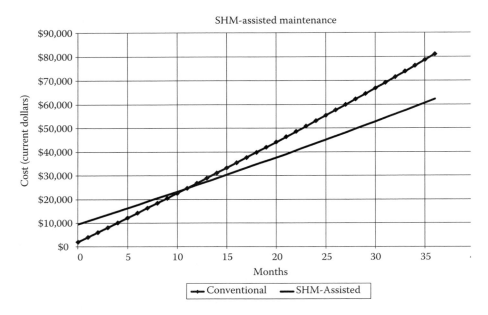

FIGURE 9.46 SHM-assisted maintenance cost—$K = 4$.

the camera, the system cost will be zero, making the breakeven point much closer. (Note that total cost will not be zero since SHM maintenance and other personnel costs are not included in the illustration.)

9.5.6.2 Value (Benefit) of Cleaning-Type Maintenance Activities ($i = 1$ to $i = 7$), $\alpha < 1$

The value of periodic maintenance activity with an assumed period T_{SM} over a period T was established in Equation 9.34. Let us now propose to use a CCTV system that would indicate that the maintenance activity cycle needed is T_M. Proceeding as before, we find that $\alpha = T_{SM}/T_M < 1$. This indicates that the conventional maintenance period is too long and needs to be shortened. As mentioned earlier, if the maintenance period is shortened accordingly, it would result in a better performance. Can we quantify, in an economic sense, such a performance increase, thus justifying the expenditure incurred for purchasing an SHM system?

Following the steps of the previous section, we can express the value (benefit) of reducing the maintenance period in current dollars (thus, the value or benefit of using an SHM technique) as

$$V_{Sij} = V_{mij} \left(\sum_{k=0}^{k=M} \left(1+I_{SM}\right)^k - \sum_{k=0}^{k=\alpha M} \left(1+I_M\right)^k \right) - \frac{C_{CCTV}}{K} \tag{9.41}$$

We will use the same examples as before. Let us consider the economics of an SHM system that considers the need for debris removal and cleaning drains, expansion joints, and gratings. The total value per activity is $V_{mij} = \$1,000.00$. The discount rate is 4% and α is estimated as 0.75. The cost of installing a CCTV is $10,000.00 in current dollars. The CCTV is designed to monitor only one of the above activities ($K = 1$). Figure 9.47 shows a comparison between value or benefit (in current dollars) between a conventional maintenance schedule at one activity per month and an SHM-assisted activity which is determined to be at 0.75 month. The breakeven point is only after 27 months. After 3 years, the SHM-assisted maintenance would save almost $3,500.00 (estimated in value or benefits gained).

For a different outlook on the potential gained value or benefit of an SHM-assisted approach, assume that for a small increase in the cost of CCTV to $30,000.00, the outreach of the monitoring

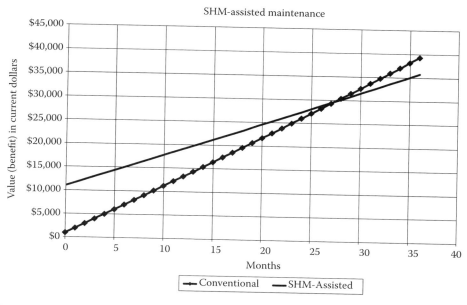

FIGURE 9.47   SHM-assisted maintenance value—$K = 1$.

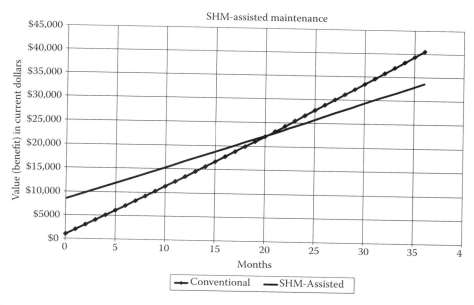

FIGURE 9.48   SHM-assisted maintenance value—$K = 4$.

system will increase to include more maintenance activities. This would mean $K = 4$. Figure 9.48 shows the comparisons. The breakeven point is sooner at 20 months. The savings (estimated in value or benefits gained) after 36 months will increase to almost $8,800.00.

It should be noted that the above CCTV example is for illustration only. All the costs, frequency of activities with and without SHM use, scope of the activities, associated benefits, and so on are all assumed to illustrate the quantitative approach, which can be used and, which does not represent the real data from experience. Note that user costs, and maintenance of SHM costs are also not mentioned, but should be included in the approach.

The same concept can be used to quantitatively evaluate the use of appropriate SHM methods for various activities such as the use of fiber-reinforced polymer (FRP) wrapping for columns and capbeams of bridges or use of smart sensors to monitor chloride content in the decks.

9.5.6.3 Corrosion of Bridge Deck Surfaces

Reinforced concrete bridge decks are susceptible to corrosion owing to the use of deicing chemicals and exposure to salty environments. Corrosion affects different spots, based on the chloride content, and is hard to detect in the initial stages. The usual methods of trying to locate the hidden damaged spots are by chain dragging, hammering, or taking cores and testing them. The costs of these activities are relatively high, and the rate of return of the activities (damaged spots discovered for a given monetary unit) is relatively low (Ettouney and Alampalli 2006). An SHM solution of this problem is the use of passive chloride detection sensors (Watters et al. 2001 and Watters 2003), which can be embedded during the concrete pour. Chloride contents can be measured periodically. When the chloride content at a given location is judged to be sufficient to start the corrosion process, appropriate maintenance activities can be undertaken before it becomes a serious problem. The cost will be the number of sensors and cost of collecting data. If this activity can increase the deck replacement/overlay cycle, it may provide a benefit. For example, if a conventionally maintained bridge deck lasts 10 years, but deck with maintenance assisted by sensors lasts 20 years, then there is a benefit due to extending the cycle. Additional benefits will result from not using routine methods used to evaluate the decks. Cost–benefit analysis will be very useful in making a decision on installing such sensors.

9.5.6.4 Maintenance of Icy Conditions

Icy condition on pavements is a frequent occurrence on roads and bridges during winter. Maintenance departments must estimate such occurrences and then decide whether or not to apply the chemicals to avoid the dangerous icy conditions. Such a process is time-consuming and costly. A real-time monitoring solution was used by Michigan DOT (see FHWA 2005) for the problem. The SHM-sensing solution uses a Road Weather Information System (RWIS). The RWIS stations include pavement sensors which measure the temperature of the road surface and then determine whether the pavement is wet or dry. It also determines the amount of chemicals needed to resolve the icing situation. The information is then transmitted to a central site, where the maintenance crew can take accurate decisions on where and when to send the deicing equipment and how much chemicals are needed. Such accurate decisions can be cost-effective. The breakeven point can be determined using the method outlined above.

Mixing the RWIS stations with the chloride detection sensors (such as those described in the previous section) offers a dual use. Such a solution promises to address icing and corrosion problems simultaneously. This is equivalent to $K = 2$ in Equations 9.40 or 9.41.

9.5.6.5 FRP Bridge Decks

FRP materials have been used for bridge decks instead of conventional reinforced concrete or steel decks. The use of FRP materials for bridge decks has numerous advantages, including light weight and ease of construction (Alampalli et al. 1999, 2003). Unfortunately, because of the sandwich-type topology of FRP decks, it is difficult to ascertain the deck condition by visual inspection. Traditionally, load testing is needed to evaluate the deck condition (Alampalli et al. 2004). Unfortunately, by the time load testing gives the required information, it may be too late, as maintenance (or rehabilitation/replacement) cost may be too high. Ettouney and Alampalli (2006) suggested the use of strategically placed sensors, during the manufacturing of the FRP deck, to resolve the issue and to initiate maintenance. This would eliminate the need for more costly repairs if the problems in the FRP deck are not detected in time.

9.5.7 Closing Remarks

This section illustrates the possible role SHM can play for bridge maintenance. Cost/benefit (value) approach for use of SHM in maintenance applications was presented, thus quantifying the decision making process involved in maintenance. A problem that illustrates the use on cost/benefit method is offered. Finally, case studies of the use of SHM techniques are discussed. In all, the value (cost/benefit) of using SHM techniques for bridge maintenance is demonstrated.

9.6 REPAIR

9.6.1 Economies of Repair in Bridges

For the purpose of this work, we will define "repair" as any bridge activity that is not a periodic maintenance (which was covered in the previous section). Bridge rehabilitation projects on a major scale are not covered in this section. Our main task is to explore the ways in which structural health techniques can help in repair. A cost–benefit approach would be a suitable place to start. Structural health techniques can help in either reducing repair costs or increasing repair value. To simplify the discussion, we subdivide the bridge into components similar to those in Table 9.17. Since possible repair activities for each component vary, depending on the nature of the component, we cannot use a "repair activity" as an analogy to the maintenance activity of Table 9.17 (this is, in fact, due to the differing natures of maintenance and repair). Instead, we use a qualitative approach widely used in BMS: subdivide repair activities into several discrete levels, ranging from minimal repair to replacement of the component under consideration. We choose, for the purpose of presentation, five levels of repair, as shown in Table 9.17. The repair levels remind us immediately of deterioration levels (states) used in the Markovian modeling of deterioration (see Section 5.3). We note that deterioration level and repair level are fairly different entities. The differences are

- Deterioration, as a physical happening, is continuous; discrete deterioration levels (states) are only approximations that are used for the sake of simplicity (Frangopol and Yang 2004). Repair levels are discrete; they depend on the bridge owner's decision.
- Deterioration can be an impetus for repair; although not the only impetus.
- There are several reasons for embarking on a repair activity in addition to deterioration. These can be natural (floods, earthquakes, scour) or man-made (traffic accidents).

Figure 9.49 shows how deterioration and repair activities are interrelated. These distinctions between deterioration and repair levels will be used later in this section.

Cost of repair: Cost of repair can be defined as the direct cost of labor, equipment use, and materials expended. Similar to maintenance activities, another source of costs is the indirect cost from traffic routing or lane closings during maintenance. We can estimate that the total repair cost for the ith repair state (level) and jth component is C_{rij}. For N repair activities over a time period T and with a discount rate of I per activity period, $L=T/N$ the total cost of repair C_{Rij} (in current dollars) is

$$C_{Rij} = C_{rij} \left(\sum_{k=0}^{k=N} (1+I)^k \right)$$

(9.42)

Value (Benefit) of repair: The value of repair is, of course, to enable the bridge to perform as long as possible. However, estimating the value (benefit) of repair in a quantifiable fashion is not a straightforward undertaking. We can approach the subject by asking what by now is a familiar question to the reader: What is the cost of not performing the repair? There can be several consequences, such as

- The bridge will have a shorter lifespan
- The bridge will not function properly

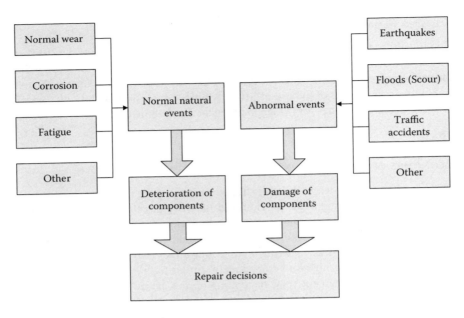

FIGURE 9.49   Deterioration and repair.

Shorter lifespan: The effects of repair on lifespan of bridge have been studied by various authors. The reader is referred to Chapter 8 for a discussion on the subject and examples. For our immediate purpose, however, we are interested in the question: What is the extended lifespan of a bridge as a result of a single, one-time repair? We submit that a particular repair activity is controlled by one of two entities:

- Time (period) controlled
- Deterioration (state) controlled

Deterioration (state) controlled repair is done on an as-needed basis. In deterioration-controlled repair, a repair activity occurs when a particular component deteriorates from a particular state to a preset state. For example, when the cracks in a wearing surface reach a preset size and density*, an activity for repairing the cracks is initiated. Figure 9.50 shows the deterioration-controlled repair model. The repair activity is begun when the components deteriorate ΔD. We note that after the ith single repair activity, the life of the component is increased by ΔL_i. We note that as the component ages, the extended life is ΔL_i naturally, reduced, so that $\Delta L_i > \Delta L_{i+1}$. Deterioration-controlled repair has the advantage of efficient utilization of repair costs and/or value. It has the obvious disadvantage of difficult management, scheduling, and manpower allocation.

Time-controlled repair occurs when the repair event occurs at a preset time. For example, repairing corrosion at a particular column, say every 5 years, is a time-controlled repair. Figure 9.51 shows the time-controlled repair model. The repair activity is triggered at a constant time period ΔT. We note that the ith single repair activity is triggered after the component has deteriorated by ΔD_i. We note that as the component ages, the incremental deterioration ΔD_i for the constant repair period is increased such that $\Delta D_i < D_{i+1}$. As in deterioration-controlled repair, the extended component life after a single repair is ΔL_i. Time-controlled repair has the advantage of ease of management and scheduling. Since it is done without accounting for the actual deterioration or need, it can be inefficient (both cost and/or value).

---

* Number of cracks in a given surface.

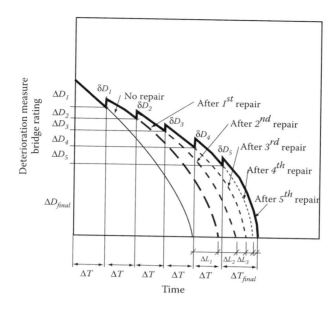

FIGURE 9.50 Repair at constant deterioration.

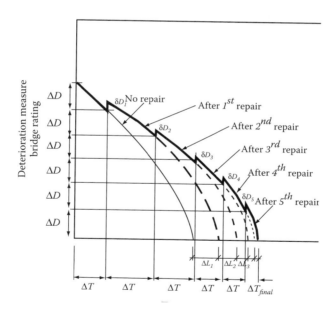

FIGURE 9.51 Repair at constant period.

Value because of delayed replacement: We can now identify one of the values of a single repair activity: an increase in component life of ΔL_i. Expressing this life extension in monetary terms is easy, Let us assume that the current cost of the new component is CC. The value of single ith repair (because of life extension, which leads to a delayed replacement) is

$$V_i = CC\left(\left(1+I\right)^{N_{i-1}} - \left(1+I\right)^{N_i} \right)$$

(9.43)

Where I is the expected discount rate in a given L period. The total estimated component life after the ith repair is

$$L_i = L_0 + \sum_{k=1}^{k=i} \Delta L_k \qquad (9.44)$$

The component life, with no repair activity is L_0. From the above, $N_i = L_i/L$

After i repairs, the total cumulative value of all repairs is

$$V_i = CC\left((1+I)^{N_i} \right) \qquad (9.45)$$

Value because of lowered rate of depreciation: Another source of value for repairs is due to lowering the rate of depreciation as a direct outcome of extending the life of the component. Figure 9.52 illustrates this concept. Let us assume that the value (direct cost, CC) of a new component will be depreciated on a straight line, and there is no salvage value for the component when it is replaced. We realize that this depreciation model can be improved. For our current purpose, the linear assumption is adequate.

From Figure 9.52, the value of a single, ith repair activity can be expressed as

$$V_i = CC\left(\frac{1}{L_i} - \frac{1}{L_{i+1}} \right) T_i \qquad (9.46)$$

We assume that T_i is the time of ith repair. To gain some insight into Equation 9.46, let us assume

$$\alpha_i = \frac{T_i}{L_{i+1}} \qquad (9.47)$$

$$\beta_i = \frac{\Delta L_i}{L_i} \qquad (9.48)$$

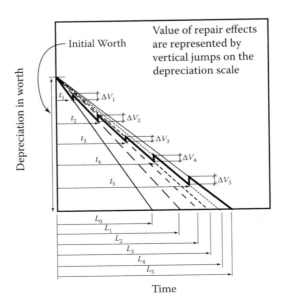

FIGURE 9.52   Effects of repair on depreciation.

Equation 9.46 can be written as

$$V_i = CC \, \gamma_i \tag{9.49}$$

with

$$\gamma_i = \beta_i \, \alpha_i \tag{9.50}$$

We note that the dimensionless factor α_i represents the timing of the ith repair. The dimensionless factor β_i represents the effects of the ith repair on extending the life of the component. The dimensionless parameter γ_l can be evaluated in the realistic ranges of $0 < \alpha_i < 1$ and $0 < \beta_i < 0.5^*$. Table 9.20 shows the values of γ_i in these realistic ranges. Clearly, the timing of repair is an important factor in the value of repair.

Repair value and functionality: When a bridge or a bridge component is repaired, it is understood that the functionality of the repaired item is either maintained or improved. Considering deterioration as an indication of functionality, Figures 9.50 and 9.51 show clearly the relationship between repair activities and improvements in the deterioration state; hence the functionality of components/bridges. We can be less qualitative if we remember that Chase and Gáspár (2000) proved that there is a relationship between average load rating R and condition rating which, in turn, represents deterioration state D (see Figure 9.53). Let us assume that the relationship can be expressed as

$$R = a + b \, D \tag{9.51}$$

Typical values of regression parameters a or b can be found in Chase and Gáspár (2000). The interpretation of Equation 9.51 is clear: when the bridge condition changes from D_i to D_j, the capacity also changes from R_i to R_j, if we use the capacity as a functionality measure. We have just related condition to functionality. We must now find a way to estimate the value of function changes. Such a seemingly formidable task can be simplified to a great degree if we try to find, as we have done before, the cost of functionality *loss* instead.

Two interrelated but different relationships are needed. First, the rating versus time relationship, T–D diagram, is needed. Such a relationship has been discussed earlier and displayed in Figures 9.50 and 9.51. The other relationship is the functionality value of repair versus rating, V_f–D, space is established. To do so, we need to define the cost of total loss of function as V_{00}. Note that we use the cost of functionality loss as a measure of functional value. We can estimate V_{00} using the methods

---

TABLE 9.20
Effects of Repairs on Depreciation of Components

β_i	Repair Timing Factor: α_i				
	0.1	**.25**	**.5**	**.75**	**1.00**
0.10	0.010	0.025	0.050	0.075	0.100
0.20	0.020	0.050	0.100	0.150	0.200
0.33	0.033	0.083	0.165	0.248	0.330
0.43	0.043	0.108	0.215	0.323	0.430
0.50	0.050	0.125	0.250	0.375	0.500

---

* We assume that a single repair will not increase the component's life by more than 50%. However, the equations are valid for more general situations.

from several references. Perhaps the most detailed reference is offered by Hawk (2003). We only have to relate V_{00} to either a condition rating or a load capacity rating. To keep our evaluation applicable to both bridge components and the overall bridge, we will use the condition rating. Thus, we have to populate the functional value-condition rating space. In that space, we identify two condition-rating points: the nearly new condition, D_{00} and the total loss of function condition D_{min}. Clearly V_{00} and D_{00} represent one point in the functional value-condition rating space (at the nearly new condition, the functional value is equal to the cost of total loss of functionality). The total loss of functionality condition D_{min} corresponds naturally to a functional value of 0. For the sake of simplicity, we will assume that a straight line can simulate the behavior of the component between these two points. Figure 9.53 shows the functional value condition-rating space. Analytically, the relationship can be expressed as

$$V_f = a_f + b_f D \tag{9.52}$$

$$b_f = \frac{V_{00}}{D_{00} - D_{min}} \tag{9.53}$$

$$a_f = -b_f D_{min} \tag{9.54}$$

Let us consider a simple example that shows how we can build a functional value condition-rating space. Assume that a bridge rating of 3 would be considered a complete loss of functionality. A bridge rating of 7 would be considered new. Thus,

$$D_{min} = 3 \tag{9.55}$$

$$D_{00} = 7 \tag{9.56}$$

If the bridge official estimates that the cost of total functionality loss is $1,000,000, then

$$V_{00} = 1,000,000 \tag{9.57}$$

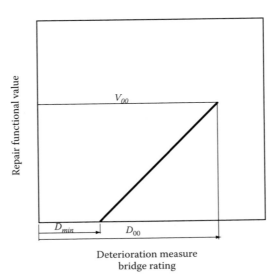

FIGURE 9.53 V_f–D space.

The V_f–D relationship in such a case is

$$V_f = 250,000(-3+D) \tag{9.58}$$

We proceed now to generalize the functionality value of repair versus time diagram, T–V_f for repair activities. We have seen earlier (Figure 9.52) that a single repair activity would generate two results: (1) an increase in rating δD_i and such an increase would lead to (2) an increased life of the component ΔL_i. The relationship can be expressed as

$$\Delta L_i = f(\delta D_i) \tag{9.59}$$

Or inversely

$$\delta D_i = g(\Delta L_i) \tag{9.60}$$

The repair activity will have an effect on the V_f–D space. Starting from a nearly new state (D_{00}, V_{00}), point 0 in Figure 9.54, the condition of the component deteriorates, as well as its functional value, until it reaches point 1 (D_{11}, V_{11}). It is decided that such a condition is not acceptable, and a repair activity is initiated. This repair activity (1st) will result in a new point 2 (D_{12}, V_{12}). After another time period, the condition and functional value will drop to point 3 (D_{21}, V_{21}). Again, it is decided that such a condition is not acceptable, and another repair activity (2nd) is initiated, which results in a new point 4 (D_{22}, V_{22}). The repair will continue until the component or the bridge is replaced or decommissioned. Thus, the ith repair would change the component from (D_{i1}, V_{i1}) to (Di_2, Vi_2). Such a repair state would result in changes in both condition rating δD_i and functional value δV_{fi} such that

$$\delta D_i = D_{i2} - D_{i1} \tag{9.61}$$

$$\delta V_{fi} = V_{i2} - V_{i1} \tag{9.62}$$

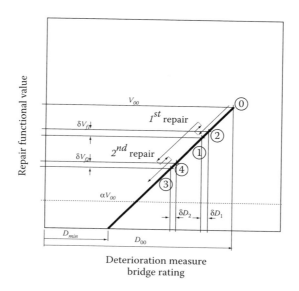

FIGURE 9.54 Functional value of repair—bridge rating (deterioration).

Note that $V_{i1} \geq \alpha V_{00}$ while $V_{i2} \leq V_{00}$. This is practical, since the value after any repair activity V_{i2} cannot exceed the nearly new value V_{00}. Similarly, there must be a minimum functional value αV_{00}, where $\alpha \leq 1$ that the component or the bridge cannot be permitted to go below, for safety reasons.

Since it is assumed that the V_f–D relationship is linear, we can state

$$\delta V_{fi} = b_f \, \delta D_i \tag{9.63}$$

When a repair activity is performed, the resulting δD_i should have an effect on V_f as shown in Figure 9.54.

We arrive at an important observation: the functional repair value needs to increase for single repair activity as the component or the total bridge ages.

To continue with the above example, if after some time, the condition rating of the component is estimated at 5, then the functional value is now $V_f (D = 5) = 250,000 \, x (-3 + 5) = 500,000$. The manager decided that a repair activity is needed to improve its rating. Two possible repair strategies would bring the rating back to 6 and 6.5, respectively. The functional value for each strategy would be \$750,000 and \$875,000, respectively. The choice between the two strategies would depend on other cost–benefits (values) of the two repair strategies.

Before we close the discussion on the functional value of repair, some observations need to be made. The reader will note that the basis of Equation 9.52 is a linearity assumption. It is fair to say that such an assumption is made only for the sake of simplicity; the actual relationship is probably a nonlinear one. It can be convex or concave or even a more complex configuration, as shown schematically in Figure 9.12. There is room for improving Equation 9.52 for a given bridge component or for the bridge as a whole. However, the current linearity assumption does not affect the general accuracy of the approach.

Value (benefit) of repair activities: The total value of a single repair activity can then be expressed as

$$V_{Fi} = \sum_{j=1}^{j=3} V_{fij} \tag{9.64}$$

This value is expressed in current dollars. The first two terms in Equation 9.64 relate to the direct cost of the component (or the bridge), while the third relates to the actual function of the component (or the bridge). In many situations, the magnitude of V_{fi3} will be much higher than the sum of V_{fi1} and V_{fi2}. For example, the function of a support is much more valuable than the direct cost of replacing it. Evaluating V_{fi1} and/or V_{fi2} is much easier than evaluating V_{fi3}. However, because of the importance of V_{fi3} in Equation 9.64, its magnitude must be evaluated with care.

Cost-value (benefit) of repair: As in maintenance, a breakeven point is when

$$V_{Fi} = C_{Fi} \tag{9.65}$$

Again, $V_{Mi} < C_{Mi}$ is not desirable, since it indicates that the cost exceeds the value. Also, if $V_{Mi} > C_{Mi}$, the repair efficiency is high, no changes are needed. Note that in the latter case, the above methods can still be used to prioritize different repair strategies, based on minimizing the C_{Mi}/V_{Mi} (cost to benefit or value) ratio.

9.6.2 Repair and Structural Health Monitoring

We can identify the different ways in which SHM techniques can be of help in bridge repair:

- Damage detection
- Condition assessment

- Optimizing the repair timing and scheduling
- Obtaining and verifying transition probabilities
- Obtaining and verifying parameters needed for bridge reliability studies

We discussed the last two subjects in Section 9.3 of this chapter. For now, we discuss the first three subjects.

9.6.2.1 Damage Detection

Traditionally, SHM has been used mostly for predicting damage to structural components. For example, Tables 9.21 and 9.22 show the many factors that can damage bridge components; they are varied indeed. They are varied in nature, effect, rate, and magnitude. This prompts the edict: different SHM techniques are needed for different causes of hazards that prompt bridge repairs.

We use the factor $Mshm_{ij}$ which is a SHM measure for the ith hazard and the jth bridge component as introduced earlier. Tables 9.21 and 9.22 show different values of $Mshm_{ij}$ for different hazards and bridge components.

To discuss in detail the role of SHM in damage and repair, we develop Tables 9.21 and 9.22 further. First, the types of damage that can occur at a particular bridge component, due to a particular hazard, is shown in Tables 9.23 and 9.24. The types of measurements that can detect such damage in an SHM are discussed at length in Chapters 1 through 5. Additional detailed discussion of damage detection, sensing, and measurements can be found in chapters 5 and 7 by Ettouney and Alampalli (2012).

TABLE 9.21
Relationship between SHM, Reasons for Repair and Bridge Components (a)

			1	2	3	4	5	6	7
i		j	Bearings	Bearing Wall	Abutments	W. wall	Seats	Primary Members	Secondary Members
1	Abnormal	Earthquakes	2	2	2	2	2	2	2
2		Flood (Scour)	2	1	1	1	3	3	3
3		Traffic accidents	1	1	1	1	1	1	1
4		Wind	2	2	2	2	2	2	2
5	Normal	Wear and tear	1	1	1	1	1	1	1
6		Corrosion	1	2	2	2	1	2	2
7		Fatigue	2	2	2	2	2	2	2

TABLE 9.22
Relationship between SHM, Reasons for Repair and Bridge Components (b)

			8	9	10	11	12	13
i		j	Curbs	Sidewalk	Deck	Wearing Surface	Piers	Joints
1	Abnormal	Earthquakes	2	2	2	2	2	2
2		Flood (Scour)	1	1	1	1	2	1
3		Traffic accidents	1	1	2	1	1	1
4		Wind	2	2	2	2	2	2
5	Normal	Wear and tear	1	1	1	1	1	1
6		Corrosion	1	1	2	2	2	1
7		Fatigue	NA	NA	1	NA	1	2

TABLE 9.23
Examples of Types of Damage that can be Detected by SHM (a)

i	j		1 Bearings	2 Bearing Wall	3 Abutments	4 W. Wall	5 Seats	6 Primary Members	7 Secondary Members
1	Abnormal	Earthquakes	Excessive relative motions or rotations	Large displacements, tilt, or deformations	Large displacements, tilt, or deformations		Unseating, loss of anchoring	Large deformations, excessive cracks	Large deformations, excessive cracks
2		Flood (Scour)	Loss of support	Soil erosion below or behind the structural components	Soil erosion below or behind the structural components	Soil erosion below or behind the structural components	Loss of support		
3		Traffic accidents	Direct impact can have severe damage effects. Damage includes component failure, cracking, unseating, etc.						
4		Wind	Excessive relative motions or rotations	Large displacements, tilt, or deformations	Unseating, loss of anchoring	Large deformations, excessive cracks	Excessive relative motions or rotations	Large displacements, tilt, or deformations	Unseating, loss of anchoring
5	Normal	Wear and tear	Frozen bearings. Loss of section	Delamination, spalling, cracking			Loss of support, loss of anchoring	Loss of section. Cracking, delamination	
6		Corrosion	Loss of section						
7		Fatigue	Fatigue cracks						

TABLE 9.24
Examples of Types of Damage that can be Detected by SHM (b)

i		j	8 Curbs	9 Sidewalk	10 Deck	11 Wearing Surface	12 Piers	13 Joints
1	Abnormal	Earthquakes	Cracks, spalling					Large deformations, excessive cracks
2		Flood (Scour)		If submerged, moisture can cause delamination and unwanted cracks			Loss of support	Secondary effects include abnormal deformation due to loss of support elsewhere
3		Traffic accidents		Direct impact, spilled fuel can cause cracking and delamination			Failed components, severe deformations, local failure	
4		Wind	Cracks, spalling	Large deformations, excessive cracks	Cracks, spalling	Large deformations, excessive cracks	Cracks, spalling	Large deformations, excessive cracks
5	Normal	Wear and tear	Spalling, delamination and cracks. Also rusting and loss of strength					
6		Corrosion						
7		Fatigue	Limited effects, but cracks can occur owing to fatigue				Fatigue cracks can occur	

9.6.2.2 Condition Assessment

We discussed in the previous section the role of SHM in damage detection as a part of repair of bridge components. One outcome of assessing damage is that it entails condition assessment of components. We discussed condition assessment in Section 9.2 and showed the several ways in which it can be used in bridge management. Verifying accurately different conditions for components is another way that SHM can be of help in repair of bridges. We try to discuss the following questions:

1. Can SHM techniques help in automating and/or quantifying condition assessment?
2. Which bridge components are amenable for SHM condition assessment?
3. What are the cost–benefit techniques that can be used?

The first step in answering any of these questions is to link the condition rating of different components to types and severity of damage. As an example, we show in Figure 9.55 a possible algorithm to relate damage, as measured in crack size, and condition assessment of pavement (deck). Thus, if an SHM technique is developed to monitor the crack size of the pavement (through embedded fiber optics or high-resolution cameras, for example), then it is possible to have a continuous condition rating for the pavement, both in time (through continuous monitoring) and space (through monitoring most or all of the pavement).

We now discuss which bridge component can benefit from an SHM for condition rating. From Table 9.23, which relates the types of damage, different components, and different hazards, another table that relates condition-rating metrics to damage of components and hazards can be established. Note that Table 9.23 is fairly simple and is only for illustration. The point is that SHM can be used almost universally for all types of hazard, for all components to assess condition ratings.

Finally, the all-important cost–benefit issue is discussed. This can be best explained by exploring a practical example. A bridge deck is covered by a 2-in bituminous pavement. The total area of the pavement is 100 yards^2. Conventionally, the pavement is manually (visually) inspected every 2 years and, if needed, a repair is initiated. The cost of manual inspection is $1,500 or $1,200 for experienced or inexperienced inspectors, respectively. As a part of the inspection, the inspectors issue a condition-rating report for the pavement. Let us assume that the condition-rating system is similar to that of Figure 9.55, that is, it has five possible discrete conditions; thus, the outcome of a single inspection will be a qualitative single condition rating for the pavement.

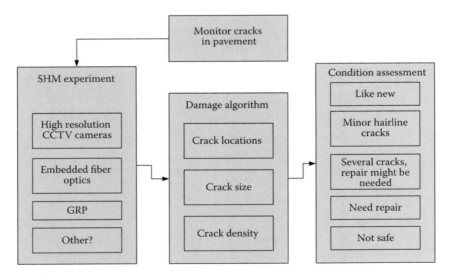

FIGURE 9.55 SHM role in pavement condition rating.

Consider now an SHM technique, by using a high-resolution CCTV camera (the same logic can be applied if other technologies are used). The initial cost of the system is $6,000 (including installation, training, etc.); it has a lifespan of 6 years. We ignore the operating costs since they are negligible. The CCTV system is trained to continually take pictures of the pavement and then estimate crack sizes as they occur. Since it is high-resolution system, it can detect the smallest crack size. For consistency in comparison with manual inspection, let us assume that the detection algorithm is activated only once every 2 years (the time optimization issue will be discussed in the next section). When turned on, the SHM system will measure the information about all the cracks in the pavement, including the number of cracks and the size (length, width) of each. Also, the exact location of each crack will be measured.

One of the differences between the output of manual condition rating and SHM condition rating is obvious: the amount of information. For manual condition rating, only the qualitative condition will be the output. For SHM condition rating, all crack information will be present. Statistical operations will help in reducing the information into a single condition rating. However, the information gathered can be saved in a database and used for other tasks (transition probabilities in a deterioration estimation project, for example). Now we have two condition ratings as results from manual inspection and SHM operations. Is there an advantage that either of them has over the other? Remember that the manual inspection rating is a qualitative rating, while the SHM rating is a quantitative rating. Each of these ratings has an implied coefficient of variation. The SHM quantitative coefficient of variation can be easily estimated using statistical methods from the measurements of the CCTV.

To compare the cost of manual versus SHM condition ratings, the problem can be stated as a Bayesian probabilistic model (see Ettouney and Alampalli 2012 for the theoretical basis of Bayesian modeling). For now, we assign five physical states of the pavement condition rating θ_i with $i = 1 \rightarrow 5$ as shown in Table 9.25.

The historical deterioration probabilities of each physical state are needed. We assume that historical probabilities $P(\theta_i)$ are as shown in Table 9.26. These probabilities means: the probability that a given pavement will be in the condition state i at any given time. The conditional probabilities $P(X=j|\theta_i)$ are evaluated next. They indicate the probability that a particular evaluation activity will predict a physical pavement state j given a pavement actual state i. For the purpose of this discussion, three evaluation factors are compared: experienced inspectors, inexperienced inspectors, and SHM technique. For inexperienced inspectors, $j = 3$ and $i = 2$, the expression $P(X = 3|\theta_2)X$ means: the probability that inexperienced inspectors will evaluate the pavement condition to be "Several Cracks, Repair Might be needed" when the actual pavement condition is "Minor Hairline Cracks." The magnitudes of these conditional probabilities can easily and accurately be computed by simple observation studies. For our immediate purpose, and because of the lack of such information as of the writing of this chapter, we make the realistic assumptions of those conditional probabilities, as shown in Table 9.26. We study only the case of $X = 3$: the case when the condition assessor indicates that the pavement condition is "Several Cracks, Repair Might be needed."

We are interested in evaluating the conditional probabilities $P(\theta_i|X=j)$. This indicates the probability that the actual pavement condition is θ_i given that the assessor indicated that the condition

TABLE 9.25

Physical States of Pavement Condition Rating

Physical State θ_i	Condition Rating
θ_1	Like new
θ_2	Minor hairline cracks
θ_3	Several cracks, repair might be needed
θ_4	Need repair
θ_5	Not safe

is j. Bayes theorem can be used to evaluate these conditional probabilities. The results for $X = 3$ are shown in Table 9.27. The probability that SHM effort will accurately predict the actual pavement condition is 77%—much higher than the probability that the experienced inspector would make an accurate prediction (53%). The probability that the inexperienced inspector would make the correct call is only 46%, as expected. These results are not unexpected. However, to continue with our cost–benefit analysis, we need to estimate the cost of these assessments. We use a weighted cost approach (see Chapter 8 by Ettouney and Alampalli 2012) to estimate the expected costs of the three assessment methods. The total costs are shown in Tables 9.28 and 9.29.

The results of Table 9.29 indicated that the added cost of using an inexperienced inspector, when compared with the cost of using an experienced inspector, is more than $500. The cost saving when using an SHM, when compared with using an experienced inspector, is in excess of $1,000. For a complete picture, we need to include the inspection costs and the CCTV costs. The cost model is shown in Table 9.29 for a single inspection cycle (2 years). The value of using SHM is clear (savings of almost 12%). If SHM is not possible, then the value of using experienced inspectors is also self-evident (savings of about 1.5%). We remind the reader that in this comparison, we assumed a fixed inspection/repair cycle (2 years). We investigate in the next section the implications of releasing this assumption. It should be noted that all costs shown are arbitrary and thus are only for illustration purposes.

We start first with evaluating the cost.

TABLE 9.26
Different Conditional Probabilities

θ_i		θ_1 (%)	θ_2 (%)	θ_3 (%)	θ_4 (%)	θ_5 (%)
$P(\theta_i)$		10	20	30	30	10
Experienced inspector	$P(X = 3 \mid \theta_i)$	5	20	70	40	20
Inexperienced inspector	$P(X = 3 \mid \theta_i)$	5	25	60	45	25
SHM	$P(X = 3 \mid \theta_i)$	2	10	85	15	10

TABLE 9.27
Conditional Probabilities for $X = 3$

	$P(X = 3)$ (%)	$P(\theta_1 \mid X = 3)$ (%)	$P(\theta_2 \mid X = 3)$ (%)	$P(\theta_3 \mid X = 3)$ (%)	$P(\theta_4 \mid X = 3)$ (%)	$P(\theta_5 \mid X = 3)$ (%)
Experienced inspector	40	1	10	53	30	5
Inexperienced inspector	40	1	13	46	34	6
SHM	33	1	6	77	14	3

TABLE 9.28
Construction Costs of Different Pavement Condition Evaluation Techniques

Cost of repair/yard²	$0.00	$1.00	$10.00	$50.00	$200.00	Total Cost/ Yard²	Total Cost
Experienced inspector	$0.00	$0.10	$5.32	$15.19	$10.13	$30.73	$3,073.42
Inexperienced inspector	$0.00	$0.13	$4.56	$17.09	$12.66	$34.43	$3,443.04
SHM	$0.00	$0.06	$7.68	$6.78	$6.02	$20.54	$2,054.22

TABLE 9.29
Total Cost Comparison for a Single Assessment

	Total Cost
Experienced inspector	$4,573.42
Inexperienced inspector	$4,643.04
SHM	$4,054.22

Note: The cost of the CCTV system is discounted over three inspection cycles ($6,000/3 = $2,000 per cycle).

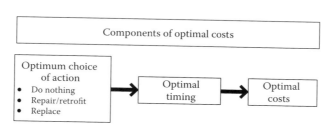

FIGURE 9.56 Components of optimal costs.

9.6.2.3 Optimum Repair Timing and Scheduling

It is fair to assume that when damage is predicted, a repair activity will (should?) follow on time. So, in a generic sense, we can say that optimizing repair timing and scheduling (the first bullet, above) is at the heart of SHM. In this section, the cost–benefit (value) of SHM is explored. We can use the economic theory of repair presented earlier as our vehicle to find this SHM cost–benefit (value).

9.7 BRIDGE MANAGEMENT TOOLS AND SHM

9.7.1 Overview

The main objective of any successful asset management is to ensure user safety at reasonable costs. A logical process of achieving this goal in bridge management is the optimal choice of actions, timing, and costs, as shown in Figure 9.56.

Available to bridge managers are several tools that help them achieve their objectives. Guidelines and manuals form the basic tool in bridge management. There are various guides and manuals at every step in the management process. We explore how the use of guidelines and manuals can be improved by employing different SHCE/SHM techniques and methods. BMSs offer valuable tools for the bridge manager, and several packages are available. Of those we discuss two in detail: Pontis and BRIDGIT. While exploring the main blocks of the systems, we emphasize the important role of SHCE/SHM in improving the quality of information and thereby the results used by those systems. We should also mention that other BMS are available. An overview of other systems can be found in a study by PENNDOT (2003). We end this section by presenting details of some SHCE techniques that can directly improve planning in bridge management. STRID, DMID and decision making techniques can be successfully used in many ways to aid planning efforts. Figure 9.57 shows the interrelationship of different BMS tools.

9.7.2 Guidelines and Manuals

Many State DOTs have specific guidelines that accommodate different situations and hazard mitigation efforts. For example, NYSDOT (2003) has a detailed manual on scour hazard (see Figure 9.58).

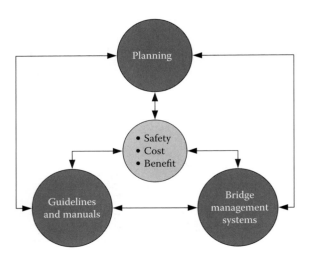

FIGURE 9.57   Elements of bridge management.

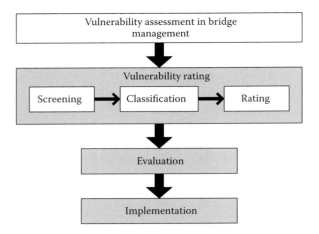

FIGURE 9.58   Hydraulic vulnerability program (NYSDOT).

TABLE 9.30
Role of SHCE in Improving Bridge Guides and Manuals

Managing General Hazards		SHCE Role
Vulnerability Assessment	Screening	Sensing technologies; STRID and DMID techniques
	Classifying	Neural networks
	Rating	Neural networks
Evaluation		STRID and DMID
Implementation		Decision making processes

It specifies clearly three steps: (1) vulnerability assessment, (2) evaluation, and (3) implementation. In each of these steps, SHM/SHCE can have a role that would improve accuracy, as shown in Table 9.30.

We note that the format of hazard management of Figure 9.58 and Table 9.30 are fairly similar for other hazards such as overload hazard (NYSDOT 2001). We discuss in more detail this approach and how SHCE/SHM can improve the assessment.

To assess the vulnerability of bridges to overload $V_{OVERLOAD}$ a risk-based approach is followed. It is based on estimating bridge classification and consequences as follows

Classification: Classification of the bridge is a qualitative measure of the potential vulnerability of the structure to overload failure. It is used in the overall vulnerability-rating process. It uses a set of qualitative and quantitative measures to compute the bridge classification. First, it computed the overload expectancy factor as

$$CL = E_{OVERLOAD} \cdot E_{DETOUR} \tag{9.66}$$

The overload expectancy factor estimates the likelihood of an overload on the bridge. It depends on two factors: $E_{OVERLOAD}$, the posting factor that estimates the probability of exceeding the posted load and the detour factor E_{DETOUR} that accounts for the likelihood of an overload due to a particular detour length. In addition, the structural capacity assessment factor CS is made of two parts: the resistance assessment factor RS that evaluates structural resistance to applied loads and the condition assessment factor CS that accounts for deterioration effects on load-carrying capacity. These factors are related by

$$CS = RS + CS \tag{9.67}$$

A normalized classification factor C_{CL} is then computed using

$$C_{CL} = 65 \frac{CL}{CS} \tag{9.68}$$

Classification factor C_{CL} is then adjusted using a data quality review and a local condition assessment. The qualitative bridge classification is then assigned using the computed values of C_{CL} as in Table 9.31.

Consequence: Consequences of bridge overload depend on two factors: failure type and bridge exposure. Failure type FAIL is assigned values of 5, 3, or 1 for catastrophic (sudden and complete failure), partial collapse (large deformation), or localized structural damage, respectively.

Bridge exposure EXP is a function of traffic volume and functional class. Traffic volume T_{VOL} is assigned factors ranging from 0 to 2, depending on the estimated volume. Functional class F_{CL} is assigned values 0, 1, 2, and 3 for local roads, collector bridges, arterial bridges, and interstates, respectively. Thus

$$EXP = T_{VOL} + F_{CL} \tag{9.69}$$

The consequence value is then computed as

$$CON = FAIL + EXP \tag{9.70}$$

TABLE 9.31
Classifications of Bridge Overload Vulnerability

C_{CL} Value	Classification
$10 > CL$	High
$11 \geq CL \geq 5$	Medium
$CL < 6$	Low
$CL = 0$	Not vulnerable to overload

TABLE 9.32
Vulnerability to Bridge Overload

$V_{OVERLOAD}$ Score	Rating	Comment
>15	1	Safety priority
13–16	2	Safety program
9–14	3	Capital program
<15	4	Inspection program
<9	5	No action
-	6	Not applicable

TABLE 9.33
SHCE interaction with Vulnerability to Overload

Issue	SHCE Role
Overload expectancy factor	Accurate monitoring of traffic volume and loads. Better statistical data
Structural resistance	Tools of STRID. Accurate load testing. Accurate vibration methods for structural parameters
Deterioration effects	See Section 9.3 of this chapter
Failure type	See Chapter 3 by Ettouney and Alampalli (2012)
Overall method	The method is based on relative risk approach. It produces deterministic results. A probabilistic estimation (similar to the methods of Chapter 8 by Ettouney and Alampalli 2012) can account for the inherent uncertainties, thus leading to a more informative decision making process

Vulnerability Rating: The final bridge vulnerability to overload $V_{OVERLOAD}$ score is then computed as

$$V_{OVERLOAD} = CL + CON \tag{9.71}$$

The scores are subdivided into six ratings as shown in Table 9.32.

We recognize Equation 9.71 as a form of relative risk expression. It contains the hazard level and damage represented in *CL*. The consequences are represented in CON. Upon inspecting Equations 9.66 through 9.71, we recognize that the qualitative nature of the approach may be improved by using SHCE in evaluating one or more of the components (Table 9.33).

9.7.3 PONTIS

9.7.3.1 Overall

One of the popular tools for BMSs is the Pontis system (Pontis 2005a, 2005b). Owned by AASHTO and licensed to several State DOTs for managing their bridge networks, Pontis has been under development since 1989. It was first developed for FHWA. In 1994, FHWA transferred the ownership to AASHTO. The Pontis system can help bridge managers in optimizing the allocation of funding for protecting and preserving their infrastructures. It also helps in archiving information (parameters and inspection results) about bridge networks. In addition, Pontis includes several modeling techniques that can quantify decision making processes and accommodate uncertainties and knowledge gaps in many bridge management protocols.

9.7.3.2 Preservation (Maintenance, Rehabilitation, and Repair) versus Improvement

Pontis system differentiates between preservation and improvement in its overall management logic. Preservation includes maintenance, rehabilitation, and repair (MR&R). Improvements include activities such as widening, raising, strengthening, or replacements. During computation of the optimal course of action, Pontis computes preservation (MR&R) and improvement efforts independently. Actually, the mathematical models for both efforts are fairly different, as explained below:

Preservation (MR&R): Pontis uses deterioration models for different components, such as deterioration of paint. As a response to the computed deterioration, the possible decisions that can be made to accommodate it (in the form of MR & R) are modeled, including their costs. Those annual events (deterioration and improvements) are modeled using the theory of discounted programming. The main outcome of the preservation models is optimal preservation solutions available to the bridge official.

Improvements: Improvements in Pontis include four specific actions available to the bridge official. These actions are consistent with applicable functional standards and improvement feasibility. They are (1) widening of lanes, (2) raising the bridge level, (3) strengthening components, and (4) replacing components. The improvements are mathematically modeled using bridge information as well as other information such as cost and traffic. In all, both cost and benefits of the improvements are computed. Figure 9.59 illustrates these interrelationships conceptually.

Simulation: The preservation models are simulated with the improvement needs using a dynamic nonlinear programming technique. The results of the analysis are complete optimal planning solutions for as long as 99 years.

9.7.3.3 Technical Components

Pontis is a wide-ranging and comprehensive management tool. It includes a rich content of technical tools. Covering all of those tools is beyond the scope of this chapter. Most of those technical tools

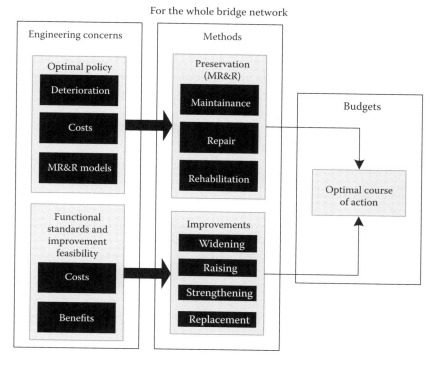

FIGURE 9.59 Components of Pontis system.

can be enhanced by SHCE/SHM techniques. We mention a few of these tools below, and then we end this section by exploring how SHCE/SHM can help those techniques.

Databases: Pontis contains a wide array of databases. Perhaps the most visible database is the physical inventory database, which includes, but is not limited to (1) bridge information, (2) component inspection results, (3) inspection events, (4) roadway information, and (5) structural properties.

Default parameters: Several default parameters used in building different mathematical models are included in Pontis. These default parameters can be overridden by the user if more accurate information is available.

Inspection-based computations: There are several parameters that are computed within Pontis which utilize inspection records. They are (1) components and overall structural health index, (2) NBI translator, and (3) sufficiency rating and (structural deficient/functionally obsolete) classifications. Results of inspection efforts are used to compute these three parameters, which are then used during the Pontis search for optimal strategies.

Deterioration models: Pontis uses Markov processes to estimate deterioration states of different bridge components. The transition probability matrices are built-in. They are evaluated using two sources:

- Available inspection data
- Experience and judgment of officials

Cost models: There are cost models for potential actions that can be made by bridge managers. These models can be adjusted to fit different economic conditions.

9.7.3.4 Possible SHM Utilization

Estimations of different technical parameters can be improved by SHCE techniques as shown in Table 9.34

9.7.4 BRIDGIT

Another BMS tool is BRIDGIT (see Hawk and Small 1998). It models inspection results for individual bridge components that are compatible with NBI specifications. In addition, protective systems (e.g., paint) are treated independently from underlying elements. The analytical approach performs

TABLE 9.34
SHCE/SHM Role in Pontis Parameters

Technical Parameter	SHCE/SHM Role
Databases	Agrawal et al. (2008b) provided a methodology for establishing a realistic and accurate bridge database. The method is summarized in Section 9.5
Default parameters	Measuring different metrics using SHM projects can improve estimates of the probabilities of the transition matrices
Inspection-based computations	See Section 9.4
Deterioration models	Measuring different deterioration metrics can improve estimates of the probabilities of the transition matrices (see Section 9.5)
Cost models	Chapter 10 explores in detail the relationship between life cycle analysis (LCA) and SHM. Also, see Section 9.4.3: Economic Theory of Inspection

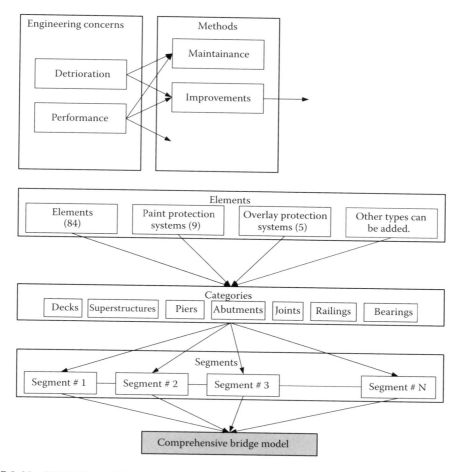

FIGURE 9.60 BRIDGIT building blocks.

multiperiod optimizations as contrasted with sequential-period optimization (as in Pontis system). Similar to Pontis, it uses Markovian processes for estimating rates of deterioration of different components. One of the features of BRIDGIT is that it focuses on life cycle cost analysis of the structure. The life cycle computations span over 20 years. Figure 9.60 shows the schematics of BRIDGIT building blocks.

The role of SHCE/SHM in improving the quality of information used by BRIDGIT is similar to the SHCE-Pontis role as described previously.

9.7.5 SHM AND PLANNING

Structural health in civil engineering methods can aid in improving the accuracy and usability of different bridge management tools. We discuss how some STRID, DMID, and decision making techniques can be related to BMS tools.

9.7.5.1 STRID, Neural Networks, and Classification Applications

Neural networks were shown to be valuable assets in many structural and damage identification situations. Ettouney and Alampalli (2012) mentioned in Chapter 6 another type of application of neural network in the SHM field, which is the classification application. A potential neural network classification application is bridge rating. The conventional practice of bridge rating is based on a set of observations about the condition of the bridge and its components. This set of observations determines the rating of a bridge. The observations are usually a preset series of safety and performance issues that are given weights. The weights are assigned to each of these issues by an inspector during or following a visual inspection of the bridge. A preset algorithm will then integrate all the weights into a final bridge rating. Neural networks can be used for bridge rating by training the network to observe the preset performance and safety issues and the corresponding classification (bridge rating). The bridge rating as a classification problem is shown schematically in Figure 9.61. The observations are represented by the small circles in the observation space. The different bridge ratings are highly nonlinear functions of the observations. Neural networks are then trained to classify (rate) the bridge based on preset conditions. The input layer is the observations, while the output layer is the bridge ratings (classification). After training, the neural network can be used to rate the bridge, given a new set of observations. A sample neural network topology for bridge-rating classification is shown in Figure 9.62. The sample includes an input layer (observations), and output layer (bridge rating), and a hidden layer or layers. Generally, the number of input neurons in such an application would be large, since each of these input neurons represents a single observation point. The observation points needed to produce a bridge rating are numerous. On the output side, the number of output neurons will be limited. The bridge-rating classifications are usually limited. They range from four to nine ratings in most practical cases. The neural network example of Figure 9.62 shows four output rating classifications and an arbitrary number of observation input neurons. We suggest that, due to the highly nonlinear and complex nature of the bridge-rating problem, more than one hidden layer might be needed.

The use of neural networks for bridge-rating application and for classification applications in general offers several advantages over the conventional bridge-rating approach. Neural networks are capable of expanding and pruning the input parameters in an automatic and accurate manner. They can also modify easily the bridge-rating definition. Note that some rating systems are in the range from 1 to 9, while some others use a rating system from 1 to 7. It is not easy to correlate one rating mechanism to the other. The neural network approach can retrain observations to readjust bridge rating as desired. For example, let us assume that a particular bridge authority decides to add two observation issues that affect bridge ratings, for example, a new restraint condition assessment following a change of bridge restraint system and a new scour riprap condition assessment after placing a new synthetic riprap near the bridge abutment. Let us assume also that the rating levels of Figure 9.62 have increased from four levels to seven levels to be consistent with other bridge-rating systems. These two new observation issues and the three additional rating levels can be handled easily by adding two input neurons and three output neurons to the original

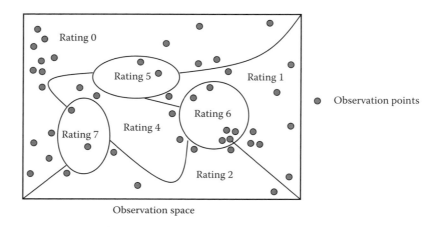

FIGURE 9.61   Neural network model for bridge rating.

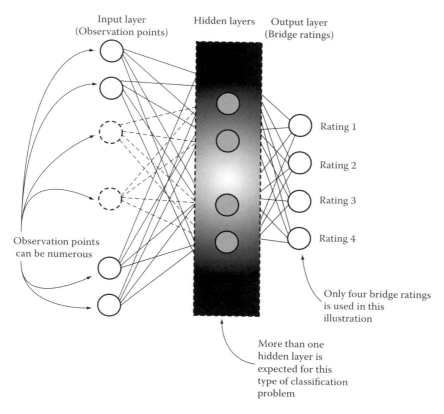

Input layer
(Observation points)

Hidden layers

Output layer
(Bridge ratings)

Rating 1

Rating 2

Rating 3

Rating 4

Observation points
can be numerous

Only four bridge ratings
is used in this
illustration

More than one
hidden layer is
expected for this
type of classification
problem

FIGURE 9.62   Neural network example for bridge rating.

network of Figure 9.62. The new neural network topology will then be retrained so that it can be used for the new rating system. It can also produce bridge ratings for previous conditions that can be used as needed by bridge owners. Figure 9.63 shows the new neural network topology. Obviously, the neural network application simplified and automated an otherwise very complex and time-consuming situation.

9.7.5.2   Decision Making: Deterministic versus Probabilistic Dynamic Programming

Different BMS tools utilize the methods of dynamic programming in finding an optimal course of action. Dynamic programming is based on Bellman's principle of optimality: "An optimal policy has the property that, whatever be the initial state and initial decision, the remaining decisions must constitute an optimal with regard to the state resulting from the first decision." There are four types of issues that are needed to solve a dynamic programming problem in bridge management. First, the problem is subdivided into stages that are usually annual or biannual. Second, we define the states of the bridge which are different information that completely define pertinent parameters at each stage. A policy is then defined that describes decision making rules at every stage, given the current states. Finally, an optimal policy that optimizes any objective that the decision maker desires; usually cost or value is defined.

The problem is formalized (Pontis 2005b) as a minimization process of the expected value

$$C(i) = E\left[\sum_{n=1}^{n=\infty} \left(c\left(i_n, a_n\right)I^n\right)\middle| x_0 = i\right]$$

(9.72)

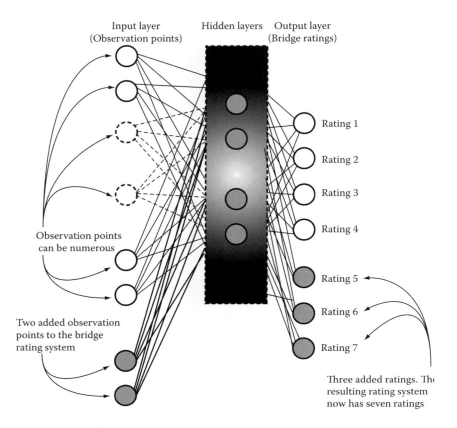

FIGURE 9.63   Modified bridge-rating system and its effect on the neural network model.

This minimization problem tries to find the minimum conditional expectation of the cost $C(i)$, given an initial state of the system of $x_0 = i$. The cost $c(i_n, a_n)$ is the cost at time stage n of a decision a and a system state i. The discount rate is I. Assuming stationary policies, the above minimization problem can be restated as

$$C(i) = \min\left(c(i_n, a_n) + I\sum_j p_{ij} C(j) \right) \tag{9.73}$$

The transition probabilities, in a Markov processes sense, are p_{ij}. Using a nonlinear minimization solver, different BMS programs solve Equations 9.72 and 9.73 to find optimal decisions a_n. For simple problems, we recognize the process as a simplified decision tree technique.

Even though the dynamic programming models as described above do include uncertainties of parameters, in the form of transition probabilities, p_{ij}, they produce single-value results, without the benefit of producing variances that can add to the usefulness of the models. Recall that Chapter 8 by Ettouney and Alampalli (2012) addressed this situation for risk-based models. Expanding the deterministic nature of dynamic programming into more useful probabilistic applications was discussed, and a few methods of accommodating such a situation were presented.

The use of Taylor series in certain closed-form situations was also explored in Chapter 8 by Ettouney and Alampalli (2012). Elsayed and Ettouney (1994), by using the Taylor series, have introduced a closed-form approach that analyzes programming problems with uncertain parameters. Such an approach, if used with the dynamic programming methods in BMS, would produce variances of random parameters in an efficient manner.

Optimization methods in bridge management systems

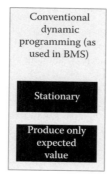
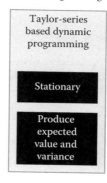
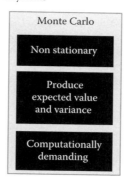

FIGURE 9.64   Comparison of dynamic programming methods in BMS.

Another basic problem with the dynamic programming approach as used in BMS systems is the stationary assumption: transition probabilities and policies are stage-independent. The use of Monte Carlo simulation can overcome this shortcoming. The Monte Carlo approach will also produce the required variances. Unfortunately, the Monte Carlo technique can be computationally inefficient, especially when different decision scenarios are explored.

The three techniques of dynamic programming method in BMS are compared in Figure 9.64.

REFERENCES

AASHTO. (1964). "Information Guide for Maintenance Personnel," American Association for State Highway and Transportation Officials, Washington, DC.

ASCE, ASCE/SEI-AASHTO. (2009). "White paper on bridge inspection and rating," Ad Hoc Group on Bridge Inspection, Rating, Rehabilitation, and Replacement, *Journal of Bridge Engineering*, 14.

Agrawal, A., Kawaguchi, A., Chen, Z., Lagace, S., and Delisle, R., (2008b). "Case Study on Factors Affecting Deterioration of Bridge Elements in New York," NDE/NDT for Highways and Bridges: Structural Materials Technology (SMT), ASNT, Oakland, CA.

Agrawal, A., Kawaguchi, A., and Qian, G. (2008a). "Bridge Element Deterioration Rates: Phase I, Report," TIRC/NYSDOT Project # C-01–51, NYSDOT, Albany, NY.

Alampalli, S., Ettouney, M., and Gajer, R. (2003). "Structural Health Issues for FRP Bridge Decks," 2003 New York City Bridge Conference, New York, NY.

Alampalli, S. and Jalinoos, F. (2009). *Use of NDT Technologies in US Bridge Inspection Practice*, Materials Evaluation, ASNT, Columbus, OH.

Alampalli, S., O'Connor, J. S., and Yannotti, A. P. (1999). "Advanced Composites for Cost-Effective Rehabilitation of Bridges," *Advances in Composite Materials and Mechanics*, ASCE Special Publication, pp. 76–84.

Alampalli, S., Schongar, G., and Greenberg, H. (2004). "In-Service Performance of an FRP Superstructure." Special Report 141, Transportation Research and Development Bureau, New York State Department of Transportation, Albany, NY.

Baker, D. V. and Chen, S-E. (2000). "Visual Inspection Enhancement via. Virtual Reality," Structural Materials Technology: an NDT Conference, ASNT, Atlantic City, NJ.

Bakht, B., Jaeger, L., and Mufti, A. (2002). "Use of in-service Performance of a Bridge to Modify its Safety Assessment: An Introduction," *Proceedings of 1st International Workshop on Structural Health Monitoring of Innovative Civil Engineering Structures*, ISIS Canada Corporation, Manitoba, Canada.

Benjamin, J. R. and Cornell, C. A. (1970). *Probability, Statistics, and Decision-Making for Civil Engineers*. McGraw-Hill, New York, NY.

Chase, S. B. and Gáspár, L. (2000). "Modeling the reduction in load capacity of highway bridges with age." *Journal of Bridge Engineering*, ASCE, 5(4), 331–336.

DeLisle, R., Shufon, J., and Adams, L. (2003). "Development of Network-Level Bridge Deterioration Curves for Use in NYSDOT's Asset Management Process," NYS-DOT, Albany, NY.

DeLisle, R., Shufon, J., and Adams, L. (2004). "Development of Network-Level Bridge Deterioration Curves for Use in NYSDOT's Asset Management Process." Presented at the 2004 Transportation Research Board Annual Meeting, Washington, DC.

DeLisle, R., Sullo, P., and Grivas, D. (2002). "A Network-Level Pavement Performance Model That Incorporate Censored Data," TRP preprint Number 03–4281.

Elsayed, A. E. and Ettouney, M. M. (1994). "Perturbation analysis of linear programming problems with random parameters," *Computers and Operation Research*, 21, 211–224.

Ettouney, M. M. and Alampalli, S. (2000). "Engineering Structural Health." ASCE Structures Congress 2000, Philadelphia, PA.

Ettouney, M. M. and Alampalli, S. (2006). *Engineering of Structural Health*, CRC Press, Boca Raton, FL.

Ettouney, M. M. and Alampalli, S. (2012). *Infrastructure Health in Civil Engineering: Theory and Components*, CRC Press, Boca Raton, FL.

FHWA. (2003). *Bridge Inspector's Reference Manual*, (BIRM - 2003).

FHWA. (1995). *Recording and Coding Guide*.

FHWA. (2005). http://www.fhwa.dot.gov/winter/roadsvr/CS037.htm. Accessed April 6, 2011.

Frangopol, D. and Yang, C. (2004). "Health and Safety of Civil Infrastructures: A Unified Approach," *Proceedings of 2nd International Workshop on Structural Health Monitoring of Innovative Civil Engineering Structures*, ISIS Canada Corporation, Manitoba, Canada.

Furuta, H., Hirokane, M., Muraki, H., Tanaka, S., and Frangopol, D. (2004). "Bridge Management System Using Digital Photos," *Proceedings of 2nd International Workshop on Structural Health Monitoring of Innovative Civil Engineering Structures*, ISIS Canada Corporation, Manitoba, Canada.

Hawk, H. (2003). "Life-Cycle Cost Analysis," NCHRP Report 483, Transportation Research Board, Washington, DC.

Hawk, H. and Small, E. P. (1998). "The BRIDGIT bridge management system." *Structural Engineering International*, (4) 309–314. International Association of Bridge and Structural Engineering (IABSE), Zurich, Switzerland.

Jalinoos, F. (2008). "Use of Nondestructive Evaluation (NDE) Tools in the U.S. Bridge Inspection Practice," NDE/NDT for Highways and Bridges: Structural Materials Technology (SMT), ASNT, Oakland, CA.

Kong, J. S. and Frangopol, D. M. (2003). "Evaluation of expected life-cycle maintenance cost of deteriorating structures," *Journal of Structural Engineering*, ASCE, 129(5), 682–691.

Krieger, J. and Friebel, W. D. (2000). "NDT Methods for the Inspection of Road Tunnels," Structural Materials Technology: an NDT Conference, ASNT, Atlantic City, NJ.

Livingston, R. A., Ceesay, J., Amden, A. M., and Newman, J. W. (2006). "Development of a Portable Laser Shearography System for Crack Detection in Bridges," NDE Conference on Civil Engineering, ASNT, St. Louis, MO.

Madanat, S. and Ibrahim, W. (1995). "Poisson regression models of infrastructure transition probabilities," *ASCE Journal of Transportation Engineering*, 121(3), 267–272.

Medina, E., Aldrin, J., Allwin, D., Fisher, J., Ahmed, M., and Knopp, J. (2005). "Simulation-Based Design and Tradeoff Analysis with Probabilistic Risk Assessment for NDE and SHM," *Proceedings of the 2005 ASNT Fall Conference*, Columbus, OH.

Mishlani, R. and Madanat, S. (2002). "Computation of infrastructure transition probabilities using stochastic duration models," *ASCE Journal of Infrastructure Systems*, 8(4), 756–765.

Miyamoto, A. and Motoshita, M. (2004). "An integrated Internet Monitoring System for Bridge Maintenance," *Proceedings of 2nd International Workshop on Structural Health Monitoring of Innovative Civil Engineering Structures*," ISIS Canada Corporation, Manitoba, Canada.

NBIS (2004). Federal Regulations: *National Bridge Inspection Standards*, Federal Highway Administration, Washington, DC.

NYSDOT. (2001). "Overload Vulnerability Manual," Report by New York State Department of Transportation, Albany NY.

NYSDOT. (2003). "Hydraulic Vulnerability Manual," Report by New York State Department of Transportation, Albany NY.

NYSDOT (2007). "SFY 2006–07: Annual Report of Bridge Management and Inspection Programs," New York State Department of Transportation, Albany, NY

PENNDOT. (2003). "Software Analysis Report," Report by Pennsylvania Department of Transportation, PENNDOT Harrisburg, PA.

Pontis 4.4. (2005a). "Pontis Release 4.4: User's Manual," Prepared for AASHTO by Cambridge Systematics, Inc., Cambridge, MA.

Pontis 4.4. (2005b). "Pontis Release 4.4: Technical Manual," Prepared for AASHTO by Cambridge Systematics, Inc., Cambridge, MA.

Sathantip, C., Rens, K., and Neimann, Y. "Deterioration Model for Bridge Management Systems: User's Manual," City and County of Denver, Department of Public Works, Denver, CO.

Shoukouhi, P., Gucunski, N., and Wiggenhauser, H. (2006). "Using Impact Echo for Non-Destructive Detection of Delamination in Concrete Bridge Decks," NDE Conference on Civil Engineering, ASNT, St. Louis, MO.

Söderqvist, M-K. and Veijola, M. (1998). "The Finnish Bridge Management System," Structural Engineering International.

Takeda, T., Yasue, S., Oshima, T., Tamba, I., and Mikami, S. (2004). "Bridge Management System (BMS) for Multi-Span Long Bridges," *Proceedings of 2nd International Workshop on Structural Health Monitoring of Innovative Civil Engineering Structures*, ISIS Canada Corporation, Manitoba, Canada.

Taljsten, B., Hejll, A., and Olofsson, T. (2002). "Structural Health Monitoring of Two Railway Bridges," *Proceedings of 1st International Workshop on Structural Health Monitoring of Innovative Civil Engineering Structures*," ISIS Canada Corporation, Manitoba, Canada.

Testa, R. B. and Yanev, B. S. (2002). "Bridge maintenance level assessment," *Computer-Aided Civil and Infrastructure Engineering*, 17(5), 358–367.

Thompson, -P. D., Small, E. P., Johnson, M., and Marshall, A. R. (1998). "The Pontis bridge management system." *Structural Engineering International*. International Association of Bridge And Structural Engineering (IABSE), Zurich, Switzerland 8(4), 303–308.

Ulku, A. E., Attanayake, U., and Aktan, H. M. (2006). "Remaining Service Life Estimation of Bridge Components by Knowledge Discovery Techniques," NDE Conference on Civil Engineering, ASNT, St. Louis, MO.

Watters, D. G. (2003). "Wireless Sensors Will Monitor Bridge Decks," Better Roads Magazine, February 2003.

Watters, D. G., Jayaweera, P., Bahr, A. J., and Huestis, D. L. (2001). "Design and Performance of Wireless Sensors for Structural Health Monitoring," Review of Progress in Quantitative Nondestructive Evaluation, Brunswick, ME, 21A, 969–976.

Yanev, B. (1996). "Optimal Management and Rehabilitation Strategy for the Bridges of New York City." *Proceedings of the International Conference on Retrofitting of Structures*, March 11-13, 1996, Columbia University, NY, pp. 311-327.

Yanev, B. (1997). "Life-Cycle Performance of Bridge Components in New York City." *Recent Advances in Bridge Engineering, Zurich*, July 1997, pp. 385-392.

Yanev, B. (1998). "Bridge Management for New York City." *Structural Engineering International*. 1998/03. 8(4), 211–215. International Association of Bridge and Structural Engineering (IABSE), Zurich, Switzerland.

10 Life Cycle Analysis and Infrastructure Health

10.1 INTRODUCTION

10.1.1 OVERVIEW

Life cycle analysis (LCA) is an essential step in any long-term planning or decision making for infrastructures, both existing and new. Without an accurate LCA, it is difficult to prioritize projects over a long term, impossible to integrate varying infrastructures with different estimated lifespan, or make an accurate long-term economic projection on a project at the local, regional, or national level. The LCA is perhaps one of the few logical ways to relate different infrastructure sectors; thus, it is important to promote different methodologies and technologies that help in accurate evaluation of LCA. In developing an accurate LCA, discount rates as well as potential uncertainties need to be included.

Other technologies that can benefit and contribute to LCA are enhancement of databases that contain a vast amount of information about infrastructures (both new and existing), IT technologies, monitoring and sensing, and well-understood materials behavior (both conventional and advanced). Decision making tools and techniques such as risk (assessment, treatment, and/or communication) need to be an integral part of LCA. We should note that infrastructure resiliency is a contributor to LCA and should be considered as such. Other important factors that contribute to LCA and hence can benefit from a deeper understanding of LCA are

1. Durability issues: such as designs for infrastructure/material durability
2. Multihazards considerations (man-made and or natural)
3. Environmental effects; sustainability issues
4. Energy use/conservation effects
5. Potential long-term climate changes: this is a specially important subject for infrastructures with long lifespan such as bridges and/or tunnels
6. Balancing funds between short-term and long-term needs

One of the major emerging roles of LCA is its utilization by the performance-based design (PBD) paradigm. Ettouney and Alampalli (2012, Chapter 3) explained that PBD can quantify the consequences of a design. If we use monetary metrics to define consequences, and integrate the consequences over all possible hazards for a particular design, then we can easily show that the computed PBD consequence is simply the life cycle cost (LCC) of the particular design.

Given the emerging importance of LCA, we note that it does come with a price. For an accurate LCA, we need

1. High-quality information about the structure and its environment
2. Accurate life cycle analytical basis for the structural system, hazard demands, future trends, and past lessons

In the rest of this section, we explore some of the previous research works in the field. We then introduce a new outlook on LCA.

10.1.2 LITERATURE REVIEW

The dependence of life cycle estimate of structures on accurate information was recognized by Peil and Mehdianpour (2000). They observed that conventional models in estimating life cycle estimate hazards or design loads, use an objective system transfer model that can estimate the system response to hazards, and establish an acceptance model (through knowledge of acceptable damage), estimate conditions of the system and estimate expected life cycle of the model. The authors opined that this conventional model is not accurate, especially in evaluating the system response or the damage steps. They suggested increasing the accuracy of life cycle estimation by relying on structural health monitoring (SHM), particularly by measuring strains at critical locations. They suggested that from the strain measurements an accurate estimation of fatigue-accumulated damage can be computed. Then the life cycle of the system can be predicted.

El-Diraby and Rasic (2002) devoted their work on LCC to the use of smart material. They presented theoretical arguments on LCC and the different categorizations of the aggregates of LCC. Oshima et al. (2004) related (damage identification DMID) to life cycle management of bridges. Yagi et al. (2004) showed use of the reaction force method by measuring in real time the axle weight of trucks, truck speeds, and truck types. The information can then be fed into an algorithm to estimate LCCs. Yoneda and Edamoto (2004) presented a method to estimate the gross weight of trucks passing on a bridge. Again, such information can be fed into an algorithm to estimate LCCs.

10.1.3 NEW PHILOSOPHY OF LCA

Traditional LCA concentrated heavily on estimating life cycle cost analysis (LCCA). In many situations, LCA was usually intertwined with LCCA. However, we observe that such a view is limited and can lead ultimately to inaccurate decisions. We submit that there are other important aspects to LCA. LCA contains three essential components: costs, benefits, and lifespan. All the three components need to be accurately estimated for LCA evaluation, as shown in Figure 10.1. They are explained below:

Costs: These can include costs of management, labor, retrofits, initial construction, operations, decommissioning, and so on. Traditionally, estimating costs is the easiest component of LCA.

Benefits: These include cost savings from different activities (lifespan extension, safer and secure environment, etc.), direct benefits and revenue, and indirect social, economic, and political benefits. Estimating benefits is much more difficult than estimating costs; yet, it is essential to accurately estimate benefits to ensure a proper decision making process.

Lifespan: Estimating the expected lifespan of an infrastructure, or a project relating to an infrastructure, is as essential as estimating costs or benefits. Yet, it is the least understood or studied component of many infrastructures. Currently, lifespan estimation is mostly qualitative and usually does not have a solid technological basis.

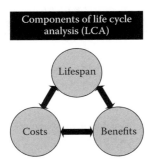

FIGURE 10.1 Components of LCA.

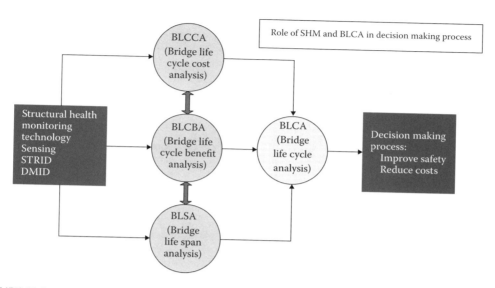

FIGURE 10.2   Role of SHM in BLCA and decision making.

Since accuracy of LCA depends mainly on the quality of information and the assumptions made, it is clear that SHM can play an essential role in all of LCA components (Figure 10.2). We have seen throughout this book, the important intersections between LCA and SHM components and fields. This chapter aims at providing a central view and some quantitative tools that can be of help to decision makers in all aspects of infrastructure health. We use bridge life cycle analysis (BLCA) as basis for discussion; however, most of the concepts are applicable to other types of civil infrastructures.

10.2   BRIDGE LIFE CYCLE COST ANALYSIS

10.2.1   Overview

Bridge life cycle cost analysis (BLCCA) is needed as a part of bridge management operations. It can be applied to the cost of specific projects or to the overall cost of ownership. The units of cost are usually monetary units. In many situations, costs can be nonmonetary, such as social or environmental. In those situations, identifying costs must be made by using qualitative estimates. Another technique is to use a lesser qualitative approach such as the utility theory approach (Ettouney and Alampalli 2012, Chapter 8). We also note that since estimating costs depends on future events, good estimates of future discount and inflation rates are needed. We introduce here some basic components of costs that occur during the lifespan of a bridge. For more comprehensive bridge cost presentation, see Hawk (2003), Chapman (2003), and Chapman and Leng (2004).

10.2.2   Components of Costs

Bridge life cycle costs, BLCC, can be defined as

$$BLCC = BLCC|_A + BLCC|_U + BLCC|_H \tag{10.1}$$

with $BLCC|_A$, $BLCC|_U$, and $BLCC|_H$ as the agency costs, user costs, and hazards costs, respectively. The basic relationship of Equation 10.1 is shown in Figure 10.3. The components of $BLCC|_A$,

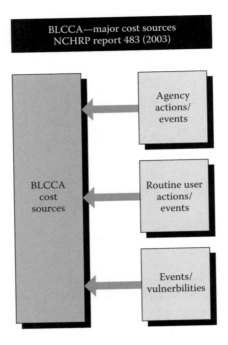

FIGURE 10.3 Major sources of cost.

$BLCC|_U$, and $BLCC|_H$ are shown in Figures 10.4, 10.5, and 10.6, respectively. A general model for evaluating components of Equation 10.1 is

$$BLCC|_A = \sum_{j=1}^{N_{COMPONENT}} \sum_{i=i_{START}}^{N_{SPAN}} C_{ij} \left(1 + I_{ij}\right)^i \qquad (10.2)$$

Similar expressions can be used for $BLCC|_U$, and $BLCC|_H$.

In Equation 10.2, we define

$N_{COMPONENT}$ = Number of cost components
N_{SPAN} = Lifespan of the bridge (or the project)
i_{START} = Year that component j will become functional
C_{ij} = Cost of component j during year i
I_{ij} = Estimate of discount rate pertaining to component j during year i

Use of Equations 10.1 and 10.2 forms the basic skeleton of BLCCA as used by most decision makers. We offer the following additional points in this regard.

Interdependencies of parameters: Note that Equation 10.2 is presented as a simple sum. This implies that the estimate of costs C_{ij} are independent. Such an assumption is not accurate in many situations. For example, the cost of an earthquake, if it occurs, might reduce the cost of overload or collision. This is because by retrofitting against earthquake damage, the effects of overload or collision might be eliminated or reduced. The decision maker is advised to study cost components carefully and decide on how to modify Equations 10.1 and/or 10.2 to accommodate any possible interdependency. An example of analysis of interdependencies of future events is given in Ettouney and Alampalli (2012, Chapter 8).

Cost of hazards: General form of costs of any hazard can be expressed as

$$C_H = \int_S H(S) C_D(S) dS \qquad (10.3)$$

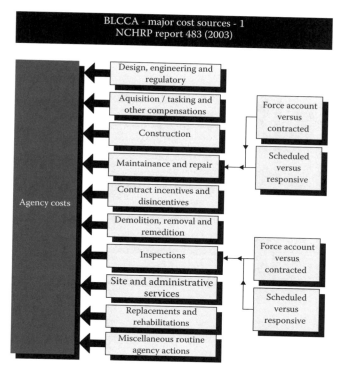

FIGURE 10.4   Details of major sources of agency costs.

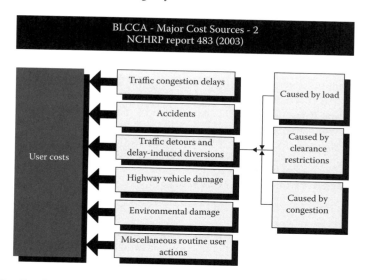

FIGURE 10.5   Details of major sources of user costs.

where

S = Hazard space
$H(S)$ = Hazard level
$C_D(S)$ = Cost of $H(S)$

Details of costs of several specific hazards (scour, earthquake, fatigue, and corrosion) are given in Chapters 1 through 5.

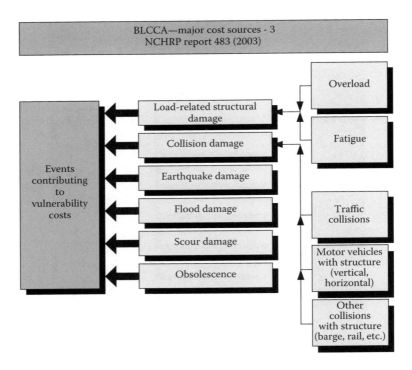

FIGURE 10.6   Details of major sources of vulnerability costs.

Uncertainties of parameters: Note that most components in Equations 10.1 through 10.3 are uncertain. Thus, BLCC should be treated as an uncertain (random) entity. Any variance within any component in BLCC should be reflected in the decision making process. Ignoring uncertainties can lead to wrong decisions based on accurate estimation of BLCC. Chapter 8 by Ettouney and Alampalli (2012) contains several examples that explore this point.

Use of SHM for increased accuracy: One of the main drawbacks of using BLCCA as a tool of decision making is that it requires accurate estimation of many unknown components. We argue that the use of SHM can help in reducing uncertainties and improving accurate estimates of BLCC components. Several modes of SHM-BLCCA interrelationships are discussed in Section 10.7.

10.3   BRIDGE LIFE CYCLE BENEFIT ANALYSIS

10.3.1   Overview

Bridge life cycle benefit analysis (BLCBA) is the second part of bridge BLCA management operations. It can be applied to the benefits of specific projects or to the overall benefits of ownership. The units of benefit are not as easily defined in monetary terms as the cost counterpart. In some situations, benefits can be defined as cost savings. In many other situations, benefits are nonmonetary, such as social or environmental. In those situations, identifying benefits must be made by using qualitative estimates. Another technique is to use a lesser qualitative approach such as the utility theory approach (Ettouney and Alampalli 2012, Chapter 8). As in cost estimates, since estimating costs depends on future events, good estimates of future discount and inflation rates are needed. We introduce here some basic components of benefits that occur during the lifespan of a bridge. Unfortunately, there are not many reports or studies devoted to benefits. In most situations, however, a benefit can be regarded as the opposite of cost. As such, cost-only references, such as Hawk (2003), Chapman (2003), and Chapman and Leng (2004), can be used to explore BLCBA.

10.3.2 COMPONENTS OF BENEFITS

Since there is no widely used formal definition of bridge life cycle benefits BLCB, we define it as

$$BLCB = BLCB|_C + BLCB|_S \tag{10.4}$$

With $BLCB|_C$ and $BLCB|_S$ as the benefits resulting from cost savings and increased safety, respectively. Note that this definition of BLCB is general and consistent with accepted universal bridge management objectives. The basic relationship of Equation 10.4 is shown in Figure 10.7. The components of $BLCB|_C$ and $BLCB|_S$ are shown in Figure 10.8. The models for evaluating components of Equation 10.4 are

$$BLCB|_C = \sum_{j=1}^{N_{COMPONENT}} \sum_{i=i_{START}}^{N_{SPAN}} \left(\overline{C}_{ij} + \overline{B}_{ij}\right)\left(1 + I_{ij}\right)^i \tag{10.5}$$

$$BLCB|_S = \sum_{j=1}^{N_{COMPONENT}} \sum_{i=i_{START}}^{N_{SPAN}} B_{ij}\left(1 + I_{ij}\right)^i \tag{10.6}$$

In Equations 10.2 and 10.6, we define

$N_{COMPONENT}$ = Number of benefit components
N_{SPAN} = Lifespan of the bridge (or the project)
i_{START} = Year that component j will become functional
$\overline{C}_{ij}$ = Cost savings for component j during year i
$\overline{B}_{ij}$ = Indirect benefits for cost savings for component j during year i
B_{ij} = Safety benefits for component j during year i
I_{ij} = Estimate of discount rate pertaining to component j during year i

Equations 10.4 through 10.6 form the proposed skeleton of BLCBA. They can be used in several decision making situations. We offer the following additional points in this regard.

Interdependencies of parameters: Note that Equations 10.2 and 10.6 are presented as simple sums. This implies that the estimate of benefits $(\overline{C}_{ij}, \overline{B}_{ij}, B_{ij})$ are independent. As in cost situations, such an assumption is not accurate in many situations. For example, the benefit of an earthquake retrofit project (performance improvement) will also produce the benefits of reducing costs of overload or collision hazards. This is because by retrofitting against earthquake damage, the effects of overload or collision might be eliminated or reduced. In addition, the potential for progressive collapse failure of the bridge will be reduced. The decision maker is advised to study the benefit

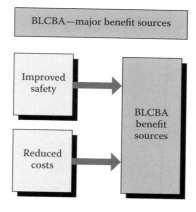

FIGURE 10.7 Benefit categories for infrastructures (bridges).

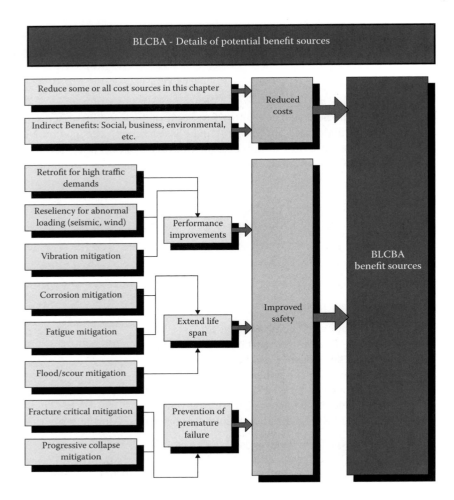

FIGURE 10.8 Details of benefit sources.

components carefully and decide on how to modify Equations 10.4 through 10.6 to accommodate any possible benefit interdependency.

Cost reduction of hazards retrofits: General form of cost reduction of any hazard retrofits can be expressed as

$$\bar{C}_H = \int_S H(S)(C_D(S) - \bar{C}_D(S))\,dS \tag{10.7}$$

where
S = Hazard space
$H(S)$ = Hazard level
$C_D(S)$ = Cost of $H(S)$, without any retrofit
$\bar{C}_D(S)$ = Cost of $H(S)$, including retrofit effects

Note that in Equation 10.7

$$C_D(S) > \bar{C}_D(S) \geq 0 \tag{10.8}$$

The limits of Equation 10.8 can be exceeded in certain multihazard conditions: when retrofitting for a hazard might have a negative effect on the response of the structure to another hazard (see

Ettouney, Alampalli, and Agrawal 2005). Details of costs reductions of several specific hazards (scour, earthquake, fatigue, and corrosion) are given in Chapters 1 through 5.

Uncertainties of parameters: Similar to costs, we note that most components in Equations 10.4 through 10.7 are uncertain. Thus, BLCB should be treated as an uncertain (random) entity. Any variance within any component in BLCB should be reflected in the decision making process.

Use of SHM for increased accuracy: BLCBA requires accurate estimation of many unknown or uncertain components. Similar to cost estimations, we argue that the use of SHM can help in reducing uncertainties and improving the accuracy of estimates of BLCB components. Several modes of SHM-BLCBA interrelationships are discussed in Section 10.7.

10.4 BRIDGE LIFESPAN ANALYSIS

10.4.1 OVERVIEW

Estimating the lifespan of a bridge is definitely an important task but not an easy one. It is important since it is an integral part of almost all decisions that bridge owners make. The main reasons for the importance of having a good estimate of lifespan are (1) cost (reduction, or increase) implications, and (2) its safety implications. Actually, it is in lifespan estimate where both cost and safety issues intersect and become inseparable. Consider, for example, the deterioration rate of a bridge, which is a basic parameter in estimating bridge lifespan. Overestimating the deterioration rate can lead to an underestimation of lifespan, as shown in Figure 10.9. Such an overestimation can lead to decisions that might lead to financial loss. On the other hand, underestimating deterioration rate (Figure 10.10) might lead to highly undesirable, unsafe decisions.

Estimating lifespan involves several issues that are difficult to quantify. The very topic of estimating lifespan in itself is an attempt to find a quantitative solution to a problem that is highly qualitative, or, at the very least, a problem that includes numerous qualitative as well as quantitative parameters. Some of these difficulties are

- Bridges are composed of numerous components. It is far easier to estimate the lifespan of components than the lifespan of component assemblage (bridge).
- Difficulty in quantifying the metrics that govern the decision of the exact point of life termination of the bridge or its components.
- On a component level, it is not easy to extrapolate damage information to lifespan information.
- The highly uncertain parameters of the whole process.

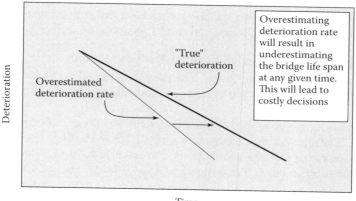

FIGURE 10.9 Effects of overestimating deterioration rates.

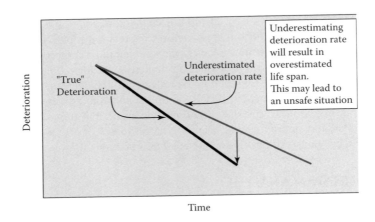

FIGURE 10.10   Effects of underestimating deterioration rates.

In the rest of this section we explore some definitions of lifespan. We then discuss some of the methods that can be used to estimate lifespan.

10.4.2   Lifespan, Remaining Life, and Useful Life

We use the term "lifespan" in this chapter to identify the time used for computing LCC and life cycle benefits. As such, we are leaving the decision of termination to the decision maker. This is done in recognition of the fact that such a termination point is highly subjective, with the possible exception of a situation of bridge failure. We intentionally do not use one of the two more popular terms in this field: "remaining life" or "useful life."

Remaining life: This term is usually used when professionals study fatigue (see Alampalli and Lund 2006; also, see Chapter 5). As such, it indicates the remaining life to fatigue failure of the bridge or components. Obviously, remaining life, following this definition, is much longer than the time that should be used in LCCs/benefits.

Useful life: This definition is fairly similar to our concept of lifespan. Note that the termination points in situations when the useful life concept is used vary a great deal. To keep our definition focused narrowly on BLCCA and BLCBA, we chose a different expression of lifespan to differentiate it from the useful life expression.

10.4.3   Methods for Evaluating Bridge Lifespan Analysis

Figure 10.11 shows some methods that can help the decision maker in evaluating lifespan. We subdivide the methods into general/qualitative and specific/quantitative methods, as follows:

10.4.3.1   General/Qualitative Bridge Lifespan Analysis Methods

Qualitative methods include National Bridge Inventory (NBI) trend methods. The NBI methods in estimating the useful life of bridges are generic in nature and are based on statistical observations of several bridges.

Deterioration rates have been used to estimate the useful life of bridges (see Agrawal, Kawaguchi, and Qian 2009). The method uses qualitative metrics for evaluating deterioration of bridges. Statistical observations and regression are then used to estimate the useful life of the bridge.

Another method that can be used to estimate lifespan is bridge ratings. Utilizing the ratings of a specific bridge and the way it varies as time progresses might give a qualitative indication of its lifespan. Remember that our definition of lifespan is not the same as useful or remaining life; we intend "lifespan" to be used exclusively during the estimation of BLCCA/BLCBA. As such, it is possible to use bridge ratings to estimate lifespan.

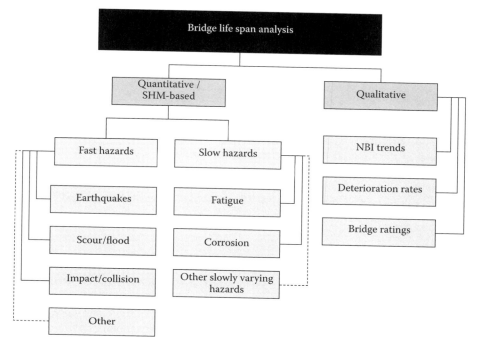

FIGURE 10.11 BLSA methods.

10.4.3.2 Specific/Quantitative Bridge Lifespan Analysis Methods

Specific/quantitative methods for estimating bridge lifespan analysis (BLSA) can be further subdivided into fast and slow events. Fast events include earthquakes, scour, flood, impact, collision, and so on. Slow events include fatigue, corrosion, etc. For both types of events, a general framework for estimating lifespan can be as follows:

1. For a given structural state, estimate a reasonable level of hazard at which the structural lifespan comes to an end (termination point). Note that such a point does not necessarily mean failure; it can mean an unacceptable damage state.
2. Use available information about the hazard of interest to estimate the time T_{SPAN} at which the hazard level at #1, above, will occur, with a reasonable confidence level.
3. Time T_{SPAN} is the required lifespan.

10.5 INTERRELATIONSHIP OF BLCCA, BLCBA, AND BLSA

10.5.1 Overview

One of the major weaknesses in the current state of computing BLCCA is the presumed independence of sources of costs during the lifespan of the bridge. As such, it is an accepted practice to assume that a cost due to a cost source is independent from the cost due to a cost source. Such a basic assumption is clearly not accurate. For example, a cost of seismic retrofit would be affected by cost of scour (foundation retrofit) since changes in foundation detailing might affect seismic behavior of the bridge.

Accommodating such interrelationships in BLCCA (BLCBA or BLSA) is not an easy task. However, ignoring the interrelationship would result in an inaccurate result for the LCA on hand.

This section explores several aspects of these interrelationships between life cycle sources of bridge costs, bridge benefits, and bridge lifespan.

10.5.2 INTERRELATIONSHIP OF COSTS: MULTIHAZARD CONSIDERATIONS

Following the general concepts of Hawk (2003), it is possible to express the cost of the jth hazard per T unit of time, C_j as

$$C_j = \int h_j(x)c_{aj}(x)dx \qquad (10.9)$$

where $h_j(x)$ is the probability of the occurrence of jth hazard of intensity x during time period T. The cost of such an occurrence is $c_{aj}(x)$.

We note that at the limit, if we are interested in computing the cost of failure of a particular system, Equation 10.9 would need the computations of that system reliability (Melchers 2002; Bazovsky 2004).

If there are multihazard situations, then the total cost of all hazards for a period T, C_T is

$$C_T = \sum_{j=1}^{j=NH} C_j \qquad (10.10)$$

Where NH is the number of hazards under consideration. Equations 10.9 and 10.10 presume, for the sake of simplicity, that occurrences of hazards are independent of each other, which is a common assumption. This is a fairly reasonable assumption and implies that the occurrence of earthquake at a given site for a given time period T is independent of a flood or a bomb blast during the same period. A major implication of Equations 10.9 and 10.10 is that costs of different hazards are not interrelated. We need to examine such an implication more carefully. When some hazards coincide, such as blizzards that combine wind, snow, and/or ice hazards, Equation 10.10 should be modified to include this combination; however, such a situation is beyond the scope of this work.

The aforementioned theory of multihazard states that for any given system there is an inherent resilience to any type of hazard. The qualitative interrelationship of hazards was discussed in Chapter 1 by Ettouney and Alampalli (2012). This means that there is an interrelationship between earthquake hazard resiliency and flood hazard resiliency, for example. This interrelationship is a function of the intensity of the hazard. For example, supplying seismic restraints might not increase medium flood costs, but it can definitely have a major effect in situations when the floods are extreme—preventing the bridge span from dislodging. This interrelated resiliency between different hazards at different levels can be directly expressed in hazard costs. On the basis of this, we modify Equation 10.10 such that

$$C_j = \sum_{k=1}^{k=NH} \int h_k(x)c_{ajk}(x)dx \qquad (10.11)$$

Factor $c_{ajk}(x)$ is the cost of hazard $h_k(x)$ at intensity x given hazard j.

Let us express Equation 10.11 in its discrete form

$$C_j = \sum_{k=1}^{k=NH} \sum_{i=1}^{i=NS} H_{ik} C_{aikj} \qquad (10.12)$$

The hazard intensity space x is subdivided into NS subdivisions. The probability of ith hazard intensity is thus H_{ik}, while the corresponding cost is C_{aikj}.

Equation 10.12, in combination with Equation 10.9, can be used to assess the total cost of multi-hazards for any system. The interaction between the hazards is accounted for by the terms $k \neq j$. Note that the sign of C_{aikj} controls how different hazards at different intensities can affect the total cost. If the sign of C_{aikj} is positive, then the jth and kth hazards have conflicting demands on the system, thus increasing the total cost. If C_{aikj} is negative, then mitigating the j and k hazards, which have consistent demands, can end up reducing the total cost.

A final observation about Equations 10.10 through 10.12 is that all the costs in these equations are assumed to take place during a unit period of time T. The equations can easily be generalized to total LCC equations by introducing a rate of return for T and performing a present value analysis. The reader is referred to Hawk (2003) for such generalized LCC analysis.

10.5.3 Interrelationship of Benefits: Multihazard Considerations

Similar considerations should be extended when evaluating BLCBA. Interactions between benefits from different components can be quantified using equations similar to Equation 10.12. Again, the interaction terms of benefits can have either positive or negative signs, depending on whether the components enhance or conflict with each other's benefits.

10.5.4 Interrelationship of Costs, Benefits, and Lifespan

The interrelationship of costs, benefits, and lifespan should be considered when performing BLCA. The cost–benefit interrelationship is partially accommodated in Equation 10.9, where cost reductions are considered benefits. However, additional interactions must be accommodated. For example, rehabilitation measures would be included in BLCCA. Their effects on BLSA should also be considered. Since performing rehabilitation will have an effect on extending bridge lifespan, thus have an effect on BLCB.

10.5.5 Closing Remarks

The interrelationships discussed in this section are difficult to account for. On the other hand, it is clear that these interrelationships need to be accommodated; otherwise, the results might not be accurate. A simple way to account for the complex nature of interrelationships is by using a Monte Carlo simulation technique (see Ettouney and Alampalli 2012). Within Monte Carlo, complex models of relationships can be built, thus avoiding the confines of the simple format of Equations 10.9 through 10.12. Chapter 8 by Ettouney and Alampalli (2012) gave an example of using the Monte Carlo technique with interdependencies within the model. Another advantage of the Monte Carlo technique is that it would produce probabilistic results for BLCC, BLCB, and BLS automatically, thus making the decision making process more realistic and accurate.

10.6 USE OF BLCA IN DECISION MAKING

10.6.1 General

The main purpose of any BLCA effort is to compute the cost BLCC and benefit BLCB over the lifespan, T_{SPAN} of the asset. We need to ask ourselves: Why are these important to owners? We can offer two reasons:

- When embarking on any effort (maintenance, repair, retrofit, etc.) on an existing construction, it is essential to compare the cost of the effort C_e with the benefit of the effort B_e over the length of time T_{SPAN}. Again, based on the resulting value, $V_e = B_e/C_e$, a reasonable decision on the new construction can be made. We call this issue *cost–lifespan* issue.
- When embarking on a new construction, it is essential to compare C with the benefit B over the length of time T_{SPAN}. Based on the resulting value, $V = B/C$, a reasonable decision on the new construction can be made. We call this issue *cost–benefit* issue.

It is clear that BLCA can be a valuable decision making tool in the structural health in civil engineering (SHCE) field. Both uses of BLCA will be explored in more detail next.

10.6.2 Cost–Lifespan of BLCA

There are three ways to compute the cost of an effort C_e. The most obvious and least desired method is to assume that

$$C_e = C_0 \tag{10.13}$$

With C_0 is the initial cost of the effort. The limitation of this approach is that it does not account for the discount rate over time T_{SPAN}. An improved estimate would be

$$C_e = PV(C_0, T, I) \tag{10.14}$$

Where $PV()$ is the present value of C_0 discounted of T_{SPAN} with a rate of I. See Ettouney and Alampalli (2012, Chapter 8) for details of present value computations.

There are two limitations to this approach:

- It does not permit quantifiable computations of the benefit B_e of the effort. Thus the value V_e of the effort can only be estimated qualitatively, not quantitatively.
- It does not account for the effects of the effort on other costs (current and future) that might be needed for the structure on hand.

Accommodating the second limitation is straightforward:

$$C_e = C_2 - C_1 \tag{10.15}$$

Note that C_2 and C_1 are the BLCCA of the bridge with and without the effort under consideration. Obviously, direct cost C_0 is included in the computations of C_2. It is not clear if $C_2 > C_1$, since the effects of spending C_0 might actually result in an overall reduction of BLCCA over time span T_{SPAN}. We can then expand Equation 10.15 into three scenarios as follows:

$$
\begin{aligned}
&\text{For } C_2 > C_1, \quad C_e = C_2 - C_1 \\
&\text{For } C_2 = C_1, \quad C_e = 0 \\
&\text{For } C_2 < C_1, \quad B_{direct} = C_1 - C_2
\end{aligned}
\tag{10.16}
$$

The direct benefit B_{direct} of the effort is self-evident.

There is another benefit of an effort that will result from expected changes in the lifespan of the project that needs to be quantified. It is reasonable to expect that the effort under consideration will have an effect on the lifespan T_{SPAN}. Assume that the lifespan is T_1 if the effort under consideration is not performed. If the effort is performed, then the lifespan will change to T_2. It is also expected that $T_2 > T_1$. We now have two pairs of cost–lifespan scenarios. The do-nothing pair, C_1 and T_1, and the performed-effort pair, C_2 and T_2, need to be compared.

A simple way to compare the two scenarios is to compare the rate of cost for each scenario, δ_{CT}. We identify it as

$$\delta_{CT} = \frac{C}{T_{SPAN}} \tag{10.17}$$

The rate of cost is a measure of the total cost per time unit. Higher δ_{CT} indicates a higher cost of ownership; it is considered a nondesirable solution. The implicit assumption is a constant cost rate during the lifespan of the bridge. A more realistic solution that allows for an in-depth analysis of the time-dependent rate of cost is possible, but it will not be discussed in this chapter.

We can now compare the two scenarios in a quantitative manner. First, we compute the rate of cost for each of the scenarios as

$$\delta_{CT}\big|_1 = \frac{C_1}{T_1} \qquad (10.18)$$

$$\delta_{CT}\big|_2 = \frac{C_2}{T_2} \qquad (10.19)$$

We then compute the benefit-cost, or value, V_e, of the increased lifespan as

$$V_e = \frac{\delta_{CT}\big|_1}{\delta_{CT}\big|_2} = \frac{C_1}{T_1} \cdot \frac{T_2}{C_2} \qquad (10.20)$$

In Equation 10.20, we considered only the rate of cost δ_{CT} in estimating V_e. We observe that there are benefits that are gained while the bridge is in service. We can identify a rate of benefit δ_{BT} as

$$\delta_{BT} = \frac{B}{T_{SPAN}} \qquad (10.21)$$

Where B is the BLCB for lifespan T_{SPAN}.

Similar to the cost developments, the rate of benefit is a measure of total benefit per time unit. Higher δ_{BT} indicates a higher benefit from the bridge of ownership; it is considered a desirable solution. Again, the implicit assumption is a constant benefit rate during the lifespan of the bridge. A more realistic solution that allows for a time-dependent rate of benefit is possible, but it will not be discussed in this chapter.

We can now compute the rate of benefit for each of the scenarios as

$$\delta_{BT}\big|_1 = \frac{B_1}{T_1} \qquad (10.22)$$

$$\delta_{BT}\big|_2 = \frac{B_2}{T_2} \qquad (10.23)$$

We can now generalize our estimate of V_e as

$$V_e = \frac{\delta_{CT}\big|_1}{\delta_{CT}\big|_2} \cdot \frac{\delta_{VT}\big|_2}{\delta_{VT}\big|_1} = \frac{C_1}{B_1} \cdot \frac{B_2}{C_2} \qquad (10.24)$$

If $V_e > 1.0$ then the effort on hand has a better value than the do-nothing scenario. If $V_e < 1.0$, then it is less costly to do nothing. If $V_e \cong 1.0$ then more detailed considerations need to be investigated. These considerations include the many assumptions used in estimating the four basic parameters C_1, T_1, C_2, and T_2.

10.6.3 COST–BENEFIT OF BLCA

Another use of BLCCA is for evaluating the economic merits of new constructions. Assume that for a desired new bridge lifespan T_{SPAN} the overall cost and benefit of a planned bridge are C and B, respectively. The value of the new construction can be estimated as

$$V = \frac{B}{C}$$ (10.25)

The decision making process is then simple:

1. If $V > 1.0$ then proceed with the new construction.
2. if $V < 1.0$ then the construction is not advisable.
3. if $V \cong 1.0$ then more detailed considerations need to be investigated. These considerations include the many assumptions used in estimating both C and B.

For example, if the cost of a new project over a desired lifespan of 50 years is estimated at $30 million and its estimated benefits are $45 million, the value of such a project is $V = 1.5$. Clearly, it is a great value, and the project should proceed.

Note that V is a function of the desired lifespan T_{SPAN}. This makes it possible to compare two new construction scenarios with two different lifespans, T_1 and T_2. The comparison follows similar steps as before. The relative value, V_R of the two scenarios can be shown to be

$$V_R = \frac{V_1}{V_2} = \frac{C_2}{B_2} \cdot \frac{B_1}{C_1}$$ (10.26)

The cost, benefit, and value of the first construction scenario are C_1, B_1, and V_1, respectively. For the second scenario, the cost, benefit, and value are C_2, B_2, and V_2, respectively.

Suppose there is another scenario for the above project with a longer lifespan of 60 years. The cost of the project will be increased to $35 million, while its benefits will increase to $55 million. Using Equation 10.26 shows that the relative value of the two scenarios is 0.95. This indicates that the second scenario offers a better value than the first.

10.7   SHM ROLE IN BLCCA, BLCBA, AND BLSA

10.7.1   NEED FOR ACCURACY

Accurate input is needed to produce accurate BLCA. SHM techniques can help in producing more accuracy in needed inputs. In this section we explore some specific methods that SHM techniques can aid in producing accurate information for BLCA.

10.7.2   SHM AND BLCCA

Tables 10.1 through 10.3 show specific situations where SHM techniques might produce accurate estimates of bridge LCC. Tables 10.1 through 10.3 also show the relative importance of SHM components in each situation.

10.7.3   SHM AND BLCBA

Accurate estimating of benefits can be difficult. This is perhaps one area where SHM can help to produce reasonable results. Table 10.4 shows specific situations where SHM techniques might produce accurate estimates of bridge benefits. Table 10.4 also shows the relative importance of SHM components in each situation.

10.7.4   SHM AND BLSA

SHM techniques can play major role in accurate estimation of BLSA. Already, SHM has been used to estimate fatigue remaining life (Lund and Alampalli 2004, for example). Also, corrosion

TABLE 10.1
SHM Techniques for Better Cost Estimation of Bridge Agency Costs

Routine User Action	Comments	Importance of SHM Component		
		Sensing	STRID	DMID
Design, engineering, and regulatory	STRID will aid in producing more efficient designs	M	M	L
	Sensing and damage identification can aid in accurate estimation of rehabilitation costs			
Construction	Construction costs can be accurately estimated using decision making techniques	H	H	M
	Accurate sensing and STRID can result in better construction sequencing			
Maintenance and repair	By accurately quantifying and predicting damage (using sensing, STRID, and DMID) accurate maintenance and repair costs can result	H	M	H
Contract incentives and disincentives	The metrics incentives/disincentives can be better measured using STRID and sensing	L	M	L
Demolition, removal, and remediation	By sensing deterioration and other damage, accurate scenarios for demolition processes will result. This will produce accurate cost estimates	L	M	L
Inspection	Using different SHM/SHCE techniques can help focus the inspection process, producing more accurate estimates of inspection manpower and costs	H	M	M
Site and administrative services	Using different SHM/SHCE techniques can help focus the site administration services, producing more accurate estimate of manpower and costs	L	M	M
Replacements and rehabilitations	**Replacement:** By accurately quantifying and predicting damage (using sensing, STRID, and DMID) accurate rehabilitation costs can result	M	M	M
	Rehabilitation: STRID will aid in producing more efficient designs			
	Sensing and damage identification can aid in accurate estimations of rehabilitation costs			

Note: Only Sensing, STRID, DMID are explicitly included since decision making is the main focus of the table. H, High; M, Medium; l, Low; NA, Not applicable.

TABLE 10.2
SHM Techniques for Accurate Cost Estimation of User Costs

Routine User Events	Comments	Importance of SHM Component		
		Sensing	STRID	DMID
Traffic congestion delays	Monitoring traffic demands can produce accurate cost estimation	H	L	L
Accidents		H	L	L
Traffic detours and delays		M	L	L
Highway vehicle damage		L	M	M
Environmental damage		M	M	M
Miscellaneous routine user action		L	L	L

Note: Only Sensing, STRID, DMID are explicitly included since decision making is the main focus of the table. H, High; M, Medium; L, Low; NA, Not applicable.

TABLE 10.3
SHM Techniques for Accurate Cost Estimation of Vulnerability Costs

Other Events	Specific Events	Comments	Importance of SHM Component		
			Sensing	STRID	DMID
Load-related structural damage	Overload	STRID, DMID, and sensing can produce accurate estimations of rehabilitation costs	M	M	H
	Fatigue	STRID, DMID, and sensing can produce accurate estimations of rehabilitation costs	H	M	H
		Also, continues fatigue monitoring can result in accurate estimation of probabilities of occurrences			
Collision damage	Traffic collisions	Monitoring traffic demands can produce accurate cost estimation using different decision making techniques	L	L	L
	Motor vehicles with structure	Pre-, during-, and postevent sensing, STRID, DMID can aid in accurate cost estimation	M	M	H
	Other collision with structure (barge, rail, etc.)		M	M	H
Earthquake damage			H	H	H
Flood damage			M	M	M
Scour damage			M	M	M

Note: Only Sensing, STRID, DMID are explicitly included since decision making is the main focus of the table.
H, High; M, Medium; L, Low; NA, Not applicable.

rate measurements have been used to estimate lifespan (Bentur, Diamond, Berke 1997). In general, SHM as an aid for an accurate BLSA follows the following methods:

1. **Measure a trend:** a varying metric as a function of time. For example, strain measurements for fatigue remaining life or corrosion rates for corrosion remaining life. Measuring deterioration rates (Agrawal, Kawaguchi, Qian 2009) can also produce trends. For fast hazards, the process is a bit less direct. However, measuring trends of river bottom level can aid in estimating scour threat. Similarly, measuring traffic statistics can produce accurate trends for estimating effects of collisions on lifespan.
2. **Assess structural capacity:** This is done using structural identification (STRID) tools. Since estimating lifespan is directly related to as-built structural capacity, different STRID tools can be used in conjunction with field measurements to assess structural capacities.
3. **Assess damage effects:** This is done using DMID techniques. By identifying damage, location, and extent, it is possible to estimate the effects of such damage on the lifespan of the structure.

In all of the above methods, decision making processes of Ettouney and Alampalli (2012, Chapter 8) can be used in combination with SHM methods.

Table 10.5 shows specific situations where SHM techniques might produce accurate estimates of bridge lifespan. Table 10.5 also shows the relative importance of SHM components in each situation.

TABLE 10.4
SHM Techniques for Accurate Benefit Estimation

| Management Objective | Potential Benefit Sources | | SHM Technique for Better Benefit Estimation |
	Category	Source	
Reduce costs	Direct cost reduction	Indirect benefits	Accurate bridge rating that might reduce costs by producing higher ratings
Improve safety	Performance improvements	Retrofit for high traffic demands Resiliency for abnormal loadings	Measure volume and trends in traffic conditions
		Vibration mitigation	Vibration monitoring. Modal identification. See Chapter 6 by Ettouney and Alampalli (2012)
	Extend lifespan	Corrosion mitigation Fatigue mitigation Flood/Scour	See Chapters 3 and 4 See Chapter 5 See Chapter 1
	Prevent premature failure	Fracture critical Progressive collapse mitigation	Estimate propensity for brittle failure and loss of stability

Note: Only Sensing, STRID, DMID are explicitly included since decision making is the main focus of the table.

TABLE 10.5
SHM Techniques for Accurate Estimation of Bridge Lifespan

Factors Affecting Lifespan		SHM Technique for Better Lifespan Estimation
Qualitative	Deterioration rates	Improve inspection integration with NDT methods. Accurate STRID and DMID methods that relate deterioration estimate and damage identification (see Chapter 9)
	Bridge ratings	See Chapter 8
Fast hazards	Earthquakes	See Chapter 2
	Scour/Flood	See Chapter 1
Slow hazards	Fatigue	See Chapter 5
	Corrosion	See Chapters 3 and 4

Note: Only Sensing, STRID, DMID are explicitly included since decision making is the main focus of the table.

10.7.5 Deterministic versus Probabilistic Information

In Ettouney and Alampalli (2012) and throughout this chapter, we discussed deterministic and probabilistic assessments of information. We showed that deterministic values in many situations are not satisfactory in producing appropriate decisions. We argued that the knowledge of both mean and variances are needed for solid decision making. The BLCA situation is no different. We observe that the vast majority of BLCCA are based on deterministic methods. As such, the computed values of BLCA can be either inaccurate or biased. The decision maker is advised to perform probabilistic BLCA that recognizes the uncertainties of input values.

This makes even a stronger case for using SHM as an aid in performing BLCA. SHM measurements can be designed to produce any statistical measure needed for probabilistic BLCA. If the decision maker decides to embark on probabilistic BLCA, it is strongly advised that proper planning for needed information is undertaken before the actual project starts.

10.7.6 Closing Remarks

We showed the potential of using SHM techniques to accurately produce information for BLCA. Of course, this raises an important question: Can the value of SHM information for producing accurate BLCA outweigh the cost of SHM experiments? The answer of course is case dependent. We suggest that the decision maker embark on a study to evaluate the value of information from SHM. The cost of SHM experiment is then compared with the value of information. Such a comparison should help in deciding whether or not to use SHM in producing accurate information for BLCA. A detailed example that illustrates this process is provided in Section 10.7.

10.8 GENERALIZED APPROACH TO LCA

10.8.1 Introduction

Extending bridge service life is one of the main aims of any bridge owner. The reasons are self-evident: by extending service life, society can enjoy considerable economic benefits. Of course, extending useful life must be done while maintaining or enhancing the safety of the stakeholders.

The major problem that many bridge officials have while estimating BLCCA (and also BLCBA and BLSA) is the lack of information needed for accurate results. We argue that SHM can be used as a vehicle for estimating BLCA and all of its subcomponents. Specific steps of using SHM to help in estimating accurate BLCA are given below.

10.8.2 Generalized Step-by-Step Approach

10.8.2.1 General

A basic and important relationship between BLCA and SHM was shown in this chapter. We propose that SHM can be used with BLCA in one or more of several modes

- Mode I: BLCA needs would initiate an SHM activity (or activities).
- Mode II: SHM activity (or activities) is initiated for another objective. However, the results and data collected during the SHM effort can be used later if a BLCA effort is initiated.

Figure 10.12 shows the two modes of BLCA-SHM interrelationship.

In Mode I, the step-by-step approach to relate BLCA and SHM are

1. Identify pertinent cost, benefits and/or lifespan sources for the case on hand. To execute this step, the BLCA Equations 10.1 through 10.8 are used.
2. Decide on whether SHM activities can be used to estimate the cost sources.

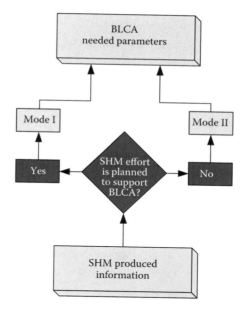

FIGURE 10.12 Modes of BLCA—and SHM interrelationship.

3. If it is decided that an SHM activity can be used to estimate BLCCA cost(s), a suitable SHM effort can then be designed.

Figure 10.13 summarizes the steps of Mode I.

Mode II is essentially the inverse of Mode I. It is assumed that an SHM activity is already in existence, or warranted for some goal other than BLCA. In either of these situations, the additional benefit of obtaining an accurate estimate of BLCA from the SHM activity can be accomplished. Mode II is also the basis of the Serendipity Principle in SHM as described by Ettouney, Alampalli, and Agrawal (2005). The steps of Mode II are as follows:

1. For an existing or new SHM effort, identify the collected data and the goal of the project.
2. Decide whether the collected data is suitable for an accurate estimate of
    a. Cost measure in a BLCA evaluation and/or
    b. Benefit measure in a BLCBA evaluation and/or
    c. Lifespan measure in a BLSA evaluation process.
3. Process the information for any BLCCA, BLCBA, and/or BLSA evaluation effort(s).

Figure 10.14 summarizes the steps of Mode II. Figure 10.15 shows an example of the use of Mode II.

10.8.2.2 Example—Mode I

As an example to Mode I BLCA, consider a bridge official who has two options for capital improvements of the bridge. Option A is to decommission the bridge and build a new one. Option B is to retrofit the existing bridge. The cost/benefit analysis that the bridge official has embarked on is based on an estimation of the lifespan of the existing bridge (see Hawk 2003). Obviously, the longer lifespan of the existing bridge, the more attractive option B would be. Thus, it is clear that an accurate estimation of bridge lifespan is warranted.

The next step would be for the official to decide on the parameters that affect the lifespan of the bridge. Some of these parameters are societal: traffic needs, for example. Other factors can be

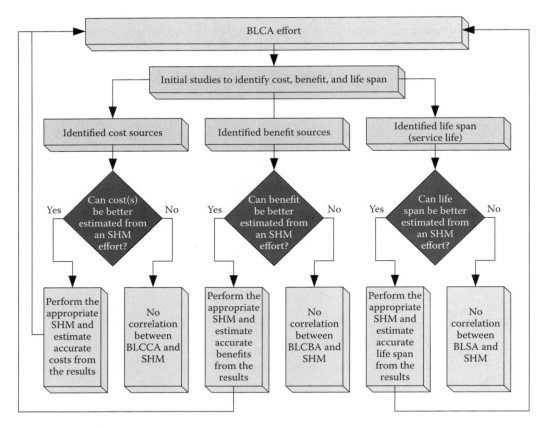

FIGURE 10.13 Mode I BLCA—SHM process.

rate of normal wear and tear and vulnerability to abnormal hazards such as earthquakes or scour. On the basis of this, the official embarked on an SHM study to evaluate the lifespan of the bridge. The details of the study are shown in Table 10.6. Obviously, the costs of the SHM study should be included in any BLCA analysis. The method of accommodating the value of such information has been described elsewhere (Benjamin and Cornell 1970).

10.8.2.3   Example—Mode II

Let us consider a situation where an SHM-based modal identification project is under way. The purpose of this project is to estimate number of fatigue cycles in certain parts of the bridge (see Lund and Alampalli 2004). The decision maker is planning to embark on a BLCA to estimate seismic hazard effects and the needed mitigation measures. It is well known that an accurate estimation of modal damping during any seismic evaluation is important. Overestimating damping can lead to unsafe seismic mitigation solutions. Underestimating damping can lead to unnecessary and costly mitigation measures. Because of this, the decision maker might decide to monitor also enough information so that as-built modal damping can be measured. The knowledge of accurate modal damping can help the analysts in estimating appropriate damping values during seismic events.

10.8.3   Advanced Concepts of the LCA Process

Life cycle analysis is a powerful decision making and management tool that is still in its infancy as of the writing of this chapter. Ultimately, LCA should include all important attributes of infrastructure

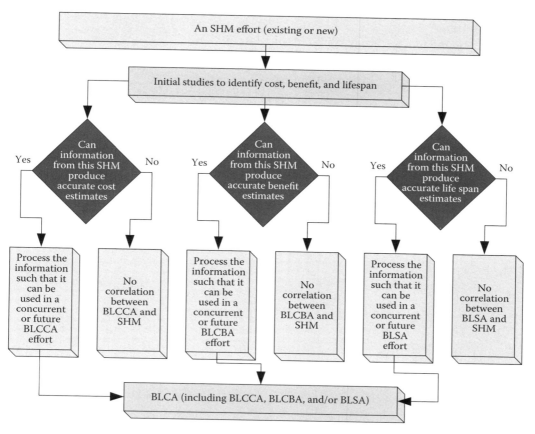

FIGURE 10.14   Mode II BLCA—SHM process.

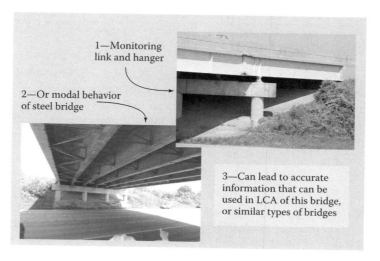

FIGURE 10.15   Illustration of Mode II.

TABLE 10.6
SHM Scheme for Accurate Estimation of Lifespan

Factor Affecting Life span Estimation	SHM Experiment
Future traffic needs	Install traffic volume monitors
Normal wear and tear	Use existing data for statistical analysis. Monitor corrosion, fatigue, and other conditions
Seismic hazard	Perform structural identification analysis to estimate potential and extent of damage during different earthquake scenarios
Scour hazard	Install scour monitors to estimate potential and extent of damage during different flood/scour scenarios

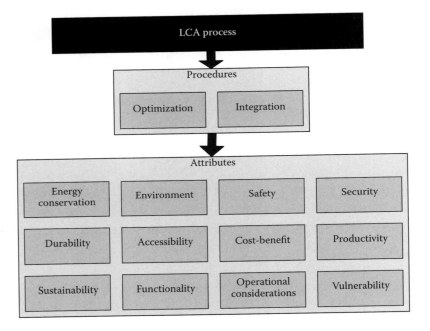

FIGURE 10.16   Advanced LCA process.

behavior. These attributes include safety, security, environmental concerns, energy, durability, and operability. All these attributes should be addressed by LCA from two procedural viewpoints: integration and optimization. Obviously, SHM will play an important role in achieving the complex objectives of LCA. This is due to two factors: (1) SHM will provide accurate information to the LCA process, and (2) SHM will verify many of the analytical assumptions built into different LCA methodologies. Figure 10.16 illustrates this advanced LCA process.

REFERENCES

Agrawal, A., Kawaguchi, A., and Qian, G. (2009). "Bridge Element Deterioration Rates: Phase I Report," New York State Department of Transportation, Project # C-01–51, Albany, NY.
Alampalli, S. and Lund, R. (2006). "Estimating fatigue life of bridge components using measured strains." *Journal of Bridge Engineering, ASCE*, 11(6), 725–736.
Bazovsky, I. (2004). *Reliability Theory and Practice*, Dover Publications, Mineola, NY.
Benjamin, J. R. and Cornell, A. C. (1970). *Probability, Statistics, and Decision for Civil Engineers*, McGraw-Hill, New York.

Bentur, A., Diamond, S., and Berke, N. (1997). *Steel Corrosion in Concrete: Fundamentals and Civil Engineering Practice*, E&FN Spon, an imprint of Chapman & Hall, London, UK.

Chapman, R. (2003). "Applications of Life-Cycle Cost Analysis to Homeland Security Issues in Constructed Facilities: A Case Study," National Institute of Standards and Technology, Gaithersburg, MD.

Chapman, R. and Leng, C. (2004). *Cost-Effective Responses to Terrorist Risks in Constructed Facilities*, National Institute of Standards and Technology, Gaithersburg, MD.

El-Diraby, T. and Rasic, I. (2002). "A Framework for Life-Cycle Costing of Smart Materials in Civil Infrastructure," *Proceedings of 1st International Workshop on Structural Health Monitoring of Innovative Civil Engineering Structures*, ISIS Canada Corporation, Manitoba, Canada.

Ettouney, M. and Alampalli, S. (2012). *Infrastructure Health in Civil Engineering: Theory and Components*, CRC Press, Boca Raton, FL.

Ettouney, M., Alampalli, S., and Agrawal, A. (2005). "Theory of Multihazards for Bridge Structures," *Journal of Bridge Structures: Assessment, Design and Construction*, Taylor and Francis, 1(3), 281–291.

Hawk, H. Life-Cycle Cost Analysis (2003), "Life-Cycle Cost Analysis," NCHRP Report 483, Transportation Research Board, Washington, DC.

Lund, R. and Alampalli, S. (2004). "Estimating Fatigue Life of Patroon Island Bridge Using Strain Measurements," Special Report No. 142, New York State Department of Transportation, Albany, NY.

Melchers, R. (2002). *Structural Reliability Analysis and Prediction*, John Wiley and Sons, New York, NY.

Oshima, T., Takeda, T., Beskhyroun, S., Yasue, S., and Tamba, I. (2004). "Structural Health Monitoring for Life Cycle Management (LCM) of Bridges," *Proceedings of 2nd International Workshop on Structural Health Monitoring of Innovative Civil Engineering Structures*, ISIS Canada Corporation, Manitoba, Canada.

Peil, U. and Mehdianpour, M. (2000). "Life Cycle Prediction via Monitoring," *Proceedings of 2nd International Workshop on Structural Health Monitoring*, Stanford University, Stanford, CA.

Yagi, T., Nakano, T., Yamada, K., and Ojio, T. (2004). "Service Load Monitoring by BWIM Using Reaction Force Method," *Proceedings of 2nd International Workshop on Structural Health Monitoring of Innovative Civil Engineering Structures*, ISIS Canada Corporation, Manitoba, Canada.

Yoneda, M. and Edamoto, K. (2004). "Development of Bridge Weigh-in-Motion System Using Genetic Algorithm on Highway Bridges," *Proceedings of 2nd International Workshop on Structural Health Monitoring of Innovative Civil Engineering Structures*, ISIS Canada Corporation, Manitoba, Canada.

11 Role of Structural Health Monitoring in Enhancing Bridge Security

11.1 INTRODUCTION

Bridge security has now emerged as an important subject. Protecting our bridges, which represent key components of transportion infrastructure, is essential for national security, mobility, and economic vitality. Direct attacks on critical structures can lead to casualties and seriously damage regional and national economies. However, because each bridge is unique and complex, defining and securing vulnerable components against varied threats pose a challenge. Securing structural components of a bridge is only one part. Overall site conditions and lifelines often carried by bridges also need to be protected (see Figure 11.1). The threats to our bridges are complex and can vary significantly in severity. These, coupled with the numerous stakeholders who must interact efficiently in times of crisis, make bridge security a complex subject, often not well understood, although its importance in national well-being is highly acknowledged. Thus, ensuring bridge security for public safety is a technically challenging task, potentially involving immense resources.

Structural health monitoring (SHM) has also emerged recently as a viable engineering field with the potential of helping bridge owners enhance safety while reducing operating and maintenance costs and thereby increasing the service life of bridges. SHM has been shown to be of help in normal bridge operations; and also in monitoring corrosion, fatigue, and structural behavior under abnormal hazards such as earthquakes, high winds, and scour. Several SHM methodologies, techniques, hardware, and software have been developed and used to achieve different bridge management goals, thus improving informed decision making processes, enhancing safety, and reducing costs. As discussed earlier, structural health in civil engineering (SHCE) includes both SHM and decision making. Since the use of decision making tools is important for bridge security, we will explore both SHM and SHCE in this chapter.

11.1.1 This Chapter

As of the writing of this chapter, the role of SHM in bridge security is not well defined or understood. This chapter discusses several issues that correlate SHM and bridge security. This includes various SHM technologies, hazards that affect bridge security, the temporal nature of security (before, during, and after event), interaction between hazards, bridge components, and disciplines, and interaction among stakeholders.

Perhaps the most effective introduction to the subject is to consider the different strategies for enhancing bridge security. These strategies are known as 4-D strategies. We explore the role of SHM/SHCE within each of these 4-Ds. This will lead us to different techniques used by professionals in electronic security. We observe that there are commonalities between electronic security and SHM sensing. Exploring SHM techniques to enhance bridge security comes next. There are various angles to the SHM role in enhancing bridge security. This includes detailed discussion of bridge components, SHM components (sensors, structural identification, and damage identification),

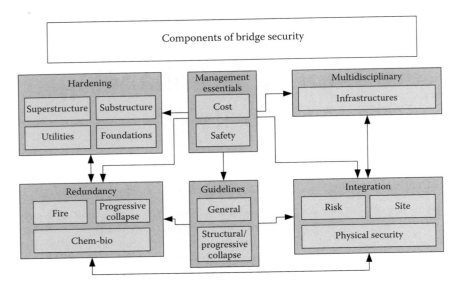

FIGURE 11.1   Components of bridge security.

multihazard and multidisciplinary factors, sequence of events during a security hazard, cost–benefit/value, and finally the all-important subject of stakeholder interaction.

Recognizing the importance of decision making in bridge security, this chapter presents two case studies that explore the use of SHM technologies for bridge security. An example of a risk-based prioritization of mitigation options is investigated. The second example shows how to use a cost–benefit analysis technique to ensure optimum value of the project on hand.

This chapter will conclude with an exhaustive list of observations and recommendations. We believe that the subjects in this chapter can be of help to SHM and security communities in understanding the role of SHM in enhancing bridge security and in focusing/prioritizing their efforts to reduce costs while improving safety and security.

11.2   CONCEPT OF 4DS

Security of bridges or any other type of infrastructures can be subdivided into four principles: deterrence, denying, detection, and defense. These four principles are referred simply as 4Ds. Each of the 4Ds contains set of philosophies that might be applied to enhance the security of the bridge. There are situations where some of the philosophies of the 4Ds can be used for SHM applications, or vice versa. Obviously, such situations can be exploited to enhance both the security of the bridge and to improve SHM potential of the bridge simultaneously. In what follows, we describe each of the 4Ds and some situations where its philosophies intersect with SHM applications.

11.2.1   SHM AND DETERRENCE

Deterrence is a security concept that allows potential aggressors to realize that the bridge structure is being monitored. Deterrence includes noticeable security procedures such as security patrols and guards, improved lighting, and noticeable closed circuit video and TV (CCTV). Table 11.1 shows potential intersections between security deterrence methods and SHM.

11.2.2   SHM AND DENYING

The concept of denying implies that different measures are taken to prevent the aggressor from accessing the bridge site. Measures that enhance denying include fencing, barriers. Controlled parking in the area can improve bridge security immensely.

11.2.3 SHM AND DETECTION

Surveillance and monitoring activities of potential aggressors is the basis of detection. Also, reporting mechanisms are part of detection systems. As the name implies, detection of unwanted activities would include several potential intersections with SHM. Table 11.2 shows some of these potential intersections.

11.2.4 SHM AND DEFENSE

Defense in bridge security includes using physical measures that enhances bridge capability to resist blast pressures. These measures include increasing standoff distance as shown in Figure 11.2, increasing strength of the bridge structure, and increasing redundancy of bridge structure. Obviously, there are several intersections between defense and SHM activities as shown in Table 11.3.

11.2.5 GENERAL CONCEPTS FOR SITE SECURITY

There are several important concepts that need to be considered when designing site security as shown in Table 11.4.

11.2.6 GENERAL CONCEPTS FOR PHYSICAL BRIDGE SECURITY

Some considerations that can enhance bridge physical security are shown in Table 11.5. The table also shows the needed coordination with pertinent SHM applications.

11.3 SECURITY-SPECIFIC TECHNOLOGY AND SHM UTILIZATION

Infrastructure security has been a field of study for as long as infrastructures have existed. Technologies that enhance security have also evolved over the years. Currently, there are several

TABLE 11.1
Deterrence and SHM

Activity	Deterrence Application	Other SHM Uses
CCTV	Used to monitor suspicious activity	Observe traffic flow and statistics, impact/accidents
Lighting	Used to observe potential nonwanted activities	Helps in clarity of SHM-related video

TABLE 11.2
Detection and SHM

SHM/detection activity	Wireless sensing
Surveillance	Optical technologies
Pattern (physical) recognition	Remote monitoring/sensing
Substance recognition	Imaging techniques
Embedded sensing	Biometric sensing

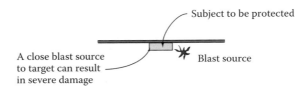

Baseline: Target with no protection

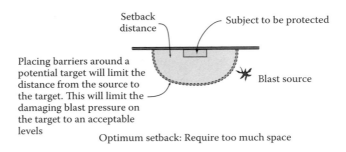

Optimum setback: Require too much space

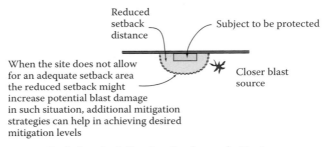

Realistic setback: Require other forms of mitigation measures

FIGURE 11.2 Denying strategy example: use of barriers to increase setback.

TABLE 11.3
Defense and SHM

SHM Activity	Defense Application	Other SHM Uses
Measuring strains, stresses	Used to aid in designing hardening and redundancy retrofit.	Used in structural and damage identification
Measuring displacements, velocities, and accelerations	Used to aid in designing hardening and redundancy retrofits	Used in structural and damage identification

technologies specifically designed to enhance infrastructure security. Most of those methods were developed from the security viewpoint. We can thus claim that they aim at protecting structural health from the potential ills that security hazards might inflict on the infrastructure. In this section we will explore these techniques and present an overview of the vast array of security applications these technologies can offer. In addition, we discuss the potential of using some of these security-specific technologies to enhance other structural health needs, thus enhancing the value of these applications by increasing the benefits while maintaining the cost level.

TABLE 11.4
Bridge Site Security Concepts

Concept

Include security and protection measures in the calculation of land area requirements

Provide a sitewide public address system and emergency call boxes at readily identified locations

Provide intrusion detection systems for all lifelines within the bridge site

TABLE 11.5
Bridge Physical Security and SHM

Concept	SHM Application
Secure manholes and access hatches	Need coordination with the normal bridge inspection activities
Install redundant communication mechanisms	Need to coordinate with any SHM activity to reduce any potential interference between security and SHM goals
Harden or limit use of nonstructural components	SHM can be utilized to observe any potential degradation in the system worthiness as time progresses

11.3.1 ELECTRONIC SECURITY SYSTEMS

Electronic security systems can be used to enhance protection of bridge structures (or any other type of civil infrastructures). Electronic security systems can be categorized based on four zones within the site of the bridge, or infrastructure, of interest. Starting from the outside, the four security zones can be identified as

Exterior intrusion detection zone: This zone covers open areas around the infrastructure. It can be a single zone or multiple zones separated by a hard fence or barrier. The zones can be concentric or adjacent to each other.

Boundary-penetrating zone: This area constitutes a hard barrier, such as a fence, wall, or glass barrier.

Volumetric motion zone: This is an interior zone where it is of interest to detect any motion within the zone.

Electronic entry zone: This zone constitutes doors or gates and the different measures that govern them.

Figure 11.3 shows schematically the different security zones. Obviously, the type of sensors that can be used in each of these zones depends on the characteristic of the zone itself and the type of intrusion/threat that might affect it. For an in-depth description of these concepts, the reader is referred to FEMA 426 and FEMA 459.

11.3.2 Security-Sensing Technologies

Security-sensing systems can be subdivided into several categories according to the security zone layer topology.

Exterior intrusion detection sensors: These types of sensors include fence sensors, buried line sensors, microwave sensors, infrared sensors, and video motion sensors. See Table 11.6 for more descriptions of the sensors.

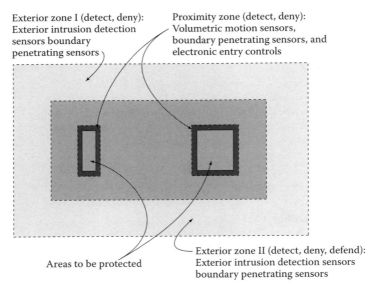

Exterior zone I (detect, deny): Exterior intrusion detection sensors boundary penetrating sensors

Proximity zone (detect, deny): Volumetric motion sensors, boundary penetrating sensors, and electronic entry controls

Areas to be protected

Exterior zone II (detect, deny, defend): Exterior intrusion detection sensors boundary penetrating sensors

Zones of protection of electronic security measures

FIGURE 11.3   Zones of protection when using electronic security.

TABLE 11.6
Categories of Exterior Intrusion Detection Sensors

General Category	General Description of Category	Specific Sensor Type
Fence sensors	Detect breakage in fences	Strain sensitive cables
		Taught wire Sensors
		Fiber optic cable sensors
		Capacitance proximity sensors
Microwave sensors	Detect motion with greater accuracy.	Bi-Static System
Microwave and infrared sensors	Detect motion with greater accuracy. Detects movements in absence of visible light waves	Mono-Static System
Video motion sensors	Turns on / off depending if motion is detected	

Boundary-penetrating sensors: There are several types of sensors that can be used to monitor this security zone. They include structural vibration sensors, glass-breaking sensors, ultrasonic sensors, balanced magnetic switches, and grid wire sensors. Table 11.7 shows the basic properties of the sensors.

Volumetric motion sensors: This category of sensors includes microwave sensors, passive infrared sensors, video motion sensors, point sensors, capacitance sensors, pressure mats, and pressure switches. Also, dual-technology sensors are used for this category of sensors. Table 11.8 shows the properties of each sensor category.

Electronic entry controls: This category of sensors includes coded devices, credential devices, and biometric devices. See Table 11.9 for description.

11.4   SHM-SPECIFIC TECHNIQUES AND BRIDGE SECURITY

11.4.1   OVERVIEW

This section discusses different issues that relate SHM/SHCE to bridge security. We first relate bridge components to both SHM and security. In the next four sections, we discuss the four

TABLE 11.7
Categories of Boundary-Penetrating Sensors

General Category
Structural vibration sensors
Glass breakage sensors
Ultrasonic sensors
Balanced magnetic switches
Grid wire sensors

TABLE 11.8
Categories of Volumetric Motion Sensors

General Category
Microwave sensors
Passive infrared motion sensors
Video motion sensors
Point sensors
Capacitance sensors
Pressure mats
Pressure switches
Dual technology sensors

TABLE 11.9
Categories of Electronic Entry Controls

General Category	General Description of Category	Specific Sensor Type
Coded Devices	These devices offer entry to secure areas by allowing for entry of per-coded passcode	Electronic keypad Computer-controlled Keypad
Credential Devices	These devices recognize the user through a personal code.	Magnetic strip card Wiegrand-effect card Proximity card Smart card Barcode
Biometric Devices	These devices recognize the user through specific bio- recognition	Fingerprints Hand geometry Retinal patterns

SHCE components (sensing, structural identification, damage identification, and decision making) related to bridge security. We then present an overview of the SHM techniques of remote sensing. Recognizing that multihazards and multidisciplinary issues are essential in the coupling of SHM and bridge security, we discuss those issues next. A brief description of sequences of security hazard effects on the use of SHM methods follows. We end this section with brief discussions on the management subjects of cost–benefits and interaction among stakeholders.

11.4.2 BRIDGE COMPONENTS

Bridge security is a broad and complex subject. When we try to address it from an SHM viewpoint, it becomes even more complex. To simplify matters, we divide the security components into five broad areas: hardening, redundancy, integration, guidelines, and management goals. Note that all these components are not physical.

The security hardening component relates vulnerability and mitigation measures to strength and stiffness of physical subcomponents such as superstructures (decks, girders, bracings), substructures (piers, bents, columns), utilities (water, electric, and gas lines), and foundations. Redundancy relates to physical issues such as progressive collapse, fire resistance, and mechanical equipment that resist chem-bio hazards. Integration is another important component, and it includes risk management and assessment. This is because several factors need careful consideration when applying any risk methodology. Site consideration from a security viewpoint is another subject that needs the integration of various issues such as topology, proximity to residential or office buildings, landscaping features, pedestrian patterns, and so on. Physical and electronic security is another security integration subject. Many parameters need to be integrated to achieve optimum security, such as personnel qualifications, types of sensors used, and optimal use of human and financial resources. Another general component that has a direct impact on bridge security is guidelines. The guidelines can be general (FEMA 2006, 2008, 2010) or specific, such as structural (FEMA 2009). Finally, perhaps the most important security component is management. The component includes different evaluations and techniques that address cost/benefits and safety/security of bridges.

We note that SHM tools have a direct impact on some components. In addition, some components are related to vulnerabilities and/or mitigation of the bridge as it is affected by security hazard. Table 11.10 discusses these interrelationships.

11.4.3 SENSORS AND INSTRUMENTATIONS

Several sensing techniques used in SHM were discussed in Chapter 5 by Ettouney and Alampalli (2012). Many of them can be used to enhance bridge security; some of these are detailed in Table 11.11.

Wireless sensing and dense sensor arrays: One of the most important recent advances in SHM is the emergence of wireless-sensing technologies (see Chapter 5 by Ettouney and Alampalli 2012). Wireless sensing in turn has permitted the use of dense sensor arrays. The wide-sensing array has obvious benefits in SHM applications, as discussed by Ettouney and Alampalli (2012, Chapter 5). When used to enhance bridge security, the wireless sensing/dense sensor arrays can have even more benefits, such as

- Due to the intensity of blast near the source, the near-damage sensors might malfunction; however, thanks to the dense sensor array, many sensors will remain functional, and appropriate identification of the damage can still be performed (see Figure 11.4).
- Some wireless-sensing technologies can be miniaturized and embedded, thus providing better protection from intense blast pressures.

Instrumentation: SHM instrumentation network can be planned to be used as a backup instrumentation network for physical security measures. If such a redundancy is beyond the available budget, then both SHM and physical security instrumentation networks can be used to complement each other.

11.4.4 STRUCTURAL IDENTIFICATION

Ettouney and Alampalli (2012, Chapter 6) presented four uses of structural identification techniques. We also discussed the three major categories of structural identification methods. We discuss in

TABLE 11.10
SHM Relationships To Security of Bridge Components

Component	Issues	SHM Roles		
		General	Vulnerability	Mitigation
Hardening	Superstructure	Sensing responses. Structural and damage identifications	Yes: assess points of vulnerabilities	Yes: evaluate different mitigation methods
	Substructure	Sensing responses. Structural and damage identifications		
	Utilities	Physical security technologies: detection methods and techniques	Detection and deterrence strategies	
	Foundations	Sensing responses. Structural and damage identification	Yes: assess points of vulnerabilities	Yes: evaluate different mitigation methods
	Progressive Collapse	Sense extent of collapse and safety of remaining structure. Structural and damage identification methods can be used	Yes: assess potential of progressive collapse	Yes: assess adequacy of mitigation measures
	Fire	Sense occurrence of fire and safety of remaining structure. Structural and damage identification methods can be used	NA	NA
	Chem-bio	Sense occurrence of attack	NA	NA
Integration issues	Risk management/ assessment	SHM measurement database can help in accurate risk assessments. Decision-Making Toolbox (DMTB)` can aid in risk management		
	Site considerations	Physical security technologies: detection methods and techniques	Detection and deterrence strategies	
Guidelines	Physical security General Structural	See Section 11.3 of this chapter		
Management goals	Safety/security Cost/benefit/value	SHM measurement database can help in accurate risk assessments. DMTB can aid in risk management		

TABLE 11.11
Role of SHM Sensors and Instrumentations in Enhancing Bridge Security

SHM Sensor Type (Ettouney and Alampalli 2012, Chapter 3)	Use in Bridge Security
Strain	Direct strain measurements can aid in damage and structural identification. Also, for both mitigation and vulnerability assessment (see Figure 11.5)
Position	Can help in postdisaster effects in assessing safety of structures (see Figure 11.6)
Displacement, velocity, and accelerations	Can aid in damage and structural identification. Also for both mitigation and vulnerability assessment
Pressure	Measurements of blast pressures, thus aiding in structural identification efforts
Temperature	Near-field sensing can help in identifying blast source and any structural degradation due to temperature
Force	Can be used to measure behavior of bearings
Inclinometers	Can help in postdisaster effects in assessing safety of structures
Corrosion Sensors	NA
Fiber optics sensors	Can be used in several pre- and post-event measurements, thus aiding in structural and damage identification efforts. Can be very effective if embedded, since the reduced direct blast pressure will make it possible to sense closer to the blast source than surface mounted sensors (see Figure 11.7)

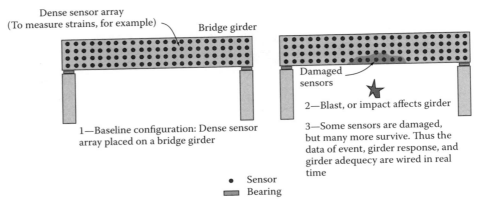

FIGURE 11.4 Dense sensor array utilization for enhancing bridge security.

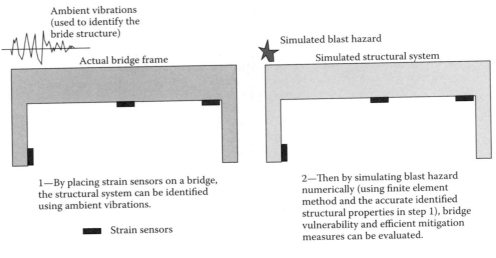

FIGURE 11.5 Use of strain sensors for accurate vulnerability and mitigation evaluation.

Table 11.12 the applicability of each of the three categories for the four uses of structural identification as applied to blast loading.

11.4.4.1 Time, Amplitude, and Spatial Scale Issues

Blast and impact/ramming hazards are short-duration loading conditions with high amplitudes. Therefore, any accurate structural modeling must include high-resolution (small size) finite elements that can simulate the resulting strain localization and high-excited natural modes. For example, Figure 11.8 shows a finite-element bridge model built to analyze the response of the bridge to bomb blast (see Agrawal, A., pers. comm.). In the same study (Agrawal, A., pers. comm.), built a detailed finite-element model for the bridge elastomer bearing, as shown in Figures 11.9 through 11.11. Note

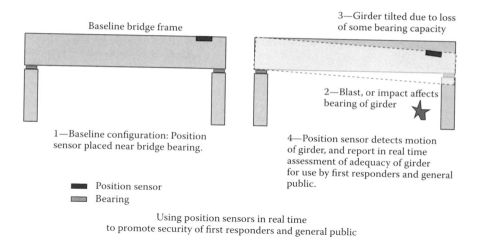

Baseline bridge frame

3—Girder tilted due to loss of some bearing capacity

2—Blast, or impact affects bearing of girder

1—Baseline configuration: Position sensor placed near bridge bearing.

4—Position sensor detects motion of girder, and report in real time assessment of adequacy of girder for use by first responders and general public.

■ Position sensor
▨ Bearing

Using position sensors in real time to promote security of first responders and general public

FIGURE 11.6 Position sensor can help in real time.

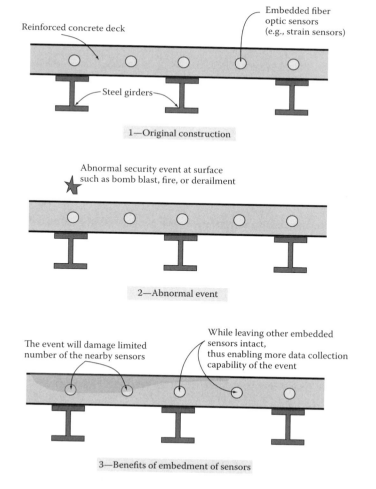

Reinforced concrete deck

Embedded fiber optic sensors (e.g., strain sensors)

Steel girders

1—Original construction

Abnormal security event at surface such as bomb blast, fire, or derailment

2—Abnormal event

The event will damage limited number of the nearby sensors

While leaving other embedded sensors intact, thus enabling more data collection capability of the event

3—Benefits of embedment of sensors

FIGURE 11.7 Use of embedded fiber optics to enhance bridge security.

TABLE 11.12
Use of Conventional Structural Identification Methods for Bridge Blast Hazard

Different Uses of Structural Identification	Modal Identification	Parameter Identification	Nonphysical Methods—ANN
Condition assessment of existing structures	Can be used to detect shifts in structural modes, which is a subjective way of assessing structural condition after a severe blast	Can be used to detect changes in physical structural properties, which is a subjective way of assessing structural condition after a severe blast	Needs baseline as well as several databases of similar events to ensure accurate results
Design validation of new structures	These tasks would require nonlinear analysis; modal identification is not capable of accurate nonlinear modeling of structures	These tasks would require nonlinear analysis; linear parameter identification methods would result in inaccurate parameters. Nonlinear methods can be adequate	
Analytical model updating for new and existing structures			
Damage identification of new and existing structures	Since modal identification can use only a few lower structural modes, it might not be adequate for accurate detection of damage resulting from blast loading	High-resolution nonlinear models can detect damage case by case. No general automated methods are available for this	Lack of physical significance of the ANN weights limits the use of ANN for this purpose

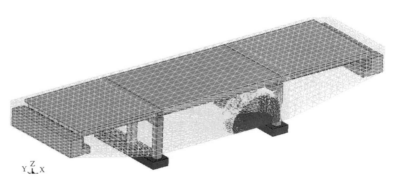

FIGURE 11.8   Finite-element modeling of blast event on a 3-span bridge. (Courtesy of Dr. Anil Agrawal.)

the high resolution of the finite elements that were needed to obtain accurate results from the blast event.

11.4.5 DAMAGE IDENTIFICATION

Identification of damage resulting from blasts can be done using conventional nondestructive testing (NDT) tools or the more general SHM tools. Many of these tools can be beneficial in identifying damage during the three time sequence of the event as follows:

- Before event: Estimate potential damage so that mitigation measures can avoid those estimated vulnerabilities. Also, measurements are needed for use as a baseline for comparison with damaged condition.
- During event: Generally possible only for using SHM real-time tools. In general, NDT methods cannot be used during event.
- After event: Most methods can be used for damage identification.

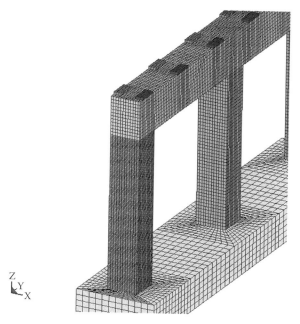

FIGURE 11.9 Finite-element model of bridge bent and elastomer bearings. (Courtesy of Dr. Anil Agrawal.)

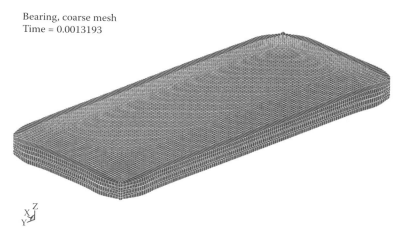

FIGURE 11.10 Details of elastomer bearing. (Courtesy of Dr. Anil Agrawal.)

Table 11.13 shows the applicability of different NDT methods for use in damage identification of bridges susceptible to blasts. The applicability is scored on a scale 0–10, with score of 1 being the least applicable and score of 10 being the most applicable.

11.4.6 REMOTE SENSING

The general concepts and applications of remote sensing in SHM was defined earlier as the potential of structural or damage identification of infrastructures using remotely placed sensors/instrumentation. Using this definition, remote sensing is obviously a suitable application in bridge security. There are several bridge security applications that can use the concepts of

Bearing, coarse mesh
step 1 time: 0.000000

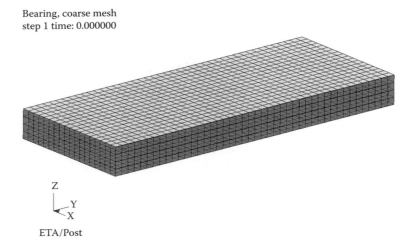

Z
Y
X

ETA/Post

FIGURE 11.11   Elastomer bearing finite-element model. (Courtesy of Dr. Anil Agrawal.)

TABLE 11.13
Applicability of NDT Methods for Identification of Damage Caused by Bomb Blast

NDT Methods	Before Event	During Event	After Event
Ultrasonic	7	2	9
Electromagnetic	5	3	7
Optical	7	6	8
Chemical	1	1	1
Visual	8	8	9
Thermal	6	7	7
Penetrating radiation	2	1	5
Acoustic emission	7	3	7
Imaging techniques	9	5	8

remote sensing. For example, in a bomb blast that affects a particular bridge or any other type of infrastructure, it is of interest to evaluate the degree of structural damage in real time. If there is no built-in SHM sensors on the damaged system, or if the built-in sensors did exist but were damaged during the blast, a remote-sensing resource can help in assessing the damaged system. Other applications of remote sensing to bridge security are ramming (barges or cars/trucks) or severe fire of bridge components that are difficult to reach by conventional NDT. This concept is shown in Figure 11.12.

As discussed by Ettouney and Alampalli (2012), a remote-sensing system for bridge security would involve the following components (see Figure 11.12):

Remote transmitter/receiver: This will be a portable or stationary sensing device. Sensing techniques can be laser-based, thermography-based, vibration-based, or ultrasonic. The sensors will measure displacements, strains, or rotations of the structural components.

Analysis component: The sensor data would be analyzed to determine location, type, extent, and severity of structural damage.

Decision making algorithm: Upon detecting damage information, decision making algorithms would help managers or decision makers choose the optimal course of action.

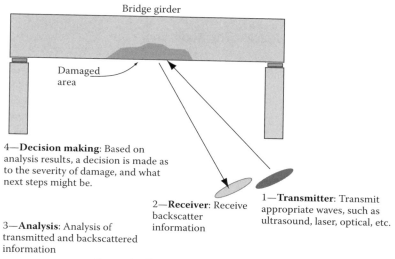

Bridge girder

Damaged area

4—**Decision making**: Based on analysis results, a decision is made as to the severity of damage, and what next steps might be.

3—**Analysis**: Analysis of transmitted and backscattered information

2—**Receiver**: Receive backscatter information

1—**Transmitter**: Transmit appropriate waves, such as ultrasound, laser, optical, etc.

Concepts of remote sensing after a blast event

FIGURE 11.12 Remote-sensing concepts after a blast.

11.4.7 Multihazards Factors

The concept of multihazards as applied to infrastructures was explored earlier. The subject of multihazards consideration is of special importance when considering bridge (or any other infrastructure) security. The reason is the combination of the high cost of bridge security measures and the relatively low probability of occurrence of security events. Because of this, it is beneficial to use mitigation measures of other hazards as much as possible as part of bridge security projects. Thus, any successful use of multihazard consideration will result in an efficient use of funding while achieving the desired level of safety and security. Table 11.14 shows different hazards and the way they interact with bomb blast hazards for different components of SHM/SHCE.

11.4.8 Multidiscipline Factors

The fields of bridge security and SHCE/SHM require the coordination and participation of many disciplines. These include police/law enforcement, fire departments, emergency managers, and engineering professionals. Each of these disciplines has a different outlook on the combined subjects of SHCE/SHM and bridge security. Therefore, it is extremely important to have all these disciplines involved in all phases of the strategies to enhance bridge security by using SHCE/SHM techniques. Table 11.15 shows the different disciplines and the way they can interact for different components of SHM/SHCE.

11.4.9 Sequence of Events during Hazards

Most security-related hazards involve a temporal sequence. Such a sequence can be simply subdivided into three parts:

- Before event
- During event
- After event

TABLE 11.14
Multihazards in Bridge SHM/SHCE Security Applications

Hazard	Sensors and Measurements	Structural Identification	Damage Identification	Decision Making
Seismic	Motion and deformation sensors can be used for blast applications	Nonlinear structural identification methods are common for blast and seismic fields	Seismic behavior extends to post yield regime, similar to structural response to blast	DMTB should be similar to these hazards. In particular, risk and reliability models
Wind	Motion and deformation sensors can be used for blast applications. Also, wind pressure sensors can be used, assuming that the pressure range is wide enough to encompass both wind and blast pressure levels	Structural identification models that apply to wind might need adjustment before being used for blast. This is because wind responses are usually elastic, while blast responses are usually inelastic	Blast and wind damage modes might be different, since the load and response characteristics are different for each hazard	All these hazards share being dynamic and relatively short duration
Flood-scour	Some flood/scour sensors might be used to sense foundation/soil response to blast	Soil and foundation identification models can be used for both scour/flood and blast situations	Soil and foundation damage identification models can be used for both scour/flood and blast situations	
Impact— traffic	Motion and deformation sensors can be used for blast applications	Nonlinear structural identification methods are common for blast and impact fields	Behavior of bridge components to impact effects (cars, trucks and or barges) extends to post yield regime, similar to structural response to blast	
Normal wear and tear	Long-term or continuous monitored motion or deformation sensors can be used to monitor blast events	Linear structural identification models used for normal wear and tear might not be useful for near-field situations. However, for structural locations away from blast source, such models can be useful	Damage models and types are different from blast damage	The continuous nature of the wear/tear, fatigue and corrosion hazards might require some DMTB different from the short-duration and low- probability blast hazard
Corrosion	Half-cell, pH, or humidity sensors are not directly useful in blast events	NA	NA	
Fatigue	Fatigue strain sensors near welds or at structural connections are suitable for sensing blast response. This is assuming that the sensors are not damaged during blast. Embedded sensors are more protected than exposed sensors	Local fatigue structural models tend to be nonlinear static; they are not useful for the highly dynamic blast situation	Even though both hazards would produce highly localized and nonlinear damage, each damage is different in nature. Different damage models are needed	

TABLE 11.15
Multidisciplinary Issues in Bridge SHM / SHCE Security Applications.

Stakeholders	Sensors and Measurements	Structural Identification	Damage Identification	Decision Making
Police – law enforcement, fire department, and emergency management	Sensing displacements, tilt, and other global building performance for damaged structures	Hidden areas that might have injured or trapped people in it.	Potential of additional collapse of the remaining structure.	Risk-reward decisions during emergencies. Shoring needs.
Engineers	In addition to the above, utilize more specific non-destructive testing methods for more specific information about damaged buildings.	Analytical models of damaged structures	Levels of damage to the system or sub-systems.	Retrofit possibilities. Demolition or reconstruction decisions.

TABLE 11.16
Use of SHCE Tools for Different Sequences of Bomb Blasts

SHCE Components	Before Event	During Event	After Event
Measurement sensors	Enhance deterrence, detection, and denying mitigation measures. See Sections 11.2 and 11.3	Measure responses in real time	Continued sensing would help in detecting any incipient failures
Structural Identification	Use for analysis of potential hardening or increasing redundancy of bridge	Evaluate extent of damage in real time	Estimate extent of damage to plan retrofit and public safety
Damage identification	Use in a multihazard mode for other abnormal hazards (wind, earthquakes, etc.)	Evaluate extent of damage in real time	Estimate extent of damage to plan retrofit and public safety
Decision making	Use DMTB (Chapter 8 by Ettouney and Alampalli 2012) to design mitigation measures. Also, see Sections 11.4.7 through 11.4.11 and 11.5 of this chapter for examples	Provides guidance for emergency response	Use DMTB (Chapter 8 by Ettouney and Alampalli 2012) to design retrofit measures

Note: Table 11.16 applies also to other security-related hazards such as car/truck or barge ramming. For fire hazard, this table is also fairly applicable, albeit with some minor modifications.

The role of SHM will vary depending on the sequence. Table 11.16 shows the different roles that SHM/SHCE can play during the time sequence of a bomb blast.

11.4.10 COST–BENEFIT AND VALUE

The cost of several SHM activities is usually measured in monetary units and sometimes in non-monetary units (such as social and psychological costs). Benefits of the SHM are generally measured in terms of increased security.

All parameters (cost, benefit, and value) are qualitative. Two types of value are considered: perceived value and actual value. A perceived value is the value as perceived by the public as a whole. Actual value is the value as judged by experts or based on real data available from other sources.

11.4.11 INTERACTION AMONG STAKEHOLDERS

Interaction among stakeholders in the SHM-bridge security fields is of utmost importance. Without such interaction, the efficiency of using SHM methods to enhance bridge security is reduced, which might result in lesser security or higher costs. The following stakeholders (see Figure 11.13) are identified as having an interest in SHM-bridge security:

- Bridge owners
- Federal government
- Research community
- Consultants (engineering)
- Manufacturers (both SHM and physical security technologies)
- Security professionals (police, federal agencies, security consultants, etc.)

Figure 11.13 shows both the complexity and importance of interaction among stakeholders. The lines of communication must be clear and well defined among all the stakeholders. If some communication lines are not well established, it might lead to inefficient results: higher costs and lesser safety. The center of communication among stakeholders is the bridge owner. Some pertinent issues in establishing efficient communication among stakeholders are

- The applicability of interaction among specific stakeholders in the security-SHM fields.
- The importance of interaction among specific stakeholders.
- The cost/benefit of maintaining interaction.
- Given that the different stakeholders have different styles of practice and different modes of operation (security and engineering consultants, for example), it is important to set a common communication baseline among the stakeholders.

In a recent workshop (Alampalli and Ettouney 2007), it was concluded that the interaction of owners and security professionals with other stakeholders seemed more applicable and important than others. See Section 11.8 for more results of the workshop.

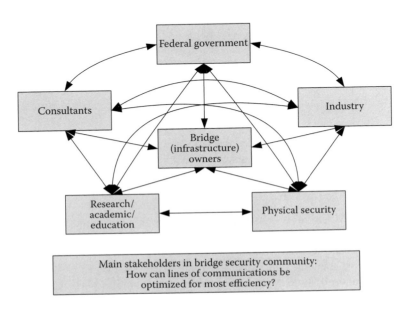

FIGURE 11.13 Interaction among bridge security stakeholders.

11.5 DECISION MAKING: PRIORITIZATION

The value of using SHM technologies to enhance bridge security can be illustrated using decision making processes. For example, consider a situation where a particular part in the system is to be protected from an impact-type hazard. This impact-type hazard can be a terrorist bomb blast. To study the protection projects, the decision maker chooses a risk-based approach. The main advantages of such an approach are (1) it will help compare the different protection or mitigation schemes, and (2) it can be as simple or as complex as needed. By using a simple risk-based analysis, the computed risks are mostly qualitative. As the computation of the risk become more complex, the computed risk becomes more quantitative. For this example, the decision maker decides to use a simple risk-computing method. Since only prioritizing the protection methods from blast hazard is needed, the qualitative nature of the results is acceptable.

The problem facing the decision maker is the need to mitigate the potential vulnerability of a particular target that might be subjected to a bomb blast. The blast source is located at a distance L from the area to be protected. If the size of the blast threat is W, then the resulting blast pressure p can be expressed as

$$p = f(W, L, G) \tag{11.1}$$

where G represent the properties and the geometry of the structure.

Equation 11.1 is a fairly complex equation that has been studied by various authors (see, for example, Mays and Smith 1995). What concerns us in this example is the direct (albeit not linear) relationship between the blast pressures and the size of the blast source and the distance of the source from the area to be protected. Moreover, it is well known that the latter relation is an inverse relation, that is, the blast pressures p decrease as the distance L increases. Figure 11.14 shows the

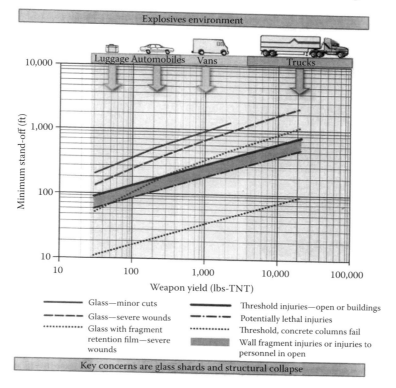

FIGURE 11.14 Bomb blast effects.

relationship between the size of the bomb blast threat, the distance to the target, and the relative damage of the target (FEMA 452).

Let us make the reasonable assumption that protecting the structure (or any mitigation measure) is achieved by reducing p and/or its effects on the structure. This can be done by inspecting Equation 11.1 as follows:

- By reducing or eliminating the size of the blast threat W. This can be done by increasing surveillance, adding manpower, and so on (see FEMA 426)
- By increasing the distance L. Note that as $L \to \infty$, $p \to 0$. This can be done by placing barriers (both vehicular and/or personnel) (see FEMA 430).
- By changing G, the effects of p on the structure can be reduced. This can be done by either hardening the structure or increasing its redundancies.

Clearly, several possibilities are available to reduce p. Some basic mitigation strategies are shown in Figure 11.2. After further analysis of all those possibilities, the decision maker concluded that there are seven viable options. A summary of the options can be found in Table 11.17.

Recall that the expression for risk R due to any hazard as introduced earlier is

$$R = H \cdot V \cdot I \tag{11.2}$$

where H, V, and I are values for hazards, vulnerabilities, and impact. The values are such that

$$1 \leq H \leq 10 \tag{11.3}$$
$$1 \leq V \leq 10 \tag{11.4}$$
$$1 \leq I \leq 10 \tag{11.5}$$

The assigned values to H, V, and I are qualitative; they represent the severity of each of the three parameters for the condition on hand, as shown in Tables 11.18 through 11.20, for evaluating impact, vulnerability, and hazard ratings, respectively. Note that in Table 11.18 asset value is used instead of impact. There are differing views on the use of asset value versus impact in the risk equation. However, such differences are beyond the scope of this chapter. The reader should refer to FEMA 426 and FEMA 452 for a detailed discussion on the differences between the two topics. For now, we use both of them interchangeably. Additionally, the reader is referred to FEMA 452 for more details on the logic and use of Tables 11.18 through 11.20.

To use the risk, Equation 11.2, the decision maker assigned a hazard score of 7. For such a hazard level, the vulnerability and impact scores for different mitigation options were estimated. Finally, using Equation 11.2, the total risk score was computed. Table 11.21 shows the scores and the risk for each mitigation option. Table 11.21 also shows the ranking of different methods, according to their risk scores.

It is of interest to note that in the above example the combinations of mitigation methods did produce much better risk scores than the single mitigation method solution. Among the single mitigation methods, the SHM-only approach produced a risk score that is in a tie with the setback-only score. When coupled with other methods, its risk scores produced the topmost risk score: the least risk is when the decision maker uses a mitigation solution that couples SHM methods with an increased setback method.

The above risk-based method resulted in a simple prioritization scheme for available mitigation methods. The simplicity of the method, along with its capacity to include all three performance measures (hazard, vulnerability, and impact), makes it a powerful tool for decision makers. However, we need to make the following observations:

- While computing the different vulnerability and impact scores for different mitigation solutions, the decision maker needed to accommodate the available mitigation budget. This means that if the available budget is $C_{\text{AVAILABLE}}$ and the required budget of each of the

TABLE 11.17
Different Mitigation Options

ID	Mitigation/Protection Measure	Strategy	Description
1	Increase L	Denying	Setback solution: The distance to the blast threat is increased by placing barriers and fences around the protected space (see Figure 11.2). Due to the limited space available at the site, the increased distance is not as effective in reducing the threat to adequate levels
2	Limit W	Deterrence + Detection	SHM solution: This solution relies mostly on SHM technologies. It calls for added surveillance and electronic equipment. Add detection equipment at appropriate locations to ensure timely detection. Ensure that some, but not all, SHM devices are visible. This is a cost-effective solution, but technologies might not be well tested
3	Improve G	Defense	Hardening/structural redundancy solution: To improve (harden) the structure, additional stiffness and strength measures are needed. This includes adding steel jackets to columns, adding FRP wraps to wall, and providing better foundation anchoring. Due to cost implications, this solution might not be as effective as desired
4	Improve G + increase L	Denying + defense	Combined solution of setback and hardening
5	Improve G + limit W	Defense + deterrence + detection	Combined solution of SHM and hardening
6	Increase L + limit W	Denying + deterrence + detection	Combined solution of setback and SHM
7	Do nothing	NA	The decision maker decides not to make any improvements/changes

TABLE 11.18
Impact Rating, I. (FEMA 452)

		Asset Value
Very High	10	Loss or damage of the bulding's assets would have exceptionally grave consequences, such as exetensive loss of life, widespread severe injuries, or total loss of primary services, core processes, and functions.
High	8–9	Loss or damage of the building's assets would have grave consequences, such as loss of life, severe injuries, loss of primary services, or major loss of core processes and functions for an extended period of time.
Medium High	7	Loss or damage of the building's assets would have serious consequences, such as serious injuries or impairment of core processes and functions for an extended period of time.
Medium	5–6	Loss or damage of the building's assets would have moderate to serious consequences, such as injuries or impairment of core functions and processes.
Medium Low	4	Loss or damage of the building's assets would have moderate consequences, such as minor injuries or minor impairment of core functions and processes.
Low	2–3	Loss or damage of the building's assets would have minor consequences or impact, such as a slight impact on core functions and processes for a short period of time.
Very Low	1	Loss or damage of the building's assets would have negligible consequences or impact.

TABLE 11.19
Vulnerability Rating, V. (FEMA 452)

		Criteria
Very High	10	Very High—One or more major weaknesses have been indentified that make the asset extremely susceptible to an aggressor or hazard. The building lacks redundancies/ physical protection and the entire building would be only functional again after a very long period of time after the attack.
High	8–9	One or more major weaknesses have been identified that make the asset highly susceptible to an aggressor or hazard. The building has poor redundancies/physical protection and most parts of the building would be only functional again after a long period of time after the attack.
Medium High	7	An Important weakness has been identified that makes the asset very susceptible to an aggressor or hazard. The building has inadequate redundancies/physical protection and most critical functions would be only operational again after a long period of time after the attack.
Medium	5–6	A weakness has been indentified that makes the asset fairly susceptible to an aggressor or hazard. The building has insufficient redundancies/physical protection and most part of the building would be only functional again after a considerable period of time after the attack.
Medium low	4	A weakness has been identified that makes the asset somewhat susceptible to an aggressor or hazard. The building has incorporated a fair level of redundancies/ physical protection and most critical functions would be only operational again after a considerable period of time after the attack.
Low	2–3	A minor weakness has been identified that slightly increases the susceptibility of the asset to an aggressor or hazard. The building has incorporated a good level of redundancies/physical protection and the building would be operational within a short period of time after an attack.
Very Low	1	No weaknesses exist. The building has incorporated excellent redundancies/physical protection and the building would be operational immediately after an attack.

TABLE 11.20
Hazard Rating, H. (FEMA 452)

		Threat/Hazard Rating
Very High	10	There is an extremely high likelihood of one or more threats/hazards impacting the site and a history of numerous damages at the site from post events.
High	8–9	It is high likelihood of one or more threats/hazards impacting the site and there is a history of at least one event causing significant damages.
Medium High	7	There is a high likelihood of one or more threats/hazards impacting the site and a history of some damages from past events.
Medium	5–6	There exists a significant possibility of one or more threats/natural hazards impacting the site. There may or may not be a damage history at the site.
Medium low	4	There exists a moderate to low possibility of one or more threats/hazards impacting the site. There may or may not be a damage history at the site.
Low	2–3	There exists a slight possiblity of one or more threats/natural hazards impacting the site. There is little or no history of damages from past events.
Very Low	1	There is little or no likelihood of one or more threats/natural hazards impacting the site. There is no history of damages from past events.

TABLE 11.21
Risk Scores

ID	H	V	I	R	Prioritization
1	7	5	6	210	3
2	7	6	5	210	3
3	7	7	7	343	4
4	7	4	6	168	2
5	7	6	5	210	3
6	7	3	5	105	1
7	7	9	8	504	5

mitigation solutions is $C^i_{REQUIRED}$, where $i = 1,2...7$, then the actual budget used in estimating the risk scores is $C^i_{REQUIRED} = \text{MINIMUM} (C_{AVAILABLE}, C^i_{REQUIRED})$. This means that the risk prioritizations of Table 11.21 do not account for mitigation costs in an equitable manner if $C^i_{REQUIRED}$ varies for different mitigation solutions. This limitation can be overcome by including mitigation costs in the risk score computations.

- At the outset of the process, the decision maker needs to assume a hazard score H. In this case it was $H = 7$. Obviously, the risk scores of Table 11.21 will change if the assumed value of H changes. In many situations, the decision maker might not have accurate information to warrant such a deterministic hazard score. In those situations, utilizing the probability of hazard occurrences might provide a more accurate solution. This is a life cycle analysis approach, which is discussed next.

11.6   LIFE CYCLE ANALYSIS

We explore in this section the longer view of the bridge security decision making process: life cycle cost analysis (LCCA). The various tools of LCCA presented in the previous chapters can be used. However, for the sake of simplicity, we make some assumptions. First we assume that inflation i and discount rate i_{inf} are zero, that is,

$$i = 0\% \tag{11.6}$$

and

$$i_{inf} = 0\% \tag{11.7}$$

We also assume that the lifespan L_{span} of the security projects, expressed in years, is known. Finally, no benefit analysis will be considered in this example. The decision will be based solely on life cycle cost (LCC) considerations.

Based on the above assumptions, a simplified version of total LCC is

$$LCC = C_0 + L_{span} C_{HAZARD} \tag{11.8}$$

The initial cost, C_0 can be expressed as

$$C_0 = C_{CONSTRUCTION} + C_{OTHER} \tag{11.9}$$

While the cost of hazard C_{HAZARD} per year can be expressed as

$$C_{HAZARD} = \int_{-\infty}^{+\infty} f(x) C(x) \, dx \tag{11.10}$$

where $f(x)$ is the probability density function of the occurrence of the hazard per year. The cost of hazard occurrence is $C(x)$.

A discrete version of Equation 11.10 can be expressed as

$$C_{\text{HAZARD}} = \sum_{i=1}^{i=N_{DIV}} p_i \cdot C_i \qquad (11.11)$$

Cost C_i is the cost incurred if a hazard with an annual probability of occurrence of p_i occurs. Note that C_i includes all possible costs (damage to systems, business interruption, social costs, and other indirect costs). We observe that Equation 11.8 includes all three risk parameters: hazard, vulnerability, and impact.

We can now have a more detailed analysis of the security example of the previous section using LCC as the basis of the decision making process.

To start the process of computing LCC, the decision maker should estimate the annual probability of hazard occurrences. The process will result in a table similar to Table 11.22. Note that the decision maker decides to subdivide the severity of hazards into six qualitative measures that range from none to severe. More elaborate and quantitative descriptions are possible; however, the simple descriptions of Table 11.22 are adequate for our current example.

The next step in LCCA is to estimate the initial costs of different mitigation measures. For the purpose of this example, we use the same measures of the previous section. Table 11.23 shows an estimate of the initial costs of each of the measures. The do-nothing option costs nothing.

TABLE 11.22
Annual Probability of Blast Hazard

ID	Hazard Severity	Annual Probability
1	Severe	1.00E-06
2	Medium-high	1.00E-05
3	Medium	1.00E-04
4	Medium-low	1.50E-03
5	Low	3.00E-02
6	None	9.68E-01
	Total	1.00E + 00

TABLE 11.23
Initial Costs

ID	Mitigation/Protection Measure	Initial Cost
1	Increase L	1000
2	Limit W	500
3	Improve G	1400
4	Improve G + increase L	2200
5	Improve G + limit W	1900
6	Increase L + limit W	800
7	Do nothing	0

The next step is to estimate the costs that will be incurred if a hazard of a severity in Table 11.22 occurs while one of the seven measures of Table 11.23 is used by the decision maker. Note that the estimated costs should include the costs of damaged systems, costs of business interruptions, detours, or causalities. In addition, any indirect costs must be accommodated. This step will require some analysis (system, economic, and/or social) of the situation on hand. These analyses can be advanced or simplified, depending on the type and importance of the system. The results of such analyses would be damage cost estimates as in Table 11.24.

Assuming that $L_{span} = 20$ years, and utilizing Equations 11.8 and 11.11, the hazard cost per year, C_i, the total hazard costs, C_{HAZARD}, and LCC for each of the mitigation choices can be computed as in Table 11.25. The table also includes prioritization of the methods based on cost only. On closer inspection, it seems that the prioritization follows closely the relative initial cost of different methods. This can be explained by observing that the relative initial costs, Table 11.22, are high when compared with the costs of hazards as estimated in Table 11.24. The LCC for different methods are displayed in Figure 11.15. Because of the relatively high initial costs compared to hazard costs, it is of interest to note that the trivial do-nothing case is still not the least desirable; it actually is more cost-effective than several mitigation options.

It is of interest to see the effects of relatively higher hazards costs. Let us assume that the hazards costs are as displayed in Table 11.26 instead. Following a similar procedure, the hazard cost per year, C_i, the total hazard costs, C_{HAZARD}, and LCC for each of the mitigation choices are computed as in Table 11.27. The LCC for different methods are displayed in Figure 11.16. The prioritizations have changed, with the new least expensive method as method # 6. The change reflects the high cost of hazards relative to the initial mitigation costs: the least expensive mitigating solution is not the

TABLE 11.24
Costs of Hazards for Different Mitigation Measures

ID	Mitigation/Protection Measure	Severity of Hazard					
		Severe	Medium-High	Medium	Medium-Low	Low	None
1	Increase L	5000	3000	2000	500	100	0
2	Limit W	8000	5000	3000	1000	500	0
3	Improve G	7000	4000	2500	800	300	0
4	Improve G + increase L	4000	2500	1800	300	100	0
5	Improve G + limit W	6000	3500	2200	600	400	0
6	Increase L + limit W	3500	2200	1700	200	100	0
7	Do nothing	12000	8000	5000	3000	2000	0

TABLE 11.25
Life Cycle Costs

ID	Mitigation/Protection Measure	Hazard Cost per Year, C_i	Total Hazard Costs, C_{HAZARD}	Life Cycle Cost (LCC)	Prioritization
1	Increase L	3.99E + 00	7.97E + 01	1.08E + 03	3
2	Limit W	1.69E + 01	3.37E + 02	8.37E + 02	1
3	Improve G	1.05E + 01	2.10E + 02	1.61E + 03	5
4	Improve G + increase L	3.66E + 00	7.32E + 01	2.27E + 03	7
5	Improve G + limit W	1.32E + 01	2.63E+02	2.16E + 03	6
6	Increase L + limit W	3.50E + 00	6.99E + 01	8.70E + 02	2
7	Do nothing	6.51E + 01	1.30E + 03	1.30E + 03	4

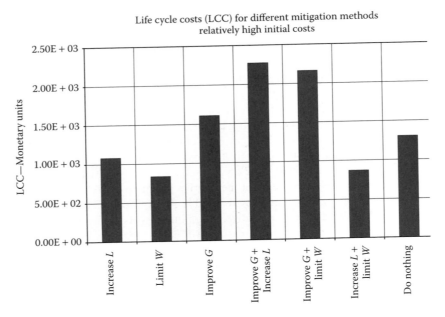

FIGURE 11.15 Life cycle costs for different mitigation methods—relatively high initial costs.

TABLE 11.26
Costs of Hazards for Different Mitigation Measures: High Relative Hazard Costs

	Mitigation/Protection Measure	Severity of Hazard					
ID		Severe	Medium-High	Medium	Medium-Low	Low	None
1	Increase L	500,000	300,000	200,000	50,000	10,000	0
2	Limit W	800,000	500,000	300,000	100,000	50,000	0
3	Improve G	700,000	400,000	250,000	80,000	30,000	0
4	Improve G + increase L	400,000	250,000	180,000	30,000	10,000	0
5	Improve G + limit W	600,000	350,000	220,000	60,000	40,000	0
6	Increase L + limit W	350,000	220,000	170,000	20,000	10,000	0
7	Do nothing	120,0000	800,000	500,000	300,000	200,000	0

TABLE 11.27
Life Cycle Costs: High Relative Hazard Costs

ID	Mitigation/Protection Measure	Hazard Cost per Year, C_i	Total Hazard Costs, C_{HAZARD}	Life cycle Cost (LCC)	Prioritization
1	Increase L	3.99E + 02	7.97E + 03	8.97E + 03	2
2	Limit W	1.69E + 03	3.37E + 04	3.42E + 04	6
3	Improve G	1.05E + 03	2.10E + 04	2.24E + 04	4
4	Improve G + Increase L	3.66E + 02	7.32E + 03	9.52E + 03	3
5	Improve G + limit W	1.32E + 03	2.63E + 04	2.82E + 04	5
6	Increase L + limit W	3.50E + 02	6.99E + 03	7.79E + 03	1
7	Do nothing	6.51E + 03	1.30E + 05	1.30E + 05	7

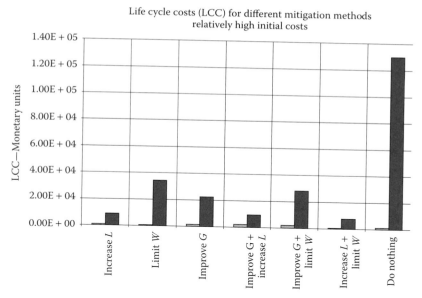

Life cycle costs (LCC) for different mitigation methods
relatively high initial costs

FIGURE 11.16   Life cycle costs for different mitigation methods—relatively low initial costs.

solution with the least expensive initial costs. Due to the relative high cost of hazard, note that the trivial do-nothing case is now the least desired case, with much higher costs than any of the methods under consideration.

The above example shows how LCCA can help decision makers in prioritizing mitigation measures that enhance bridge security. It also shows that SHM methods can play a significant role in enhancing bridge security while reducing overall mitigation costs. The simplifying assumptions introduced at the start of this example can easily be removed, if desired, and a more accurate, yet more complex LCCA would result.

11.7   CONCLUDING REMARKS

We have seen that the merging of the two important and evolving paradigms of SHM (and SHCE in general) with bridge security holds great promise. The technology transfer between the two fields is ongoing, and in many cases it seems almost like force-fitting them at times. Of course, we should not try to force-fit them. Immense steps have been taken, as we have seen in this chapter, but we have a long way to go. Some important observations can be listed as

- A multidisciplinary approach must be followed at all steps. This includes involving the law enforcement agencies and others from the beginning. Engineers should change their culture—from i.e., accept uncertainty. SHM and security overlap. We should try to take advantage of the cooperation between different disciplines and try to get more benefits from security work.
- Practicality of solutions must be accommodated. We need to have an objective approach to the problem of bridge security. A cost–benefit analysis can help reach such a goal.
- Opportunities in sensor technologies can be used for after events, where events may not just be security threats. Transportation is still the key to recovery operations, and so be knowledgeable about various scenarios, such as primary and secondary threats.
- New technologies hold immense promise. There is a need to be open-minded about them and to evaluate them in the future as needed.

- Complexity of models should be dependent on the type of problem on hand. Structural identification options from other fields can be applied to civil engineering issues. Civil engineers should not be afraid of exploring other options.
- Security threat is another hazard. SHM is an optimization of two contradictory demands. SHM should also lead to bridges that do not need monitoring (and does not require monitoring always).
- Need more stakeholders' interaction to get better perspectives.
- Sensor network should be tied to decision making process. Hundred percent safety is an unreal objective.
- For security, visual inspection is still safe and reliable. Systems should be economical, redundant, and reliable.
- There is a need for screening and prioritization.
- Safety versus security: Security should complement safety measures.
- All hazard threat and all hazard mitigation should be considered.
- Performance, safety, security, cost–benefits should be looked into seriously.
- Security is a small but important component of SHM.
- Incorporate smart sensors that address multiple hazards.
- There is no security without monitoring—can be simple to complex. Challenge here is how we approach it.
- Security, vulnerability, mitigation, and ultimately value can be realized by serious consideration of all aspects of the problem.
- Security is one another vulnerability bridge engineers should be aware of. Multihazards approach is important as it helps bottomline.

In addition to these, we offer the following recommendations:

- A balance between safety, cost, and cost/benefit must be reached.
- Event sequencing (before, during, and after) in bridge security can use current state of the art in SHM as applied to earthquake hazard.
- Fire, ramming (impact), and radiation hazards must be considered together in the bridge security field.
- Multihazard considerations can improve efficiency, that is, improve safety while reducing costs.
- The need for a multidisciplinary approach should be highlighted; all stakeholders must be present when planning or designing a security project.
- Perceived value versus actual value of bridge security projects are different.
- Although interaction among different stakeholders in the SHM-bridge security field was considered important, the interaction between bridge owners and security professionals was considered more important.

11.8 APPENDIX: A WORKSHOP ON SECURITY AND SHM

Given the emerging complex needs of bridge security and the tools and techniques of SHM, we can ask the following questions:

- How pertinent is the use of SHM tools to the subject of bridge security?
- Even if SHM tools are pertinent for bridge security, how important are they in resolving bridge security issues?
- What is the current availability of SHM tools that can enhance bridge security?
- Is there a value (cost–benefit) in using SHM for enhancing bridge security?

- What are the needs of various stakeholders for efficient interaction of SHM-bridge security demands (multidisciplinary issues)?
- Can SHM tools developed for other hazards be used efficiently to enhance bridge security (multihazard issues)?

It is obvious that finding detailed answers to these questions calls for immense effort and research. A workshop sponsored by NYSDOT and FHWA was convened at the offices of Weidlinger Associates, New York, NY, on January 12, 2006. The one-day workshop was limited to ten people and was attended by representative stakeholders. The deliberations addressed several issues that pertain to SHM in enhancing bridge security. Numerous interrelationships were discussed. The relative importance of issues as agreed upon by the participants was documented by Alampalli and Ettouney (2007). We report here some results of the workshop. For a complete documentation of the issues, the reader should refer to the workshop report. This report should help decision makers, who are responsible for enhancing bridge security and prioritizing available resources, to get optimal value.

11.8.1 GENERAL CONSIDERATIONS

General considerations included bridge components, hardening, redundancy, guidelines, and management goals as related to bridge security and SHM technologies. The scores in Figures 11.17 through 11.30 in this appendix range from 0 to 10, with higher scores indicating higher validity. Figure 11.17 shows the relative importance of SHM technologies as they relate to hardening of different bridge components when they are exposed to blast hazard. Figure 11.18 shows the relative importance of SHM technologies as they relate to increasing bridge redundancy as the bridge is exposed to different hazards. The availability of guidelines relating to bridge security is scored in Figure 11.19. Finally, Figure 11.20 shows how the workshop participants have scored the importance of SHM tools (as related to bridge security) in meeting management goals (safety and cost). It can be seen that safety is more pertinent than cost in this subject.

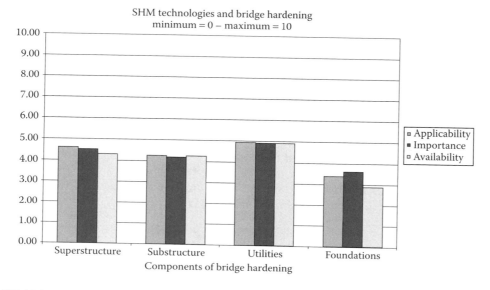

FIGURE 11.17 Scores for SHM technologies and bridge hardening.

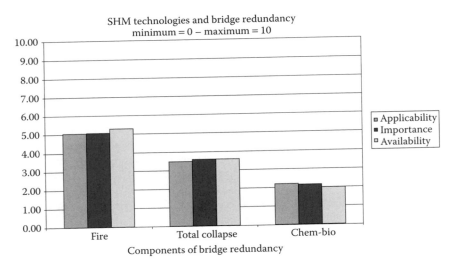

FIGURE 11.18   Scores for SHM technologies and bridge redundancy.

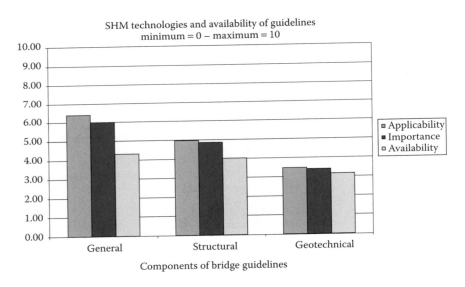

FIGURE 11.19   Scores for SHM technologies and bridge guidelines.

11.8.2   Sequence of Events during Hazards

The workshop also deliberated the use of SHM technologies before, during, and after events such as bomb blasts, fire, ramming, and radiation. The scoring of the applicability of SHM technologies is shown in Figures 11.21 through 11.24 for bomb blast, fire, ramming, and radiation, respectively. These relative scoring results can help decision makers and researchers in prioritizing their procurement and research needs.

11.8.3   Multihazards and Multidiscipline Factors

The importance of multihazards considerations in using SHM technologies for bridge security was also deliberated at the workshop. The hazards considered were seismic, wind, flood/scour, impact/

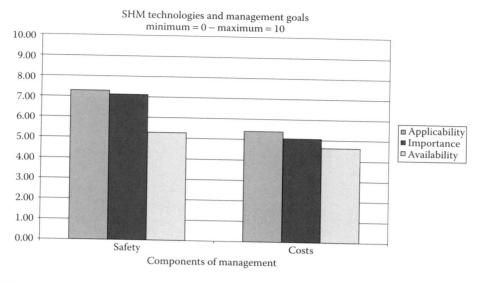

FIGURE 11.20 Scores for SHM technologies and bridge management goals.

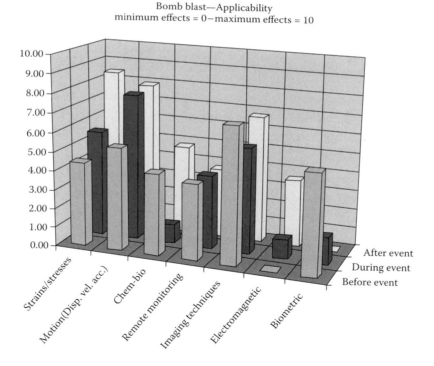

FIGURE 11.21 Bomb blast time sequence applicability of SHM technologies.

traffic, normal deterioration, and corrosion. The scoring of the applicability and importance of SHM technologies is shown in Figures 11.25 and 11.26.

The importance and applicability of a multidisciplinary approach in using SHM technologies for bridge security was also deliberated. The disciplines considered were police, fire, emergency management, and different engineering disciplines. The scoring of the applicability and importance of SHM technologies is shown in Figures 11.27 and 11.28.

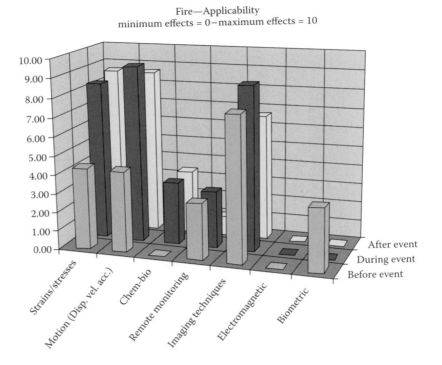

FIGURE 11.22   Fire time sequence applicability of SHM technologies.

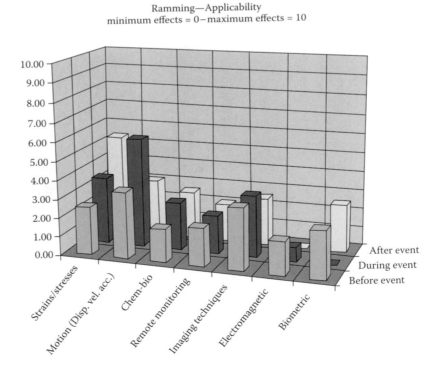

FIGURE 11.23   Ramming time sequence applicability of SHM technologies.

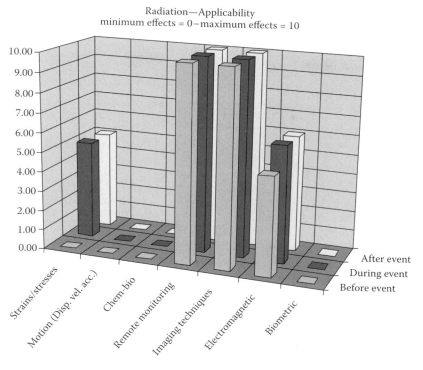

FIGURE 11.24 Radiation time sequence applicability of SHM technologies.

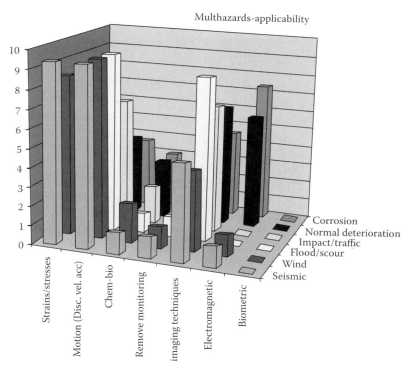

FIGURE 11.25 Scores for applicability of SHM for multihazards issues.

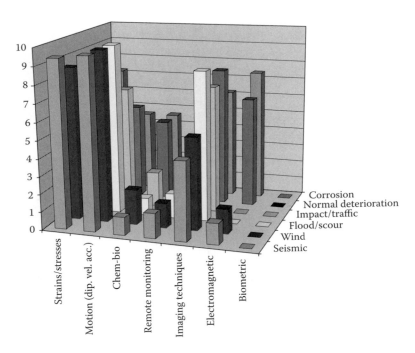

FIGURE 11.26   Scores for importance of SHM for multihazards issues.

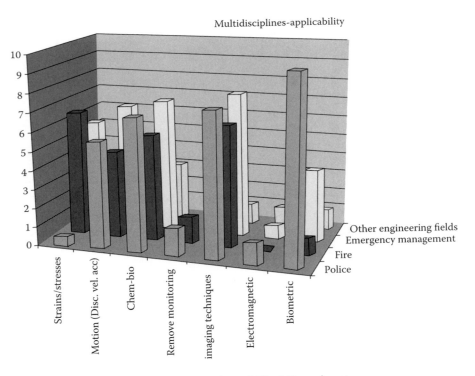

FIGURE 11.27   Scores for applicability of SHM for multidisciplinary issues.

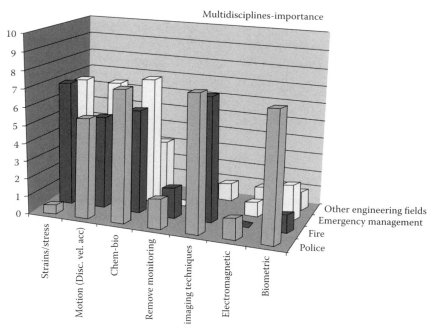

FIGURE 11.28   Scores for importance of SHM for multidisciplinary issues.

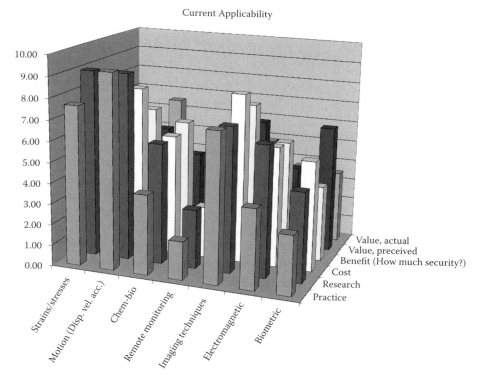

FIGURE 11.29   Importance of SHM technologies to different bridge security issues.

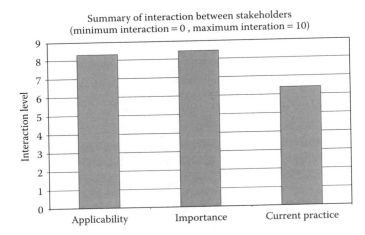

FIGURE 11.30   Scores for interaction among stakeholders: SHM and bridge security issues (overall).

11.8.4   Current and Future Use of SHM Technologies

Several issues require the confluence of SHM and bridge security fields. These include practice versus research needs, cost and benefit considerations, and perceived versus actual value of the issues. The applicability of different SHM technologies in relation to those issues was deliberated and scored as shown in Figure 11.29.

11.8.5   Interaction among Stakeholders

One of the most important factors in the successful execution of a project is smooth and efficient interaction among stakeholders. Figure 11.30 shows how the workshop participants scored the applicability, importance, and current practice (as of the date of the workshop) of interaction among stakeholders.

REFERENCES

Alampalli, S. and Ettouney, M. (2007). *Proceedings of the Workshop on Structural Health Monitoring and Its Role in Enhancing Bridge Security*, New York State Department of Transportation, Albany, NY.

Ettouney, M. and Alampalli, S. (2012). *Infrastructure Health in Civil Engineering: Theory and Components*, CRC Press, Boca Raton, FL.

FEMA. (2006). "Risk Assessment: A How-To Guide to Mitigate Potential Terrorist Attacks," FEMA 452, Federal Emergency Management Agency, FEMA Publication, Washington, DC.

FEMA. (2008). "Site and Urban Design for Security: Guidance against Potential Terrorist Attacks," FEMA 430, Federal Emergency Management Agency, FEMA Publication, Washington, DC.

FEMA. (2009). "Handbook for Rapid Visual Screening of Buildings to Evaluate Terrorism Risk," FEMA 455, Federal Emergency Management Agency, FEMA Publication, Washington, DC.

FEMA. (2010). "A How-To Guide to Mitigate Potential Terrorist Attacks," FEMA 426, Federal Emergency Management Agency, FEMA Publication, Washington, D.C.

FHWA. (2003). *Recommendations for Bridge and Tunnel Security*, Federal Highway Administration, Washington, D.C.

Mays, G. C. and Smith, P. D. (1995). *Blast Effects on Buildings*, Thomas Telford, London, UK.

Appendix

Unit Conversion

Table A.1 contains the units of major engineering metrics used in this volume and the relationships between the units in both SI and US Customary units. The conversion factors are rounded off to a reasonable decimal point.

Table A.1
Conversion Table between SI and US Customary Units

Engineering Metric	To Convert From ... SI	to ... US Customary	Divide by
Linear acceleration	m/s^2	in/s^2	0.0254
	m/s^2	ft/s^2	0.3048
Angle	radian	radian	1
Angular acceleration	$radian/s^2$	$radian/s^2$	1
Angular velocity	$radian/s$	$radian/s$	1
Area	m^2	in^2	0.000645
	m^2	ft^2	0.0929
	mm^2	in^2	645.2
Energy	J	ft.lb	1.356
Force	N or $kg.m/s^2$	lbf	4.448
	kN	kipf	4.448
Frequency	Hz	Hz	1
Impulse	N.s or kg.m/s	lb . s	4.448
Distance/length	m	in	0.0254
	mm	in	25.40
	m	ft	0.3048
	km	mi	1.609
Mass	kg	lb (mass)	0.4536
	G	oz	28.35
	kg	slug	14.59
	kg	ton	907.2
Moment of a force	N.m	lb.in	0.112867
Power	W or J/s	ft.lbf/s	1.356
	W or J/s	hp	745.7
Pressure or stress	Pa	psi or lb/in^2	6895
	Pa	lb/ft^2	47.88
	kPa	psi or lb/in^2	6.895
Work	J	lb.ft	1.356
Velocity	m/s	in/s	0.0254
	m/s	mi/h	0.447
	km/h	mi/h	1.609
Volume: solid	m^3	in^3	1.64E-05
	m^3	ft^3	0.02832
	cm^3	in^3	16.39

continued

TABLE A.1 (continued)
Conversion Table between SI and US Customary Units

Engineering Metric	To Convert From ... SI	to ... US Customary	Divide by
Volume: liquid	l	gal (US)	3.785
	l	qt	0.946
Time	s	s	1

Notes: m = meter, s = seconds, kg = kilogram, J = joule, kN = kilo newton, qt = quart, N = newton, Hz = hertz, W = watt, Pa = pascal, l = liter, kipf = kilo pounds (force), in = inches, lb = pounds (force), gal = gallon, ft = feet, mm = millimeter, oz = ounce, mi = mile, km = kilo meter, g = gram.

Index

Note: *Italicized* page references denote figures and tables.